Lecture Notes on Data Engineering and Communications Technologies

Volume 47

Series Editor

Fatos Xhafa, Technical University of Catalonia, Barcelona, Spain

The aim of the book series is to present cutting edge engineering approaches to data technologies and communications. It will publish latest advances on the engineering task of building and deploying distributed, scalable and reliable data infrastructures and communication systems.

The series will have a prominent applied focus on data technologies and communications with aim to promote the bridging from fundamental research on data science and networking to data engineering and communications that lead to industry products, business knowledge and standardisation.

** **Indexing: The books of this series are submitted to SCOPUS, ISI Proceedings, MetaPress, Springerlink and DBLP** **

More information about this series at http://www.springer.com/series/15362

Leonard Barolli · Yoshihiro Okada ·
Flora Amato
Editors

Advances in Internet, Data and Web Technologies

The 8th International Conference on Emerging Internet, Data and Web Technologies (EIDWT-2020)

 Springer

Editors
Leonard Barolli
Department of Information
and Communication Engineering
Fukuoka Institute of Technology
Fukuoka, Japan

Yoshihiro Okada
Innovation Center for Educational
Resources, University Library
Kyushu University
Fukuoka, Japan

Flora Amato
Department of Electrical Engineering
and Information Technology
University of Naples "Frederico II"
Naples, Italy

ISSN 2367-4512   ISSN 2367-4520   (electronic)
Lecture Notes on Data Engineering and Communications Technologies
ISBN 978-3-030-39745-6   ISBN 978-3-030-39746-3   (eBook)
https://doi.org/10.1007/978-3-030-39746-3

This Springer imprint is published by the registered company Springer Nature Switzerland AG
The registered company address is: Gewerbestrasse 11, 6330 Cham, Switzerland

Welcome Message of EIDWT-2020 International Conference Organizers

Welcome to the 8th International Conference on Emerging Internet, Data and Web Technologies (EIDWT-2020), which will be held from February 24 to 26, 2020, at Kitakyushu International Convention Center, Kitakyushu, Japan.

The EIDWT is dedicated to the dissemination of original contributions that are related to the theories, practices and concepts of emerging Internet and data technologies yet most importantly of their applicability in business and academia toward a collective intelligence approach.

In EIDWT-2020 will be discussed topics related to Information Networking, Data Centers, Data Grids, Clouds, Crowds, Mashups, Social Networks, Security Issues and other Web 2.0 implementations toward a collaborative and collective intelligence approach leading to advancements of virtual organizations and their user communities. This is because current and future Web and Web 2.0 implementations will store and continuously produce a vast amount of data, which if combined and analyzed through a collective intelligence manner will make a difference in the organizational settings and their user communities. Thus, the scope of EIDWT-2020 includes methods and practices which bring various emerging Internet and data technologies together to capture, integrate, analyze, mine, annotate and visualize data in a meaningful and collaborative manner. Finally, EIDWT-2020 aims to provide a forum for original discussion and prompt future directions in the area. For EIDWT-2020 International Conference, we accepted for presentation 57 papers (about 30% acceptance ratio).

An international conference requires the support and help of many people. A lot of people have helped and worked hard for a successful EIDWT-2020 technical program and conference proceedings. First, we would like to thank all authors for submitting their papers. We are indebted to Program Area chairs, Program Committee members and reviewers who carried out the most difficult work of carefully evaluating the submitted papers. We would like to give our special thanks to Honorary Chair of EIDWT-2020 Prof. Makoto Takizawa, Hosei University, Japan, for his guidance and support. We would like to express our appreciation to our keynote speakers for accepting our invitation and delivering very interesting keynotes at the conference.

We would like as well to thank the Local Arrangements Chairs for making excellent local arrangements for the conference. We hope you will enjoy the conference and have a great time in Kitakyushu, Japan.

EIDWT-2020 International Conference Organizers

EIDWT-2020 Steering Committee Chair

Leonard Barolli   Fukuoka Institute of Technology (FIT), Japan

EIDWT-2020 General Co-chairs

Yoshihiro Okada   Kyushu University, Japan
Flora Amato   University of Naples "Frederico II", Italy
Wenny Rahayu   La Trobe University, Australia

EIDWT-2020 Program Committee Co-chairs

Tomoya Enokido   Rissho University, Japan
Zahoor Khan   Higher Colleges of Technology, UAE
Juggapong Natwichai   Chiang Mai University, Thailand

EIDWT-2020 Organizing Committee

Honorary Chair

Makoto Takizawa   Hosei University, Japan

General Co-chairs

Yoshihiro Okada   Kyushu University, Japan
Flora Amato   University of Naples "Frederico II", Italy
Wenny Rahayu   La Trobe University, Australia

Program Co-chairs

Tomoya Enokido   Rissho University, Japan
Zahoor Khan   Higher Colleges of Technology, UAE
Juggapong Natwichai   Chiang Mai University, Thailand

International Advisory Committee

David Taniar   Monash University, Australia
Janusz Kacprzyk   Polish Academy of Sciences, Poland
Vincenzo Loia   University of Salerno, Italy
Arjan Durresi   IUPUI, USA

Publicity Co-chairs

Santi Caballé   Open University of Catalonia, Spain
Pruet Boonma   Chiang Mai University, Thailand
Elis Kulla   Okayama University of Science, Japan
Farookh Hussain   University of Technology Sydney, Australia
Nadeem Javaid   COMSATS University Islamabad, Pakistan

International Liaison Co-chairs

Fang-Yie Leu	Tunghai University, Taiwan
Admir Barolli	Aleksander Moisiu University of Durres, Albania
Kin Fun Li	University of Victoria, Canada
Akio Koyama	Yamagata University, Japan
Omar Hussain	University of New South Wales, Australia

Local Organizing Committee Co-chairs

Keita Matsuo	Fukuoka Institute of Technology, Japan
Tomoyuki Ishida	Fukuoka Institute of Technology, Japan
Donald Elmazi	Fukuoka Institute of Technology, Japan

Web Administrators

Kevin Bylykbashi	Fukuoka Institute of Technology, Japan
Miralda Cuka	Fukuoka Institute of Technology, Japan

Finance Chair

Makoto Ikeda	Fukuoka Institute of Technology, Japan

Steering Committee Chair

Leonard Barolli	Fukuoka Institute of Technology, Japan

PC Members

Akimitsu Kanzaki	Shimane University, Japan
Akio Koyama	Yamagata University, Japan
Akira Uejima	Okayama University of Science, Japan
Akshay Uttama Nambi S. N.	Microsoft Research India, India
Alba Amato	National Research Council (CNR)–Institute for High-Performance Computing and Networking (ICAR), Italy
Alberto Scionti	LINKS Foundation, Italy
Albin Ahmeti	TU Wien, Austria
Alex Pongpech	National Institute of Development Administration, Thailand
Ali Rodan	Higher Colleges of Technology, UAE
Alfred Miller	Higher Colleges of Technology, UAE
Amelie Chi Zhou	Shenzhen University, China
Amin M. Khan	Pentaho, Hitachi Data Systems, Japan
Ana Azevedo	ISCAP, Porto, Portugal

Mazin Abuharaz	Higher Colleges of Technology, UAE
Minghu Wu	Hubei University of Technology, China
Mingwu Zhang	Hubei University of Technology, China
Minoru Uehara	Toyo University, Japan
Mirang Park	Kanagawa Institute of Technology, Japan
Monther Tarawneh	Higher Colleges of Technology, UAE
Morteza Saberi	UNSW Canberra, Australia
Mohsen Farid	University of Derby, UK
Motoi Yamagiwa	University of Yamanashi, Japan
Mouza Alshemaili	Higher Colleges of Technology, UAE
Muawya Aldalaien	Higher Colleges of Technology, UAE
Mukesh Prasad	University of Technology, Australia
Muhammad Iqbal	Higher Colleges of Technology, UAE
Naeem Janjua	Edith Cowan University, Australia
Naohiro Hayashibara	Kyoto Sangyo University, Japan
Naonobu Okazaki	University of Miyazaki, Japan
Neha Warikoo	Academia Sinica, Taiwan
Nobukazu Iguchi	Kindai University, Japan
Nobuo Funabiki	Okayama University, Japan
Olivier Terzo	LINKS Foundation, Italy
Omar Al Amiri	Higher Colleges of Technology, UAE
Omar Hussain	UNSW Canberra, Australia
Osama Alfarraj	King Saud University, Saudi Arabia
Osama Rahmeh	Higher Colleges of Technology, UAE
P. Sakthivel	Anna University, Chennai, India
Panachit Kittipanya-Ngam	Electronic Government Agency, Thailand
Paolo Bellavista	University of Bologna, Italy
Pavel Smrž	Brno University of Technology, Czech Republic
Peer Shah	Higher Colleges of Technology, UAE
Pelle Jakovits	University of Tartu, Estonia
Per Ola Kristensson	University of Cambridge, UK
Philip Moore	Lanzhou University, China
Pietro Ruiu	LINKS Foundation, Italy
Pornthep Rojanavasu	University of Phayao, Thailand
Pruet Boonma	Chiang Mai University, Thailand
Raffaele Pizzolante	University of Salerno, Italy
Ragib, Hasan	The University of Alabama at Birmingham, USA
Rao Mikkilineni	C3dna, USA
Richard Conniss	University of Derby, UK
Ruben Mayer	University of Stuttgart, Germany
Sachin Shetty	Old Dominion University, USA
Sajal Mukhopadhyay	National Institute of Technology, Durgapur, India
Salem Alkhalaf	Qassim University, Saudi Arabia
Salvatore Ventiqincue	University of Campania "Luigi Vanvitelli", Italy
Samia Kouki	Higher Colleges of Technology, UAE

EIDWT-2020 Reviewers

Ali Khan Zahoor
Amato Flora
Amato Alba
Barolli Admir
Barolli Leonard
Bista Bhed
Caballé Santi
Chellappan Sriram
Chen Hsing-Chung
Cui Baojiang
Di Martino Beniamino
Embarak Ossama
Enokido Tomoya
Ficco Massimo
Fiore Ugo
Fun Li Kin
Gotoh Yusuke
Hussain Farookh
Hussain Omar
Javaid Nadeem
Ikeda Makoto
Ishida Tomoyuki
Kayes Asm
Kikuchi Hiroaki
Kolici Vladi

Koyama Akio
Kouki Samia
Kulla Elis
Leu Fang-Yie
Matsuo Keita
Moore Philip
Koyama Akio
Kryvinska Natalia
Ogiela Lidia
Ogiela Marek
Okada Yoshihiro
Palmieri Francesco
Paruchuri Vamsi Krishna
Rahayu Wenny
Sato Fumiaki
Spaho Evjola
Sugawara Shinji
Takizawa Makoto
Taniar David
Terzo Olivier
Uehara Minoru
Venticinque Salvatore
Wang Xu An
Woungang Isaac
Xhafa Fatos

EIDWT-2020 Keynote Talks

Delay Tolerant Networking Technology and Disaster Management—Theoretical and Practical Aspects of DTN Technology

Hiroyoshi Miwa

Kwansei Gakuin University, Sanda, Hyogo Prefecture, Japan

Abstract. Immediately after a large-scale disaster such as the Great East Japan Earthquake in 2011 struck, both wired and wireless communications, do not work at all in the affected area. However, in such an environment, keeping communications and sharing information is absolutely imperative. Delay/Disruption/Disconnect Tolerant Networking (DTN) is the technology that establishes communications in an environment characterized by lack of continuous connectivity, high loss rates and long propagation delays. A routing scheme, store-carry-forward, in which a mobile node first stores a message, carries it while moving, and then forwards it to either an intermediate node or the destination node, is essential for DTN. The store-carry-forward routing scheme makes use of opportunistic communication based on human serendipitous encounters which human mobility patterns cause. Recently, interesting knowledge about human mobility patterns and serendipitous encounters was found. A human mobility model which is consistent with the found properties was proposed. We can design an efficient algorithm for the store-carry-forward routing scheme by considering the mathematical mobility model. This is an example that theoretical knowledge and results can solve a practical problem. The optimization theory and the probability theory are useful also in the disaster management. In this talk, we introduce our theoretical and practical knowledge for DTN technology.

Opportunities in IoT Data Processing Research

David Taniar

Monash University, Melbourne, Australia

Abstract. IoT research is not a new topic. It started long ago with Distributed Sensor Network (or Wireless Sensor Network), where the focus was much on communication networking, wireless ad-hoc network, connectivity, spontaneous formation of networks or self-discovery. Recently, there is a lot of IoT research focusing on data analytics of sensor data. Data analytics is very human nature because human likes to know patterns, to find association, to explain behavior, as well as to predict the future based on data that comes from various sensors. Therefore, IoT research spans from sensor as hardware and networks, to data analytics for end users. It is well known in data analytics that up to 80% of data analytics projects is dedicated to data management, including data preparation, cleansing and dealing with missing and inaccurate data. This is even more so in IoT data research since sensors produce noise, anomalies, missing data, etc. Therefore, there is a growing IoT research focusing on data management, which can be considered the middle part of the IoT research spectrum. In this talk, I will discuss opportunities in IoT data management research, including sensor data storage and management, data cleansing and missing data, data retrieval, as well as the user of spatial techniques. I will use several illustrations from our IoT projects for various industries, such as railway, manufacturing, environment, AgTech, utility, as well as healthcare. By the end of this talk, the audience would appreciate and understand the opportunities that IoT research could give, beyond networking or data analytics. The focus is on IoT data management and processing.

Contents

Judging Students' Learning Style from Big Video Data Using Neural Network

Noriyasu Yamamoto(✉)

Department of Information and Communication Engineering,
Fukuoka Institute of Technology, 3-30-1 Wajiro-Higashi, Higashi-Ku,
Fukuoka 811-0295, Japan
nori@fit.ac.jp

Abstract. In general, at the universities the lecturers use the grades to evaluate the students' performance. However, the grades do not show the accumulated knowledge. Thus, the students can't understand whether they have studied enough to understand all the topics of the lecture. Recently, with the advancement of IoT technology and applications, we can record students' study for every lecture using video images in all directions (360-degree). We also can display high quality image for students. During the lecture, by the response from the students' smart phone, the students' learning style was recorded in the database. However, the learning style jugged by the data received from the students' smartphone had not a good accuracy. In this paper, in order to accurately judge the students' learning style from big video data recorded in the database, we used a neural network.

1 Introduction

In general, at the universities the lecturers use the grades to evaluate the students' performance. However, the grades do not show the accumulated knowledge. Thus, the students can't understand whether they have studied enough to understand all the topics of the lecture. Recently, with the advancement of IoT technology and applications, we can record students' study for every lecture using sharp video image in all directions (360-degree). Also, we can display high quality image for students. However, by recording many lectures, we will have big data. Therefore, we need to convert these data into short video and show to the students in order that students easily understand their study style.

For reducing video time, we can use a fixed point camera. We need some 360-degree cameras and we should consider the location point for cameras. Then, a target student can be automatically searched using these big (long) video data. In addition, we should convert the data into short video in order that a student can easily understand his learning style. For implementation of the proposed system [1–10], we will use 360-degrees cameras for searching the students' location and judge their learning style.

During the lecture, by the response from the students' smart phone, the students' learning style was recorded in the database [11]. However, the learning style jugged by the data received from the students' smartphone had not a good accuracy. In this paper, in order to accurately judge the students' learning style from big video data recorded in the database, we used a neural network.

© Springer Nature Switzerland AG 2020
L. Barolli et al. (Eds.): EIDWT 2020, LNDECT 47, pp. 1–6, 2020.
https://doi.org/10.1007/978-3-030-39746-3_1

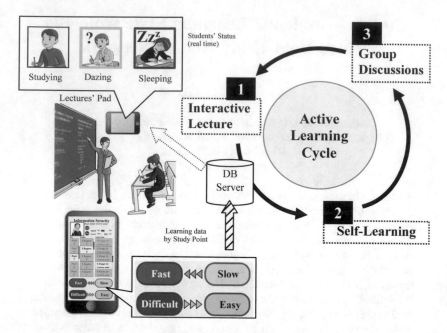

Fig. 1. Structure of ALS.

The paper structure is as follows. In Sect. 2, we introduce the implemented Active Learning System (ALS). In Sect. 3, we present the judging algorithm for the students' learning style from big video data used by the neural network. Finally, in Sect. 4, we give some conclusions and future work.

2 Implemented Active Learning System (ALS)

In this section, we present our implemented Smartphone-based Active Learning System (ALS), which was used to increase the students learning motivation [1–10]. Figure 1 shows the structure of ALS. By using ALS, the lecturer performs the interactive lecture by confirming the understanding degree of the student using their smartphone in real time. Prior to the lecture, the lecturer prepares "study points". "Small examination" refers to a mini quiz prepared for each study point. A mini quiz consists of simple multiple-choice questions. "Understanding level" is set by the result of the "Small examination". "Lecture speed" suggests whether students find the lecture progress too fast or too slow. By "Understanding level" and "Lecture speed" students' understanding degree can be judged. These two functions are used as feedbacks through the application on students' smartphone. During the lecture, these data are recorded in the database. By using by this database, the students' learning status is classified in three types as follows.

[Type 1] Studying: The student concentrates on study.
[Type 2] Dazing: The student does not concentrate on study.
[Type 3] Sleeping: The student is sleeping without studying.

In ALS, 360-degrees cameras are used for searching the student location and judging the student learning style. By using 360-degree camera, a target student can be automatically searched by using the recorded video data. During the lecture, by the response from the students' smart phone, the students' learning style was recorded in the database [11]. However, the learning style jugged by the data received from the students' smartphone had not a good accuracy.

3   Judging Students' Learning Style from Big Video Data Using Neural Network

In this section, in order to accurately judge the students' learning style from big video data recorded in the database, we used a neural network. Figure 2 shows three video cameras for students' 3D data from three directions. The following sequences show how to judge the students' learning style from three video camera data using a neural network.

Fig. 2. Target students' video which taken by 360-degree Camera from the front and the side.

[Sequence 1] Student are detected by the data received from three video cameras.
[Sequence 2] The face, head and hand are detected by the students' action on the video. As a result, the students' frame is decided by Feature Detection System as shown in Fig. 3.

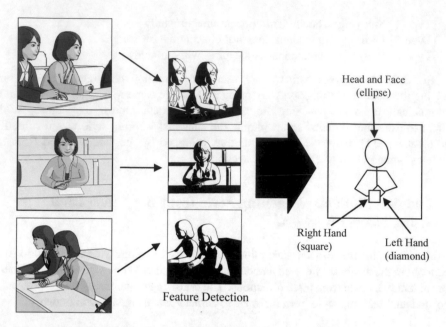

Fig. 3. Detecting students' body frame, head (face) and hand.

[Sequence 3] Then, the neural network learns the students' frame input and the training
data which students used during learning as shown in Fig. 4.

[Sequence 4] The students' learning style is judged by using the neural network.

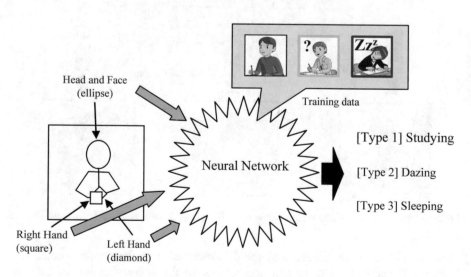

Fig. 4. Judging students' learning style.

4 Conclusions

Using our implemented ALS, during the lecture, by the response from the students' smart phone, the students' learning style was recorded in the database. However, the learning style jugged by the data received from the students' smartphone had not a good accuracy.

In this paper, in order to accurately judge the students' learning style from big video data recorded in the database, we used a neural network. In the future, we plan to perform extensive experiments with integrated ALS and proposed neural network judging system.

References

1. Yamamoto, N.: An interactive e-learning system for improving students motivation and self-learning by using smartphones. J. Mobile Multimedia (JMM) **11**(1&2), 67–75 (2015)
2. Yamamoto, N.: New functions for an active learning system to improve students self-learning and concentration. In: Proceeding of the 18th International Conference on Network-Based Information Systems (NBiS-2015), pp. 573–576, September 2015
3. Yamamoto, N.: Performance evaluation of an active learning system to improve students self-learning and concentration. In: Proceeding of the 10th International Conference on Broadband and Wireless Computing, Communication and Applications (BWCCA-2015), pp. 497–500, November 2015
4. Yamamoto, N.: Improvement of group discussion system for active learning using smartphone. In: Proceeding of the 10th International Conference on Innovative Mobile and Internet Services in Ubiquitous Computing (IMIS-2016), pp. 143–148, July 2016
5. Yamamoto, N.: Improvement of study logging system for active learning using smartphone. In: Proceedings of the 11th International Conference on P2P, Parallel, Grid, Cloud and Internet Computing (3PGCIC-2016), pp. 845–851, November 2016
6. Yamamoto, N., Uchida, N.: Improvement of the interface of smartphone for an active learning with high learning concentration. In: Proceeding of the 31st International Conference on Advanced Information Networking and Applications Workshops (AINA-2017), pp. 531–534, March 2017
7. Yamamoto, N., Uchida, N.: Performance evaluation of a learning logger system for active learning using smartphone. In: Proceeding of the 20th International Conference on Network-Based Information Systems (NBiS-2017), pp. 443–452, August 2017
8. Yamamoto, N., Uchida, N.: Performance evaluation of an active learning system using smartphone: a case study for high level class. In: Proceeding of the 6th International Conference on Emerging Internet, Data & Web Technologies (EIDWT-2018), pp. 152–160, March 2018
9. Yamamoto, N., Uchida, N.: Dynamic group formation for an active learning system using smartphone to improve learning motivation. In: Proceedings of the 12th International Conference on Innovative Mobile and Internet Services in Ubiquitous Computing (IMIS-2018), pp. 183–189, March 2018

10. Yamamoto, N., Uchida, N.: Performance evaluation of a smartphone-based active learning system for improving learning motivation during study of a difficult subject. In: Proceeding of the 21th International Conference on Network-Based Information Systems (NBiS-2018), pp. 531–539, August 2018

11. Yamamoto, N.: Converting big video data into short video: using 360-degree cameras for searching students location and judging students learning style. In: Proceeding of The 22nd International Conference on Network-Based Information Systems (NBiS-2019), pp. 459–464, September 2019

A Multi-modal Interface for Control of Omnidirectional Video Playing on Head Mount Display

Yusi Machidori[1], Ko Takayama[1], and Kaoru Sugita[2]([✉])

[1] Graduate School of Engineering, Fukuoka Institute of Technology,
3-30-1 Wajiro-Higashi, Higashi-Ku, Fukuoka 811-0295, Japan
{mgm19106, mgm18102}@bene.fit.ac.jp
[2] Department of Information and Communication Engineering,
Fukuoka Institute of Technology, 3-30-1 Wajiro-Higashi, Higashi-Ku,
Fukuoka 811-0295, Japan
sugita@fit.ac.jp

Abstract. In recent years, the omnidirectional video has been used for various purposes. It can be placed on a Head Mount Display (HMD) to give immersive experiences on a head tracking function. In general, the HMD only support a controller to operate a viewpoint, but it can't see the controller on a virtual world over the HMD. In this paper, we introduce a multi-modal interface to control a spatio-temporal position of an omnidirectional video on HMD application software. We also show an implementation to use a general HMD attached on a Leap motion controller. The implemented multi-modal interface can operate an omnidirectional video used for motions of hands and voice through the Leap Motion controller and a general microphone.

1 Introduction

In recent years, the omnidirectional video has been used for various purposes such as remote monitoring, promotion of tourist sites, entertainments and so on. These videos can usually play an application software and a web application attached on the omnidirectional camera controlling both the direction and size of view by some user interface components such as buttons, text box/fields and menus. Also, a Head Mount Display (HMD) is used to give immersive experiences on these videos supporting the head tracking. On the other hands, in general the HMD only support a controller to operate a viewpoint, but it can't see the controller on a virtual world over the HMD. For this reason, we have introduced a multi-modal interface [1] to a VR application on HMD attached a Leap motion controller [2], which voice commands and gesture inputs are supposed to become one of natural input methods on the virtual world [3].

In this paper, we introduce a multi-modal interface to control a spatio-temporal position of an omnidirectional video on HMD application software. We also show the implementation and the use of a general HMD attached on a Leap motion controller.

This paper is organized as follows. We introduce the multi-modal interface to control the omnidirectional video on HMD application software in Sect. 2. Section 3

© Springer Nature Switzerland AG 2020
L. Barolli et al. (Eds.): EIDWT 2020, LNDECT 47, pp. 7–11, 2020.
https://doi.org/10.1007/978-3-030-39746-3_2

shows implementation of the windows application supporting a mixed reality (MR) headset and a Leap motion controller. Finally, Sect. 4 concludes the paper.

2   Multi-modal Interface for Control of Spatio-Temporal Position of Omnidirectional Video

The omnidirectional video can operate both: a view position and a playback position. In general, in an application software, the view position is moved by drag operations of mouse and the playback position is controlled by the SeekBar. In HMD application software, the view position is operated by a controller over the HMD supporting the head tracking function. Simultaneously, the playback position is also operated by a controller through the SeekBar. These operations are not inconvenience because the controller is blinded by the HMD and the SeekBar is caused by increasing the number of operations and unintuitive operations on the HMD application software. For this reason, we introduce a multi-modal interface to control both the view position and playback position of omnidirectional video playing on the HMD as shown in Fig. 1.

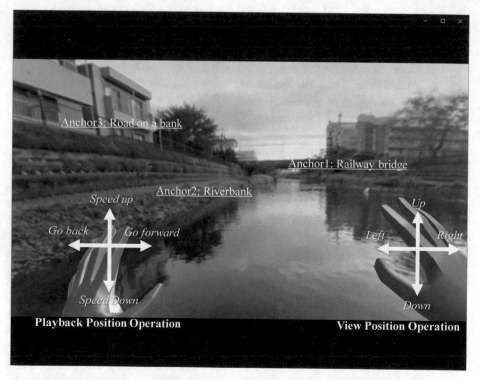

Fig. 1. Multi-modal interface to control a spatio-temporal position of an omnidirectional video.

In the multi-modal interface, the right hand and voice is used to move a view point, and a left hand can operate the playback position and speed of the omnidirectional video. The view point can be moved by the direction of motion by the right hand and the voice corresponding to an anchor (e.g. Railway bridge, Riverbank and Road on a bank). The playback position is moved to a direction of horizontal motion in left hand. The playback speed can be changed to a high speed in the upward direction and to low speed in the downward direction of left hand.

3   Implemented Multi-modal Interface

We implemented our approach on the Windows 10, which can be displayed on the windows Mixed Reality (MR) Headset. For the implementation of multi-modal interface, a laptop PC is connected to a Leap motion controller (i.e. a gesture recognition device), a microphone and a HMD as shown in Fig. 2 and Table 1. The PC specifications are shown in Table 2. We developed the implemented multi-modal interface by using the software environments shown in Table 3. We used Unity over C# language for the HMD application software development platform, the Web Speech API for a voice input, the WebSocketSharp for enabling the Web speech API, the Leap Motion Interaction Engine for hand UI and the Leap Motion Unity Core Assets for a gesture recognition.

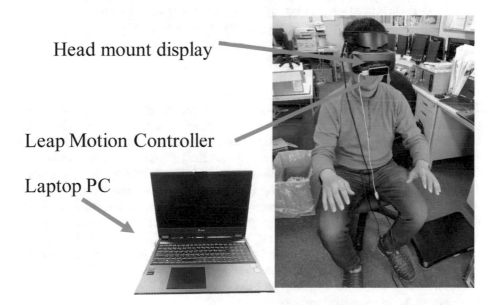

Head mount display

Leap Motion Controller

Laptop PC

Fig. 2. HMD and leap motion controller.

Table 1. Input and output devices.

Device	Maker	Product
Windows MR headset	Lenovo	G0A20002JP
Leap motion controller	Leap motion	LM-C01-JP
Microphone	NASUM	Capacitor microphone

Table 2. PC environment.

Device	Maker	Product
CPU	Intel	Core i7-8750H
Memory	–	DDR 8 GB
GPU	NVIDIA	GeForce RTX 2070

Table 3. Software used for implementation of proposed system.

Type of software	Software	Version
OS	Windows 10 Education	1803
IDE	Visual Studio 2017	15.7.4
Development language	C#	2018.1.6f1
Voice recognition	Web Speech API/WebSocket Sharp	–
Hand user interface	Leap Motion Interaction Engine	1.2.0
Gesture recognition	Leap motion unity core assets	4.4.0

4   Conclusions

In this paper, we presented a multi-modal interface for the control a spatio-temporal position of an omnidirectional video on HMD application software. Our multi-modal interface can control both: the view position and playback position of omnidirectional video using motions of hands and voice. We also have shown the implementation of the multi-modal interface and its use to a general HMD attached on a Leap motion controller. The implemented multi-modal interface is running on the Windows 10 and can be displayed on windows MR Headset. It can operate an omnidirectional video used for motions of hands and voice through the Leap Motion controller and a general microphone. Currently, we are carrying out some experiments to find operating problems of our multi-modal interface in order to improve the effectiveness of the spatio-temporal position of an omnidirectional video on the HMD and VR.

In the future work, we will carry out extensive experiments to evaluate the performance of the implemented system.

References

1. Vo, M.T., Waibel, A.: A multimodal human-computer interface: combination of speech and gesture recognition. In: Proceedings of InterCHI-1993, pp. 69–70 (1993)
2. Leap Motion: Ultraleap Ltd (2019). https://www.leapmotion.com/
3. Machidori, Y., Takayama, K., Sugita, K.: Implementation of multi-modal interface for VR application. In: Proceedings of the iCAST-2019, pp. 370–373 (2019)

Evaluation of a TBOI (Time-Based Operation Interruption) Protocol to Prevent Late Information Flow in the IoT

Shigenari Nakamura[1]($\boxtimes$), Tomoya Enokido[2], and Makoto Takizawa[1]

[1] Hosei University, Tokyo, Japan
nakamura.shigenari@gmail.com, makoto.takizawa@computer.org
[2] Rissho University, Tokyo, Japan
eno@ris.ac.jp

Abstract. In the IoT (Internet of Things), the CapBAC (Capability-Based Access Control) model is proposed to make devices secure. Here, an owner of a device issues a capability token, i.e. a set of access rights to a subject. The subject is allowed to manipulate the device according to the access rights authorized in the capability token. In the CapBAC model, a subject sb_i can get data of a device d_1 brought to another device d_2 by getting the data from the device d_2 even if the subject sb_i is not allowed to get data from the device d_1. Here, the data of the device d_1 illegally flow to the subject sb_i. In addition, a subject sb_i can get data of the device d_1 generated at time τ even if the subject sb_i is not allowed to get the data at time τ. In this case, the data come to the subject sb_i later than expected by the subject sb_i to get the data, i.e. the data flow late to the subject sb_i. In our previous studies, OI (Operation Interruption) and TBOI (Time-Based OI) protocols are proposed. In the OI and TBOI protocols, only illegal operations and both types of illegal and late operations are interrupted, respectively. In this paper, we evaluate the TBOI protocol in terms of the number of operations interrupted. In the evaluation, we show the late information flow is prevented in addition to the illegal information flow in the TBOI protocol differently from the OI protocol.

Keywords: IoT (Internet of Things) · Device security · CapBAC (Capability-Based Access Control) model · Illegal information flow · Information flow control · Late information flow · TBOI (Time-Based Operation Interruption) protocol

1 Introduction

In order to make secure information systems, access control models are used [1]. Here, only an authorized subject is allowed to manipulate an object in an authorized operation. However, even if a subject is not allowed to get data in an object o_i, the subject can get the data by accessing another object o_j [1]. Here, illegal information flow from the object o_i via the object o_j to the subject occurs.

© Springer Nature Switzerland AG 2020
L. Barolli et al. (Eds.): EIDWT 2020, LNDECT 47, pp. 12–23, 2020.
https://doi.org/10.1007/978-3-030-39746-3_3

Hence, we have to prevent illegal information flow among subjects and objects in the access control models. In the LBAC (Lattice-Based Access Control) model [13], each entity like subjects and objects are assigned a security class. Illegal information flow is defined based on the relations among classes and every operation implying the illegal information flow is prohibited. In our previous studies, various types of protocols to prevent illegal information flow are proposed. In the paper [6], protocols to prevent illegal information flow occurring in database systems are proposed based on the RBAC (Role-Based Access Control) model [14]. In the papers [7,9], protocols to prevent illegal information flow occurring in P2PPSO (Peer-to-Peer Publish/Subscribe with Object concept) systems are proposed based on the TBAC (Topic-Based Access Control) model [11].

The IoT (Internet of Things) is composed of various types and millions of nodes including not only computers but also devices like sensors and actuators [4,12]. Here, the traditional access control models such as the RBAC [14] and ABAC (Attribute-Based Access Control) [15] models are not adopted for the IoT due to the scalability of the IoT. Hence, the CapBAC (Capability-Based Access Control) model is proposed [3]. Here, an owner of a device issues a capability token to a subject like users and applications. The capability token is defined to be a set of access rights, $\langle d, op \rangle$ for a device d and an operation op. The subject is allowed to manipulate the device d in an operation op only if the capability token including the access right $\langle d, op \rangle$ is issued to the subject. In addition, capability token includes the validity period which shows the capability token is valid from when to when. Each subject can manipulate devices according to the capability token during the validity period of the capability token.

Suppose a subject sb_i is issued a capability token including a pair of access rights $\langle d_1, get \rangle$ and $\langle d_2, put \rangle$ of a pair of devices d_1 and d_2 by owners of the devices. Suppose the device d_1 is a sensor and the device d_2 is equipped with a pair of sensor and actuator. A sensor just gives sensor data to a subject. On the other hand, an actuator supports an action to store data to the device. A subject sb_j is issued a capability token including an access right $\langle d_2, get \rangle$ by an owner of the device d_2. First, the subject sb_i gets sensor data obtained by the sensor d_1 and then gives the data to the device d_2. Next, the subject sb_j gets the data from the device d_2. Here, the subject sb_j can obtain the data of the device d_1 via the device d_2 although the subject sb_j is not issued a capability token including the access right $\langle d_1, get \rangle$. Here, the device d_1 is a source one of information flow. This is illegal information flow from the device d_1 to the subject sb_j. In our previous studies, an OI (Operation Interruption) protocol is proposed to prevent illegal information flow in the IoT based on the CapBAC model [8]. First, the legal and illegal information flow relations among subjects and devices are defined based on the CapBAC model. Then, it is checked whether or not the illegal information flow occurs based on the information flow relations. If the illegal information flow occurs, the operation is interrupted at the device. Hence, every illegal information flow is prevented from occurring.

In the OI protocol, every subject sb_i can not get data of a device d_1 in another device d_2 even if the subject sb_i is allowed to get data of the device d_2. However, every subject sb_i can get data of a device d_1 generated at time τ even if the subject sb_i is not allowed to get the data at time τ. This means, every illegal information flow is prevented but every information flow of data generated in the device d_1 before the validity period of the access right $\langle d_1, get \rangle$ to a subject is ignored in the OI protocol because information flow relations are based on only devices. Here, the older information comes to a subject than what the subject expects to get. Since the information comes to a subject *late*, the information flow is referred to as *late information flow* [10]. In order to prevent late information flow in addition to illegal information flow, a TBOI (Time-Based OI) protocol is proposed [10]. Here, the operations occurring illegal and late information flow are interrupted. In this paper, we evaluate the TBOI protocol in terms of the number of operations interrupted compared with the OI protocol. In the evaluation, the more number of operations are interrupted in the TBOI protocol than the OI protocol. However, every late information flow is prevented in addition to illegal information flow in the TBOI protocol differently from the OI protocol.

In Sect. 2, we discuss the system model. In Sect. 3, we define types of information flow relations based on the CapBAC model. In Sect. 4, we overview the OI and TBOI protocols. In Sect. 5, we evaluate the TBOI protocol in terms of the number of operations interrupted compared with the OI protocol.

2   System Model

In order to make information systems secure, types of access control models [2,14,15] are widely used. Here, a system is composed of two types of entities, subjects and objects. A subject s issues an operation op to an object o. Then, the operation op is performed on the object o. Here, only an authorized subject s is allowed to manipulate an object o in an authorized operation op. Most of the access control models are based on ACLs (access control lists) such as RBAC (Role-Based Access Control) [14] and ABAC (Attribute-Based Access Control) [15] models. An ACL is a list of access rules specified by an authorizer. Each access rule $\langle s, o, op \rangle$ is composed of a subject s, an object o, and an operation op. This means, a subject s is granted an access right of an object o and an operation op. In the ACL system, if a subject s tries to access the data of an object o in an operation op, a service provider has to check whether or not the subject is authorized to manipulate the object o in the operation op by using the ACL, i.e. $\langle s, o, op \rangle$ in the ACL. In scalable systems like the IoT, the ACL gets also scalable and it is difficult to maintain and check the ACL. Hence, the RBAC and ABAC models are not suitable for the scalable distributed systems where there is no centralized coordinator and each node is an autonomous process which makes a decision by itself.

In the paper [3], the CapBAC (Capability-Based Access Control) model is proposed as an access control model for the IoT. Here, there is an owner of each device. An owner of a device first issues a capability token CAP_i to a subject

sb_i. The capability token CAP_i issued to the subject sb_i is defined to be a set of access rights. An access right is a pair $\langle d, op \rangle$ of a device d and an operation op. A subject sb_i is allowed to manipulate a device d in an operation op only if a capability token CAP_i including an access right $\langle d, op \rangle$ is issued to the subject sb_i, i.e. $\langle d, op \rangle \in CAP_i$. Otherwise, the access request is rejected.

In the distributed CapBAC model [5], there is no intermediate entity between each pair of a subject and a device to implement the access control. Each capability token includes the validity period. Here, an access request to manipulate a device d in an operation op of a subject sb_i whose capability token CAP_i expires is rejected even if the capability token CAP_i includes the access right $\langle d, op \rangle$. Let $gt_i^s.st$ be the first time when a subject sb_i is allowed to issue a get operation to a device s. On the other hand, $gt_i^s.et$ is the end time when a subject sb_i is allowed to issue a get operation to a device s. That is, the subject sb_i is allowed to issue a get operation to a device s from $gt_i^s.st$ to $gt_i^s.et$.

In the IoT, a set D of devices $d_1, \ldots, d_{dn}$ $(dn \geq 1)$ are interconnected in networks. In this paper, we consider three types of devices, sensor, actuator, and hybrid device. A sensor device just collects data by sensing events which occur in physical environment. An actuator device acts according to the action request from a subject. Let SB be a set of subjects $sb_1, \ldots, sb_{sbn}$ $(sbn \geq 1)$. A subject sb_i gets and puts data from a sensor s and to an actuator a, respectively. In addition, a subject sb_i issues both get and put operations to a hybrid device h like robots and cars.

Through manipulating devices, data are exchanged among devices and subjects. If data of a device d flow into an entity e like subjects and devices, the device d is referred to as a *source* device of the entity e. Let $sb_i.D$ and $d.D$ be sets of source devices whose data flow into a subject sb_i and a device d, respectively, which are initially ϕ. For example, if data of a device d_1 flow into a subject sb_i and a device d_2, the sets $sb_i.D$ and $d_2.D$ include the device d_1, i.e. $sb_i.D = d_2.D = \{d_1\}$. Let $minT_e^s$ be the oldest generation time of the data which are generated in the device s and flow into an entity e.

3 Information Flow Relations

In this section, we define types of information flow relations on devices and subjects based on the CapBAC model. Let $IN(sb_i)$ be a set of devices whose data a subject sb_i is allowed to get, i.e. $IN(sb_i) = \{d \mid \langle d, get \rangle \in CAP_i\}$ $(\subseteq D)$. A subject sb_i can get data from a device d $(\in D)$ only if the device d is included in the set $IN(sb_i)$.

Definition 1. A device d *flows* to a subject sb_i $(d \rightarrow sb_i)$ iff (if and only if) $d.D \neq \phi$ and $d \in IN(sb_i)$.

If $d \rightarrow sb_i$ holds, data brought to the device d may be brought to the subject sb_i. Otherwise, no data flow from the device d into the subject sb_i.

Definition 2. A device d *legally flows* to a subject sb_i $(d \Rightarrow sb_i)$ iff $d \rightarrow sb_i$ and $d.D \subseteq IN(sb_i)$.

The condition "$d.D \subseteq IN(sb_i)$" means that the data in the device d are allowed to be brought into the subject sb_i.

Definition 3. A device d *illegally flows* to a subject sb_i ($d \mapsto sb_i$) iff $d \rightarrow sb_i$ and $d.D \nsubseteq IN(sb_i)$.

The condition "$d.D \nsubseteq IN(sb_i)$" means that the data in some device of $d.D$ are not allowed to be brought into the subject sb_i.

In the legal information flow relation ($\Rightarrow$), only devices are checked. In the following information flow relations, not only devices but also time are checked.

Definition 4. A device d *timely flows* to a subject sb_i ($d \Rightarrow_t sb_i$) iff $d \Rightarrow sb_i$ and $\forall s \in d.D$ ($gt_i^s.st \leq minT_d^s \leq gt_i^s.et$).

The condition "$\forall s \in d.D$ ($gt_i^s.st \leq minT_d^s \leq gt_i^s.et$)" means that only the data generated in the device s while the subject sb_i is allowed to issue a get operation to the device s flow into the subject sb_i.

Definition 5. A device d *flows late* to a subject sb_i ($d \mapsto_l sb_i$) iff $d \Rightarrow sb_i$ and $\exists s \in d.D$ ($\neg(gt_i^s.st \leq minT_d^s \leq gt_i^s.et)$).

The condition "$\exists s \in d.D$ ($\neg(gt_i^s.st \leq minT_d^s \leq gt_i^s.et)$)" means that the data generated in the device s while the subject sb_i is not allowed to issue a get operation to the device s flow into the subject sb_i. This means, the subject sb_i can get data even if the data are not generated while the subject sb_i is allowed to get the data.

4   Protocols

In this paper, we consider three types of devices, sensor, actuator, and hybrid devices.

Sensor and hybrid devices collect data by sensing events occurring around them. Once a device d gets data by sensing events, the device d is added to the set $d.D$ of the device d. There are two types, full and partial types of sensing of sensor and hybrid devices. If a device collects data in a full sensing at time τ, the whole data of the device are overwritten. On the other hand, if the device collects data in a partial sensing at time τ, only some data are overwritten.

Actuators and hybrid devices get data which are collected by other devices like sensors and hybrid devices. Subjects issue get operations to sensors and hybrid devices to get data collected by the sensors and hybrid devices. On the other hand, subjects issue put operations to actuators and hybrid devices to store data which the subjects get from sensors and hybrid devices. There are two types, full and partial types for each get and put operation of subjects. If a subject sb_i issues a full get operation to a device d, the whole data obtained by the subject sb_i are overwritten. On the other hand, if the subject sb_i issues a partial get operation to the device d, only some data are overwritten. If a subject sb_i issues a full put operation to a device d, the whole data stored in the device d are overwritten. On the other hand, if the subject sb_i issues a partial put operation to the device d, only some data are overwritten.

4.1 An OI (Operation Interruption) Protocol

In the IoT with the CapBAC model, a capability token CAP_i which is a set of access rights is issued to a subject sb_i. If the capability token CAP_i includes an access right $\langle d, op \rangle$, the subject sb_i is allowed to manipulate the device d in an operation op. If data of a device d_1 brought into a device d_2 are brought into a subject sb_i which is not allowed to get the data of the device d_1, the data of the device d_1 illegally flow into the subject sb_i. In Sect. 3, the illegal information flow is defined based on the CapBAC model. In this section, we discuss the OI (Operation Interruption) protocol [8] to prevent illegal information flow based on the information flow relations among subjects and devices.

$IN(sb_i)$ is a set of devices whose data are allowed to be got by a subject sb_i, i.e. $IN(sb_i) = \{d \mid \langle d, get \rangle \in CAP_i\}$. $sb_i.D$ is a set of source devices whose data are brought to the subject sb_i. $d.D$ is a set of source devices whose data are brought to the device d. For each subject sb_i and device d, the sets $sb_i.D$ and $d.D$ are manipulated, which are initially ϕ. If a subject sb_i issues a get operation to a device d, it is checked whether or not the subject sb_i is allowed to get all the data in the device d. If at least one device whose data are not allowed to be brought into the subject sb_i exists in the device d, i.e. $d.D \nsubseteq IN(sb_i)$, the get operation is interrupted to prevent illegal information flow.

The OI protocol is shown as follows:

[OI (Operation Interruption) Protocol]

1. A device d gets data by sensing events occurring around the device d.
   a. $d.D = d.D \cup \{d\}$;
   b. Then, the device d collects data;
2. A subject sb_i issues a get operation to a device d, i.e. $d \rightarrow sb_i$ holds.
   a. If $d \Rightarrow sb_i$,
      i. If a full get operation is issued, $sb_i.D = d.D$;
      ii. If a partial get operation is issued, $sb_i.D = sb_i.D \cup d.D$;
      iii. Then, the subject sb_i gets data from the device d;
   b. Otherwise, the get operation is interrupted at the device d;
3. A subject sb_i issues a put operation to a device d.
   a. If a full put operation is issued, $d.D = sb_i.D$;
   b. If a partial put operation is issued, $d.D = d.D \cup sb_i.D$;
   c. Then, the subject sb_i puts data to the device d;

4.2 A TBOI (Time-Based OI) Protocol

In the OI protocol, every illegal information flow is prevented by interrupting operations implying illegal information flow. However, a subject can get data generated at time τ although the subject is not allowed to get data at time τ. In this case, the data come to the subject later than expected by the subject to get the data, i.e. the data flow late to the subject.

In this section, we discuss a TBOI (Time-Based OI) protocol [10] to prevent late information flow in addition to illegal information flow based on the information flow relations among subjects and devices. Here, time is considered differently from the OI protocol. In the OI protocol, each device makes an authorization decision based on sets of devices whose data flow into each entity. On the other hand, in the TBOI protocol, the time when data are generated and validity period of an access right to get the data are used in addition to sets of devices. Every subject sb_i is allowed to issue an get operation to a device s at time between $gt_i^s.st$ and $gt_i^s.et$. $sb_i.D$ and $d.D$ are sets of source devices whose data flow into a subject sb_i and a device d, respectively. $minT_e^s$ is the oldest generation time of the data which are generated in the device s and flow into an entity e. For each subject sb_i and device d, the variable $minT_e^s$ is manipulated. If a subject sb_i issues a get operation to a device d, it is checked whether or not every data of a device s in the device d which may be brought into the subject sb_i are generated during the validity period of the access right $\langle s, get \rangle$ granted to the subject sb_i by using the variables $minT_d^s$, $gt_i^s.st$, and $gt_i^s.et$. If the data generated at time when the subject sb_i is not allowed to get the data exist in the device d, i.e. $\exists s \in d.D$ $(\neg(gt_i^s.st \leq minT_d^s \leq gt_i^s.et))$, the get operation is interrupted to prevent late information flow.

The TBOI protocol is shown as follows:

[TBOI (Time-Based OI) Protocol]

1. A device d gets data by sensing events occurring around the device d at time τ.
   a. $d.D = d.D \cup \{d\}$;
   b. If $minT_d^d = NULL$, $minT_d^d = \tau$;
   c. If a full sensing is performed, $minT_d^d = \tau$;
      /* If a partial sensing is performed, $minT_d^d$ is not changed */
   d. Then, the device d collects data;
2. A subject sb_i issues a get operation to a device d, i.e. $d \rightarrow sb_i$ holds.
   a. If $d \Rightarrow_t sb_i$,
      i. If a full get operation is issued,
         A. $sb_i.D = d.D$;
         B. For each device s such that $s \in d.D$, $minT_i^s = minT_d^s$;
      ii. If a partial get operation is issued,
         A. $sb_i.D = sb_i.D \cup d.D$;
         B. For each device s such that $s \in (sb_i.D \cap d.D)$, $minT_i^s = min(minT_i^s, minT_d^s)$;
         C. For each device s such that $s \notin sb_i.D$ and $s \in d.D$, $minT_i^s = minT_d^s$;
      iii. Then, the subject sb_i gets data from the device d;
   b. Otherwise, the get operation is interrupted at the device d;
3. A subject sb_i issues a put operation to a device d.
   a. If a full put operation is issued,
      i. $d.D = sb_i.D$;
      ii. For each device s such that $s \in sb_i.D$, $minT_d^s = minT_i^s$;

b. If a partial put operation is issued,
  i. $d.D = d.D \cup sb_i.D$;
  ii. For each device s such that $s \in (d.D \cap sb_i.D)$, $minT_d^s = min(minT_d^s, minT_i^s)$;
  iii. For each device s such that $s \notin d.D$ and $s \in sb_i.D$, $minT_d^s = minT_i^s$;
c. Then, the subject sb_i puts data to the device d;

Example 1. Figure 1 shows an example of the TBOI protocol. There are two devices, a sensor s and a hybrid device h. Suppose a capability token including a pair of access rights $\langle s, get \rangle$ and $\langle h, put \rangle$ is issued to a subjects sb_i. On the other hand, a capability token including the access rights $\langle s, get \rangle$ and $\langle h, get \rangle$ is issued to another subjects sb_j. We also suppose $gt_i^s.st = t_1$, $gt_i^s.et = t_2$, $gt_j^s.st = t_3$, $gt_j^s.et = t_4$, $gt_j^h.st = t_3$, and $gt_j^h.et = t_4$. Here, $IN(sb_i) = \{s\}$ and $IN(sb_j) = \{s, h\}$.

First, the sensor s collects data by sensing events occurring around itself at time τ, i.e. $s.D = \{s\}$ and $minT_s^s = \tau$. Next, the subject sb_i gets data from the sensor s because $s \Rightarrow_t sb_i$ holds. Hence, $sb_i.D = sb_i.D (= \phi) \cup s.D (= \{s\}) = \{s\}$ and $minT_i^s = minT_s^s = \tau$. Then, the hybrid device h collects data by sensing events occurring around itself at time τ', i.e. $h.D = \{h\}$ and $minT_h^h = \tau'$. After that, the subject sb_i puts the data to the hybrid device h. Here, $h.D = h.D (= \{h\}) \cup sb_i.D (= \{s\}) = \{h, s\}$ and $minT_h^s = minT_i^s = \tau$. Finally, the subject sb_j tries to get the data from the hybrid device h. Here, $h \mapsto_l sb_j$ holds because $minT_h^s (= \tau) < gt_j^s.st (= t_3)$. Hence, the get operation of the subject sb_j is interrupted at the hybrid device h to prevent late information flow.

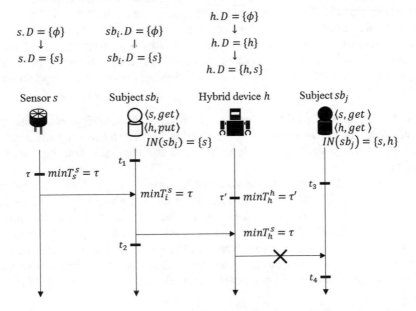

Fig. 1. TBOI protocol.

5 Evaluation

We evaluate the TBOI protocol on a subject set $SB = \{sb_1, \ldots, sb_{sbn}\}$ and a device set $D = \{d_1, \ldots, d_{dn}\}$ in terms of the number of operations interrupted.

In the evaluation, we assume that every subject issues only get and put operations to sensors and actuators, respectively. On the other hand, every subject issues get and put operations to hybrid devices. This means, for each sensor s and actuator a, there is one access right $\langle s, get \rangle$ and $\langle a, put \rangle$ and for each hybrid device h, there are a pair of access rights $\langle h, get \rangle$ and $\langle h, put \rangle$. Let sn, an, and hn be the numbers of sensors, actuators, and hybrid devices, respectively. Here, a set AC of access rights includes $sn + an + 2hn$ access rights. Every capability token CAP_i issued to a subject sb_i is composed of access rights. Let $mxan$ show the maximum number of access rights to be granted to every subject. acn_i is the number of access rights granted to a subject sb_i. Here, the number acn_i is randomly selected out of numbers $1, \ldots, mxan$. The number acn_i of access rights are randomly granted to each subject sb_i.

Let $ac_{i,j}$ be an access right in a capability token CAP_i issued to a subject sb_i. Every access right $ac_{i,j}$ has validity period $vp_{i,j}$ [tu] which shows how long the access right $ac_{i,j}$ is valid. $mxvp$ shows maximum validity period [tu] of every access right. If an access right $ac_{i,j}$ is granted to a subject sb_i, the validity period $vp_{i,j}$ of the access right $ac_{i,j}$ is randomly selected out of numbers $1, \ldots, mxvp$ [tu].

In the evaluation, we consider twenty five subjects ($sbn = 25$) and fifty devices ($dn = 50$). First, sbn subjects and dn devices are randomly generated, i.e. $SB = \{sb_1, \ldots, sb_{25}\}$ and $D = \{d_1, \ldots, d_{50}\}$. Every subject is issued capability tokens which are composed of access rights randomly selected from a set AC. The number of access rights granted to each subject in capability tokens is randomly selected out of numbers $1, \ldots, 30$, i.e. $mxan = 30$. The validity period $vp_{i,j}$ of each access right $ac_{i,j}$ is randomly selected out of numbers $1, \ldots, 30$, i.e. $mxvp = 30$. After the generation of subjects and devices, the following procedures are performed in the TBOI protocol:

1. Every sensor and hybrid device d collects data by sensing events with probability 0.2. Here, the sensing type is randomly selected so that the sensing is full one with probability 0.5. If a device d collects data, the data stored in the device d.
2. Every subject sb_i issues an operation with probability 0.5. Then, the operation type is randomly selected in get and put with same probability. If a subject sb_i issues a get operation to a device d, data in the device d flow into the subject sb_i. On the other hand, if a subject sb_i issues a put operation to a device d, data in the subject sb_i flow into the device d. For each operation, full and partial types are randomly selected with same probability. For example, a subject sb_i issues a full get operation with probability 0.125 ($= 0.5 \times 0.5 \times 0.5$).
3. For every subject sb_i, the validity period $vp_{i,j}$ of every access right $ac_{i,j}$ is decremented by one. If the validity period $vp_{i,j}$ gets 0, the access right $ac_{i,j}$

is revoked from the subject sb_i. If every access right $ac_{i,j}$ included in the capability token CAP_i is revoked from the subject sb_i, the capability token CAP_i is also revoked from the subject sb_i.

4. Every subject sb_i which has no capability token is issued capability tokens which are composed of access rights randomly selected from a set AC. acn_i is randomly selected out of numbers $1, \ldots, mxan$.

In the simulation, we make the following assumptions:

- If $\exists s \in d.D \ (minT_d^s < minT_i^s)$, a subject sb_i does not issue a full get operation to a device d.
- If $\exists s \in sb_i.D \ (minT_i^s < minT_d^s)$, a subject sb_i does not issue a full put operation to a device d.
- Every subject sb_i does not issue a get operation to a device d such that $d.D = \phi$.
- Every subject sb_i such that $sb_i.D = \phi$ does not issue a put operation to a device d.

The above procedures are assumed to be performed in one time unit [tu]. Let st be a simulation time. This means, the above procedures are iterated st times. Here, $st = 0, 50, 100, 150,$ or 200. We randomly create a subject set SB and a device set D one hundred times. For a pair of given sets SB and D, the above procedures for st [tu] are iterated one hundred times. Finally, we calculate the average number of illegal and late get operations interrupted.

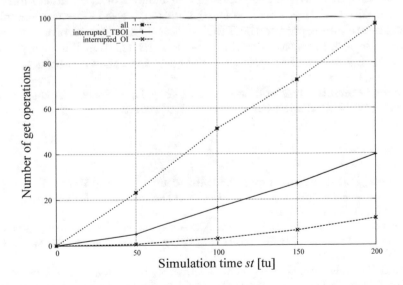

Fig. 2. Number of get operations.

Figure 2 shows the numbers of get operations for the simulation time st [tu] in the OI and TBOI protocols. The line with the label "all" shows the total number of get operations issued by sbn subjects. The other lines with the labels "interrupted_TBOI" and "interrupted_OI" show the number of get operations interrupted in the TBOI and OI protocols, respectively. Here, the more number of get operations are interrupted in the TBOI protocol than the OI protocol because late information flow is prevented in addition to illegal information flow in the TBOI protocol differently from the OI protocol.

6 Concluding Remarks

In order to make IoT (Internet of Things) secure, we take the CapBAC (Capability-Based Access Control) model. In this paper, we consider three types of devices, sensor, actuator, and hybrid devices. In the IoT, a subject sb_i can get data of a device d_1 from another device d_2 to which the data of the device d_1 are brought although the subject sb_i is not allowed to get the data from the device d_1. Here, illegal information flow occurs. In our previous studies, the OI (Operation Interruption) protocol where operations implying illegal information flow are interrupted is proposed to prevent illegal information flow in the IoT based on the CapBAC model. However, in the OI protocol, late information flow such that a subject sb_i can get data of a device d_1 generated at time τ even if the subject sb_i is not allowed to get the data at time τ occurs. Hence, the TBOI (Time-Based OI) protocol where operations implying illegal and late information flow are interrupted is proposed in our previous studies. In this paper, we evaluate the TBOI protocol in terms of the number of operations interrupted compared with the OI protocol. In the evaluation, we show the more number of operations are interrupted in the TBOI protocol than the OI protocol because not only illegal information flow but also late information flow are prevented from occurring in the TBOI protocol differently from the OI protocol.

Acknowledgements. This work was supported by Japan Society for the Promotion of Science (JSPS) KAKENHI Grant Number JP17J00106.

References

1. Denning, D.E.R.: Cryptography and Data Security. Addison Wesley, Boston (1982)
2. Fernandez, E.B., Summers, R.C., Wood, C.: Database Security and Integrity. Adison Wesley, Boston (1980)
3. Gusmeroli, S., Piccione, S., Rotondi, D.: A capability-based security approach to manage access control in the internet of things. Math. Comput. Modell. **58**(5–6), 1189–1205 (2013)
4. Hanes, D., Salgueiro, G., Grossetete, P., Barton, R., Henry, J.: IoT Fundamentals: Networking Technologies, Protocols, and Use Cases for the Internet of Things. Cisco Press, Indianapolis (2018)
5. Hernández-Ramos, J.L., Jara, A.J., Marín, L., Skarmeta, A.F.: Distributed capability-based access control for the internet of things. J. Internet Serv. Inf. Secur. **3**(3/4), 1–16 (2013)

6. Nakamura, S., Duolikun, D., Enokido, T., Takizawa, M.: A flexible read-write abortion protocol to prevent illegal information flow among objects. J. Mobile Multimedia **11**(3&4), 263–280 (2015)
7. Nakamura, S., Enokido, T., Takizawa, M.: Causally ordering delivery of event messages in P2 PPSO systems. Cogn. Syst. Res. **56**, 167–178 (2019)
8. Nakamura, S., Enokido, T., Takizawa, M.: Information flow control based on the CapBAC (capability-based access control) model in the IoT. Int. J. Mobile Comput. Multimedia Commun. **10**(4), 13–25 (2019)
9. Nakamura, S., Enokido, T., Takizawa, M.: Protocol to efficiently prevent illegal flow of objects in P2P type of publish/subscribe (ps) systems. SOCA **13**(4), 323–332 (2019)
10. Nakamura, S., Enokido, T., Takizawa, M.: A TBOI (time-based operation interruption) protocol to prevent late information flow in the IoT. In: Proceedings of the 14th International Conference on Broad-Band Wireless Computing, Communication and Applications, pp. 93–104 (2019)
11. Nakamura, S., Ogiela, L., Enokido, T., Takizawa, M.: An information flow control model in a topic-based publish/subscribe system. J. High Speed Netw. **24**(3), 243–257 (2018)
12. Oma, R., Nakamura, S., Duolikun, D., Enokido, T., Takizawa, M.: An energy-efficient model for fog computing in the internet of things (IoT). Internet Things Eng. Cyber Phys. Hum. Syst. **1-2**, 14–26 (2018)
13. Sandhu, R.S.: Lattice-based access control models. IEEE Comput. **26**(11), 9–19 (1993)
14. Sandhu, R.S., Coyne, E.J., Feinstein, H.L., Youman, C.E.: Role-based access control models. IEEE Comput. **29**(2), 38–47 (1996)
15. Yuan, E., Tong, J.: Attributed based access control (ABAC) for web services. In: Proceedings of the IEEE International Conference on Web Services (ICWS 2005) (2005)

A Dynamic Tree-Based Fog Computing (DTBFC) Model for the Energy-Efficient IoT

Ryuji Oma[1]([✉]), Shigenari Nakamura[1], Tomoya Enokido[2], and Makoto Takizawa[1]

[1] Hosei University, Tokyo, Japan
ryuji.oma.6r@stu.hosei.ac.jp, nakamura.shigenari@gmail.com,
makoto.takizawa@computer.org
[2] Rissho University, Tokyo, Japan
eno@ris.ac.jp

Abstract. In order to increase the performance of the IoT (Internet of Things), types of the FC (Fog Computing) models are proposed. Here, subprocesses of an application process to handle sensor data are performed on fog nodes in addition to servers. Fog nodes are hierarchically structured where a root node shows a cluster of servers and a leaf node indicates an edge node which communicates with sensors and actuators. As sensor data increases and decreases, some fog node gets overloaded and underloaded, respectively. In this paper, we propose a dynamic tree model for fog computing. A fog node supports a subsequence of subprocesses of an application process. If a fog node is overloaded, a fog node is splitted to a pair of nodes which support a prefix subsequence and the other postfix subsequence of the subprocesses, respectively. In another way, a fog node is replicated to a pair of nodes each of which supports the same subprocesses while receiving different input data. We propose a model to give the energy consumption of nodes splitted and replicated. By using the model, we newly consider a DTBFC (Dynamic TBFC) model where fog nodes are dynamically created and dropped depending on traffic of nodes so that the energy consumption and execution time of fog nodes can be reduced.

Keywords: TBFC (Tree-Based Fog Computing) model · DTBFC (Dynamic TBFC) model · Eco IoT

1 Introduction

In the IoT (Internet of Things) [8], not only computers like servers and clients but also millions of sensors and actuators are interconnected in networks. The FC (Fog Computing) model [15] is proposed to reduce the communication and processing traffic to handle sensor data in the IoT. Sensor data is processed by a fog node and the output data is sent to another fog node. Thus, servers in clouds finally receive data processed by fog nodes.

In order to reduce the energy consumption and execution time of fog nodes and servers, types of the TBFC (Tree-Based FC) models [6,7,9–14] are proposed. Here, fog nodes are hierarchically structured in a tree, where a root node is a server and a

© Springer Nature Switzerland AG 2020
L. Barolli et al. (Eds.): EIDWT 2020, LNDECT 47, pp. 24–34, 2020.
https://doi.org/10.1007/978-3-030-39746-3_4

leaf node is an edge node which communicates with sensors and actuators. Each node receives input data from child nodes, obtains output data by processing the input data, and sends the output data to a parent node. Here, each subprocess is supported by some fog node and the tree structure of fog nodes is not changed even if the amount of sensor data is so changed that nodes are overloaded or underloaded.

In this paper, we newly propose a DTBFC (Dynamic TBFC) model where fog nodes are dynamically created and dropped as the traffic of each node increases and decreases, respectively, so that the energy consumption and execution time of nodes can be reduced. In this paper, we assume an application process is realized as a sequence of subprocesses. An edge subprocess receives input data from sensors and sends output data obtained by processing the input data to a succeeding subprocess. Thus, each subprocess receives input data from a preceding subprocess and sends output data to a succeeding subprocess. Initially, every subprocess of an application process is supported by a root fog node f like the CC (Cloud Computing) model [2]. As the amount of sensor data increases, the node f is more heavily loaded since the node f receives more amount of input data from child nodes. A node f supporting subprocesses is splitted or replicated to a pair of nodes f and f_1 so that the energy consumption of the nodes can be reduced. In the splitting way, the nodes f and f_1 support different prefix and the other postfix subsequences of the subprocesses, respectively. Here, the processing load can be distributed to the nodes f and f_1. In the replication way, the node f_1 supports the same subsequences of the subprocesses as the node f. The input data sent by the child nodes can be distributed to the nodes f and f_1. We propose a model to give the energy consumption and execution time of each fog node splitted and replicated.

In Sect. 2, we propose the DTBFC model. In Sect. 3, we discuss the energy consumption and execution time of a fog node. In Sect. 4, we discuss algorithms to split and replicate fog nodes.

2 Dynamic Tree-Based Fog Computing (DTBFC) Model

An application process P to handle data from sensors and to issue actions to actuators is assumed to be a sequence of subprocesses $p_1, ..., p_m$ $(m \geq 1)$ in this paper. The subprocesses p_1 and p_m are a root and edge ones, respectively. Here, the edge subprocess p_m receives input data from sensors. The subprocess p_m obtains output data d_m by processing the input data from sensors. For example, the subprocess takes an average value of all the input data as output data. Then, the subprocess p_m sends the output data d_m to the preceding subprocess p_{m-1}. Thus, each subprocess p_i receives input data d_{i+1} from a succeeding subprocess p_{i+1} and sends output data d_i to a preceding subprocess p_{i-1}. The output ratio r_i of the subprocess p_i is $|d_i|/|d_{i+1}|$. For example, if d_{i+1} is a collection of ten values $v_1, ..., v_{10}$ and an average value v of the ten values is obtained as output data d_i by a subprocess p_i, the output ratio r_i is 0.1. A sequence $P' = \langle p_i, p_{i+1}, ..., p_j \rangle$ $(1 \leq i \leq j \leq m)$ of subprocesses is a *subsequence* of a sequence $P = \langle p_1, ..., p_i, ..., p_j, ..., p_m \rangle$ $(P' \sqsubset P)$ of subprocesses. Subsequences $p_1, ..., p_i$ and $p_j, ..., p_m$ $(1 \leq i, j \leq m)$ are a prefix and postfix of a sequence $p_1, ..., p_m$.

Initially, all the subprocesses $p_1, ..., p_m$ are supported by one fog node f which is a root node, e.g. servers in a cloud. Every sensor initially sends sensor data to the root node f where the data is processed by the subprocesses $p_1, ..., p_m$ of the application process P. As the amount of sensor data increases, the execution time and energy consumption of the root node f increase and network traffic to the root node f also increases. In order to improve the performance and energy consumption, some subprocesses of the node f are distributed to another node f_1. There are two ways to distribute subprocesses, *splitting* and *replication* of the node f. In one way, the node f is *splitted* to a pair of nodes f and f_1, where a prefix subsequence $p_1, ..., p_l$ $(l \leq m)$ of the subprocesses are supported by the node f and the postfix subsequence $p_{l+1}, ..., p_m$ are supported by the node f_1 as shown in Fig. 1(1). The node f_1 receives all the data from the child nodes and processes the data by the subprocess $p_{l+1}, ..., p_m$. The node f_1 then sends the output data d_{l+1} of the subprocess p_{l+1} to the node f. Here, the node f is a parent node of the node f_1 and the node f_1 is a child node of the node f. The processing load of the node f thus decreases since the subprocesses $\langle p_{l+1}, ..., p_m \rangle$ are not performed. If the amount of input data of the nodes f_1 and f_2 decreases, the node f_1 is merged into the node f, i.e. the node f_1 is dropped and the node f supports all the subprocesses $p_1, ..., p_l, p_{l+1}, ..., p_m$.

Next, if the node f_1 gets too heavily loaded, the node f_1 is *replicated* to a node f_2, i.e. the subsequence $p_{l+1}, ..., p_m$ of the subprocesses are supported by both the nodes f_1 and f_2 as shown in Fig. 1(2). Each of the child nodes sends data to one of the nodes f_1 and f_2. Here, the nodes f_1 and f_2 process smaller amount of sensor data. The nodes f_1 and f_2 send the output data to the parent node f. The nodes f_1 and f_2 are child nodes of the node f. If the amount of sensor data to be processed by the nodes f_1 and f_2 decreases so that one node can process all sensor data, one node, say the node f_2 is dropped and every sensor data is sent to the node f_1.

In the DTBFC model, a root node f has child nodes $f_1, ..., f_c$ $(c \geq 1)$. Here, each node f_i has also child nodes $f_{i1}, ..., f_{i,c_i}$ $(c_i \geq 1)$. Thus, a fog node f_R has child nodes $f_{R1}, ..., f_{R,c_R}$ $(c_R \geq 1)$ where the label R is a sequence of numbers. The label R of a node f_R shows a path from the root node f to the node f_R. Here, f_R is a parent node of each node f_{Ri}. Let $C(f_R)$ be a set of child nodes and $pt(f_R)$ be a parent node of a node f_R. Each node f_R supports a subsequence $SP(f_R)$ $(\sqsubseteq P)$ of subprocesses in an application process P. The output ratio ρ_R of a node f_R is $\Pi_{p_i \in SP(f_R)} r_i$. For each edge node f_R where $R = i_1 ... i_h$ $(1 \leq h \leq m)$, a sequence $SP(f_{i_1}), SP(f_{i_2}), ..., SP(f_{i_1...i_h})$ of subprocesses $f_{i_1}, f_{i_2}, ..., f_{i_1...i_h}$ is a sequence $p_1, ..., p_m$ of the application process P.

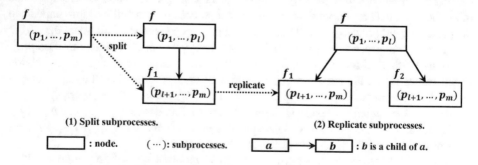

(1) Split subprocesses.   (2) Replicate subprocesses.

☐ : node.   (···): subprocesses.   [a] → [b] : b is a child of a.

Fig. 1. Splitting and replication of subprocesses.

3   Computation and Energy Consumption Models of a Fog Node

3.1   Computation Model

An application process P is a sequence $\langle p_1, ..., p_m \rangle$ ($m \geq 1$) of subprocesses. A node f_R takes input data $D_R = \{d_{R1}, ..., d_{R,c_R}\}$ of size i_R (= $|D_R|$) from child nodes $f_{R1}, ..., f_{R,c_R}$, respectively. The input data D_R is processed and the output data d_R of size o_R (= $|d_R|$) is generated by the subprocesses $SP(f_R)$. Here, $SP(f_R)$ is a subsequence $\langle p_{s_R}, p_{s_R+1}, ..., p_{e_R} \rangle$ ($\sqsubseteq P$) of subprocesses supported by a node f_R where $1 \leq s_R \leq e_R \leq m$. Then, the node f_R sends the output data d_R to a parent node $pt(f_R)$, where $o_R = \rho_R \cdot i_R$ for the output ratio ρ_R.

Let CR_R show the computation rate of a fog node f_R to the root node, i.e. server f. We assume the computation rate CR of a root node f is 1 and $CR_R \leq 1$ for every node f_R. This means, the root node f is $1/CR_R$ times faster than a node f_R. $TP_i(x)$ and $TP_{iR}(x)$ show the execution time [sec] of a subprocess p_i to process data of size x by a root node f and a node f_R, respectively. The execution time $TP_i(x)$ of each subprocess p_i on a root node f to process input data of size x is $ct_i \cdot C_i(x)$ where ct_i is a constant and $C_i(x)$ is assumed to be x or x^2. The size of the output data of the subprocesses p_i is $r_i \cdot x$ where r_i is the output ratio of the subprocess p_i. If a subprocess p_i is performed on a fog node f_R, the execution time $TP_{iR}(x)$ is $TP_i(x)/CR_R$. $TC_R(SP(f_R), x)$ shows the execution time [sec] of a node f_R to process data of size x by the subprocesses $SP(f_R)$. The total execution time $TC_R(SP(f_R), x)$ of the subprocesses $SP(f_R)$ on a node f_R to process input data D_R of size x is given as follows:

$$TC_R(SP(f_R), x) = \Sigma_{i=s_R}^{e_R} TP_{iR}(rf_i \cdot x) = \Sigma_{i=s_R}^{e_R} TP_i(rf_i \cdot x)/CR_R. \qquad (1)$$

Here, $rf_{e_R} = 1$ and $rf_i = r_{e_R} \cdot r_{e_R-1} \cdot ... \cdot r_{i-1}$ for $s_R \leq i \leq e_R$. The output ratio ρ_R of the node f_R is $rf_{s_R} \cdot r_{s_R} = r_{e_R} \cdot r_{e_R-1} \cdot ... \cdot r_{s_R}$. The size of the output data d_{s_R} of the subprocess p_{s_R}, i.e. the size of the output data d_R of the node f_R is $\rho_R \cdot x$.

The execution time $TI_R(x)$ and $TO_R(x)$ of a node f_R to receive and send data of size x are proportional to the data size x, i.e. $TI_R(x) = rc_R + rt_R \cdot x$ and $TO_R(x) = sc_R + st_R \cdot x$. Here, rc_R, rt_R, sc_R, and st_R are constants. A node f_R receives input data $d_{R1}, ..., d_{R,c_R}$ from child nodes $f_{R1}, ..., f_{R,c_R}$, respectively, where $x_i = |d_{Ri}|$ and $x = x_1 + \cdots + x_{c_R}$. It takes $TTI_R(x) = TI_R(x_1) + \cdots + TI_R(x_{c_R})$ [sec] to receive the input data $d_{R1}, ..., d_{R,c_R}$.

It takes totally $TF_R(x)$ (= $TTI_R(x) + TC_R(SP(f_R), x) + \delta_R \cdot TO_R(\rho_R \cdot x)$) [sec] to process input data $D_R = \{d_{R1}, ..., d_{R,c_R}\}$ of size x in each node f_R. Here, if f_R is a root, $\delta_R = 0$, else $\delta_R = 1$. The total execution time $TT_R(\langle p_{s_R}, ..., p_{e_R} \rangle, x)$ [sec] of a node f_R with the subprocesses $SP(f_R) = \langle p_{s_R}, ..., p_{e_R} \rangle$ for input data D_R of size x is given as follows:

$$TT_R(\langle p_{s_R}, ..., p_{e_R} \rangle, x) = TTI_R(x) + TC_R(\langle p_{s_R}, ..., p_{e_R} \rangle, x) + \delta_R \cdot TO_R(\rho_R \cdot x) \qquad (2)$$
$$= TTI_R(x) + \Sigma_{i=s_R}^{e_R} TP_i(rf_i \cdot x)/CR + \delta_R \cdot TO_R(\rho_R \cdot x).$$

In a Raspberry Pi 3 Model B [1] node f_R which is connected with a 10 [Gbps] network, $rc_R = 0$, $sc_R = st_R \cdot 7$, $rt_R = 9 \cdot st_R$, $rt_R = ct_R$, $st_R = 0.00001$, and $rt_R = 0.00009$ [11].

3.2 Energy Consumption Model

$EI_R(x)$, $EC_R(SP(f_R), x)$, and $EO_R(x)$ show electric energy [J] consumed by a node f_R to receive, process, and send data of size x, respectively. In this paper, we assume each node f_R follows the SPC (Simple Power Consumption) model [3–5] for simplicity. The power consumption of a node f_R to process data is $maxE_R$ [W]. A node f_R consumes the power $maxE_R$ [W] for time $TC_R(SP(f_R), x)$ [sec] to process data of size x. Hence, the energy $EC_R(SP(f_R), x)$ [J] consumed by a node f_R to perform subprocesses $SP(f_R)$ to process input data of size x (> 0) is $maxE_R \cdot TC_R(SP(f_R), x)$.

The power consumption PI_R and PO_R [W] of a node f_R to receive and send data are $re_R \cdot maxE_R$ and $se_R \cdot maxE_R$, respectively. In the Raspberry Pi 3 Model B node f_R, $re_R = 0.729$, $se_R = 0.676$, and $maxE_R = 3.7$ [W] [11]. The energy consumption $EI_R(x)$ and $EO_R(x)$ [J] to receive and send data of size x (> 0), respectively, are $EI_R(x) = PI_R \cdot TTI_R(x) = re_R \cdot maxE_R \cdot (c_R \cdot rc_R + rt_R \cdot x)$ and $EO_R(x) = PO_R \cdot TO_R(x) = se_R \cdot maxE_R \cdot (sc_R + st_R \cdot x)$.

A node f_R consumes the energy $EF_R(SP(f_R), x)$ to receive and process input data $d_{R1}, ..., d_{R,c_R}$ of size $x_1, ..., x_{c_R}$, respectively, where $x = x_1 + \cdots + x_{c_R}$ by the subprocesses $SP(f_R) = \langle p_{s_R}, ..., p_{e_R} \rangle$ and to send the output data d_R of size $\rho_R \cdot x$ to a parent node $pt(f_R)$:

$$
\begin{aligned}
EF_R(SP(f_R), x) &= EI_R(x) + EC_R(SP(f_R), x) + \delta_R \cdot EO_R(\rho_R \cdot x) \\
&= \{re_R \cdot TTI_R(x) + TC_R(SP(f_R), x) + \delta_R \cdot se_R \cdot TO_R(\rho_R \cdot x)\} \cdot maxE_R \\
&= \{(re_R \cdot (c_R \cdot rc_R + rt_R \cdot x)) + \Sigma_{i=s_R}^{e_R} ct_i \cdot C_i(rf_i \cdot x)/CR_R \\
&\quad + \delta_R \cdot se_R \cdot (sc_R + st_R \cdot \rho_R \cdot x)\} \cdot maxE_R.
\end{aligned}
\tag{3}
$$

Here, $\delta_R = 0$ if f_R is a root, else $\delta_R = 1$.

4 Algorithms to Split and Replicate Nodes

4.1 Splitting and Replicating Nodes

An application process P is a sequence $\langle p_1, ..., p_m \rangle$ ($m \geq 1$) of subprocesses. Suppose a node f_R supports a subsequence $SP(f_R) = \langle p_{s_R}, ..., p_l \rangle$ ($\sqsubseteq P$) of subprocesses and a child node f_{Ri} supports a subsequence $SP(f_{Ri}) = \langle p_{l+1}, ..., p_{e_R} \rangle$ ($\sqsubseteq P$) as shown in Fig. 2. Suppose a subprocess p_l of the node f_R moves to the child node f_{Ri} and the size of input data of D_R is x. Here, the execution time of the node f_R decreases since the subprocess p_l is additionally performed on the child node f_{Ri}. Hence, the execution time of the node f_R is decreased by $TI_R(r_l \cdot x) - (TI_R(x) + TP_l(x)/CR_R)$ while the execution time of the node f_{Ri} is increased by $(TO_{Ri}(r_l \cdot x) + TP_l(x)/CR_{Ri}) - TO_{Ri}(x)$. Here, r_l is the output ratio of the subprocess p_l. Thus, if $(TI_R(r_l \cdot x) - TI_R(x)) + (TO_{Ri}(r_l \cdot x) - TO_{Ri}(x)) + (TP_l(x)/CR_{Ri} - TP_l(x)/CR_R) < 0$, the total execution time of the node f_R and f_{Ri} can be reduced by migrating the subprocess p_l to the node f_{Ri}.

We discuss how much energy nodes consume by splitting and replicating the nodes. Suppose a node f_R supports a subsequence $SP(f_R) = \langle p_{s_R}, p_{s_R+1}, ..., p_{e_R} \rangle$ ($\sqsubset P$) of subprocesses of the application process P which has child nodes $cf_{R1}, ..., cf_{R,q_R}$ ($q_R \geq 0$). Let $CF(f_R)$ be a set $\{cf_{R1}, ..., cf_{R,q_R}\}$ ($q_R \geq 0$) of child nodes of a node f_R. First,

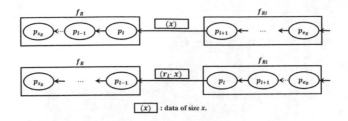

Fig. 2. Split subprocess p_l.

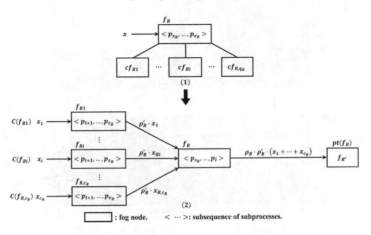

Fig. 3. Splitting and replication of a node.

we consider that the node f_R is splitted and replicated to the nodes $f_R, f_{R1}, ..., f_{R,c_R}$ ($c_R \geq 1$) as shown in Fig. 3. Here, the node f_R supports a subsequence $\langle p_{s_R}, ..., p_l \rangle$ of subprocesses and each node f_{Ri} supports a subsequence $\langle p_{l+1}, ..., p_{e_R} \rangle$ of subprocesses ($s_R \leq l < e_R$). Each child node $c f_{Ri}$ is connected to a node f_{Rj}. Here, $q_R \geq c_R$. Let $C(f_{Ri})$ be a set of child nodes which are connected to the node f_{Ri}. Here, $C(f_{Ri}) \subseteq CF(f_R)$, $C(f_{Ri}) \neq \phi$, and $C(f_{Ri}) \cap C(f_{Rj}) = \phi$ for every pair of different f_{Ri} and f_{Rj}, and $C(f_{R1}) \cup ... \cup C(f_{R,c_R}) = CF(f_R)$. In Fig. 3(1), the node f_R receives the input data D_R of size x. In Fig. 3(2), each node f_{Ri} receives the input data D_{Ri} of size x_i from child nodes in $C(f_{Ri})$. Here, $x = x_1 + ... + x_{c_R}$. The input data D_{Ri} of size x_i is received and processed by the subprocesses $p_{e_R}, p_{e_R-1}, ..., p_{l+1}$. The output ratio ρ_{Ri} of each child node f_{Ri} is the same ρ'_R since each node f_{Ri} supports the same subprocesses $SP(f_{Ri}) = \langle p_{l+1}, ..., p_{e_R} \rangle$. Then, the output data d_{Ri} of size $\rho'_R (= r_{e_R} \cdot ... \cdot r_{l+1}) \cdot x_i$ is sent to the node f_R as shown in Fig. 3. The node f_R receives the input data $D_R = \{d_{R1}, ..., d_{R,c_R}\}$ from all the child nodes $f_{R1}, ..., f_{R,c_R}$. The size of the input data D_R of the node f_R is $\rho'_R \cdot (x_1 + ... + x_{c_R})$. Then, the input data D_R is processed by the subprocesses $p_l, p_{l-1}, ..., p_{s_R}$. The node f_R sends the output data d_R of size $\rho_R \cdot \rho'_R \cdot (x_1 + ... + x_{c_R}) = \rho_R \cdot \rho'_R \cdot x$ to the parent node $f_{R'} (= pt(f_R))$.

Figure 4 shows how the nodes $f_R, f_{R1}, ..., f_{R,c_R}$ consumes energy to receive, process, and send data. First, at time τ, every child node f_{Ri} receives the input data d_{Ri}. In Fig. 4,

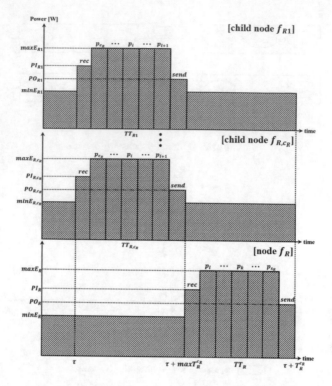

Fig. 4. Energy consumption of subprocesses.

rec shows the energy consumed by a node f_{Ri} to receive the input data d_{Ri}. The node f_{Ri} consumes the power PI_{Ri} [W] for $TI_{Ri}(x_i)$ [sec]. Then, the subprocesses $p_{e_R}, p_{e_R-1}, ..., p_{l+1}$ are performed. The node f_R consumes the power $maxE_R$ [W] to perform the subprocesses for $TC_{Ri}(\langle p_{l+1}, ..., p_{e_R} \rangle, x_i)$ [sec]. The node f_{Ri} sends the output data d_{Ri} to the node f_R. Here, the node f_R consumes the power PO_{Ri} [W] for $TO_{Ri}(\rho'_R \cdot x)$ [sec]. *send* shows the energy consumed by f_{Ri} to send the output data d_{Ri}.

The execution time $TT_{Ri}(\langle p_{l+1}, ..., p_{e_R} \rangle, x_i)$ [sec] of the node f_{Ri} for input data D_{Ri} of size x_i is $TI_{Ri}(x_i) + TC_{Ri}(\langle p_{l+1}, ..., p_{e_R} \rangle, x_i) + TO_{Ri}(\rho'_R \cdot x_i)$ [sec] since the node f_{Ri} receives the input data D_{Ri} until f_{Ri} sends the output data d_{Ri} to the node f_R as shown in Fig. 4. The node f_{Ri} consumes the energy $EF_{Ri}(\langle p_{l+1}, ..., p_{e_R} \rangle, x_i) = EI_{Ri}(x_i) + EC_{Ri}(\langle p_{l+1}, ..., p_{e_R} \rangle, x_i) + EO_{Ri}(\rho'_R \cdot x_i)$ [J]. Let $maxT_R^{CR}$ be the largest execution time $max(\{TT_{Ri}(\langle p_{l+1}, ..., p_{e_R} \rangle, x_i) \mid i = 1, ..., c_R\})$ of the nodes $f_{R1}, ..., f_{R,c_R}$. The node f_R starts receiving input data d_{Ri} from every child node f_{Ri} at time $\tau + maxT_R^{CR}$. Here, the node f_R consumes the minimum power $minE_R$ [W] for $maxT_R^{CR}$ [sec] from time τ, i.e. consumes the energy $maxT_R^{CR} \cdot minE_R$. Then, the node f_R starts processing the input data. Here, the node f_R consumes the energy $EF_R(\langle p_{s_R}, ..., p_l \rangle, \rho'_R \cdot (x_1 + ... + x_{c_R}))$ [J] for the input data $d_{R1}, ..., d_{R,c_R}$ from the child nodes $f_{R1}, ..., f_{R,c_R}$. The response time ST_R^{CR} and the total energy consumption SE_R^{CR} of the node f_R and child nodes $f_{R1}, ..., f_{R,c_R}$ are given as follows:

$$ST_R^{CR} = maxT_R^{CR} + TT_R(\langle p_{s_R}, ..., p_l \rangle, \rho_R' \cdot x). \tag{4}$$

$$\begin{aligned} SE_R^{CR} = \Sigma_{i=1}^{CR} [&EF_{Ri}(\langle p_{l+1}, ..., p_{e_R} \rangle, x_i) + \{maxT_R^{CR} - TT_{Ri}(\langle p_{l+1}, ..., p_{e_R} \rangle, x_i) \\ &+ TT_R(\langle p_{s_R}, ..., p_l \rangle, \rho_R' \cdot x)\} \cdot minE_{Ri}] \\ &+ (EF_R(\langle p_{s_R}, ..., p_l \rangle, \rho_R' \cdot x) + maxT_R^{CR} \cdot minE_R). \end{aligned} \tag{5}$$

On the other hand, if the subsequence $\langle p_{s_R}, p_{s_R+1}, ..., p_{e_R} \rangle$ of subprocesses are supported by one node f_R, the node f_R consumes the energy SE_R [J] for time ST_R [sec] to handle the input data D_R of size x;

$$ST_R = TT_R(\langle p_{s_R}, ..., p_{e_R} \rangle, x). \tag{6}$$

$$SE_R = EF_R(\langle p_{s_R}, ..., p_{e_R} \rangle, x). \tag{7}$$

If $SE_R^{CR} < SE_R$ for some l and c_R, the node f_R supporting a subsequence $SP(f_R) = \langle f_{s_R}, ..., f_{e_R} \rangle$ can be splitted and replicated to the node f_R which supports a prefix $\langle p_{s_R}, ..., p_l \rangle$ and child nodes $f_{R1}, ..., f_{R,c_R}$ each of which supports a postfix $\langle p_{l+1}, ..., p_{e_R} \rangle$ as shown in Fig. 4. The child nodes $cf_{R1}, ..., cf_{R,q_R}$ of the node f_R are changed with child nodes of the nodes $f_{R1}, ..., f_{R,c_R}$ so that each child node f_{Ri} can receive the same size of input data.

Suppose every fog node f_{Ri} is homogeneous, i.e. $CR_{Ri} = CR$, $maxR_{Ri} = maxE$, and $minE_{Ri} = minE$ ($i = 1, ..., c_R$). Every child node f_{Ri} receives the input data of the same size $x_i = x/c_R$. The response time ST_R^{CR} and total energy consumption SE_R^{CR} of the node f_R and the child nodes $f_{R1}, ..., f_{R,c_R}$ are given as follows;

$$ST_R^{CR} = maxT_R^{CR} + TT_R(\langle p_{s_R}, ..., p_l \rangle, \rho_R' \cdot x). \tag{8}$$

$$\begin{aligned} SE_R^{CR} = c_R \cdot \{&EF_{Ri}(\langle p_{s_R}, ..., p_l \rangle, x/c_R) \\ &+ (TT_R(\langle p_{s_R}, ..., p_l \rangle, \rho_R' \cdot x) \cdot minE)\} \\ &+ (EF_R(\langle p_{s_R}, ..., p_l \rangle, \rho_R' \cdot x) + maxT_R^{CR} \cdot minE). \end{aligned} \tag{9}$$

4.2 Replicating Nodes

Suppose a node f_R is a child of a node $f_{R'}$ $(= pt(f_R))$ and supports a subsequence $SP(f_R) = \langle p_{s_R}, ..., p_{e_R} \rangle$ of subprocesses. The node f_R has q_R (≥ 0) child nodes $cf_{R1}, ..., cf_{R,q_R}$ as shown in Fig. 5. The node f_R is replicated to c_R nodes $f_R^1, ..., f_R^{c_R}$ $(c_R \geq 1)$ where f_R^1 shows the node f_R. Here, each replica node f_R^i supports the same subprocesses $SP(f_R) = \langle p_{s_R}, ..., p_{e_R} \rangle$. The node f_R receives the input data D_R of size x from the nodes $cf_{R1}, ..., cf_{R,q_R}$. Each replica node f_R^i has input data of size x_i. Here, $x = x_1 + ... + x_{c_R}$. $minE_R^i$ shows the minimum power consumption of a replica node f_R^i.

Each replica node f_R^i consumes the energy $EF_R^i(SP(f_R), x_i)$ [J] for time $TT_R^i(SP(f_R), x_i)$ [sec] ($i = 1, ..., c_R$):

$$TT_R^i(SP(f_R), x_i) = TTI_R^i(x_i) + TC_R^i(SP(f_R), x_i) + \delta_R^i \cdot TO_R^i(\rho_R^i \cdot x_i). \tag{10}$$

$$EF_R^i(SP(f_R), x_i) = EI_R^i(x_i) + EC_R^i(SP(f_R), x_i) + \delta_R^i \cdot EO_R^i(\rho_R^i \cdot x_i). \tag{11}$$

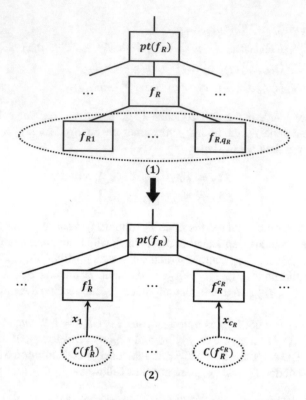

Fig. 5. Replication.

The execution time $TTT_R(SP(f_R),\ x,\ c_R)$ and total energy consumption $TEF_R(SP(f_R), x,\ c_R)$ of the replica nodes $f_R^1, ..., f_R^{c_R}$ are given as follows:

$$TTT_R(SP(f_R),\ x,\ c_R) = max(\{TT_R^i(SP(f_R),\ x_i) \mid i = 1, ..., c_R)\}) \qquad (12)$$

$$TEF_R(SP(f_R),\ x,\ c_R) = \Sigma_{i=1}^{c_R}[EF_R^i(SP(f_R),\ x_i) + (TTT(SP(f_R),\ x,\ c_R)$$
$$-TT_R^i(SP(f_R),\ x_i))\cdot minE_R^i]. \qquad (13)$$

The parent node $f_{R'}$ receives the data $d_R^1, ..., d_R^{c_R}$ whose sizes are $x_1, ..., x_{c_R}$ from the replica nodes $f_R^1, ..., f_R^{c_R}$, respectively. The execution time and energy consumption of the node $f_{R'}$ are as follows:

$$TT_{R'}(SP(f_{R'}),\ x) = TTI_{R'}(x) + TC_{R'}(SP(f_{R'}),\ x) + \delta_{R'}\cdot TO_{R'}(\rho_{R'}\cdot x). \qquad (14)$$

$$EF_{R'}(SP(f_{R'}),\ x) = EI_{R'}(x) + EC_{R'}(SP(f_{R'}),\ x) + \delta_{R'}\cdot EO_{R'}(\rho_{R'}\cdot x). \qquad (15)$$

The replica nodes $f_{R'}, f_R^1, ..., f_R^{c_R}$ totally consume the energy $RE_R^{c_R}$ [J] for time $RT_R^{c_R}$ [sec] as follows:

$$RT_R^{c_R} = TTT_R(SP(f_R),\ x,\ c_R). \qquad (16)$$

$$RE_R^{c_R} = TEF_R(SP(f_R),\ x,\ c_R) + TTI_{R'}(\rho_R\cdot x)\cdot re_{R'}\cdot maxE_{R'}. \qquad (17)$$

In the parent node $f_{R'}$, the execution time $TI_{R'}(x)$ and energy consumption $TI_{R'}(x)$ $\cdot re_{R'} \cdot maxE_{R'}$ to receive the data is changed with $TTI_{R'}(x) = TI_{R'}(x_1) + ... + TI_{R'}(x_i)$ and $TTI_{R'}(x) \cdot re_{R'} \cdot maxE_{R'}$ if the child node f_R is replicated.

If $RE_R^{CR} < EF_R(SP(f_R), x) + TI_{R'}(\rho_R \cdot x) \cdot re_{R'} \cdot maxE_{R'}$, the node f_R can be replicated to $f_R^1, ..., f_R^{CR}$. The child nodes $f_{R1}, ..., f_{R,q_R}$ are connected to the nodes $f_R^1, ..., f_R^{CR}$. Let $C(f_R^i)$ be a child nodes of each node f_R^i.

5 Concluding Remarks

In this paper, we discussed the dynamic tree model for fog computing, i.e. DTBFC model. A fog node is splitted and replicated as the traffic of the node increases in the tree. We proposed the model to give the execution time and energy consumption of nodes. By using the model, we can split and replicate nodes so that the energy consumption of nodes can be reduced.

We are now designing and evaluating algorithms to split and replicate nodes in the DTBFC model.

References

1. Raspberry pi 3 model b. https://www.raspberrypi.org/products/raspberry-pi-3-model-b/
2. Creeger, M.: Cloud computing: an overview. Queue 7(5), 3–4 (2009)
3. Enokido, T., Ailixier, A., Takizawa, M.: A model for reducing power consumption in peer-to-peer systems. IEEE Syst. J. 4(2), 221–229 (2010)
4. Enokido, T., Ailixier, A., Takizawa, M.: Process allocation algorithms for saving power consumption in peer-to-peer systems. IEEE Trans. Industr. Electron. 58(6), 2097–2105 (2011)
5. Enokido, T., Ailixier, A., Takizawa, M.: An extended simple power consumption model for selecting a server to perform computation type processes in digital ecosystems. IEEE Trans. Industr. Inf. 10(2), 1627–1636 (2014)
6. Guo, Y., Oma, R., Nakamura, S., Duolikun, D., Enokido, T., Takizawa, M.: Data and subprocess transmission on the edge node of TWTBFC model. In: Proceedings of the 11 th International Conference on Intelligent Networking and Collaborative Systems (INCoS 2019), pp. 80–90 (2019)
7. Guo, Y., Oma, R., Nakamura, S., Duolikun, D., Enokido, T., Takizawa, M.: Evaluation of a two-way tree-based fog computing (TWTBFC) model. In: Proceedings of the 13th International Conference on Innovative Mobile and Internet Services in Ubiquitous Computing (IMIS 2019), pp. 72–81 (2019)
8. Hanes, D., Salgueiro, G., Grossetete, P., Barton, R., Henry, J.: IoT Fundamentals: Networking Technologies, Protocols, and Use Cases for the Internet of Things. Cisco Press, Indianapolis (2018)
9. Oma, R., Nakamura, S., Duolikun, D., Enokido, T., Takizawa, M.: An energy-efficient model for fog computing in the Internet of Things (IoT). Internet Things 1–2, 14–26 (2018)
10. Oma, R., Nakamura, S., Duolikun, D., Enokido, T., Takizawa, M.: Energy-efficient recovery algorithm in the fault-tolerant tree-based fog computing (FTBFC) model. In: Proceedings of the 33rd International Conference on Advanced Information Networking and Applications (AINA 2019), pp. 132–143 (2019)

11. Oma, R., Nakamura, S., Duolikun, D., Enokido, T., Takizawa, M.: Evaluation of data and subprocess transmission strategies in the tree-based fog computing (TBFC) model. In: Proceedings of the 22nd International Conference on Network-Based Information Systems (NBiS 2019), pp. 15–26 (2019)
12. Oma, R., Nakamura, S., Duolikun, D., Enokido, T., Takizawa, M.: A fault-tolerant tree-based fog computing model. Int. J. Web Grid Serv. (IJWGS) **15**(3), 219–239 (2019)
13. Oma, R., Nakamura, S., Duolikun, D., Enokido, T., Takizawa, M.: Subprocess transmission strategies for recovering from faults in the tree-based fog computing (TBFC) model. In: Proceedings of the 13th International Conference on Complex, Intelligent, and Software Intensive Systems (CISIS 2019), pp. 50–61 (2019)
14. Oma, R., Nakamura, S., Enokido, T., Takizawa, M.: A nodes selection algorithm for fault recovery in the GTBFC model. In: Proceedings of the 14th International Conference on Broad-Band Wireless Computing, Communication and Applications (BWCCA 2019), pp. 81–92 (2019)
15. Rahmani, A., Liljeberg, P., Preden, J.S., Jantsch, A.: Fog Computing in the Internet of Things. Springer, Cham (2018)

Topic-Based Subgroups for Reducing Messages Exchanged Among Subgroups

Takumi Saito[1]([✉]), Shigenari Nakamura[1], Tomoya Enokido[2],
and Makoto Takizawa[1]

[1] Hosei University, Tokyo, Japan
takumi.saito.3j@stu.hosei.ac.jp, nakamura.shigenari@gmail.com,
makoto.takizawa@computer.org
[2] Rissho University, Tokyo, Japan
eno@ris.ac.jp

Abstract. In the P2PPS (P2P (peer-to-peer) type of a topic-based PS (publish/subscribe)) model, each peer can be a subscriber and publisher. Peers publishes messages to communicate with other peers. Here, messages are characterized by topics. Messages which have a common topic are considered to be related. In our previous studies, a hierarchical group model of peers is proposed, which is composed of a gateway subgroup and member subgroups of peers where related messages are causally delivered by using the linear time vector. Every member subgroup is constructed so that the member subgroup is a collection of member peers in same network in an NBG (network-based grouping) method. In this paper, we propose an HTBG (hot-topic-based grouping) method where each subgroup is *dynamically* constructed based on the hot topics which are more frequently used at that time. Here, each subgroup is constructed so that the subgroup is composed of peers which subscribe common hot topics. In the evaluation, we show the number of messages exchanged among subgroups can be reduced in the HTBG method compared with the NBG method. However, the larger number of messages are broadcast in a member subgroup in the HTBG method than the NBG method.

Keywords: Topic-based publish/subscribe system · P2P model · Hierarchical P2PPS model · Linear time vector · Hot topic

1 Introduction

The PS (publish/subscribe) model is an event-driven model of distributed systems. In topic-based PS models [13,14], contents of each message are denoted by topics. A publisher process publishes a message with publication topics. Each subscriber process subscribes messages by specifying interesting topics named subscription topics. Each message is delivered to only a subscriber process whose subscription topics include some publication topic of the message published.

In the P2PPS (P2P (peer-to-peer) type of topic-based PS) model [1,5–7], each process is a peer which can both publish and subscribe messages in a distributed manner [2,3]. A peer receives only a message whose publication topics

© Springer Nature Switzerland AG 2020
L. Barolli et al. (Eds.): EIDWT 2020, LNDECT 47, pp. 35–45, 2020.
https://doi.org/10.1007/978-3-030-39746-3_5

are also subscription topics of the peer. Here, the peer is a *target* peer of the message. A pair of messages m_1 and m_2 are considered to be *related* iff (if and only if) the messages m_1 and m_2 carry at least one common publication topic. A message m_1 causally precedes another message m_2 ($m_1 \rightarrow m_2$) iff the publication event of the message m_1 happens before the message m_2 according to the causality theory [4]. Each peer is required to causally deliver only related messages in the P2PPS model. In papers [8–10], the TBC (topic-based causally) and OBC (object-based causally) precedent relations among messages are defined to causally deliver related messages based on topics and objects carried by messages, respectively.

In papers [11,12], a hierarchical group of peers, i.e. 2P2PPS (two-layered group of a P2PPS) system is proposed. Here, a group of peers is composed of exclusive subgroups of peers. Every member subgroup is constructed in the NBG (network-based grouping) method. Here, each member subgroup is composed of member peers. Each member subgroup is composed of peers which are located in a LAN (local-area network). Member subgroups are interconnected in a WAN (wide-area network). In this sense, member subgroups are composed of peers on the basis of network addresses. In order to deliver every pair of related messages in the 2P2PPS system, a TLCO (two-layered causally ordering) protocol is proposed [11]. Here, linear time vectors are used to check causality relations among messages.

In this paper, we propose an HTBG (hot-topic-based grouping) method to more reduce the number of messages exchanged among member subgroups compared with the NBG method. For this aim, we introduce hot topics. A topic t_1 is hotter than another topic t_2 if t_1 is used by more number of messages than t_2. First, every peer is in a member subgroup G_1. Peers whose subscriptions include a hottest topic t are in a member subgroup and the other peers are grouped into another member subgroup. Here, the topic t is named grouping topic. Thus, a subgroup is a collection of peers interconnected in an overlay network, whose subscriptions include a grouping topic. Messages with a group topic are only exchanged in the member subgroup. Hence, messages exchanged among member subgroups can be reduced. In the evaluation, we show the fewer number of messages are exchanged among member subgroups in the HTBG method than the NBG method. However, the lager number of messages are broadcast in a member subgroup in the HTBG method than the NBG method.

In Sect. 2, we present a system model. In Sect. 3, we discuss the TLCO protocol and the NBG method for a 2P2PPS system. In Sect. 4, we propose the HTBG method to reduce the number of messages exchanged among subgroups in 2P2PPS system. In Sect. 5, we evaluate the HTBG method compared with the NBG method.

2   System Model

In the topic-based PS model, contents of messages are denoted by topics. Let T be a set $\{t_1, ..., t_l\}$ ($l \geq 1$) of topics in a system. Each peer p_i specifies interesting topics named *subscription* topics $p_i.S$ ($\subseteq T$). Possible topics, messages on which the peer p_i publishes, are named *publication* topics $p_i.P$ ($\subseteq T$). Each peer p_i gives a message m publication topics $m.P$ ($\subseteq T$) which denote the contents of the message m and then publishes the message m. The message m published is only received by a peer p_i where $p_i.S \cap m.P \neq \phi$. Here, the peer p_i is a target peer of the message m. A pair of messages m_1 and m_2 are *related* ($m_1 \rightleftharpoons m_2$) iff $m_1.P \cap m_2.P \neq \phi$. A pair of messages m_1 and m_2 are *independent* ($m_1 \mid m_2$) iff $m_1 \not\rightleftharpoons m_2$.

In the 2P2PPS system, two-layered groups are considered. A group G of peers is composed of a *gateway* subgroup G_0, and *member* subgroups G_1, ..., G_g ($g \geq 1$) as shown in Fig. 1. Each member subgroup G_t is composed of member peers p_{t1}, ..., p_{t,l_t} ($l_t \geq 1$) and a gateway peer p_{t0} ($t = 1$, ..., g). Each member peer belongs to only one subgroup. The gateway subgroup G_0 is composed of gateway peers p_{10}, ..., p_{g0} of subgroups G_1, ..., G_g, respectively. Each gateway peer p_{t0} belongs to both a member subgroup G_t and a gateway subgroup G_0. We assume the underlying overlay network of a subgroup is reliable. That is, a pair of messages m_1 and m_2 published by member peers are delivered to every common target member peer in the publication order without any message loss.

In each member subgroup G_t, every message is broadcast to all the member peers p_{t1}, ..., p_{t,l_t} and the gateway peer p_{t0} by taking advantage of the underlying network service. A gateway peer p_{t0} receives messages sent by a member peer p_{ti} in a subgroup G_t and forwards the messages to gateway peers of other member subgroups by using the unicast communication. On receipt of a message m from another gateway peer p_{u0}, a gateway peer p_{t0} broadcasts the message m to every member peer in the member subgroup G_t. Let $g_{t0}.S$ and $g_{t0}.P$ be sets of subscription and publication topics of member peers, respectively. A pair of member subgroups G_t and G_u are *concurrent* ($G_t \mid G_u$) iff $g_{t0}.P \cap g_{u0}.S = \phi$. A pair of member subgroups G_t and G_u are *independent* ($G_t \parallel G_u$) iff $G_t \mid G_u$ and $G_u \mid G_t$. $G_t \mid G_u$ means that no message published in the member subgroup G_t is sent to the member subgroup G_u. $G_t \parallel G_u$ means that no message is exchanged among member subgroups G_t and G_u. A pair of subgroups G_t and G_u are *concurrent for topic* t_i ($G_t \mid_{t_i} G_u$) iff $t_i \in g_{t0}.P$ and $t_i \notin g_{u0}.S$. This means, no message on the topic t_i published in the member subgroup G_t is sent to the member subgroup G_u.

A message m_1 *causally precedes* a message m_2 ($m_1 \rightarrow m_2$) iff the publication event of the message m_1 happens before the publication event of the message m_2 according to the traditional causality theory [4]. In the P2PPS system, a common target member peer of the unrelated messages m_1 and m_2 is not required to deliver a messages m_1 before another message m_2 even if $m_1 \rightarrow m_2$. A message m_1 has to be delivered before another message m_2 in a common target peer only if $m_1 \rightarrow m_2$ and the messages m_1 and m_2 are related ($m_1 \rightleftharpoons m_2$).

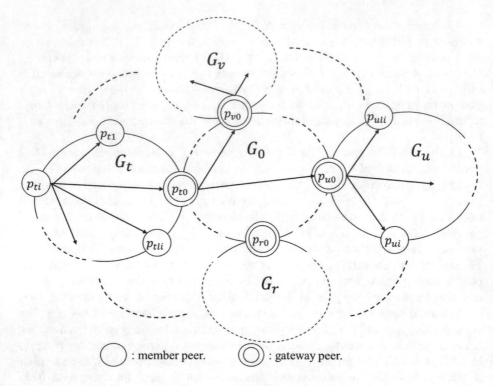

Fig. 1. Two-layered group.

3   A TLCO (Two-Layered Causally Ordering) Protocol

A 2P2PPS system G is composed of member subgroups $G_1, ..., G_g$ and a gateway subgroup G_0 as shown in Fig. 1. In papers [11,12], every member subgroup is constructed in the NBG (network-based grouping) method. Here, each member subgroup is composed of peers which are located in an LAN and the member subgroups are interconnected in a WAN. In order to causally deliver related messages to target peers in the 2P2PPS system, a TLCO (Two-Layered Causally Ordering) protocol is proposed [11]. In the TLCO protocol, each member peer p_{ti} of a member subgroup G_t manipulates a linear time vector $LV = \langle LV_1, ..., LV_g \rangle$. If a member peer p_{ti} publishes a message m, a linear time vector $LV = \langle LV_1, ..., LV_g \rangle$ of the member peer p_{ti} is carried by the message m. The tth element LV_t shows the linear time of a member subgroup G_t ($t = 1, ..., g$). $p_{ti}.LV$ stands for a linear time vector (LV) variable $LV = \langle LV_1, ..., LV_g \rangle$ of a member peer p_{ti} in a member subgroup G_t. Initially, $p_{ti}.LV = \langle 0, ..., 0 \rangle$ for every member peer p_{ti} ($i = 0, ..., l_t$). The LV field $m.LV$ of each message m shows the linear time vector LV carried by the message m. The length of each message m is $O(g)$ where g is the number of member subgroups, since the message m carries the linear time vector field $m.LV$.

Each member peer p_{ti} of a member subgroup G_t publishes and receives a message m by using topics and causally delivers messages by manipulating the LV variable $p_{ti}.LV$ and the LV field $m.LV$. Each peer publishes and receives a message as follows:

[**Member peer** p_{ti} **publishes a message** m]

1. $m.P$ = publication topics ($\subseteq T$);
2. The tth element $p_{ti}.LV_t$ in the LV variable $p_{ti}.LV = \langle LV_1, ..., LV_g \rangle$ is incremented by one in a member peer p_{ti}, i.e. $p_{ti}.LV_t = p_{ti}.LV_t + 1$;
3. For the LV field $m.LV$ of the messages m, $m.LV = p_{ti}.LV$ ($= \langle LV_1, ..., LV_g \rangle$);
4. The peer p_{ti} **publishes** the message m;

[**Member peer** p_{ti} **receives a message** m]

1. A message m arrives at a member peer p_{ti} in a member subgroup G_t;
2. $p_{ti}.LV_u = max(p_{ti}.LV_u, m.LV_u)$ (for $u = 1, ..., g$);
3. $p_{ti}.LV_t = max(p_{ti}.LV_1, ..., p_{ti}.LV_g)$;
4. **if** $m.P \cap p_{ti}.S \neq \phi$, the member peer p_{ti} **receives** the message m;
5. Otherwise, the member peer p_{ti} **neglects** the message m;

[**Gateway peer** p_{t0} **receives a message from a member peer** p_{ti}]

1. A gateway peer p_{t0} receives a message m published by a member peer p_{ti} in a member subgroup G_t;
2. The gateway peer p_{t0} **forwards** the message m to every gateway peer p_{u0} of a member subgroup G_u such that $p_{u0}.S \cap m.P \neq \phi$ in the gateway subgroup G_0;

[**Gateway peer** p_{t0} **receives a message from a gateway peer** p_{u0}]

1. A gateway peer p_{t0} receives a message m from another gateway peer p_{u0} in a gateway subgroup G_0;
2. **if** $p_{t0}.S \cap m.P = \phi$, the gateway peer p_{t0} **neglects** the message m;
3. $p_{t0}.LV_u = max(p_{t0}.LV_u, m.LV_u)$ (for $u = 1, ..., g$);
4. $p_{t0}.LV_t = max(p_{t0}.LV_1, ..., p_{t0}.LV_g)$;
5. The gateway peer p_{t0} **publishes** the message m in the member subgroup G_t;

4   An HTBG (Hot-Topic-Based Grouping) Method

In this paper, we propose an HTBG (hot-topic-based grouping) method to more reduce the number of messages exchanged among member subgroups compared with the NBG method in a 2P2PPS system. In the HTBG method, peers are dynamically grouped to member subgroups with respect to hot topics at that time. The number g of member subgroups $G_1, ..., G_g$ is invariant. In order to make the member subgroup of the LTV invariant. Initially, every member peer

p_i $(i = 0, 1, ..., l_t)$ is in the member subgroup G_1 and every other member subgroup G_t $(t > 1)$ is empty. Here, publication $p_{10}.P$ and subscription $p_{10}.S$ of the member subgroup G_1 are sets of topics published and subscribed by at least one member peer in the member subgroup G_1, respectively. Publications and subscriptions of the other member subgroups are empty. In a 2P2PPS system, hot topics which are frequently used in messages are changed through publishing and receiving messages. A topic t_1 is *hotter* than another topic t_2 $(t_1 \succ t_2)$ iff t_1 is used by more number of messages than t_2. If the hottest topic is changed, only the member peer p_i which subscribes the hottest topic ht is grouped into a same member subgroup G_t even if the peer p_i is already in another member subgroup. The hottest topic ht $(\in T)$ is named *grouping topic* of the member subgroup G_t. The publication $p_{t0}.P$ and subscription $p_{t0}.S$ of the member subgroup G_t are sets of topics published and subscribed by at least one member peer in the member subgroup G_t, respectively. A member subgroup G_s is *older* than G_t iff the grouping topic of G_s is older that G_t. For example, if the hottest topic t_1 is changed with another topic t_2. Here, a member subgroup G_s whose grouping topic is t_1 already exists and a member subgroup G_t whose grouping topic is t_2 is newly constructed. In this case, the member subgroup G_s is older than the member subgroup G_t.

[A HTBG method]

1. If the hottest topic t_1 is changed with t_2, the following steps are performed;
   a. If there exists no member subgroup which has no member peer exits, every peer in the oldest member subgroup moves into the second oldest member subgroup;
   b. Every member peer p_i which subscribes the hottest topic t_2 is grouped into a member subgroup G_t;
   c. The grouping topic of the member subgroup G_t is t_2;
   d. If a member subgroup G_s except for the new member subgroup G_t, has only one member peer p_{si}, the member peer p_{si} moves into the oldest member subgroup G_s;
   e. For every member subgroup G_u $(u = 1, \dots g)$, the publication $p_{u0}.P$ and subscription $p_{u0}.S$ are reconstructed so that $p_{u0}.P$ and $p_{u0}.S$ are sets of topics published and subscribed by at least one member peer in the member subgroup G_u, respectively;

Example 1. In Fig. 2, there are four member subgroups G_t, G_u, G_v, and G_w which are interconnected in a gateway subgroup G_0. The gateway subgroup G_0 is composed of four gateway peers p_{t0}, p_{u0}, p_{v0}, and p_{w0} of each member subgroup, respectively. Here, the total number of member peers in the system is seven.

First, every member peer is in a member subgroup G_t. The publication $p_{t0}.P$ and subscription $p_{t0}.S$ of the member subgroup G_t are sets of topics published and subscribed by at least one member peer in the member subgroup G_t, respectively. Publications and subscriptions of the other member subgroups G_u, G_v, and G_w are empty because the member subgroups have no member peer.

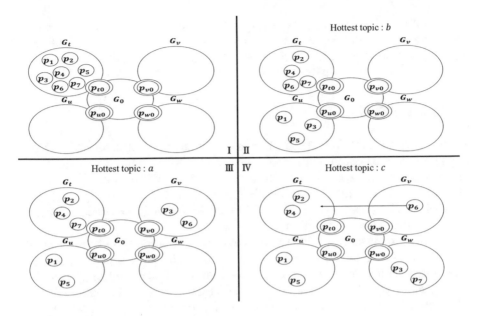

Fig. 2. HTBG method.

Next, a topic b gets a the hottest topic. Here, every member peer p_i which subscribes the topic b, i.e. $b \in p_i.S$, moves into a member subgroup G_u. In this example, member peers p_1, p_3, and p_5 subscribing the topic b move into the member subgroup G_u from the member subgroup G_t. Publications $p_{t0}.P$ and $p_{u0}.P$ and subscriptions $p_{t0}.S$ and $p_{u0}.S$ of the member subgroups G_t and G_u are reconstructed.

Then, the hottest topic b is changed with a topic a. Here, every member peer p_i which subscribes the topic a, i.e. $a \in p_i.S$ in each member subgroup moves into a member subgroup G_v. Here, member peers p_3 and p_6 subscribing the topic a move into the member subgroup G_v. Publications $p_{t0}.P$, $p_{u0}.P$, and $p_{v0}.P$ and subscriptions $p_{t0}.S$, $p_{u0}.S$, and $p_{v0}.S$ of the member subgroups G_t, G_u, and G_v are reconstructed.

Finally, the hottest topic a is changed with a topic c. Here, every member peer p_i which subscribes the topic c, i.e. $c \in p_i.S$ in each member subgroup moves to a member subgroup G_w. Here, member peers p_3 and p_7 move to the member subgroup G_w. In addition, a member peer p_6 moves from the member subgroup G_v to the member subgroup G_t because the member subgroup G_v has only one member peer p_6. Publications $p_{t0}.P$, $p_{v0}.P$, and $p_{w0}.P$ and subscriptions $p_{t0}.S$, $p_{v0}.S$, and $p_{w0}.S$ of the member subgroups G_t, G_v, and G_w are reconstructed.

In the HTBG method, every member subgroup is a collection of member peers interconnected in an overlay network, whose subscriptions include a grouping topic. Hence, messages with a grouping topic are only exchanged in the member subgroup. In the Example 1, after the topic c is decided as the hottest topic, the member subgroup G_w whose grouping topic is c has member peers. If a message m whose publication $m.P$ includes the grouping topic c is published in the member subgroup G_w, the gateway peer p_{w0} is not required to forward the message m to the other gateway peers p_{t0} and p_{u0}, i.e. $G_w \mid_c G_t$ and $G_w \mid_c G_u$.

5   Evaluation

We evaluate the HTBG method in terms of the number of messages exchanged among member subgroups compared with the NBG method. A system includes member peers $p_1, \ldots, p_n$ ($n \geq 1$). Here, $n = 30$. A 2P2PPS system G is composed of a gateway subgroup G_0 and member subgroups $G_1, \ldots, G_g$ ($g > 1$). Let T be a set of all topics $t_1, \ldots, t_l$ ($l \geq 1$) in a system. Here, $l = 30$. The number of member subgroups g is invariant. Here, $g = 5$.

In order to evaluate the HTBG protocol, we develop a time-based simulator. First, topics subscribed are randomly taken for each member peer p_i, i.e. $p_i.S$ is randomly decided from a topic set T. The number of subscription topics in $p_i.S$ of a member peer p_i is randomly selected from 6 to 10. The publication $p_i.P$ of the member peer p_i is the same as the subscription $p_i.S$ of the member peer p_i. Next, every member peer belongs to a member subgroup G_t. Then, member peers start to exchange messages in the TLCO protocol. At each time unit, a member peer p_{ti} is randomly taken in all member peers and the member peer p_{ti} publishes a message m to the other member peers p_{tj} ($j = 0, 1, \ldots, |G_t| - 1$) in the member subgroup G_t. Here, the publication $m.P$ includes one topic t which is randomly taken in the publication $p_{ti}.P$ of the member peer p_{ti}, i.e. $t \in p_{ti}.P$. The delay time $m.\delta_{t_{ij}}$ of the message m published by the member peer p_{ti} to the member peer p_{tj} ($i, j = 0, 1, \ldots, |G_t| - 1$) is randomly taken from 1 to 10 [tu (time unit)]. If the gateway peer p_{t0} receives a message m from a member peer p_{ti} in the member subgroup G_t, the gateway peer p_{t0} forwards the message m to the other gateway peers p_{u0} where $p_{u0}.S \cap m.P \neq \phi$. The delay time $m.\delta_{tu_0}$ of the message m from the gateway peer p_{t0} to a gateway peer p_{u0} in the gateway subgroup G_0 is randomly taken from 1 to 10 [tu]. If the gateway peer p_{u0} receives the message m from another gateway peer p_{t0}, the gateway peer p_{u0} forwards the message m to all the member peers in the member subgroup G_u. In the evaluation, a set of the number n of member peers is randomly created 100,000 times. For each member peer set, number pen of publication events occurs. Here, $pen = 0, 100, 200, 300, 400$, and 500. pen publication events by member peers in a member peer set is iterated 100,000 times. Finally, the average number of messages are calculated.

In the HTBG method, initially, every member peer is grouped into a member subgroup G_1 and no hottest topic exists. In the evaluation, the hottest topic is changed every fifty time units. If the hottest topic is changed, only the member peers which subscribe the hottest topic is grouped into a same member subgroup G_t. Next, the hottest topic is decided as a grouping topic of the member subgroup G_t. Then, the publication $g_{t0}.P$ and subscription $g_{t0}.S$ of the gateway peer p_{t0} are decided so that every topic published and subscribed by at least one member peer p_{ti} is included. If a member peer publishes a message m, hot topics are more frequently used as a publication topic $m.P$ of a message m.

In the NBG method, every member subgroup is constructed so that the member subgroup is a collection of peers in the same network. In the evaluation, every member peer is randomly grouped into g member subgroups so that every member subgroup is composed of six member peers and one gateway peer. All the member subgroups are not changed in the evaluation.

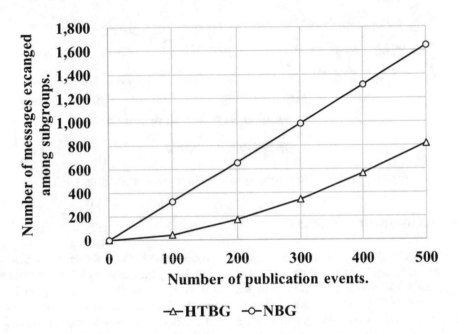

Fig. 3. Number of messages exchanged among subgroups.

Figure 3 shows the number of messages exchanged among member subgroups in the HTBG method and NBG method. The number of messages exchanged among member subgroups in the NBG method linearly increases as the number of publication events increases. The number of messages exchanged among member subgroups in the HTBG method also linearly increases from two hundred publication events because member subgroups which have no member peer exist for $pen \leq 200$. The fewer number of messages are exchanged among member subgroups in the HTBG method than the NBG method.

Figure 4 shows the number of messages broadcast in a member subgroup G_1 in the HTBG method and NBG method. The number of messages broadcast in the NBG method linearly increases as the number of publication events increases. For $pen \leq 100$, the number of messages broadcast in the HTBG method more rapidly increases because every member peer is grouped into the member subgroup G_1 initially. The lager number of messages are broadcast in the HTBG method than the NBG method.

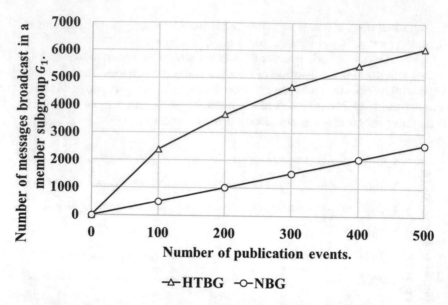

Fig. 4. Number of messages broadcast in a member subgroup G_1.

6 Concluding Remarks

The P2PPS (P2P (peer-to-peer) type of a topic-based PS (publish/subscribe)) model, messages are exchanged among peers. Here, messages are characterized by topics. Related messages have at least one common topic and every peer has to causally deliver the messages. For this aim, a hierarchical group of peers, i.e. 2P2PPS (two-layered group of a P2PPS) model and TLCO (two-layered causally ordering) protocol are proposed in our previous studies. The 2P2PPS system is composed of a gateway subgroup composed of gateway peers and member subgroups composed of a gateway peer and member peers. In the 2P2PPS system, NBG (network-based grouping) method is used to group the member peers in the same network into a same member group. In this paper, we proposed the HTBG (hot-topic-based grouping) method where each member subgroup is dynamically constructed based on the topic which is more frequently used at that time, i.e. hottest topic. Here, if the hottest topic is changed, only the member peers which subscribe the hottest topic is grouped into a same member subgroup. Since messages with the hottest topic are exchanged in only the member subgroup whose grouping topic is the hottest topic, messages exchanged among member subgroups can be reduced compared with the NBG method. In the evaluation, we show the number of messages exchanged among member subgroups in the HTBG method can be reduced compared with the NBG method. However, the larger number of messages are broadcast in a member subgroup in the HTBG method than the NBG method.

References

1. Google alert. http://www.google.com/alerts
2. Blanco, R., Alencar, P.: Event models in distributed event based systems. In: Principles and Applications of Distributed Event-Based Systems, pp. 19–42 (2010)
3. Hinze, A., Buchmann, A.: Principles and Applications of distributed Event-Based Systems. IGI Grobal (2010)
4. Lamport, L.: Time, clocks, and the ordering of event in a distributed systems. Commun. ACM **21**(7), 558–565 (1978)
5. Nakayama, H., Duolikun, D., Enokido, T., Takizawa, M.: Selective delivery of event messages in peer-to-peer topic-based publish/subscribe systems. In: Proceedings of the 18th International Conference on Network-Based Information Systems (NBiS-2015), pp. 379–386 (2015)
6. Nakayama, H., Duolikun, D., Enokido, T., Takizawa, M.: Reduction of unnecessarily ordered event messages in peer-to-peer model of topic-based publish/subscribe systems. In: Proceedings of IEEE the 30th International Conference on Advanced Information Networking and Applications (AINA-2016), pp. 1160–1167 (2016)
7. Nakayama, H., Ogawa, E., Nakamura, S., Enokido, T., Takizawa, M.: Topic-based selective delivery of event messages in peer-to-peer model of publish/subscribe systems in heterogeneous networks. In: Proceedings of the 18th International Conference on Network-Based Information Systems (WAINA-2017), pp. 1162–1168 (2017)
8. Saito, T., Nakamura, S., Duolikun, D., Enokido, T., Takizawa, M.: Object-based selective delivery of event messages in topic-based publish/subscribe systems. In: Proceedings of the 13th International Conference on Broadband and Wireless Computing, Communication and Applications (BWCCA-2018), pp. 444–455 (2018)
9. Saito, T., Nakamura, S., Duolikun, D., Enokido, T., Takizawa, M.: Evaluation of TBC and OBC precedent relations among messages in P2P type of topic-based publish/subscribe system. In: Proceedings of the Workshops of the 33rd International Conference on Advanced Information Networking and Applications (WAINA-2019), pp. 570–581 (2019)
10. Saito, T., Nakamura, S., Enokido, T., Takizawa, M.: A causally precedent relation among messages in topic-based publish/subscribe systems. In: Proceedings of the 21st International Conference on Network-Based Information Systems (NBiS-2018), pp. 543–553 (2018)
11. Saito, T., Nakamura, S., Enokido, T., Takizawa, M.: The group-based linear time causally ordering protocol in a scalable P2PPS system. In: Proceedings of the 14th International Conference on Broad-Band Wireless Computing, Communication and Applications (BWCCA-2019), pp. 471–482 (2019)
12. Saito, T., Nakamura, S., Enokido, T., Takizawa, M.: A hierarchical group of peers in publish/subscribe systems. In: Proceedings of the 11th International Conference on Intelligent Networking and Collaborative Systems (INCoS-2019), pp. 3–13 (2019)
13. Tarkoma, S.: Publish/Subscribe System: Design and Principles, 1st edn. Wiley, Hoboken (2012)
14. Tarkoma, S., Rin, M., Visala, K.: The publish/subscribe internet routing paradigm (PSIRP): Designing the future internet architecture. In: Future Internet Assembly, pp. 102–111 (2009)

Subtree-Based Fog Computing in the TWTBFC Model

Yinzhe Guo[1(✉)], Takumi Saito[1], Ryuji Oma[1], Shigenari Nakamura[1], Tomoya Enokido[2], and Makoto Takizawa[1]

[1] Hosei University, Tokyo, Japan
{yinzhe.guo.3e,takumi.saito.3j,ryuji.oma.6r}@stu.hosei.ac.jp,
nakamura.shigenari@gmail.com,makoto.takizawa@computer.org
[2] Rissho University, Tokyo, Japan
eno@ris.ac.jp

Abstract. In our previous studies, the TBFC (Tree-Based Fog Computing) model of the IoT is proposed to reduce the electric energy consumed by fog nodes and servers. Here, fog nodes are hierarchically structured in a height-balanced tree. A root node shows a cloud of servers and leaf nodes indicate edge fog nodes which communicate with sensors and actuators. First, sensor data is sent to edge fog nodes. Each fog node receives data from child fog nodes, processes the data, and sends the processed data to a parent fog node. Finally, servers receive sensor data processes by fog nodes and send actions to actuators. TWTBFC (Two-Way TBFC) model, fog nodes not only send processed data to a parent fog node but also send both the processed data to each child fog node. Fog nodes nearer to edge fog nodes decide on actions by using the data and send the actions to child fog nodes. In these two previous studies, we assume that every edge node has acquired every processed data and then makes a decision on actions to actuator. In order to process only a small amount of data scattered in various parts of the tree model and then make decision on actions to actuator, we newly propose a subtree-based fog computing in the TWTBFC model, where all edge nodes do not send the sensor data to processed and make decision on actions to actuators. Just some sensor data, i.e. a small amount of data scattered in various parts of the edge is sent to a root node, which needs to make a decision on actions to actuator.

Keywords: Energy-efficient fog computing · IoT (Internet of Things) · Two-Way TBFC (TWTBFC) model · Subtree

1 Introduction

The IoT (Internet of Things) is composed of not only computers like servers and clients but also devices like sensor and actuators installed in various things. Compared with traditional network systems like cloud computing model, the IoT is so scalable that millions of devices are included and huge amount of data

© Springer Nature Switzerland AG 2020
L. Barolli et al. (Eds.): EIDWT 2020, LNDECT 47, pp. 46–52, 2020.
https://doi.org/10.1007/978-3-030-39746-3_6

from sensors are transmitted in networks and are processed in servers. Networks and servers are too heavily loaded to timely decide on actions for all sensor data and deliver actions to actuators. In order to efficiently realize the IoT, the fog computing model [10] is proposed.

Since the IoT is scalable, huge amount of electric energy is consumed in the IoT. In order to not only increase the performance but also reduce the electric energy consumption of the IoT, the TBFC (Tree-Based Fog Computing) model is proposed in our previous studies [5,6]. Here, fog nodes are hierarchically structured in a height-balanced tree. A root node shows a cloud composed of servers and leaf nodes are edge nodes which receives sensor data from sensors and send actions to actuators. Each sensor first sends sensor data to an edge fog node. Each fog node receives input data from child fog nodes. Then, the fog node processes the data and sends the processed data i.e. output data to a parent fog node. Finally, servers in a cloud receive data processed by fog nodes. Then, the servers and deliver the actions to actuators through fog nodes. Thus, sensor data is sent upwards from sensors to a cloud of servers. Servers in the cloud send actions downward to actuators. Here, it takes time to deliver actions to actuators because servers make a decision on actions to be done by the actuator.

In our previous studies, the TWTBFC (Two-Way TBFC) model is also proposed so that actions of actuators are decided by edge fog nodes nearer to devices in order to more promptly deliver actions to actuators. Here, we propose a new TWTBFC model where only sensor data to decide on actions are collected. Accordingly, we can reduce electric energy consumption and execution time of fog nodes.

In Sect. 2, we present the new TWTBFC model of the IoT. In Sect. 3, we present the power consumption and computation module of a fog node. In Sect. 4, we evaluate the TWTBFC model.

2   Two-Way Tree-Based Fog Computing (TWTBFC) Model

The fog computing (FC) model [10] to efficiently realize the IoT [7] is composed of sensor and actuator devices, fog nodes, and clouds. Clouds are composed of servers like the cloud computing (CC) model [4]. Here, an application process p is performed on servers to process sensor data sent by sensors in networks. In the TBFC (Tree-Based Fog Computing) model [8,9], fog nodes are hierarchically structured in a tree as shown in Fig. 1. Here, the root node f is denotes a cloud of servers. Fog nodes at the bottom level are *edge* nodes which communicate with sensors and actuators.

Each node f_R has c_R (≥ 0) child nodes f_R, ..., f_{R,c_R}. Here, f_{Ri} shows the ith child node of the fog node f_R and in turn f_R is a parent node of the node f_{Ri}. $ch(f_R)$ is a set $\{f_{R1}, ..., f_{R,c_R}\}$ of child nodes of a node f_R. $pt(f_{Ri})$ shows a parent node f_R of a node f_{Ri}. An edge fog node has no child fog node and communicates with devices, i.e, receive sensor data from sensors and sends actions to actuators. An edge fog node f_R has child sensors s_{R0}, s_{R1}, ..., s_{R,sl_R-1} ($sl_R \geq 0$) and child

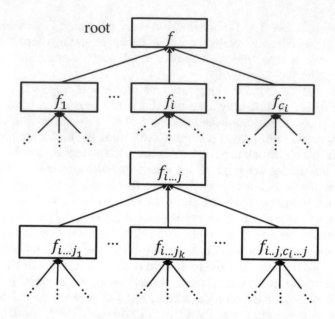

Fig. 1. TBFC model.

actuators a_{R0}, a_{R1}, ..., a_{R,al_R-1} ($al_R \geq 0$). Every edge fog node is at the same level. Each fog node f_R of the same level is equipped with a same subprocess $p(f_R)$.

In this paper, an application process p is assumed to be linear, i.e. a sequence of subprocesses p_0, p_1, ..., p_{h-1}. The edge subprocess p_{h-1} takes input data from sensors. The root subprocess p_0 is performed on a root node f, i.e. server. Each subprocess p_i takes data from a subprocess p_{i-1} and gives the processed data to a subprocess p_{i+1}. In the TBFC tree of height h, each subprocess p_i is performed on nodes of level i. Let $p(f_R)$ show a subprocess supported by in a node f_R. For each edge node f_R, a sequence of subprocess supported by node, is a path for not f to f_R is a sequence of subprocesses $p_1, ..., p_{h-1}$.

A node f_R takes input data d_{Ri} from each child node f_{Ri} ($i = 1, ..., c_R$). D_R shows a collection of the input data $d_{R1}, ..., d_{R,c_R}$ from child nodes $f_{R1}, ..., f_{R,c_R}$, respectively. The node f_R obtains output data d_R by doing the computation $f(p_R)$ on the input data D_R. Then, the node f_R sends the output data d_R to a parent node $pt(f_R)$.

A notation $|d|$ shows the size [Byte] of data d. Let i_R and o_R be the sizes of $|D_R|$ and $|d_R|$ of input data D_R and output data d_R, respectively. The ratio $|o_R| / |i_R|$ is the *output ratio* ρ_R of a node f_R. Here, $o_R = \rho_R \cdot i_R$. For example, if a fog node f_R obtains an average value of the input data $d_{R1}, ..., d_{R,l_R}$, the output ratio ρ_R is $1/l_R$.

3   Subtree-Based Computing Model

3.1   TWTBFC (Two-Way TBFC) Model

In this paper, we propose a new Two-Way TBFC (TWTBFC) model to collect only of sensor data required to decide on actions. That is, only some sensor data, i.e. a small amount of data scattered in various parts of the edge send to servers which is required to make a decision on actions of actuators.

Each sensor s_{Ri} sends sensors data ud_{Ri} to a parent edge fog node f_R. The edge fog node f_R first collects sensor data $UD_R = \{ud_{R0}, ud_{R1}, ..., ud_{R,sl_R-1}\}$ from child sensors s_{R0}, s_{R1}, ..., s_{R,sl_R-1}, respectively. Then, the edge fog node f_R sends output data ud_R obtained by processing the sensor data UD_R to root node f in a same way as the TBFC model as shown in Fig. 2.

3.2   Fog Nodes

We consider modules of each fog node f_R in the TWTBFC model. As shown in Fig. 3, an input module I_R of a node f_R receives input data ud_{Ri} from each child node f_{Ri} ($i = 0, ..., f_{R,l_R-1}$). Then, an computation module C_R does the computation on the input data UD_R and generates output data ud_R. An output module O_R sends the output data ud_R to root node f_0. The upward I_R, C_R, and O_R modules are same as the input, computation and output modules of the TBFC model.

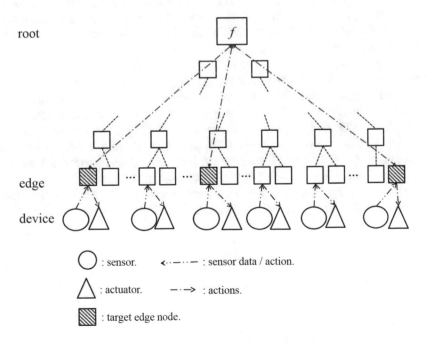

Fig. 2. Data and action transmissions.

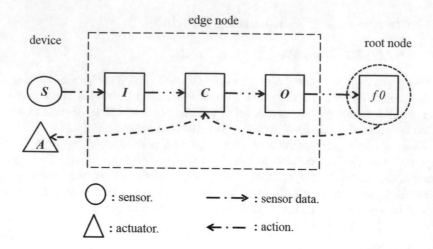

Fig. 3. Edge node.

4   Evaluation

We evaluate the TWTBFC model in terms of electric energy consumption and execution time of nodes. The TWTBFC model is composed of fog nodes structured in a tree. In this paper, we consider a height-balanced k-ary tree of fog nodes whose height is h. The reduction ratio ρ_R of each fog node f_R is assumed to be the same, i.e. $\rho_R = \rho$. We assume a root node is a server f_0 with two Inter Xeon E5-2667 CPUs [1], where the minimum electric power consumption mE_0 is 126.1 [W] and maximum electric power consumption xE_0 is 301.3 [W]. Each fog node f_R is realized by a Raspberry Pi Model B [2]. Here, the minimum electric power mE_R is 2.1 [W] and the maximum electric power xE_R is 3.7 [W] [3].

In order to make clear the computation rate of each node, we perform a same C program p which uses only CPU on the server f_0 and a fog node f_R [8,9]. It takes $mT = 0.879$ [sec] to perform the process p without any other process on the server f_0. The computation rate CR_0 of the server f_0 is assumed to be one. In a fog node f_R, it takes 4.75 [sec] to perform the same program p Hence, the computation rate CR_R of each fog node f_R is $0.879/4.75 = 0.185$.

In the evaluation, the total electric energy consumption for height of tree in TBFC model, TWTBFC model, and new-TWTBFC model as shown in Fig. 4.

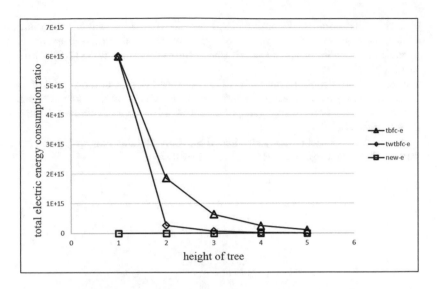

Fig. 4. Total electric energy consumption.

5 Concluding Remarks

In this paper, we proposed the modified model to efficiently realize the TWTBFC model. Here, only sensor data to make a decision on actions is collected by edge nodes. Then, edge nodes send preprocessed data to servers in cloud, only target edge nodes receive data to be processed by servers from sensors, Then, target edge nodes make a decision on actions to actuators.

References

1. Dl360p gen8. www8.hp.com/h20195/v2/getpdf.aspx/c04128242.pdf?ver=2
2. Raspberry pi 3 model b. https://www.raspberrypi.org/products/raspberry-pi-3-model-b/
3. Chida, R., Guo, Y., Oma, R., Nakamura, S., Duolikun, D., Enokido, T., Takizawa, M.: Implementation of fog nodes in the tree-based fog computing (TBFC) model of the IoT. In: Proceedings of the 7th International Conference on Emerging Internet, Data and Web Technologies (EIDWT-2019), pp. 92–102 (2019)
4. Creeger, M.: Cloud computing: an overview. Queue **7**(5), 3–4 (2009)
5. Enokido, T., Ailixier, A., Takizawa, M.: A model for reducing power consumption in peer-to-peer systems. IEEE Syst. J. **4**, 221–229 (2010)
6. Enokido, T., Ailixier, A., Takizawa, M.: An extended simple power consumption model for selecting a server to perform computation type processes in digital ecosystems. IEEE Trans. Ind. Inf. **10**, 1627–1636 (2014)
7. Oma, R., Nakamura, S., Duolikun, D., Enokido, T., Takizawa, M.: An energy-efficient model for fog computing in the internet of things (IoT). Internet Things **1–2**, 14–26 (2018)

8. Oma, R., Nakamura, S., Duolikun, D., Enokido, T., Takizawa, M.: Evaluation of an energy-efficient tree-based model of fog computing. In: Proceedings of the 21st International Conference on Network-Based Information Systems (NBiS-2018), pp. 99–109 (2018)
9. Oma, R., Nakamura, S., Enokido, T., Takizawa, M.: A tree-based model of energy-efficient fog computing systems in IoT. In: Proceedings of the 12th International Conference on Complex, Intelligent, and Software Intensive Systems (CISIS-2018), pp. 991–1001 (2018)
10. Rahmani, A.M., Liljeberg, P., Preden, J.S., Jantsch, A.: Fog Computing in the Internet of Things. Springer, Heidelberg (2018)

A Fuzzy-Based System for Actor Node Selection in WSANs Considering Task Accomplishment Time as a New Parameter

Donald Elmazi[1]([envelope]), Miralda Cuka[2], Makoto Ikeda[1], Keita Matsuo[1], and Leonard Barolli[1]

[1] Department of Information and Communication Engineering, Fukuoka Institute of Technology (FIT), 3-30-1 Wajiro-Higashi, Higashi-Ku, Fukuoka 811-0295, Japan
donald.elmazi@gmail.com, makoto.ikd@acm.org,
{kt-matsuo,barolli}@fit.ac.jp
[2] Graduate School of Engineering, Fukuoka Institute of Technology (FIT), 3-30-1 Wajiro-Higashi, Higashi-Ku, Fukuoka 811-0295, Japan
mcuka91@gmail.com

Abstract. The growth in sensor networks and importance of active devices in the physical world has led to the development of Wireless Sensor and Actor Networks (WSANs). WSANs consist of a large number of sensors and also a smaller number of actors. Whenever there is any emergency situation i.e., fire, earthquake, flood or enemy attack in the area, sensor nodes have the responsibility to sense it and send information towards an actor node. According to these data gathered, the actor nodes take a prompt action. In this work, we consider the actor node selection problem and propose a fuzzy-based system (FBS) that based on data provided by sensors and actors selects an appropriate actor node to carry out a task. We use 4 input parameters: Task Accomplishment Time (TAT), Distance to Event (DE), Power Consumption (PC) and Number of Sensors per Actor (NSA). The output parameter is Actor Selection Decision (ASD). We evaluate the proposed system by simulations. The simulation result show that the shorter the time for an actor to complete a task, the higher is the posssibility of the actor to be selected to carry out the job.

1 Introduction

Recent technological advances have lead to the emergence of distributed Wireless Sensor and Actor Networks (WSANs) which are capable of observing the physical world, processing the data, making decisions based on the observations and performing appropriate actions [1].

In WSANs, the devices deployed in the environment are sensors able to sense environmental data, actors able to react by affecting the environment or have both functions integrated. Actor nodes are equipped with two radio transmitters,

© Springer Nature Switzerland AG 2020
L. Barolli et al. (Eds.): EIDWT 2020, LNDECT 47, pp. 53–63, 2020.
https://doi.org/10.1007/978-3-030-39746-3_7

a low data rate transmitter to communicate with the sensor and a high data rate interface for actor-actor communication. For example, in the case of a fire, sensors relay the exact origin and intensity of the fire to actors so that they can extinguish it before spreading in the whole building or in a more complex scenario, to save people who may be trapped by fire [2–4].

To provide effective operation of WSAN, it is very important that sensors and actors coordinate in what are called sensor-actor and actor-actor coordination. Coordination is not only important during task conduction, but also during network's self-improvement operations, i.e. connectivity restoration [5,6], reliable service [7], Quality of Service (QoS) [8,9] and so on.

In WSANs actor nodes are heterogeneous, which is suitable for different types of tasks. For different type of tasks it is needed a different duration and difficulty to complete it. Based on this, nodes with proper capacities will be selected since they will accomplish the task faster. For this we consider the Task Accomplishment Time (TAT) parameter.

The role of Load Balancing in WSAN is to provide a reliable service, in order to have a better performance and effortness. In this paper, we also consider the Load Balancing issue. Load balancing identifies the optimal load on nodes of the network to increase the network efficiency.

In this paper, different from our previous research, we consider TAT parameter. The system is based on fuzzy logic and considers four input parameters for actor selection. We show the simulation results for different values of parameters.

The remainder of the paper is organized as follows. In Sect. 2, we describe the basics of WSANs including research challenges and architecture. In Sect. 3, we describe the system model and its implementation. The simulation results are shown in Sect. 4. Finally, conclusions and future work are given in Sect. 5.

2   WSANs

A WSAN is shown in Fig. 1. The main functionality of WSANs is to make actors perform appropriate actions in the environment, based on the data sensed from sensors and actors. When important data has to be transmitted (an event

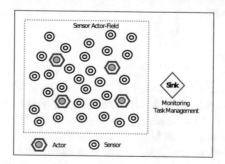

Fig. 1. Wireless Sensor Actor Network (WSAN).

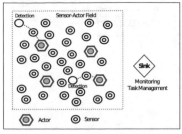 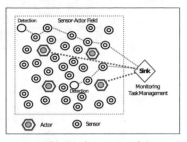

(a) Fully-Automated   (b) Semi-Automated

Fig. 2. WSAN architectures.

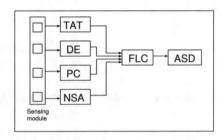

Fig. 3. Proposed system.

occurred), sensors may transmit their data back to the sink, which will control the actors' tasks from distance, or transmit their data to actors, which can perform actions independently from the sink node. Here, the former scheme is called Semi-Automated Architecture and the latter one Fully-Automated Architecture (see Fig. 2). Obviously, both architectures can be used in different applications. In the Fully-Automated Architecture are needed new sophisticated algorithms in order to provide appropriate coordination between nodes of WSAN. On the other hand, it has advantages, such as *low latency, low energy consumption, long network lifetime* [1], *higher local position accuracy, higher reliability* and so on.

3 Proposed Fuzzy-Based System

Based on WSAN characteristics and challenges, we consider the following parameters for implementation of our proposed system.

Task Accomplishment Time (TAT): Each task is represented by its duration to complete it and its difficulty. It can be accomplished by more than one actor nodes. In WSANs actor nodes are heterogeneous, which is suitable for different types of tasks. Each actor node matches its own capacity with the required capacities to complete the task. Based on this, nodes with proper capacities will be selected since they will accomplish the task faster.

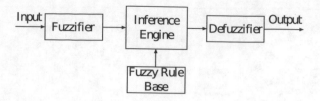

Fig. 4. FLC structure.

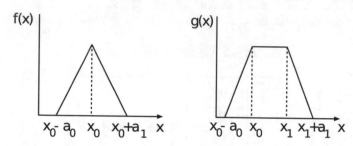

Fig. 5. Triangular and trapezoidal membership functions.

Distance to Event (DE): When an actor is called for action near an event, the distance from the actor to the event is important because when the distance is longer, the actor will spend more energy. Thus, an actor which is close to an event, should be selected.

Power Consumption (PC): As actors are active in the monitored field, they perform tasks and exchange data in different ways. Thus they have to spend energy (limited resource) based on the task and the application. It is better that the actors which consume less power are selected to carry out a task, so the network lifetime can be increased.

Number of Sensors per Actor (NSA): The number of sensors deployed in an area for sensing any event may be in the order of hundreds or thousands. So in order to have a better coverage of these sensors, the number of sensors covered by each actor node should be balanced.

Actor Selection Decision (ASD): Our system is able to decide the willingness of an actor to be assigned a certain task at a certain time.

Fuzzy sets and fuzzy logic have been developed to manage vagueness and uncertainty in a reasoning process of an intelligent system such as a knowledge based system, an expert system or a logic control system [10–22].

We use fuzzy logic to implement the proposed system. The structure of the proposed system is shown in Fig. 3. It consists of one Fuzzy Logic Controller (FLC), which is the main part of our system and its basic elements are shown in Fig. 4. They are the fuzzifier, inference engine, Fuzzy Rule Base (FRB) and defuzzifier.

Table 1. Parameters and their term sets for FLC.

Parameters	Term sets
Task Accomplishment Time (TAT)	Short (Sh), Normal (Nr), Long (Ln)
Distance to Event (DE)	Near (Ne), Moderate (Mo), Far (Fa)
Power Consumption (PC)	Low (Lo), Medium (Md), High (Hg)
Number of Sensors per Actor (NSA)	Few (Fw), Medium (Me), Many (My)
Actor Selection Decision (ASD)	Extremely Low Selection Possibility (ELSP), Very Low Selection Possibility (VLSP), Low Selection Possibility (LSP), Middle Selection Possibility (MSP), High Selection Possibility (HSP), Very High Selection Possibility (VHSP), Extremely High Selection Possibility (EHSP)

As shown in Fig. 5, we use triangular and trapezoidal membership functions for FLC, because they are suitable for real-time operation [23]. The term sets of the parameters are shown in Table 1. The x_0 in $f(x)$ is the center of triangular function, $x_0(x_1)$ in $g(x)$ is the left (right) edge of trapezoidal function, and $a_0(a_1)$ is the left (right) width of the triangular or trapezoidal function. We explain in details the design of FLC in following.

We use four input parameters for FLC:

- Task Accomplishment Time (TAT);
- Distance to Event (DE)
- Power Consumption (PC);
- Number of Sensors per Actor (NSA).

The membership functions are shown in Fig. 6 and the Fuzzy Rule Base (FRB) is shown in Table 2. The FRB forms a fuzzy set of dimensions $|T(TAT)| \times |T(DE)| \times |T(PC)| \times |T(NSA)|$, where $|T(x)|$ is the number of terms on $T(x)$. The FRB has 81 rules. The control rules have the form: IF "conditions" THEN "control action".

4   Simulation Results

The simulation results for our system are shown in Figs. 7, 8 and 9. One of the main issues in WSANs is power consumption. Lower consume of nodes energy improves the overall lifetime of the network. In Fig. 7, TAT is 0.1, PC is 0.9 and for DE from 0.1 to 0.5 we see that ASD is decreased 28% and when DE increases from 0.5 to 0.9 ASD is decreased 12%. This is because when the distance of actors to an event is far they spend more energy.

In Fig. 8, TAT is increased from 0.1 to 0.5 and we see that the probability of an actor to be selected is decreased. Because longer TAT, means that the actor node has lower probability to complete a task fast. For PC = 0.1 and for DE

Table 2. FRB of proposed fuzzy-based system.

No.	TAT	DE	PC	NSA	ASD	No.	TAT	DE	PC	NSA	ASD
1	Sh	Ne	Lo	Fw	HSP	41	Nr	Mo	Md	Me	HSP
2	Sh	Ne	Lo	Me	EHSP	42	Nr	Mo	Md	My	VLSP
3	Sh	Ne	Lo	My	HSP	43	Nr	Mo	Hg	Fw	ELSP
4	Sh	Ne	Md	Fw	MSP	44	Nr	Mo	Hg	Me	MSP
5	Sh	Ne	Md	Me	VHSP	45	Nr	Mo	Hg	My	ELSP
6	Sh	Ne	Md	My	MSP	46	Nr	Fa	Lo	Fw	LSP
7	Sh	Ne	Hg	Fw	VLSP	47	Nr	Fa	Lo	Me	HSP
8	Sh	Ne	Hg	Me	HSP	48	Nr	Fa	Lo	My	LSP
9	Sh	Ne	Hg	My	VLSP	49	Nr	Fa	Md	Fw	ELSP
10	Sh	Mo	Lo	Fw	LSP	50	Nr	Fa	Md	Me	MSP
11	Sh	Mo	Lo	Me	VHSP	51	Nr	Fa	Md	My	ELSP
12	Sh	Mo	Lo	My	LSP	52	Nr	Fa	Hg	Fw	ELSP
13	Sh	Mo	Md	Fw	VLSP	53	Nr	Fa	Hg	Me	VLSP
14	Sh	Mo	Md	Me	MSP	54	Nr	Fa	Hg	My	ELSP
15	Sh	Mo	Md	My	VLSP	55	Ln	Ne	Lo	Fw	EHSP
16	Sh	Mo	Hg	Fw	ELSP	56	Ln	Ne	Lo	Me	EHSP
17	Sh	Mo	Hg	Me	LSP	57	Ln	Ne	Lo	My	EHSP
18	Sh	Mo	Hg	My	ELSP	58	Ln	Ne	Md	Fw	VHSP
19	Sh	Fa	Lo	Fw	VLSP	59	Ln	Ne	Md	Me	EHSP
20	Sh	Fa	Lo	Me	HSP	60	Ln	Ne	Md	My	VHSP
21	Sh	Fa	Lo	My	VLSP	61	Ln	Ne	Hg	Fw	HSP
22	Sh	Fa	Md	Fw	ELSP	62	Ln	Ne	Hg	Me	EHSP
23	Sh	Fa	Md	Me	LSP	63	Ln	Ne	Hg	My	HSP
24	Sh	Fa	Md	My	ELSP	64	Ln	Mo	Lo	Fw	VHSP
25	Sh	Fa	Hg	Fw	ELSP	65	Ln	Mo	Lo	Me	EHSP
26	Sh	Fa	Hg	Me	VLSP	66	Ln	Mo	Lo	My	VHSP
27	Sh	Fa	Hg	My	ELSP	67	Ln	Mo	Md	Fw	MSP
28	Nr	Ne	Lo	Fw	VHSP	68	Ln	Mo	Md	Me	VHSP
29	Nr	Ne	Lo	Me	EHSP	69	Ln	Mo	Md	My	MSP
30	Nr	Ne	Lo	My	VHSP	70	Ln	Mo	Hg	Fw	LSP
31	Nr	Ne	Md	Fw	MSP	71	Ln	Mo	Hg	Me	HSP
32	Nr	Ne	Md	Me	EHSP	72	Ln	Mo	Hg	My	LSP
33	Nr	Ne	Md	My	MSP	73	Ln	Fa	Lo	Fw	HSP
34	Nr	Ne	Hg	Fw	LSP	74	Ln	Fa	Lo	Me	EHSP
35	Nr	Ne	Hg	Me	HSP	75	Ln	Fa	Lo	My	HSP
36	Nr	Ne	Hg	My	LSP	76	Ln	Fa	Md	Fw	LSP
37	Nr	Mo	Lo	Fw	MSP	77	Ln	Fa	Md	Me	VHSP
38	Nr	Mo	Lo	Me	VHSP	78	Ln	Fa	Md	My	LSP
39	Nr	Mo	Lo	My	MSP	79	Ln	Fa	Hg	Fw	VLSP
40	Nr	Mo	Md	Fw	VLSP	80	Ln	Fa	Hg	Me	MSP
						81	Ln	Fa	Hg	My	VLSP

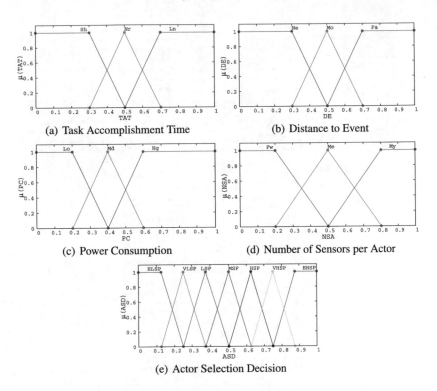

(a) Task Accomplishment Time

(b) Distance to Event

(c) Power Consumption

(d) Number of Sensors per Actor

(e) Actor Selection Decision

Fig. 6. Fuzzy membership functions.

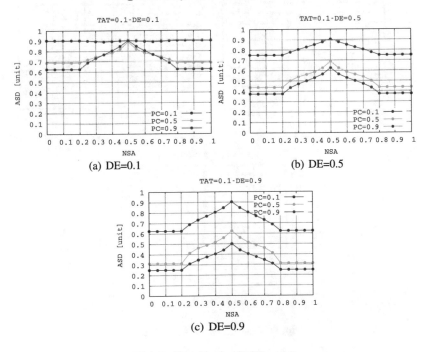

(a) DE=0.1

(b) DE=0.5

(c) DE=0.9

Fig. 7. Results for $TAT = 0.1$.

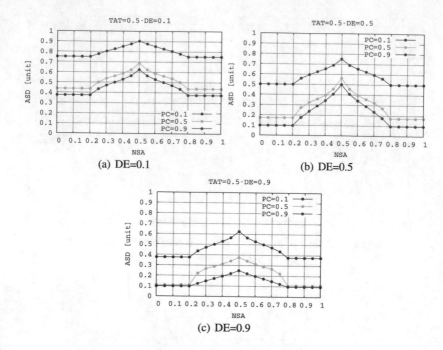

Fig. 8. Results for $TAT = 0.5$.

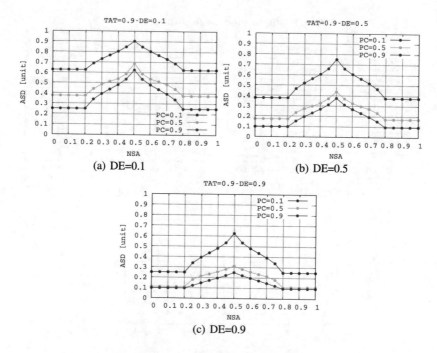

Fig. 9. Results for $TAT = 0.9$.

from 0.1 to 0.5 the ASD is decreased 15%. Also for DE from 0.5 to 0.9 the ASD is decreased 13%.

In Fig. 9 TAT is increased again, from 0.5 to 0.9 and also the actor's probability to be selected is more lower. So from NSA 0.2 to 0.5 and for PC = 0.1, the ASD increases 28%, on the other hand when NSA increases from 0.5 to 0.8, the ASD decreases also 28%.

If we compare 3 graphs for TAT from 0.1 to 0.5 (Fig. 7(b) with Fig. 8(b)) and 0.1 to 0.9 (Fig. 7(b) with Fig. 9(b)), for PC = 0.9 and DE = 0.5, the ASD is decreased 12% and 24%, respectively.

5   Conclusions and Future Work

In this paper we proposed and implemented a fuzzy-based system for actor node selection in WSANs, considering TAT as new parameter. Considering that some nodes are more reliable than the others, our system decides which actor nodes are better suited to carry out a task. From simulation results, we conclude as follows.

- When TAT parameter is increased, the ASD parameter is increased, so the probability that the system selects an actor node for the job is high.
- Actors are distributed in the network and they have different distances from an event. Actors that are further away, are less likely to be selected.
- When the DE parameter is increased, the ASD parameter is decreased, so the probability that an actor node is selected for the required task is low.
- When PC decreases, the ASD increases because the node energy is not consumed.
- If we compare 3 graphs for TAT from 0.1 to 0.5 (Fig. 7(b) with Fig. 8(b)) and 0.1 to 0.9 (Fig. 7(b) with Fig. 9(b)), for PC = 0.9 and DE = 0.5, the ASD is increased 12% and 24%, respectively.

In the future work, we will consider also other parameters for actor selection and make extensive simulations to evaluate the proposed system. Also, we will conduct experiments and compare simulation results with experimental results for different scenarios.

References

1. Akyildiz, I.F., Kasimoglu, I.H.: Wireless sensor and actor networks: research challenges. Ad Hoc Netw. J. **2**(4), 351–367 (2004)
2. Akyildiz, I., Su, W., Sankarasubramaniam, Y., Cayirci, E.: Wireless sensor networks: a survey. Comput. Netw. **38**(4), 393–422 (2002)
3. Boyinbode, O., Le, H., Takizawa, M.: A survey on clustering algorithms for wireless sensor networks. Int. J. Space Based Situated Comput. **1**(2/3), 130–136 (2011)
4. Bahrepour, M., Meratnia, N., Poel, M., Taghikhaki, Z., Havinga, P.J.: Use of wireless sensor networks for distributed event detection in disaster management applications. Int. J. Space Based Situated Comput. **2**(1), 58–69 (2012)

5. Haider, N., Imran, M., Saad, N., Zakariya, M.: Performance analysis of reactive connectivity restoration algorithms for wireless sensor and actor networks. In: IEEE Malaysia International Conference on Communications (MICC 2013), November 2013, pp. 490–495 (2013)
6. Abbasi, A., Younis, M., Akkaya, K.: Movement-assisted connectivity restoration in wireless sensor and actor networks. IEEE Trans. Parallel Distrib. Syst. **20**(9), 1366–1379 (2009)
7. Li, X., Liang, X., Lu, R., He, S., Chen, J., Shen, X.: Toward reliable actor services in wireless sensor and actor networks. In: 2011 IEEE 8th International Conference on Mobile Adhoc and Sensor Systems (MASS), October 2011, pp. 351–360 (2011)
8. Akkaya, K., Younis, M.: COLA: a coverage and latency aware actor placement for wireless sensor and actor networks. In: IEEE 64th Conference on Vehicular Technology (VTC 2006) Fall, September 2006, pp. 1–5 (2006)
9. Kakarla, J., Majhi, B.: A new optimal delay and energy efficient coordination algorithm for WSAN. In: 2013 IEEE International Conference on Advanced Networks and Telecommuncations Systems (ANTS), December 2013, pp. 1–6 (2013)
10. Inaba, T., Sakamoto, S., Kolici, V., Mino, G., Barolli, L.: A CAC scheme based on fuzzy logic for cellular networks considering security and priority parameters. In: The 9-th International Conference on Broadband and Wireless Computing, Communication and Applications (BWCCA 2014), pp. 340–346 (2014)
11. Spaho, E., Sakamoto, S., Barolli, L., Xhafa, F., Barolli, V., Iwashige, J.: A fuzzy-based system for peer reliability in JXTA-Overlay P2P considering number of interactions. In: The 16th International Conference on Network-Based Information Systems (NBiS 2013), pp. 156–161 (2013)
12. Matsuo, K., Elmazi, D., Liu, Y., Sakamoto, S., Mino, G., Barolli, L.: FACS-MP: a fuzzy admission control system with many priorities for wireless cellular networks and its performance evaluation. J. High Speed Netw. **21**(1), 1–14 (2015)
13. Grabisch, M.: The application of fuzzy integrals in multicriteria decision making. Eur. J. Oper. Res. **89**(3), 445–456 (1996)
14. Inaba, T., Elmazi, D., Liu, Y., Sakamoto, S., Barolli, L., Uchida, K.: Integrating wireless cellular and ad-hoc networks using fuzzy logic considering node mobility and security. In: The 29th IEEE International Conference on Advanced Information Networking and Applications Workshops (WAINA 2015), pp. 54–60 (2015)
15. Kulla, E., Mino, G., Sakamoto, S., Ikeda, M., Caballé, S., Barolli, L.: FBMIS: a fuzzy-based multi-interface system for cellular and ad hoc networks. In: International Conference on Advanced Information Networking and Applications (AINA 2014), pp. 180–185 (2014)
16. Elmazi, D., Kulla, E., Oda, T., Spaho, E., Sakamoto, S., Barolli, L.: A comparison study of two fuzzy-based systems for selection of actor node in wireless sensor actor networks. J. Ambient Intell. Humanized Comput. **6**, 635–645 (2015)
17. Zadeh, L.: Fuzzy logic, neural networks, and soft computing. Commun. ACM **37**, 77–84 (1994)
18. Spaho, E., Sakamoto, S., Barolli, L., Xhafa, F., Ikeda, M.: Trustworthiness in P2P: performance behaviour of two fuzzy-based systems for JXTA-overlay platform. Soft Comput. **18**(9), 1783–1793 (2014)
19. Inaba, T., Sakamoto, S., Kulla, E., Caballe, S., Ikeda, M., Barolli, L.: An integrated system for wireless cellular and ad-hoc networks using fuzzy logic. In: International Conference on Intelligent Networking and Collaborative Systems (INCoS 2014), pp. 157–162 (2014)

20. Matsuo, K., Elmazi, D., Liu, Y., Sakamoto, S., Barolli, L.: A multi-modal simulation system for wireless sensor networks: a comparison study considering stationary and mobile sink and event. J. Ambient Intell. Humanized Comput. 6(4), 519–529 (2015)
21. Kolici, V., Inaba, T., Lala, A., Mino, G., Sakamoto, S., Barolli, L.: A fuzzy-based CAC scheme for cellular networks considering security. In: International Conference on Network-Based Information Systems (NBiS 2014), pp. 368–373 (2014)
22. Matsuo, K., Elmazi, D., Liu, Y., Sakamoto, S., Mino, G., Barolli, L.: FACS-MP: a fuzzy admission control system with many priorities for wireless cellular networks and its performemance evaluation. J. High Speed Netw. 21(1), 1–14 (2015)
23. Mendel, J.M.: Fuzzy logic systems for engineering: a tutorial. Proc. IEEE 83(3), 345–377 (1995)

IoT Node Selection in Opportunistic Networks: A Fuzzy-Based Approach Considering Node's Successful Delivery Ratio (NSDR) as a New Parameter

Miralda Cuka[1]([✉]), Donald Elmazi[2], Makoto Ikeda[2], Keita Matsuo[2], and Leonard Barolli[2]

[1] Graduate School of Engineering, Fukuoka Institute of Technology (FIT),
3-30-1 Wajiro-Higashi, Higashi-Ku, Fukuoka 811-0295, Japan
mcuka91@gmail.com
[2] Department of Information and Communication Engineering, Fukuoka Institute of
Technology (FIT), 3-30-1 Wajiro-Higashi, Higashi-Ku, Fukuoka 811-0295, Japan
donald.elmazi@gmail.com, makoto.ikd@acm.org,
{kt-matsuo,barolli}@fit.ac.jp

Abstract. In opportunistic networks the communication opportunities (contacts) are intermittent and there is no need to establish an end-to-end link between the communication nodes. The enormous growth of nodes having access to the Internet, along the vast evolution of the Internet and the connectivity of objects and nodes, has evolved as Internet of Things (IoT). There are different issues for these networks. One of them is the selection of IoT nodes in order to carry out a task in opportunistic networks. In this work, we implement a Fuzzy-based System for IoT node selection in opportunistic networks. For our proposed system, we use four input parameters: Node's Free Buffer Space (NFBS), Node's Successful Delivery Ratio (NSDR), Node's Battery Level (NBL), Node Contact Duration (NCD). The output parameter is Node Selection Decision (NSD). The results show that the proposed system makes a proper selection decision of IoT nodes in opportunistic networks. The IoT node selection is increased up to 17% and 40% by increasing NBL and NSDR, respectively.

1 Introduction

Future communication systems will be increasingly complex, involving thousands of heterogeneous nodes with diverse capabilities and various networking technologies interconnected with the aim to provide users with ubiquitous access to information and advanced services at a high quality level, in a cost efficient manner, any time, any place, and in line with the always best connectivity principle. The Opportunistic Networks (OppNets) can provide an alternative way to support the diffusion of information in special locations within a city, particularly in crowded spaces where current wireless technologies can exhibit congestion issues. The efficiency of this diffusion relies mainly on user mobility. In fact,

© Springer Nature Switzerland AG 2020
L. Barolli et al. (Eds.): EIDWT 2020, LNDECT 47, pp. 64–72, 2020.
https://doi.org/10.1007/978-3-030-39746-3_8

mobility creates the opportunities for contacts and, therefore, for data forwarding [1]. OppNets have appeared as an evolution of the MANETs. They are also a wireless based network and hence, they face various issues similar to MANETs such as frequent disconnections, highly variable links, limited bandwidth etc. In OppNets, nodes are always moving which makes the network easy to deploy and decreases the dependence on infrastructure for communication [2].

In Internet of Things (IoT), the traffic is going through different networks. The IoT can seamlessly connect the real world and cyberspace via physical objects embedded with various types of intelligent sensors. A large number of Internet-connected machines will generate and exchange an enormous amount of data that make daily life more convenient, help to make a tough decision and provide beneficial services. The IoT probably becomes one of the most popular networking concepts that has the potential to bring out many benefits [3,4].

OppNets are the variants of Delay Tolerant Networks (DTNs). It is a class of networks that has emerged as an active research subject in the recent times. Owing to the transient and un-connected nature of the nodes, routing becomes a challenging task in these networks. Sparse connectivity, no infrastructure and limited resources further complicate the situation [5,6]. Routing methods for such sparse mobile networks use a different paradigm for message delivery. These schemes utilize node mobility by having nodes carry messages, waiting for an opportunity to transfer messages to the destination or the next relay rather than transmitting them over a path [7]. Hence, the challenges for routing in OppNet are very different from the traditional wireless networks and their utility and potential for scalability makes them a huge success.

In mobile OppNet, connectivity varies significantly over time and is often disruptive. Examples of such networks include interplanetary communication networks, mobile sensor networks, vehicular ad hoc networks (VANETs), terrestrial wireless networks, and under-water sensor networks. While the nodes in such networks are typically delay-tolerant, message delivery latency still remains a crucial metric, and reducing it is highly desirable [8].

The Fuzzy Logic (FL) is unique approach that is able to simultaneously handle numerical data and linguistic knowledge. The fuzzy logic works on the levels of possibilities of input to achieve the definite output. Fuzzy set theory and FL establish the specifics of the nonlinear mapping.

In this paper, we propose and implement a Fuzy-based system for selection of IoT nodes in OppNet considering four linguistic parameters: Node's Free Buffer Space (NFBS), Node's Successful Delivery Ratio (NSDR), Node's Battery Level (NBL), Node Contact Duration (NCD) for IoT node selection. We show the simulation results for different values of parameters.

The remainder of the paper is organized as follows. In the Sect. 2, we present our proposed fuzzy-based system. In Sect. 3, we evaluate our proposed system. Finally, conclusions and future work are given in Sect. 4.

66   M. Cuka et al.

2   Proposed Fuzzy-Based System

In this work, we use FL to implement the proposed system. Fuzzy sets and FL
have been developed to manage vagueness and uncertainty in a reasoning process
of an intelligent system such as a knowledge based system, an expert system or
a logic control system [9–22].

The structure of the proposed system for the node selection is shown in
Fig. 1. Based on OppNets characteristics and challenges, we consider the following parameters for implementation of our proposed system.

- IoT Node's Free Buffer Space (NFBS)
- IoT Node's Successful Delivery Ratio (NSDR)
- IoT Node's Battery Level (NBL)
- IoT Node Contact Duration (NCD)

When designing a system based on FL, the following steps must be followed:
(1) Defining the fuzzy controller input and outputs, (2) Assigning each input
and output linguistic variable and their term sets based on the level of aggregation, (3) Based on the inputs and output selected, choosing the appropriate
Fuzzy Membership Functions (FMFs) types based on the problem specifics, (4)
Building the Fuzzy Rule Base (FRB).

Our proposed system consists of one Fuzzy Logic Controller (FLC), which is
the main part of our system and its basic elements which are shown in Fig. 2.
They are the fuzzifier, inference engine, FRB and defuzzifier. The FRB forms
a fuzzy set of dimensions $|T(NFBS)| \times |T(NSDR)| \times |T(NBL)| \times |T(NCD)|$,
where $|T(x)|$ is the number of terms on $T(x)$. We have four input parameters, so
our system has 81 rules. The term sets for these parameters are shown in Table 1.
The control rules which are shown in Table 2 have the form: *IF "conditions"
THEN "control action"*.

These parameters will be represented from numerical inputs to linguistic
variables. We use FMFs to quantify the linguistic term. The FMFs of our system
our shown in Fig. 3. We have decided to use triangular and trapezoidal FMFs
for FLC due to their simplicity and computational efficiency [23]. However, the
overlap triangle-to-triangle and trapezoid-to-triangle fuzzy regions can not be

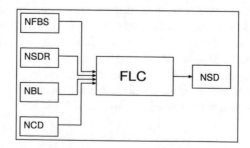

Fig. 1. Proposed system model.

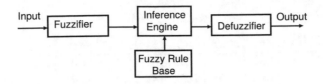

Fig. 2. FLC structure.

Table 1. Parameters and their term sets for FLC.

Parameters	Term sets
Node's Free Buffer Space *(NFBS)*	Small *(Sm)*, Medium *(Md)*, Big *(Bg)*
Node's Successful Deliviery Ratio *(NSDR)*	Low *(Lo)*, Medium *(Med)*, High *(Hg)*
Node's Battery Level *(NBL)*	Low *(Lw)*, Medium *(Mdm)*, High *(Hgh)*
Node Contact Duration *(NCD)*	Short *(Sh)*, Medium *(Medm)*, Long *(Lng)*
Node Selection Decision *(NSD)*	Extremely Low Selection Possibility *(ELSP)*, Very Low Selection Possibility *(VLSP)*, Low Selection Possibility *(LSP)*, Medium Selection Possibility *(MSP)*, High Selection Possibility *(HSP)*, Very High Selection Possibility *(VHSP)*, Extremely High Selection Possibility *(EHSP)*

addressed by any rule. It depends on the parameters and the specifics of their applications. For example, in Fig. 3(b), we see that the trapezoidal FMF for the "Lo" term set extends until 0.7. While the "Hg" term set is limited to values above 0.9. This can be explained with the tolerance the network has for a successful delivery ratio. If only 70% of the packets have been delivered, the network performance is highly affected. The only acceptable values for NSDR where QoS is ensured is for a packet loss less than 10%, hence the FMF for the "High" term set starts after 0.9.

3 Evaluation of Proposed Fuzzy-Based System

We present the simulation results in Figs. 4, 5 and 6. We show how different values of the input parameters NFBS, NSDR, NBL, NCD affect the possibility of an IoT node to be selected (NSD) for completing a task.

In Fig. 4, we show how the output parameter NSD is affected by different values of NSDR. In Fig. 4(a), NSDR = 0.1 which means IoT nodes do not have a good history of successful deliveries. In Fig. 4(b), NSDR increases to 0.9. For NSDR = 0.1 to NSDR = 0.9, NBL = 0.9 and NCD = 0.5, NSD is increased 40%. An increase in NSDR results in an increase in NSD as nodes which have a higher rate of successful delivery are preferred over others.

In Fig. 4(b), for NCD = 0.4, we see that the NSD increases with the increase of the battery level. For NBL from 0.1 to 0.5 and from 0.1 to NBL 0.9, NSD

Table 2. FRB.

No.	NFBS	NSDR	NBL	NCD	NSD	No.	NFBS	NSDR	NBL	NCD	NSD
1	Sm	Lo	Lw	Sh	ELSP	41	Md	Med	Mdm	Medm	MSP
2	Sm	Lo	Lw	Medm	ELSP	42	Md	Med	Mdm	Lng	VLSP
3	Sm	Lo	Lw	Lng	ELSP	43	Md	Med	Hgh	Sh	MSP
4	Sm	Lo	Mdm	Sh	ELSP	44	Md	Med	Hgh	Medm	VHSP
5	Sm	Lo	Mdm	Medm	LSP	45	Md	Med	Hgh	Lng	MSP
6	Sm	Lo	Mdm	Lng	ELSP	46	Md	Hg	Lw	Sh	MSP
7	Sm	Lo	Hgh	Sh	VLSP	47	Md	Hg	Lw	Medm	VHSP
8	Sm	Lo	Hgh	Medm	MSP	48	Md	Hg	Lw	Lng	MSP
9	Sm	Lo	Hgh	Lng	VLSP	49	Md	Hg	Mdm	Sh	HSP
10	Sm	Med	Lw	Sh	ELSP	50	Md	Hg	Mdm	Medm	EHSP
11	Sm	Med	Lw	Medm	VLSP	51	Md	Hg	Mdm	Lng	HSP
12	Sm	Med	Lw	Lng	ELSP	52	Md	Hg	Hgh	Sh	VHSP
13	Sm	Med	Mdm	Sh	ELSP	53	Md	Hg	Hgh	Medm	EHSP
14	Sm	Med	Mdm	Medm	LSP	54	Md	Hg	Hgh	Lng	EHSP
15	Sm	Med	Mdm	Lng	VLSP	55	Bg	Lo	Lw	Sh	VLSP
16	Sm	Med	Hgh	Sh	LSP	56	Bg	Lo	Lw	Medm	MSP
17	Sm	Med	Hgh	Medm	HSP	57	Bg	Lo	Lw	Lng	VLSP
18	Sm	Med	Hgh	Lng	LSP	58	Bg	Lo	Mdm	Sh	LSP
19	Sm	Hg	Lw	Sh	LSP	59	Bg	Lo	Mdm	Medm	HSP
20	Sm	Hg	Lw	Medm	HSP	60	Bg	Lo	Mdm	Lng	LSP
21	Sm	Hg	Lw	Lng	LSP	61	Bg	Lo	Hgh	Sh	HSP
22	Sm	Hg	Mdm	Sh	MSP	62	Bg	Lo	Hgh	Medm	VHSP
23	Sm	Hg	Mdm	Medm	VHSP	63	Bg	Lo	Hgh	Lng	HSP
24	Sm	Hg	Mdm	Lng	HSP	64	Bg	Med	Lw	Sh	VLSP
25	Sm	Hg	Hgh	Sh	VHSP	65	Bg	Med	Lw	Medm	MSP
26	Sm	Hg	Hgh	Medm	EHSP	66	Bg	Med	Lw	Lng	LSP
27	Sm	Hg	Hgh	Lng	VHSP	67	Bg	Med	Mdm	Sh	MSP
28	Md	Lo	Lw	Sh	ELSP	68	Bg	Med	Mdm	Medm	VHSP
29	Md	Lo	Lw	Medm	VLSP	69	Bg	Med	Mdm	Lng	MSP
30	Md	Lo	Lw	Lng	ELSP	70	Bg	Med	Hgh	Sh	HSP
31	Md	Lo	Mdm	Sh	VLSP	71	Bg	Med	Hgh	Medm	EHSP
32	Md	Lo	Mdm	Medm	MSP	72	Bg	Med	Hgh	Lng	VHSP
33	Md	Lo	Mdm	Lng	VLSP	73	Bg	Hg	Lw	Sh	HSP
34	Md	Lo	Hgh	Sh	LSP	74	Bg	Hg	Lw	Medm	EHSP
35	Md	Lo	Hgh	Medm	HSP	75	Bg	Hg	Lw	Lng	VHSP
36	Md	Lo	Hgh	Lng	LSP	76	Bg	Hg	Mdm	Sh	VHSP
37	Md	Med	Lw	Sh	ELSP	77	Bg	Hg	Mdm	Medm	EHSP
38	Md	Med	Lw	Medm	LSP	78	Bg	Hg	Mdm	Lng	EHSP
39	Md	Med	Lw	Lng	ELSP	79	Bg	Hg	Hgh	Sh	EHSP
40	Md	Med	Mdm	Sh	VLSP	80	Bg	Hg	Hgh	Medm	EHSP
						81	Bg	Hg	Hgh	Lng	EHSP

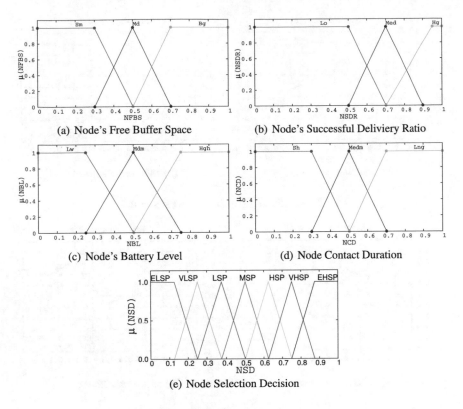

(a) Node's Free Buffer Space

(b) Node's Successful Deliviery Ratio

(c) Node's Battery Level

(d) Node Contact Duration

(e) Node Selection Decision

Fig. 3. FMFs.

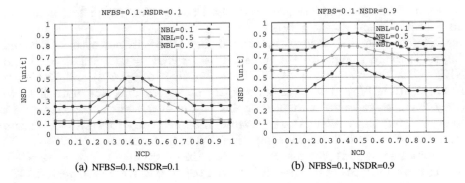

(a) NFBS=0.1, NSDR=0.1

(b) NFBS=0.1, NSDR=0.9

Fig. 4. Simulation results for NFBS = 0.1.

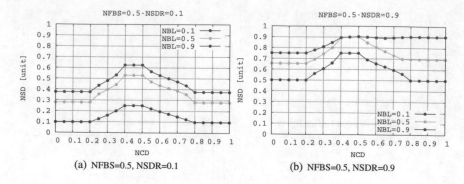

Fig. 5. Simulation results for NFBS = 0.5.

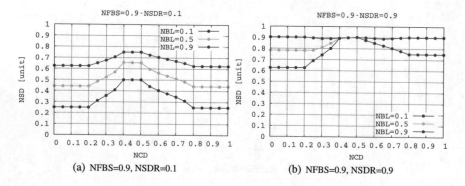

Fig. 6. Simulation results for NFBS = 0.9.

is increased to 17% and 11%, respectively. Battery level on an IoT node will increase its lifetime on the network.

Different from Fig. 4, in Figs. 5 and 6, the NFBS is increased to 0.5 and 0.9, respectively. Comparing the three figures for NCD = 0.5, NSDR = 0.1 and NBL = 0.9, the NSD increases 12% from Fig. 4(a) to Fig. 5(a) and 13% from Fig. 5(a) to Fig. 6(a). A big free buffer space on a node makes it more likely to carry the message longer without dropping it. However, some nodes may have a fully occupied buffer. Since these networks use store-carry-forward mechanism, an occupied buffer will cause a congestion due to buffer overflow.

Another important parameter that affects the selection of IoT nodes is NCD. In Fig. 5(b), for NBL = 0.9, and NCD from 0.2 to 0.5, NSD is increased 14%. When IoT nodes encounter each other, they stay in contact for different amounts of time depending on their transmission range or mobility. To ensure that the entire message has been transferred from node to node but also that nodes move across the network to expand the network connectivity, an appropriate amount of NCD is preferred.

4   Conclusions and Future Work

In this paper, we proposed a fuzzy-based IoT node selection system for Opp-
Nets considering four parameters: NFBS, NSDR, NBL, NCD. We evaluated the
proposed system by computer simulations. Our system makes the decision and
chooses which IoT node is better suited for a certain task. By increasing NFBS,
NSDR, NBL we see that NSD increases, but for NCD we need to find an optimal
time for IoT nodes to stay in contact with each other.

In the future work, we will also consider other parameters for IoT node selection and make extensive simulations and experiments to evaluate the proposed
system.

References

1. Mantas, N., Louta, M., Karapistoli, E., Karetsos, G.T., Kraounakis, S., Obaidat, M.S.: Towards an incentive-compatible, reputation-based framework for stimulating cooperation in opportunistic networks: a survey. IET Netw. **6**(6), 169–178 (2017)
2. Sharma, D.K., Sharma, A., Kumar, J., et al.: KNNR: K-nearest neighbour classification based routing protocol for opportunistic networks. In: 10-th International Conference on Contemporary Computing (IC3), pp. 1–6. IEEE (2017)
3. Kraijak, S., Tuwanut, P.: A survey on internet of things architecture, protocols, possible applications, security, privacy, real-world implementation and future trends. In: 16th International Conference on Communication Technology (ICCT), pp. 26–31. IEEE (2015)
4. Arridha, R., Sukaridhoto, S., Pramadihanto, D., Funabiki, N.: Classification extension based on IoT-big data analytic for smart environment monitoring and analytic in real-time system. Int. J. Space Based Situated Comput. **7**(2), 82–93 (2017)
5. Dhurandher, S.K., Sharma, D.K., Woungang, I., Bhati, S.: HBPR: history based prediction for routing in infrastructure-less opportunistic networks. In: 27th International Conference on Advanced Information Networking and Applications (AINA), pp. 931–936. IEEE (2013)
6. Spaho, E., Mino, G., Barolli, L., Xhafa, F.: Goodput and PDR analysis of AODV, OLSR and DYMO protocols for vehicular networks using CAVENET. Int. J. Grid Util. Comput. **2**(2), 130–138 (2011)
7. Abdulla, M., Simon, R.: The impact of intercontact time within opportunistic networks: protocol implications and mobility models. TechRepublic White Paper (2009)
8. Patra, T.K., Sunny, A.: Forwarding in heterogeneous mobile opportunistic networks. IEEE Commun. Lett. **22**(3), 626–629 (2018)
9. Inaba, T., Sakamoto, S., Kolici, V., Mino, G., Barolli, L.: A CAC scheme based on fuzzy logic for cellular networks considering security and priority parameters. In: The 9-th International Conference on Broadband and Wireless Computing, Communication and Applications (BWCCA 2014), pp. 340–346 (2014)
10. Spaho, E., Sakamoto, S., Barolli, L., Xhafa, F., Barolli, V., Iwashige, J.: A fuzzy-based system for peer reliability in JXTA-Overlay P2P considering number of interactions. In: The 16th International Conference on Network-Based Information Systems (NBiS 2013), pp. 156–161 (2013)

11. Matsuo, K., Elmazi, D., Liu, Y., Sakamoto, S., Mino, G., Barolli, L.: FACS-MP: a fuzzy admission control system with many priorities for wireless cellular networks and its performance evaluation. J. High Speed Netw. **21**(1), 1–14 (2015)
12. Grabisch, M.: The application of fuzzy integrals in multicriteria decision making. Eur. J. Oper. Res. **89**(3), 445–456 (1996)
13. Inaba, T., Elmazi, D., Liu, Y., Sakamoto, S., Barolli, L., Uchida, K.: Integrating wireless cellular and ad-hoc networks using fuzzy logic considering node mobility and security. In: The 29th IEEE International Conference on Advanced Information Networking and Applications Workshops (WAINA 2015), pp. 54–60 (2015)
14. Kulla, E., Mino, G., Sakamoto, S., Ikeda, M., Caballé, S., Barolli, L.: FBMIS: a fuzzy-based multi-interface system for cellular and ad hoc networks. In: International Conference on Advanced Information Networking and Applications (AINA 2014), pp. 180–185 (2014)
15. Elmazi, D., Kulla, E., Oda, T., Spaho, E., Sakamoto, S., Barolli, L.: A comparison study of two fuzzy-based systems for selection of actor node in wireless sensor actor networks. J. Ambient Intell. Humanized Comput. **6**(5), 635–645 (2015)
16. Zadeh, L.: Fuzzy logic, neural networks, and soft computing. Commun. ACM **37**, 77–84 (1994)
17. Spaho, E., Sakamoto, S., Barolli, L., Xhafa, F., Ikeda, M.: Trustworthiness in P2P: performance behaviour of two fuzzy-based systems for JXTA-overlay platform. Soft Comput. **18**(9), 1783–1793 (2014)
18. Inaba, T., Sakamoto, S., Kulla, E., Caballe, S., Ikeda, M., Barolli, L.: An integrated system for wireless cellular and ad-hoc networks using fuzzy logic. In: International Conference on Intelligent Networking and Collaborative Systems (INCoS 2014), pp. 157–162 (2014)
19. Matsuo, K., Elmazi, D., Liu, Y., Sakamoto, S., Barolli, L.: A multi-modal simulation system for wireless sensor networks: a comparison study considering stationary and mobile sink and event. J. Ambient Intell. Humanized Comput. **6**(4), 519–529 (2015)
20. Kolici, V., Inaba, T., Lala, A., Mino, G., Sakamoto, S., Barolli, L.: A fuzzy-based CAC scheme for cellular networks considering security. In: International Conference on Network-Based Information Systems (NBiS 2014), pp. 368–373 (2014)
21. Liu, Y., Sakamoto, S., Matsuo, K., Ikeda, M., Barolli, L., Xhafa, F.: A comparison study for two fuzzy-based systems: improving reliability and security of JXTA-overlay P2P platform. Soft Comput. **20**(7), 2677–2687 (2015)
22. Matsuo, K., Elmazi, D., Liu, Y., Sakamoto, S., Mino, G., Barolli, L.: FACS-MP: a fuzzy admission control system with many priorities for wireless cellular networks and its peroremance evaluation. J. High Speed Netw. **21**(1), 1–14 (2015)
23. Mendel, J.M.: Fuzzy logic systems for engineering: a tutorial. Proc. IEEE **83**(3), 345–377 (1995)

A Fuzzy-Based System for Admission Control in 5G Wireless Networks Considering Software-Defined Network Approach

Phudit Ampririt[1]([⊠]), Seiji Ohara[1], Yi Liu[2], Makoto Ikeda[2], Hiroshi Maeda[2], and Leonard Barolli[2]

[1] Graduate School of Engineering, Fukuoka Institute of Technology,
3-30-1 Wajiro-Higashi, Higashi-Ku, Fukuoka 811-0295, Japan
`iceattpon12@gmail.com, seiji.ohara.19@gmail.com`
[2] Department of Information and Communication Engineering, Fukuoka Institute of
Technology, 3-30-1 Wajiro-Higashi, Higashi-Ku, Fukuoka 811-0295, Japan
`ryuui1010@gmail.com, makoto.ikd@acm.org, hiroshi@gmail.com,`
`barolli@fit.ac.jp`

Abstract. The Software-Defined Network (SDN) will be the key system for efficient management and control in Fifth Generation (5G) networks. In this paper, we propose a Fuzzy-based system for admission control in 5G Wireless Networks considering SDN. The proposed system considers slice utility, delay time and grade of service to make decision on the admission control. We carried out simulations for evaluating the performance of our proposed system. From simulation results, we conclude that the considered parameters have different effects on the admission control decision. When User Request Delay Time (URDT) is increasing, Admission Decision (AD) parameter is decreased. But when Grade of Service (GS) and Slice Utility (SU) are increasing, AD is increased.

1 Introduction

Nowdays, the number of network devices are increasing and they are connected to the Internet. Especially in 5G networks, there will be billions of new devices with unpredictable traffic pattern which provide higher data rate. With the appearance of Internet of Things (IoT), these devices with generate Big Data to the Internet, which with cause the network congestion and deteriorate the Quality of Service (QoS).

The 5G network will be expected to be better than 4G. The 5G network will provide users with new experience such as Ultra High Definition Television (UHD) on Internet, support a lot of Internet of IoT devices with long battery life and high data rate on hotspot areas with high user density. Also, the routing

© Springer Nature Switzerland AG 2020
L. Barolli et al. (Eds.): EIDWT 2020, LNDECT 47, pp. 73–81, 2020.
https://doi.org/10.1007/978-3-030-39746-3_9

and switching technologies aren't important anymore or coverage area is shorter than 4G because it uses high frequency for facing higher device's volume for high user density [1–3].

Recently, there are many research work that try to build systems which are suitable to 5G era. The SDN is one of them [4]. For example, the mobile handover mechanism with SDN is used for reducing the delay in handover processing and improve the quality-of-service (QoS). Also, by using SDN the QoS can be improved by applying Fuzzy Logic (FL) on SDN controller [5–7].

This paper presents a Fuzzy-based system for admission control in 5G Wireless Networks considering three parameters: Grade of Service (GS), User Request Delay Time (URDT), Slice Utility (SU).

The rest of the paper is organized as follows. In Sect. 2 is presented an overview of SDN. In Sect. 3, we present application of Fuzzy Logic for admission control. In Sect. 4, we describe the proposed fuzzy-based system and its implementation. In Sect. 5, we explain the simulation results. Finally, conclusions and future work are presented in Sect. 6.

2   Software-Defined Networks (SDNs)

The SDN is a new networking paradigm that decouples the data plane from control plane in network. In traditional networks, the whole network is controlled by each network device, thus the management and control is difficult. Each network device has to contact each other all the time when a user want to connect or makes handover to other network devices. The SDN control plane is managed by SDN controller or cooperating group of SDN controllers. In the infrastructure layer, the network devices are receiving order from SDN controller and send data among them. The SDN structure is shown in Fig. 1 [8–10]. The SDN can manage network system while enabling new services. In congestion traffic situation, management system can be flexible. Mobility management is easier and quicker in controlling forwarding across different wireless technologies (e.g.5G, 4G, Wifi and Wimax). Also, the handover procedure is simple and the delay can be decreased.

3   Outline of Fuzzy Logic

A Fuzzy Logic System (FLS) is a nonlinear mapping of an input data vector into a scalar output, which is able to simultaneously handle numerical data and linguistic knowledge. The FL can deal with statements which may be true, false or intermediate truth-value. These statements are impossible to quantify using traditional mathematics. The FL system is used in many controlling applications such as aircraft control (Rockwell Corp.), Sendai subway operation (Hitachi), and TV picture adjustment (Sony) [11,12].

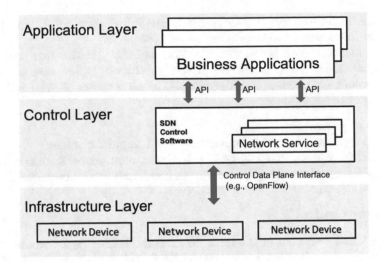

Fig. 1. Structure of SDN.

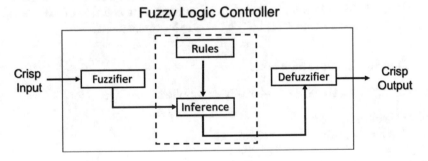

Fig. 2. FLC structure.

In Fig. 2 is shown Fuzzy Logic Controller (FLC) structure, which contains four components: fuzzifier, inference engine, fuzzy rule base and defuzzifier.

- **Fuzzifier** is needed for combining the crisp values with rules which are linguistic variables and have fuzzy sets associated with them.
- **The Rules** may be provided by expert or can be extracted from numerical data. In engineering case, the rules are expressed as a collection of IF-THEN statements.
- **The Inference engine** infers fuzzy output by considering fuzzified input values and fuzzy rules.
- **The Defuzzifier** maps output set into crisp numbers.

3.1 Linguistic Variables

A concept that plays a central role in the application of FL is that of a linguistic variable. The linguistic variables may be viewed as a form of data compression.

One linguistic variable may represent many numerical variables. It is suggestive to refer to this form of data compression as granulation.

The same effect can be achieved by conventional quantization, but in the case of quantization, the values are intervals, whereas in the case of granulation the values are overlapping fuzzy sets. The advantages of granulation over quantization are as follows:

- it is more general;
- it mimics the way in which humans interpret linguistic values;
- the transition from one linguistic value to a contiguous linguistic value is gradual rather than abrupt, resulting in continuity and robustness.

3.2 Fuzzy Control Rule

Rules are usually written in the form "IF x is S THEN y is T" where x and y are linguistic variables that are expressed by S and T that are fuzzy sets. The condition of this rule to be controlled by x, which is a control (input) variable and y is the solution (output) variable. This rule is called Fuzzy control rule. The form "IF ... THEN" is called a conditional sentence. It consists of "IF" which is called the antecedent and "THEN" is called the consequent.

3.3 Defuzzification Method

Defuzzification methods which have been applied most are:

- The Centroid Method;
- Tsukamoto's Defuzzification Method;
- The Center of Are (COA) Method;
- The Mean of Maximum (MOM) Method;
- Defuzzification when Output of Rules are Function of Their Inputs.

4 Proposed Fuzzy-Based System

In this work, we use FL to implement the proposed system. In Fig. 3, we show the overview of our proposed system. Each evolve Base Station (eBS) will receive controlling order from SDN controller and they can communicate and sending data with User Equipment (UE). On the other hand, the SDN controller will collect all data about network traffic status and controlling eBS by using the proposed fuzzy-based system. The SDN controller will be a communicating bridge between eBS and 5G core network. The proposed system is called Fuzzy-based System for Admission Control (FBSAC) in 5G wireless networks. The structure of FBSAC is shown in Fig. 4. For the implementation of our system, we consider three input parameters: Grade of Service (GS), User Request Delay Time (URDT), Slice Utility (SU) and the output parameter is Admission Decision (AD).

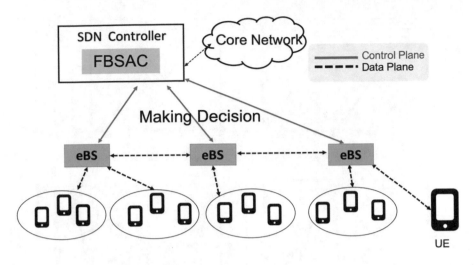

Fig. 3. Proposed system overview.

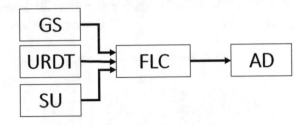

Fig. 4. Proposed system structure.

Table 1. Parameter and their term sets for FBSAC.

Parameters	Term set
Grade of Service (GS)	Low (L), Medium (M), High (H)
User Request Delay Time (URDT)	Low (Lo), Medium (Me), High (Hi)
Slice Utility (SU)	Low (Lw), Medium (Md), High (Hg)
Admission Decision (AD)	AD1, AD2, AD3, AD4, AD5, AD6, AD7

These three input parameters are fuzzified using the membership functions showed in Fig. 5(a), (b), (c). In Fig. 5(d) are shown the membership functions for the output parameter. We use triangular and trapezoidal membership functions because they are more suitable for real-time operations. We show parameters and their term sets in Table 1 and the Fuzzy Rule Base (FRB) in Table 2, which consist of 27 rules. The control rules have the form: IF "condition" THEN "control action". For example, for Rule 1: "IF GS is L, URDT is Lo and SU is Lw, THEN AD is AD3".

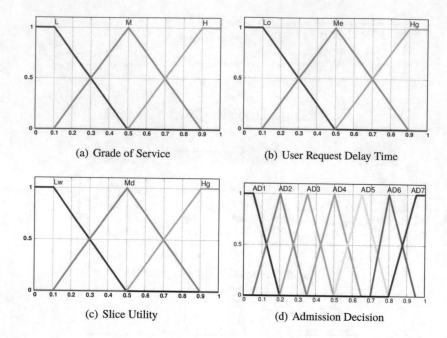

(a) Grade of Service   (b) User Request Delay Time

(c) Slice Utility   (d) Admission Decision

Fig. 5. Membership functions.

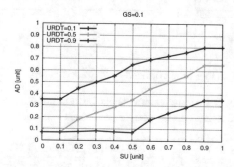

Fig. 6. Simulation results for GS $= 0.1$

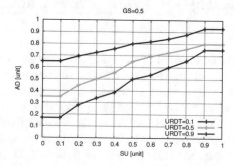

Fig. 7. Simulation results for GS $= 0.5$

Table 2. Fuzzy rule base

Rule	GS	URDT	SU	AD
1	L	Lo	Lw	AD3
2	L	Lo	Md	AD4
3	L	Lo	Hg	AD6
4	L	Me	Lw	AD1
5	L	Me	Md	AD3
6	L	Me	Hg	AD4
7	L	Hi	Lw	AD1
8	L	Hi	Md	AD1
9	L	Hi	Hg	AD3
10	M	Lo	Lw	AD4
11	M	Lo	Md	AD6
12	M	Lo	Hg	AD7
13	M	Me	Lw	AD3
14	M	Me	Md	AD4
15	M	Me	Hg	AD6
16	M	He	Lw	AD1
17	M	He	Md	AD3
18	M	He	Hg	AD4
19	H	Lo	Lw	AD6
20	H	Lo	Md	AD7
21	H	Lo	Hg	AD7
22	H	Me	Lw	AD4
23	H	Me	Md	AD6
24	H	Me	Hg	AD7
25	H	Hi	Lw	AD3
26	H	Hi	Md	AD4
27	H	Hi	Hg	AD6

5 Simulation Results

In this section, we present the simulation results of our proposed system. The simulation results are shown in Figs. 6, 7 and 8. They show the relation of AD with GS, URDT and SU. We consider the GS as constant parameter.

In Fig. 6, we consider the GS value 0.1. We change the URDT value from 0.1 to 0.9. We change the SU from 0 to 1. We see the AD is increasing when the SU is increased. When the URDT is 0.1, the AD has the best value compared with other UDRT values.

In Fig. 7, we increase the value of GS to 0.5. We see that, all AD values are increased for different URDT values. In Fig. 8, we increase the value of GS to 0.9. We see that the AD value is increased much morex compared with results in Figs. 6 and 7.

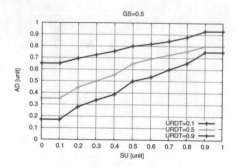

Fig. 8. Simulation results for GS = 0,9

6   Conclusions and Future Work

In this paper, we proposed a fuzzy-based system for admission control in 5G wireless networks considering SDN approach. We considered three parameters: GS, URDT and SU to decide the AD value. We used computer simulations for the performance evaluation of proposed system. From the simulation results, we see that the three parameters have different effects on the admission control decision. When URDT is increasing, AD parameter is decreased. But, when GS and SU are increasing, AD is increased.

In the future, we would like to consider other parameters and make extensive simulations to evaluate the proposed system.

References

1. Hossain, S.: 5G wireless communication systems. Am. J. Eng. Res. (AJER) **2**(10), 344–353 (2013)
2. Giordani, M., Mezzavilla, M., Zorzi, M.: Initial access in 5G mmWave cellular networks. IEEE Commun. Mag. **54**(11), 40–47 (2016)
3. Kamil, I.A., Ogundoyin, S.O.: Lightweight privacy-preserving power injection and communication over vehicular networks and 5G smart grid slice with provable security. Internet Things **8**, 100–116 (2019)
4. Hossain, E., Hasan, M.: 5G cellular: key enabling technologies and research challenges. IEEE Instrum. Measur. Mag. **18**(3), 11–21 (2015)
5. Yao, D., Su, X., Liu, B., Zeng, J.: A mobile handover mechanism based on fuzzy logic and MPTCP protocol under SDN architecture*. In: 18th International Symposium on Communications and Information Technologies (ISCIT 2018), September 2018, pp. 141–146 (2018)

6. Lee, J., Yoo, Y.: Handover cell selection using user mobility information in a 5G SDN-based network. In: 2017 Ninth International Conference on Ubiquitous and Future Networks (ICUFN 2017), July 2017, pp. 697–702 (2017)
7. Moravejosharieh, A., Ahmadi, K., Ahmad, S.: A fuzzy logic approach to increase quality of service in software defined networking. In: 2018 International Conference on Advances in Computing, Communication Control and Networking (ICACCCN 2018), October 2018, pp. 68–73 (2018)
8. Mahmoodi, T.: 5G and software-defined networking (SDN). In: 5G Radio Technology Seminar. Exploring Technical Challenges in the Emerging 5G Ecosystem, March 2015, pp. 1–19 (2015)
9. Ma, L., Wen, X., Wang, L., Lu, Z., Knopp, R.: An SDN/NFV based framework for management and deployment of service based 5G core network. China Commun. **15**(10), 86–98 (2018)
10. Li, L.E., Mao, Z.M., Rexford, J.: Toward software-defined cellular networks. In: 2012 European Workshop on Software Defined Networking, October 2012, pp. 7–12 (2012)
11. Jantzen, J.: Tutorial on fuzzy logic. Technical University of Denmark, Department of Automation, Technical report (1998)
12. Mendel, J.M.: Fuzzy logic systems for engineering: a tutorial. Proc. IEEE **83**(3), 345–377 (1995)

A Fuzzy-Based Decision System for Sightseeing Spots Considering Natural Scenery and Visiting Cost as New Parameters

Yi Liu[1](✉), Phudit Ampririt[2], Ermioni Qafzezi[2], Kevin Bylykbashi[2], Leonard Barolli[1], and Makoto Takizawa[3]

[1] Department of Information and Communication Engineering, Fukuoka Institute of Technology (FIT), 3-30-1 Wajiro-Higashi, Higashi-Ku, Fukuoka 811-0295, Japan
ryuui1010@gmail.com, barolli@fit.ac.jp
[2] Graduate School of Engineering, Fukuoka Institute of Technology (FIT), 3-30-1 Wajiro-Higashi, Higashi-Ku, Fukuoka 811-0295, Japan
iceattpon12@gmail.com, eqafzezi@gmail.com, bylykbashi.kevin@gmail.com
[3] Department of Advanced Sciences, Faculty of Science and Engineering, Housei University, Kajino-Machi, Koganesi-Shi, Tokyo 184-8584, Japan
makoto.takizawa@computer.org

Abstract. Discovering and recommending points of interest are drawing more attention to meet the increasing demand from personalized tours. In this paper, we propose and evaluate a new fuzzy-based system for decision of sightseeing spots considering four input parameters: Access Convenience (AC), Lodging Arrangements (LA), Natural Scenery (NS) and Visiting Cost (VC) to decide the sightseeing spots Visit or Not Visit (VNV). We evaluate the proposed system by computer simulations. From the simulations results, we conclude that when AC, LA and NS are increased, the VNV is increased. But, when VC is increased, the VNV is decreased.

1 Introduction

Social image hosting websites have recently become very popular. On these sites, users can upload and tag images for sharing their travelling experiences. The geotagged images are widely used in landmark recognitions and trip recommendations. Large amount of information generated from these location-based social services covers not only popular locations but also obscure ones. Since personalized tours are becoming popular, more attention is focusing on obscure sightseeing locations that are less well-known while still worth visiting. In Fig. 1 are show two dimensions of diverse sightseeing resources. The evaluation can be done using the sightseeing quality and popularity [1–5].

In this work, we use Fuzzy Logic (FL) for decision of sightseeing spots considering different conditions. The FL is the logic underlying modes of reasoning which are approximate rather then exact. The importance of FL derives from

L. Barolli et al. (Eds.): EIDWT 2020, LNDECT 47, pp. 82–88, 2020.
https://doi.org/10.1007/978-3-030-39746-3_10

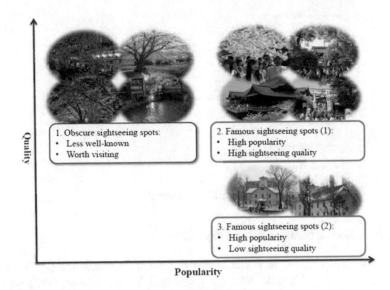

Fig. 1. Two dimensions of diverse sightseeing resources.

the fact that most modes of human reasoning and especially common sense reasoning are approximate in nature [6]. FL uses linguistic variables to describe the control parameters. By using relatively simple linguistic expressions it is possible to describe and grasp very complex problems. A very important property of the linguistic variables is the capability of describing imprecise parameters.

The concept of a fuzzy set deals with the representation of classes whose boundaries are not determined. It uses a characteristic function, taking values usually in the interval [0, 1]. The fuzzy sets are used for representing linguistic labels. This can be viewed as expressing an uncertainty about the clear-cut meaning of the label. But important point is that the valuation set is supposed to be common to the various linguistic labels that are involved in the given problem.

The fuzzy set theory uses the membership function to encode a preference among the possible interpretations of the corresponding label. A fuzzy set can be defined by examplification, ranking elements according to their typicality with respect to the concept underlying the fuzzy set [7–16].

In this paper, we propose and evaluate a fuzzy-based system for decision of sightseeing spots considering four input parameters: Access Convenience (AC), Lodging Arrangements (LA), Natural Scenery (NS) and Visiting Cost (VC) to decide the sightseeing spots Visit or Not Visit (VNV) (Fig. 2).

The structure of this paper is as follows. In Sect. 2, we present the proposed fuzzy-based system. In Sect. 3, we discuss the simulation results. Finally, conclusions and future work are given in Sect. 4.

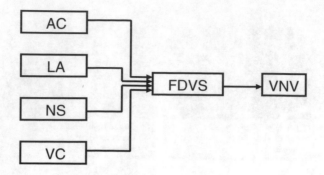

Fig. 2. FDVS structure.

2   Proposed Fuzzy-Based System

In this work, we consider four parameters: Access Convenience (AC), Lodging Arrangements (LA), Natural Scenery (NS) and Visiting Cost (VC) to decide the sightseeing spots Visit or Not Visit (VNV). These four parameters are not correlated with each other, for this reason we use fuzzy system. The membership functions for our system are shown in Fig. 3. In Table 1, we show the fuzzy rule base of our proposed system, which consists of 81 rules.

The input parameters for FVS are: AC, LA, NS and VC. The output linguistic parameter is VNV. The term sets of AC, LA, NS and VC are defined respectively as:

$$
\begin{aligned}
AC &= \{Bad,\ Normal,\ Good\} \\
&= \{Ba,\ Nor,\ Go\}; \\
LA &= \{Bad,\ Normal,\ Good\} \\
&= \{Bad,\ Normal,\ Good\}; \\
NS &= \{Bad,\ Normal,\ Good\} \\
&= \{B,\ N,\ G\}; \\
VC &= \{Low,\ Middle,\ High\} \\
&= \{Lo,\ Mi,\ Hi\}.
\end{aligned}
$$

and the term set for the output VNV is defined as:

$$
VNV = \begin{pmatrix} VisitLevel1 \\ VisitLevel2 \\ VisitLevel3 \\ VisitLevel4 \\ VisitLevel5 \\ VisitLevel6 \\ VisitLevel7 \end{pmatrix} = \begin{pmatrix} VL1 \\ VL2 \\ VL3 \\ VL4 \\ VL5 \\ VL6 \\ VL7 \end{pmatrix}.
$$

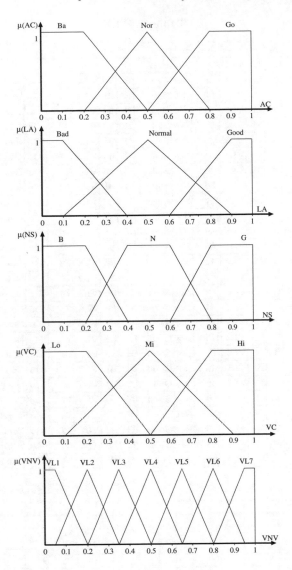

Fig. 3. Membership functions.

3   Simulation Results

In this section, we present the simulation results for our proposed fuzzy-based system. In our system, we decided the number of term sets by carrying out many simulations.

From Figs. 4, 5 and 6, we show the relation between VNV and AC, LA, NS, VC. In these simulations, we consider the LA and AC as constant parameters. In Fig. 4, we consider LA value 0.1 units. We change the AC value from 0.1 to

Table 1. FRB.

Rule	AC	LA	NS	VC	VNV	Rule	AC	LA	NS	VC	VNV	Rule	AC	LA	NS	VC	VNV
1	Ba	Bad	B	Lo	VL2	28	Nor	Bad	B	Lo	VL3	55	Go	Bad	B	Lo	VL4
2	Ba	Bad	B	Mi	VL1	29	Nor	Bad	B	Mi	VL2	56	Go	Bad	B	Mi	VL3
3	Ba	Bad	B	Hi	VL1	30	Nor	Bad	B	Hi	VL1	57	Go	Bad	B	Hi	VL2
4	Ba	Bad	N	Lo	VL3	31	Nor	Bad	N	Lo	VL4	58	Go	Bad	N	Lo	VL6
5	Ba	Bad	N	Mi	VL2	32	Nor	Bad	N	Mi	VL3	59	Go	Bad	N	Mi	VL4
6	Ba	Bad	N	Hi	VL1	33	Nor	Bad	N	Hi	VL2	60	Go	Bad	N	Hi	VL3
7	Ba	Bad	G	Lo	VL4	34	Nor	Bad	G	Lo	VL6	61	Go	Bad	G	Lo	VL7
8	Ba	Bad	G	Mi	VL3	35	Nor	Bad	G	Mi	VL4	62	Go	Bad	G	Mi	VL6
9	Ba	Bad	G	Hi	VL2	36	Nor	Bad	G	Hi	VL3	63	Go	Bad	G	Hi	VL4
10	Ba	Normal	B	Lo	VL2	37	Nor	Normal	B	Lo	VL4	64	Go	Normal	B	Lo	VL6
11	Ba	Normal	B	Mi	VL1	38	Nor	Normal	B	Mi	VL2	65	Go	Normal	B	Mi	VL4
12	Ba	Normal	B	Hi	VL1	39	Nor	Normal	B	Hi	VL1	66	Go	Normal	B	Hi	VL2
13	Ba	Normal	N	Lo	VL4	40	Nor	Normal	N	Lo	VL6	67	Go	Normal	N	Lo	VL7
14	Ba	Normal	N	Mi	VL2	41	Nor	Normal	N	Mi	VL4	68	Go	Normal	N	Mi	VL6
15	Ba	Normal	N	Hi	VL1	42	Nor	Normal	N	Hi	VL2	69	Go	Normal	N	Hi	VL4
16	Ba	Normal	G	Lo	VL6	43	Nor	Normal	G	Lo	VL7	70	Go	Normal	G	Lo	VL7
17	Ba	Normal	G	Mi	VL4	44	Nor	Normal	G	Mi	VL6	71	Go	Normal	G	Mi	VL7
18	Ba	Normal	G	Hi	VL2	45	Nor	Normal	G	Hi	VL4	72	Go	Normal	G	Hi	VL6
19	Ba	Good	B	Lo	VL3	46	Nor	Good	B	Lo	VL5	73	Go	Good	B	Lo	VL6
20	Ba	Good	B	Mi	VL2	47	Nor	Good	B	Mi	VL3	74	Go	Good	B	Mi	VL5
21	Ba	Good	B	Hi	VL1	48	Nor	Good	B	Hi	VL2	75	Go	Good	B	Hi	VL3
22	Ba	Good	N	Lo	VL5	49	Nor	Good	N	Lo	VL6	76	Go	Good	N	Lo	VL7
23	Ba	Good	N	Mi	VL3	50	Nor	Good	N	Mi	VL5	77	Go	Good	N	Mi	VL6
24	Ba	Good	N	Hi	VL2	51	Nor	Good	N	Hi	VL3	78	Go	Good	N	Hi	VL5
25	Ba	Good	G	Lo	VL6	52	Nor	Good	G	Lo	VL7	79	Go	Good	G	Lo	VL7
26	Ba	Good	G	Mi	VL5	53	Nor	Good	G	Mi	VL6	80	Go	Good	G	Mi	VL7
27	Ba	Good	G	Hi	VL3	54	Nor	Good	G	Hi	VL5	81	Go	Good	G	Hi	VL6

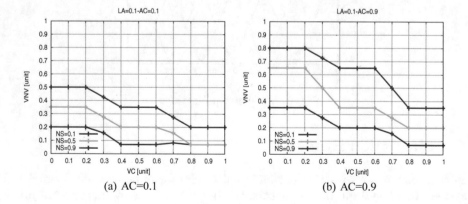

(a) AC=0.1   (b) AC=0.9

Fig. 4. Relation of VNV with VC and NS for different AC when LA = 0.1.

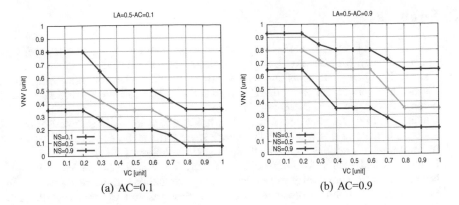

Fig. 5. Relation of VNV with VC and NS for different AC when LA = 0.5.

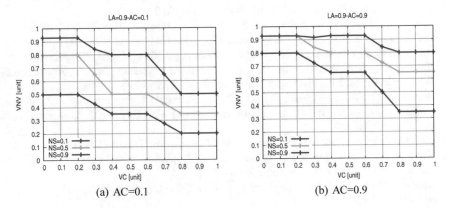

Fig. 6. Relation of VNV with VC and NS for different AC when LA = 0.9.

0.9 units and the VC value from 0 to 1 units. When the NS and AC increase, the VNV is increased. But, when the VC increased, the VNV is decreased. In Figs. 5 and 6, we increase LA values to 0.5 and 0.9 units, respectively. We see that, when the LA increases, the VNV is increased.

4   Conclusions and Future Work

In this paper, we proposed a fuzzy-based system to decide the sightseeing spots. We took into consideration four parameters: AC, LA, NS and VC. We evaluated the performance of proposed system by computer simulations. From the simulations results, we conclude that when AC, LA and NS are increased, the VNV is increased. But, by increasing VC, the VNV is decreased.

In the future, we would like to make extensive simulations to evaluate the proposed system and compare the performance of our proposed system with other systems.

References

1. Luo, J., Joshi, D., Yu, J., Gallagher, A.C.: Geotagging in multimedia and computer vision a survey. Multimedia Tools Appl. **51**(1), 187–211 (2011)
2. Chenyi, Z., Qiang, M., Xuefeng, L., Masatoshi, Y.: An obscure sightseeing spots discovering system. In: 2014 IEEE International Conference on Multimedia and Expo (ICME), pp. 1–6. IEEE (2014)
3. Cheng, Z., Ren, J., Shen, J., Miao, H.: Building a large scale test collection for effective benchmarking of mobile landmark search. In: Li, S., et al. (eds.) Advances in Multimedia Modeling, pp. 36–46. Springer, Heidelberg (2013)
4. Chen, W., Battestini, A., Gelfand, N., Setlur, V.: Visual summaries of popular landmarks from community photo collections. In: 2009 Conference Record of the Forty-Third Asilomar Conference on Signals, Systems and Computers, pp. 1248–1255. IEEE (2009)
5. Cao, X., Cong, G., Jensen, C.S.: Mining significant semantic locations from GPS data. Proc. VLDB Endow. **3**(1–2), 1009–1020 (2010)
6. Inaba, T., Obukata, R., Sakamoto, S., Oda, T., Ikeda, M., Barolli, L.: Performance evaluation of a QoS-aware fuzzy-based CAC for LAN access. Int. J. Space-Based Situated Comput. **6**(4) (2016). https://doi.org/10.1504/IJSSC.2016.082768
7. Terano, T., Asai, K., Sugeno, M.: Fuzzy Systems Theory and Its Applications. Academic Press, Inc., Harcourt Brace Jovanovich, Publishers (1992)
8. Spaho, E., Kulla, E., Xhafa, F., Barolli, L.: P2P solutions to efficient mobile peer collaboration in MANETs. In: Proceedings of of 3PGCIC 2012, November 2012, pp. 379–383 (2012)
9. Kandel, A.: Fuzzy Expert Systems. CRC Press, Boca Raton (1992)
10. Zimmermann, H.J.: Fuzzy Set Theory and Its Applications. Kluwer Academic Publishers, Dordrecht (1991). Second Revised Edition
11. McNeill, F.M., Thro, E.: Fuzzy Logic. A Practical Approach. Academic Press Inc., San Diego (1994)
12. Zadeh, L.A., Kacprzyk, J.: Fuzzy Logic for the Management of Uncertainty. Wiley, New York (1992)
13. Procyk, T.J., Mamdani, E.H.: A linguistic self-organizing process controller. Automatica **15**(1), 15–30 (1979)
14. Klir, G.J., Folger, T.A.: Fuzzy Sets, Uncertainty, and Information. Prentice Hall, Englewood Cliffs (1988)
15. Munakata, T., Jani, Y.: Fuzzy systems: an overview. Commun. ACM **37**(3), 69–76 (1994)
16. Yi, L., Kouseke, O., Keita, M., Makoto, I., Leonard, B.: A fuzzy-based approach for improving peer coordination quality in MobilePeerDroid mobile system. In: Proceedings of IMIS 2018, pp. 60–73, July 2018

Performance Evaluation of WMN-PSODGA Hybrid Simulation System for Node Placement Problem Considering Normal Distribution and Different Fitness Functions

Seiji Ohara[1](✉), Admir Barolli[2], Phudit Ampririt[1], Shinji Sakamoto[3], Leonard Barolli[4], and Makoto Takizawa[5]

[1] Graduate School of Engineering, Fukuoka Institute of Technology, 3-30-1 Wajiro-Higashi, Higashi-Ku, Fukuoka 811-0295, Japan
seiji.ohara.19@gmail.com, iceattpon12@gmail.com
[2] Department of Information Technology, Aleksander Moisiu University of Durres, L.1, Rruga e Currilave, Durres, Albania
admir.barolli@gmail.com
[3] Department of Computer and Information Science, Seikei University, 3-3-1 Kichijoji-Kitamachi, Musashino-shi, Tokyo 180-8633, Japan
shinji.sakamoto@ieee.org
[4] Department of Information and Communication Engineering, Fukuoka Institute of Technology, 3-30-1 Wajiro-Higashi, Higashi-Ku, Fukuoka 811-0295, Japan
barolli@fit.ac.jp
[5] Department of Advanced Sciences, Faculty of Science and Engineering, Hosei University, Kajino-Machi, Koganei-Shi, Tokyo 184-8584, Japan
makoto.takizawa@computer.org

Abstract. Wireless Mesh Networks (WMNs) are becoming an important networking infrastructure because they have many advantages such as low cost and increased high-speed wireless Internet connectivity. In our previous work, we implemented a Particle Swarm Optimization (PSO) based simulation system, called WMN-PSO, and a simulation system based on Genetic Algorithm (GA), called WMN-GA, for solving node placement problem in WMNs. Then, we implemented a hybrid simulation system based on PSO and distributed GA (DGA), called WMN-PSODGA. Moreover, we added in the fitness function a new parameter for the load balancing of the mesh routers called NCMCpR (Number of Covered Mesh Clients per Router). In this paper, we consider Normal distribution of mesh clients and different fitness functions. The simulation results show that WMN-PSODGA has a good performance when weight-coefficients are $\alpha = 0.6$, $\beta = 0.3$, $\gamma = 0.1$.

1 Introduction

The wireless networks and devices are becoming increasingly popular and they provide users access to information and communication anytime and anywhere

© Springer Nature Switzerland AG 2020
L. Barolli et al. (Eds.): EIDWT 2020, LNDECT 47, pp. 89–101, 2020.
https://doi.org/10.1007/978-3-030-39746-3_11

[3,8–11,14,20,26,27,29,33]. Wireless Mesh Networks (WMNs) are gaining a lot of attention because of their low-cost nature that makes it attractive for providing wireless Internet connectivity. A WMN is dynamically self-organized and self-configured, with the nodes in the network automatically establishing and maintaining mesh connectivity among itself (creating, in effect, an ad hoc network). This feature brings many advantages to WMN such as low up-front cost, easy network maintenance, robustness and reliable service coverage [1]. Moreover, such infrastructure can be used to deploy community networks, metropolitan area networks, municipal and corporative networks, and to support applications for urban areas, medical, transport and surveillance systems.

Mesh node placement in WMNs can be seen as a family of problems, which is shown (through graph theoretic approaches or placement problems, e.g. [6,15]) to be computationally hard to solve for most of the formulations [37].

We consider the version of the mesh router nodes placement problem in which we are given a grid area where to deploy a number of mesh router nodes and a number of mesh client nodes of fixed positions (of an arbitrary distribution) in the grid area. The objective is to find a location assignment for the mesh routers to the cells of the grid area that maximizes the network connectivity, client coverage and consider load balancing for each router. Network connectivity is measured by Size of Giant Component (SGC) of the resulting WMN graph, while the user coverage is simply the number of mesh client nodes that fall within the radio coverage of at least one mesh router node and is measured by Number of Covered Mesh Clients (NCMC). For load balancing, we added in the fitness function a new parameter called NCMCpR (Number of Covered Mesh Clients per Router).

Node placement problems are known to be computationally hard to solve [12, 13,38]. In previous works, some intelligent algorithms have been recently investigated for node placement problem [4,7,16,18,21–23,31,32].

In [24], we implemented a Particle Swarm Optimization (PSO) based simulation system, called WMN-PSO. Also, we implemented another simulation system based on Genetic Algorithm (GA), called WMN-GA [19], for solving node placement problem in WMNs. Then, we designed and implemented a hybrid simulation system based on PSO and distributed GA (DGA). We call this system WMN-PSODGA.

In this paper, we present the performance analysis of WMNs by WMN-PSODGA system considering Normal distribution of mesh clients and different fitness functions.

The rest of the paper is organized as follows. We present our designed and implemented hybrid simulation system in Sect. 2. The simulation results are given in Sect. 3. Finally, we give conclusions and future work in Sect. 4.

2 Proposed and Implemented Simulation System

2.1 Particle Swarm Optimization

In PSO a number of simple entities (the particles) are placed in the search space of some problem or function and each evaluates the objective function at its current location. The objective function is often minimized and the exploration of the search space is not through evolution [17].

Each particle then determines its movement through the search space by combining some aspect of the history of its own current and best (best-fitness) locations with those of one or more members of the swarm, with some random perturbations. The next iteration takes place after all particles have been moved. Eventually the swarm as a whole, like a flock of birds collectively foraging for food, is likely to move close to an optimum of the fitness function.

Each individual in the particle swarm is composed of three $\mathcal{D}$-dimensional vectors, where $\mathcal{D}$ is the dimensionality of the search space. These are the current position $\vec{x}_i$, the previous best position $\vec{p}_i$ and the velocity $\vec{v}_i$.

The particle swarm is more than just a collection of particles. A particle by itself has almost no power to solve any problem; progress occurs only when the particles interact. Problem solving is a population-wide phenomenon, emerging from the individual behaviors of the particles through their interactions. In any case, populations are organized according to some sort of communication structure or topology, often thought of as a social network. The topology typically consists of bidirectional edges connecting pairs of particles, so that if j is in i's neighborhood, i is also in j's. Each particle communicates with some other particles and is affected by the best point found by any member of its topological neighborhood. This is just the vector $\vec{p}_i$ for that best neighbor, which we will denote with $\vec{p}_g$. The potential kinds of population "social networks" are hugely varied, but in practice certain types have been used more frequently. We show the pseudo code of PSO in Algorithm 1.

In the PSO process, the velocity of each particle is iteratively adjusted so that the particle stochastically oscillates around $\vec{p}_i$ and $\vec{p}_g$ locations.

2.2 Distributed Genetic Algorithm

Distributed Genetic Algorithm (DGA) has been used in various fields of science. DGA has shown their usefulness for the resolution of many computationally hard combinatorial optimization problems. We show the pseudo code of DGA in Algorithm 2.

Population of individuals: Unlike local search techniques that construct a path in the solution space jumping from one solution to another one through local perturbations, DGA use a population of individuals giving thus the search a larger scope and chances to find better solutions. This feature is also known as "exploration" process in difference to "exploitation" process of local search methods.

Algorithm 1. Pseudo code of PSO.

/* Initialize all parameters for PSO */
Computation maxtime:= Tp_{max}, $t := 0$;
Number of particle-patterns:= m, $2 \leq m \in \mathbf{N}^1$;
Particle-patterns initial solution:= $\mathbf{P}_i^0$;
Particle-patterns initial position:= $\mathbf{x}_{ij}^0$;
Particles initial velocity:= $\mathbf{v}_{ij}^0$;
PSO parameter:= ω, $0 < \omega \in \mathbf{R}^1$;
PSO parameter:= C_1, $0 < C_1 \in \mathbf{R}^1$;
PSO parameter:= C_2, $0 < C_2 \in \mathbf{R}^1$;
/* Start PSO */
Evaluate($\mathbf{G}^0$, $\mathbf{P}^0$);
while $t < Tp_{max}$ **do**
    /* Update velocities and positions */
    $\mathbf{v}_{ij}^{t+1} = \omega \cdot \mathbf{v}_{ij}^t$
        $+C_1 \cdot \text{rand}() \cdot (best(P_{ij}^t) - x_{ij}^t)$
        $+C_2 \cdot \text{rand}() \cdot (best(G^t) - x_{ij}^t)$;
    $\mathbf{x}_{ij}^{t+1} = \mathbf{x}_{ij}^t + \mathbf{v}_{ij}^{t+1}$;
    /* if fitness value is increased, a new solution will be accepted. */
    Update_Solutions($\mathbf{G}^t$, $\mathbf{P}^t$);
    $t = t + 1$;
end while
Update_Solutions($\mathbf{G}^t$, $\mathbf{P}^t$);
return Best found pattern of particles as solution;

Fitness: The determination of an appropriate fitness function, together with the chromosome encoding are crucial to the performance of DGA. Ideally we would construct objective functions with "certain regularities", i.e. objective functions that verify that for any two individuals which are close in the search space, their respective values in the objective functions are similar.

Selection: The selection of individuals to be crossed is another important aspect in DGA as it impacts on the convergence of the algorithm. Several selection schemes have been proposed in the literature for selection operators trying to cope with premature convergence of DGA. There are many selection methods in GA. In our system, we implement 2 selection methods: Random method and Roulette wheel method.

Crossover operators: Use of crossover operators is one of the most important characteristics. Crossover operator is the means of DGA to transmit best genetic features of parents to offsprings during generations of the evolution process. Many methods for crossover operators have been proposed such as Blend Crossover (BLX-α), Unimodal Normal Distribution Crossover (UNDX), Simplex Crossover (SPX).

Mutation operators: These operators intend to improve the individuals of a population by small local perturbations. They aim to provide a component of randomness in the neighborhood of the individuals of the population. In our system, we implemented two mutation methods: uniformly random mutation and boundary mutation.

Escaping from local optima: GA itself has the ability to avoid falling prematurely into local optima and can eventually escape from them during the search process. DGA has one more mechanism to escape from local optima by considering some islands. Each island computes GA for optimizing and they migrate its gene to provide the ability to avoid from local optima (See Fig. 1).

Convergence: The convergence of the algorithm is the mechanism of DGA to reach to good solutions. A premature convergence of the algorithm would cause that all individuals of the population be similar in their genetic features and thus the search would result ineffective and the algorithm getting stuck into local optima. Maintaining the diversity of the population is therefore very important to this family of evolutionary algorithms.

Algorithm 2. Pseudo code of DGA.

/* Initialize all parameters for DGA */
Computation maxtime:= Tg_{max}, $t := 0$;
Number of islands:= n, $1 \leq n \in N^1$;
initial solution:= P_i^0;
/* Start DGA */
Evaluate(G^0, P^0);
while $t < Tg_{max}$ **do**
    for all islands **do**
        Selection();
        Crossover();
        Mutation();
    end for
    $t = t + 1$;
end while
Update_Solutions(G^t, P^t);
return Best found pattern of particles as solution;

2.3   WMN-PSODGA Hybrid Simulation System

In this subsection, we present the initialization, particle-pattern, fitness function and replacement methods. The pseudo code of our implemented system is shown in Algorithm 3. Also, our implemented simulation system uses Migration function as shown in Fig. 2. The Migration function swaps solutions among lands included in PSO part.

Initialization
We decide the velocity of particles by a random process considering the area size. For instance, when the area size is $W \times H$, the velocity is decided randomly from $-\sqrt{W^2 + H^2}$ to $\sqrt{W^2 + H^2}$.

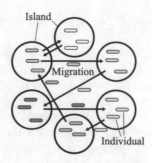

Fig. 1. Model of migration in DGA.

Algorithm 3. Pseudo code of WMN-PSODGA system.

Computation maxtime:$= T_{max}$, $t := 0$;
Initial solutions: $\boldsymbol{P}$.
Initial global solutions: $\boldsymbol{G}$.
/* Start PSODGA */
while $t < T_{max}$ **do**
    Subprocess(PSO);
    Subprocess(DGA);
    WaitSubprocesses();
    Evaluate($\boldsymbol{G}^t, \boldsymbol{P}^t$)
    /* Migration() swaps solutions (see Fig. 2). */
    Migration();
    $t = t + 1$;
end while
Update_Solutions($\boldsymbol{G}^t, \boldsymbol{P}^t$);
return Best found pattern of particles as solution;

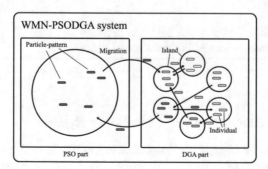

Fig. 2. Model of WMN-PSODGA migration.

Particle-Pattern

A particle is a mesh router. A fitness value of a particle-pattern is computed by combination of mesh routers and mesh clients positions. In other words, each particle-pattern is a solution as shown is Fig. 3.

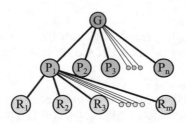

G: Global Solution
P: Particle-pattern
R: Mesh Router
n: Number of Particle-patterns
m: Number of Mesh Routers

Fig. 3. Relationship among global solution, particle-patterns and mesh routers in PSO part.

Gene Coding
A gene describes a WMN. Each individual has its own combination of mesh nodes. In other words, each individual has a fitness value. Therefore, the combination of mesh nodes is a solution.

Fitness Function
WMN-PSODGA has the fitness function to evaluate the temporary solution of the router's placements. The fitness function is defined as:

$$Fitness = \alpha \times NCMC(\boldsymbol{x}_{ij}, \boldsymbol{y}_{ij}) + \beta \times SGC(\boldsymbol{x}_{ij}, \boldsymbol{y}_{ij}) + \gamma \times NCMCpR(\boldsymbol{x}_{ij}, \boldsymbol{y}_{ij}).$$

This function uses the following indicators.

- SGC (Size of Giant Component)
  The SGC is the maximum number of the routers constructing in the same network. The SGC indicator means the connectivity of the routers.
- NCMC (Number of Covered Mesh Clients)
  The NCMC is the number of the clients belong to the network constructed by the SGC's routers. The NCMC indicator means the covering rate of the clients.
- NCMCpR (Number of Covered Mesh Clients per Router)
  The NCMCpR is the number of clients covered by each router. The NCMCpR indicator means load balancing.

WMN-PSODGA aims to maximize the value of the fitness function in order to optimize the placements of the routers using the above three indicators. The fitness function has weight-coefficients α, β, and γ for NCMC, SGC, and NCMCpR. Moreover, the weight-coefficients are implemented as $\alpha + \beta + \gamma = 1$.

Router Replacement Methods
A mesh router has x, y positions and velocity. Mesh routers are moved based on velocities. There are many router replacement methods, such as:

Constriction Method (CM)
  CM is a method which PSO parameters are set to a week stable region ($\omega = 0.729$, $C_1 = C2 = 1.4955$) based on analysis of PSO by Clerc et al. [2,5,35].

Random Inertia Weight Method (RIWM)

In RIWM, the ω parameter is changing ramdomly from 0.5 to 1.0. The C_1 and C_2 are kept 2.0. The ω can be estimated by the week stable region. The average of ω is 0.75 [28, 35].

Linearly Decreasing Inertia Weight Method (LDIWM)

In LDIWM, C_1 and C_2 are set to 2.0, constantly. On the other hand, the ω parameter is changed linearly from unstable region ($\omega = 0.9$) to stable region ($\omega = 0.4$) with increasing of iterations of computations [35, 36].

Linearly Decreasing Vmax Method (LDVM)

In LDVM, PSO parameters are set to unstable region ($\omega = 0.9$, $C_1 = C_2 = 2.0$). A value of V_{max} which is maximum velocity of particles is considered. With increasing of iteration of computations, the V_{max} is kept decreasing linearly [30, 34].

Rational Decrement of Vmax Method (RDVM)

In RDVM, PSO parameters are set to unstable region ($\omega = 0.9$, $C_1 = C_2 = 2.0$). The V_{max} is kept decreasing with the increasing of iterations as

$$V_{max}(x) = \sqrt{W^2 + H^2} \times \frac{T - x}{x}.$$

Where, W and H are the width and the height of the considered area, respectively. Also, T and x are the total number of iterations and a current number of iteration, respectively [25].

3   Simulation Results

In this section, we present some simulation results for different coefficients of the fitness function. Table 1 shows the common parameters for each simulation.

In Fig. 4 are shown the simulation results of WMN-PSODGA system when the weight-coefficients of fitness function are $\alpha = 0.8$, $\beta = 0.1$, $\gamma = 0.1$. While, Fig. 5 are shown the simulation results when the weight-coefficients are $\alpha = 0.6$, $\beta = 0.3$, $\gamma = 0.1$. Then, in Fig. 6 are shown the simulation results when the weight-coefficients are $\alpha = 0.4$, $\beta = 0.5$, $\gamma = 0.1$.

Figures 4(a), 5(a) and 6(a) are shown the visualization result after the optimization. Moreover, Figs. 4(b), 5(b) and 6(b) are shown the fitness function value while running the optimizing process in WMN-PSODGA. WMN-PSODGA executed ten times optimizing process to observe the tendency of the fitness function value.

In Fig. 4(a), some routers have not covered any client. Moreover, one of the routers could not connect with others. In Fig. 5(a), all routers are connected and all clients are covered. Also, the load balancing is better than the scenarios in Figs. 4 and 6. In Fig. 6(a), some routers have not covered any client.

In Fig. 5(b), the fitness function value has reached 100%. While in Figs. 4(b) and 6(b), the fitness function value does not reach 100%. For this reason, we conclude that for normal distribution of mesh clients the case when the coefficients of fitness function are $\alpha = 0.6$, $\beta = 0.3$, $\gamma = 0.1$ has better behavior than other cases.

Table 1. The common parameters for each simulation.

Parameters	Values
Distribution of mesh clients	Normal distribution
Number of mesh clients	48
Number of mesh routers	16
Radius of a mesh router	2.0–3.5
Number of GA Islands	16
Number of migrations	200
Evolution steps	9
Selection method	Roulette wheel method
Crossover method	SPX
Mutation method	Uniform mutation
Crossover rate	0.8
Mutation rate	0.2
Replacement method	CM
Area size	32.0×32.0

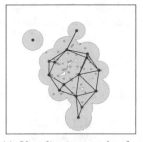 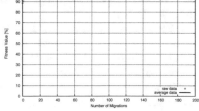

(a) Visualization result after the optimization.

(b) Fitness function value.

Fig. 4. Simulation results of when weight-coefficients are $\alpha = 0.8$, $\beta = 0.1$, $\gamma = 0.1$.

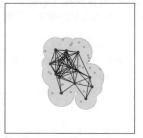

(a) Visualization result after the optimization.

(b) Fitness function value.

Fig. 5. Simulation results of when weight-coefficients are $\alpha = 0.6$, $\beta = 0.3$, $\gamma = 0.1$.

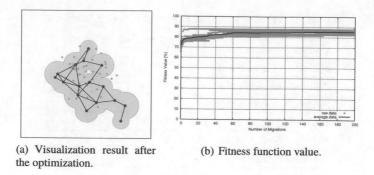

(a) Visualization result after the optimization.

(b) Fitness function value.

Fig. 6. Simulation results of when weight-coefficients are $\alpha = 0.4$, $\beta = 0.5$, $\gamma = 0.1$.

4   Conclusions

In this work, we evaluated the performance of WMNs using a hybrid simulation system based on PSO and DGA (called WMN-PSODGA). We considered the Normal distribution of mesh clients and different fitness functions. From the simulation results, we found that WMN-PSODGA has a good performance when weight-coefficients of fitness function are $\alpha = 0.6$, $\beta = 0.3$, $\gamma = 0.1$. In future work, we will consider other mesh client distributions and replacement methods.

References

1. Akyildiz, I.F., Wang, X., Wang, W.: Wireless mesh networks: a survey. Comput. Netw. **47**(4), 445–487 (2005)
2. Barolli, A., Sakamoto, S., Ozera, K., Ikeda, M., Barolli, L., Takizawa, M.: Performance evaluation of WMNs by WMN-PSOSA simulation system considering constriction and linearly decreasing Vmax methods. In: International Conference on P2P, Parallel, Grid, Cloud and Internet Computing, pp. 111–121. Springer (2017)
3. Barolli, A., Sakamoto, S., Barolli, L., Takizawa, M.: Performance analysis of simulation system based on particle swarm optimization and distributed genetic algorithm for WMNs considering different distributions of mesh clients. In: International Conference on Innovative Mobile and Internet Services in Ubiquitous Computing, pp. 32–45. Springer (2018)
4. Barolli, A., Sakamoto, S., Ozera, K., Barolli, L., Kulla, E., Takizawa, M.: Design and implementation of a hybrid intelligent system based on particle swarm optimization and distributed genetic algorithm. In: International Conference on Emerging Internetworking, Data & Web Technologies, pp. 79–93. Springer (2018)
5. Clerc, M., Kennedy, J.: The particle swarm-explosion, stability, and convergence in a multidimensional complex space. IEEE Trans. Evol. Comput. **6**(1), 58–73 (2002)
6. Franklin, A.A., Murthy, C.S.R.: Node placement algorithm for deployment of two-tier wireless mesh networks. In: Proceedings of Global Telecommunications Conference, pp 4823–4827 (2007)

7. Girgis, M.R., Mahmoud, T.M., Abdullatif, B.A., Rabie, A.M.: Solving the wireless mesh network design problem using genetic algorithm and simulated annealing optimization methods. Int. J. Comput. Appl. **96**(11), 1–10 (2014)
8. Goto, K., Sasaki, Y., Hara, T., Nishio, S.: Data gathering using mobile agents for reducing traffic in dense mobile wireless sensor networks. Mob. Inf. Syst. **9**(4), 295–314 (2013)
9. Inaba, T., Elmazi, D., Sakamoto, S., Oda, T., Ikeda, M., Barolli, L.: A secure-aware call admission control scheme for wireless cellular networks using fuzzy logic and its performance evaluation. J. Mob. Multimed. **11**(3&4), 213–222 (2015)
10. Inaba, T., Obukata, R., Sakamoto, S., Oda, T., Ikeda, M., Barolli, L.: Performance evaluation of a QoS-aware fuzzy-based CAC for LAN access. Int. J. Space Based Situated Comput. **6**(4), 228–238 (2016)
11. Inaba, T., Sakamoto, S., Oda, T., Ikeda, M., Barolli, L.: A testbed for admission control in WLAN: a fuzzy approach and its performance evaluation. In: International Conference on Broadband and Wireless Computing, Communication and Applications, pp. 559–571. Springer (2016)
12. Lim, A., Rodrigues, B., Wang, F., Xu, Z.: k-Center problems with minimum coverage. Theor. Comput. Sci. **332**(1–3), 1–17 (2005)
13. Maolin, T., et al.: Gateways placement in backbone wireless mesh networks. Int. J. Commun. Netw. Syst. Sci. **2**(1), 44–50 (2009)
14. Matsuo, K., Sakamoto, S., Oda, T., Barolli, A., Ikeda, M., Barolli, L.: Performance analysis of WMNs by WMN-GA simulation system for two WMN architectures and different TCP congestion-avoidance algorithms and client distributions. Int. J. Commun. Netw. Distrib. Syst. **20**(3), 335–351 (2018)
15. Muthaiah, S.N., Rosenberg, C.P.: Single gateway placement in wireless mesh networks. In: Proceedings of 8th International IEEE Symposium on Computer Networks, pp. 4754–4759 (2008)
16. Naka, S., Genji, T., Yura, T., Fukuyama, Y.: A hybrid particle swarm optimization for distribution state estimation. IEEE Trans. Power Syst. **18**(1), 60–68 (2003)
17. Poli, R., Kennedy, J., Blackwell, T.: Particle swarm optimization. Swarm Intell. **1**(1), 33–57 (2007)
18. Sakamoto, S., Kulla, E., Oda, T., Ikeda, M., Barolli, L., Xhafa, F.: A comparison study of simulated annealing and genetic algorithm for node placement problem in wireless mesh networks. J. Mob. Multimed. **9**(1–2), 101–110 (2013)
19. Sakamoto, S., Kulla, E., Oda, T., Ikeda, M., Barolli, L., Xhafa, F.: A comparison study of hill climbing, simulated annealing and genetic algorithm for node placement problem in WMNs. J. High Speed Netw. **20**(1), 55–66 (2014)
20. Sakamoto, S., Kulla, E., Oda, T., Ikeda, M., Barolli, L., Xhafa, F.: A simulation system for WMN based on SA: performance evaluation for different instances and starting temperature values. Int. J. Space Based Situated Comput. **4**(3–4), 209–216 (2014)
21. Sakamoto, S., Kulla, E., Oda, T., Ikeda, M., Barolli, L., Xhafa, F.: Performance evaluation considering iterations per phase and SA temperature in WMN-SA system. Mob. Inf. Syst. **10**(3), 321–330 (2014)
22. Sakamoto, S., Lala, A., Oda, T., Kolici, V., Barolli, L., Xhafa, F.: Application of WMN-SA simulation system for node placement in wireless mesh networks: a case study for a realistic scenario. Int. J. Mob. Comput. Multimed. Commun. (IJMCMC) **6**(2), 13–21 (2014)

23. Sakamoto, S., Oda, T., Ikeda, M., Barolli, L., Xhafa, F.: An integrated simulation system considering WMN-PSO simulation system and network simulator 3. In: International Conference on Broadband and Wireless Computing, Communication and Applications, pp. 187–198. Springer (2016)
24. Sakamoto, S., Oda, T., Ikeda, M., Barolli, L., Xhafa, F.: Implementation and evaluation of a simulation system based on particle swarm optimisation for node placement problem in wireless mesh networks. Int. J. Commun. Netw. Distrib. Syst. **17**(1), 1–13 (2016)
25. Sakamoto, S., Oda, T., Ikeda, M., Barolli, L., Xhafa, F.: Implementation of a new replacement method in WMN-PSO simulation system and its performance evaluation. In: The 30th IEEE International Conference on Advanced Information Networking and Applications (AINA 2016), pp. 206–211 (2016)
26. Sakamoto, S., Obukata, R., Oda, T., Barolli, L., Ikeda, M., Barolli, A.: Performance analysis of two wireless mesh network architectures by WMN-SA and WMN-TS simulation systems. J. High Speed Netw. **23**(4), 311–322 (2017)
27. Sakamoto, S., Ozera, K., Barolli, A., Ikeda, M., Barolli, L., Takizawa, M.: Implementation of an intelligent hybrid simulation systems for WMNs based on particle swarm optimization and simulated annealing: performance evaluation for different replacement methods. Soft Comput. **23**(9), 3029–3035 (2017)
28. Sakamoto, S., Ozera, K., Barolli, A., Ikeda, M., Barolli, L., Takizawa, M.: Performance evaluation of WMNs by WMN-PSOSA simulation system considering random inertia weight method and linearly decreasing Vmax method. In: International Conference on Broadband and Wireless Computing, Communication and Applications, pp. 114–124. Springer (2017)
29. Sakamoto, S., Ozera, K., Ikeda, M., Barolli, L.: Implementation of intelligent hybrid systems for node placement problem in WMNs considering particle swarm optimization, hill climbing and simulated annealing. Mob. Netw. Appl. **23**(1), 27–33 (2017)
30. Sakamoto, S., Ozeram K., Ikeda, M., Barolli, L.: Performance evaluation of WMNs by WMN-PSOSA simulation system considering constriction and linearly decreasing inertia weight methods. In: International Conference on Network-Based Information Systems, pp. 3–13. Springer (2017)
31. Sakamoto, S., Ozera, K., Oda, T., Ikeda, M., Barolli, L.: Performance evaluation of intelligent hybrid systems for node placement in wireless mesh networks: a comparison study of WMN-PSOHC and WMN-PSOSA. In: International Conference on Innovative Mobile and Internet Services in Ubiquitous Computing, pp. 16–26. Springer (2017)
32. Sakamoto, S., Ozera, K., Oda, T., Ikeda, M., Barolli, L.: Performance evaluation of WMN-PSOHC and WMN-PSO simulation systems for node placement in wireless mesh networks: a comparison study. In: International Conference on Emerging Internetworking, Data & Web Technologies, pp. 64–74. Springer (2017)
33. Sakamoto, S., Ozera, K., Barolli, A., Barolli, L., Kolici, V., Takizawa, M.: Performance evaluation of WMN-PSOSA considering four different replacement methods. In: International Conference on Emerging Internetworking, Data & Web Technologies, pp. 51–64. Springer (2018)
34. Schutte, J.F., Groenwold, A.A.: A study of global optimization using particle swarms. J. Glob. Optim. **31**(1), 93–108 (2005)
35. Shi, Y.: Particle swarm optimization. IEEE Connect. **2**(1), 8–13 (2004)
36. Shi, Y., Eberhart, R.C.: Parameter selection in particle swarm optimization. In: Evolutionary Programming VII, pp. 591–600 (1998)

37. Vanhatupa, T., Hannikainen, M., Hamalainen, T.: Genetic algorithm to optimize node placement and configuration for WLAN planning. In: Proceedings of The 4th IEEE International Symposium on Wireless Communication Systems, pp. 612–616 (2007)
38. Wang, J., Xie, B., Cai, K., Agrawal, D.P.: Efficient mesh router placement in wireless mesh networks. In: Proceedings of IEEE International Conference on Mobile Adhoc and Sensor Systems (MASS-2007), pp. 1–9 (2007)

Effect of Driver's Condition for Driving Risk Measurement in VANETs: A Comparison Study of Simulation and Experimental Results

Kevin Bylykbashi[1]([✉]), Ermioni Qafzezi[1], Makoto Ikeda[2], Keita Matsuo[2], and Leonard Barolli[2]

[1] Graduate School of Engineering, Fukuoka Institute of Technology (FIT), 3-30-1 Wajiro-Higashi, Higashi-Ku, Fukuoka 811–0295, Japan
bylykbashi.kevin@gmail.com, eqafzezi@gmail.com
[2] Department of Information and Communication Engineering, Fukuoka Institute of Technology (FIT), 3-30-1 Wajiro-Higashi, Higashi-Ku, Fukuoka 811-0295, Japan
makoto.ikd@acm.org, {kt-matsuo,barolli}@fit.ac.jp

Abstract. Vehicular Ad hoc Networks (VANETs) have gained great attention due to the rapid development of mobile Internet and Internet of Things (IoT) applications. With the evolution of technology, it is expected that VANETs will be massively deployed in upcoming vehicles. In addition, ambitious efforts are being done to incorporate Ambient Intelligence (AmI) technology in the vehicles, as it will be an important factor for VANETs to accomplish one of its main goals, the road safety. In this paper, we propose an intelligent Fuzzy-based System for Driving Risk Management (FSDRM) in VANETs. The FSDRM considers driver's vital signs data such as the driver's heart and respiratory rate, vehicle speed and vehicle's inside temperature to assess the risk level. Then, it uses the smart box to inform the driver and to provide better assistance. We aim to realize a new system to support the driver for safe driving. We evaluated the performance of the proposed system by computer simulations and experiments. From the evaluation results, we conclude that driver's heart and respiratory rate, vehicle speed, and vehicle's inside temperature have different effects on the assessment of risk level.

1 Introduction

Traffic accidents, road congestion and environmental pollution are persistent problems faced by both developed and developing countries, which have made people live in difficult situations. Among these, the traffic incidents are the most serious ones because they result in huge loss of life and property. For decades, we have seen governments and car manufacturers struggle for safer roads and car accident prevention. The development in wireless communications has allowed companies, researchers and institutions to design communication systems that provide new solutions for these issues. Therefore, new types of networks, such

© Springer Nature Switzerland AG 2020
L. Barolli et al. (Eds.): EIDWT 2020, LNDECT 47, pp. 102–113, 2020.
https://doi.org/10.1007/978-3-030-39746-3_12

as Vehicular Ad hoc Networks (VANETs) have been created. VANET consists of a network of vehicles in which vehicles are capable of communicating among themselves in order to deliver valuable information such as safety warnings and traffic information.

Nowadays, every car is likely to be equipped with various forms of smart sensors, wireless communication modules, storage and computational resources. The sensors will gather information about the road and environment conditions and share it with neighboring vehicles and adjacent roadside units (RSU) via vehicle-to-vehicle (V2V) or vehicle-to-infrastructure (V2I) communication. However, the difficulty lies on how to understand the sensed data and how to make intelligent decisions based on the provided information.

As a result, Ambient Intelligence (AmI) becomes a significant factor for VANETs. Various intelligent systems and applications are now being deployed and they are going to change the way manufacturers design vehicles. These systems include many intelligence computational technologies such as fuzzy logic, neural networks, machine learning, adaptive computing, voice recognition, and so on, and they are already announced or deployed [9]. The goal is to improve both vehicle safety and performance by realizing a series of automatic driving technologies based on the situation recognition. The car control relies on the measurement and recognition of the outside environment and their reflection on driving operation.

On the other hand, we are focused on the in-car information and driver's vital information to detect the danger or risk situation and inform the driver about the risk or change his mood [5]. Thus, our goal is to prevent the accidents by supporting the drivers. In order to realize the proposed system, we use some Internet of Things (IoT) devices equipped with various sensors for in-car monitoring.

In this paper, we propose a fuzzy-based system for driving risk measurement considering four parameters: Heart Rate (HR), Respiratory Rate (RR), Vehicle Speed (VS) and Vehicle's Inside Temperature (VIT) to determine the Driving Risk Measurement (DRM).

The structure of the paper is as follows. In Sect. 2, we present an overview of VANETs. In Sect. 3, we present a short description of AmI. In Sect. 4, we describe the proposed fuzzy-based system and its implementation. In Sect. 5, we discuss the simulation and experimental results. Finally, conclusions and future work are given in Sect. 6.

2   Vehicular Ad Hoc Networks (VANETs)

VANETs are a special case of Mobile Ad hoc Networks (MANETs) in which mobile nodes are vehicles. In VANETs nodes (vehicles) have high mobility and tend to follow organized routes instead of moving at random. Moreover, vehicles offer attractive features such as higher computational capability and localization through GPS.

VANETs have huge potential to enable applications ranging from road safety, traffic optimization, infotainment, commercial to rural and disaster scenario connectivity. Among these, the road safety and traffic optimization are considered the most important ones as they have the goal to reduce the dramatically high number of accidents, guarantee road safety, make traffic management and create new forms of inter-vehicle communications in Intelligent Transportation Systems (ITS).

The ITS manages the vehicle traffic, support drivers with safety and other information, and provide some services such as automated toll collection and driver assist systems [10]. VANETs can be defined as a part of ITS which aims to make transportation systems faster and smarter, in which vehicles are equipped with some short-range and medium-range wireless communication [12].

Despite the attractive features, VANETs are characterized by very large and dynamic topologies, variable capacity wireless links, bandwidth and hard delay constrains, and by short contact durations which are caused by the high mobility, high speed and low density of vehicles. In addition, limited transmission ranges, physical obstacles and interferences, make these networks characterized by disruptive and intermittent connectivity.

To make VANETs applications possible, it is necessary to design proper networking mechanisms that can overcome relevant problems that arise from vehicular environments.

3   Ambient Intelligence (AmI)

The AmI is the vision that technology will become invisible, embedded in our natural surroundings, present whenever we need it, enabled by simple and effortless interactions, attuned to all our senses, adaptive to users and context and autonomously acting [14]. High quality information and content must be available to any user, anywhere, at any time, and on any device.

In order that AmI becomes a reality, it should completely envelope humans, without constraining them. Distributed embedded systems for AmI are going to change the way we design embedded systems, as well as the way we think about such systems. But, more importantly, they will have a great impact on the way we live. Applications ranging from safe driving systems, smart buildings and home security, smart fabrics or e-textiles, to manufacturing systems and rescue and recovery operations in hostile environments, are poised to become part of society and human lives.

The AmI deals with a new world of ubiquitous computing devices, where physical environments interact intelligently and unobtrusively with people. AmI environments can be diverse, such as homes, offices, meeting rooms, hospitals, control centers, vehicles, tourist attractions, stores, sports facilities, and music devices.

In the future, small devices will monitor the health status in a continuous manner, diagnose any possible health conditions, have conversation with people to persuade them to change the lifestyle for maintaining better health, and communicates with the doctor, if needed [3]. The device might even be embedded into the regular clothing fibers in the form of very tiny sensors and it might communicate with other devices including the variety of sensors embedded into the home to monitor the lifestyle. For example, people might be alarmed about the lack of a healthy diet based on the items present in the fridge and based on what they are eating outside regularly.

The AmI paradigm represents the future vision of intelligent computing where environments support the people inhabiting them [1,2,21]. In this new computing paradigm, the conventional input and output media no longer exist, rather the sensors and processors will be integrated into everyday objects, working together in harmony in order to support the inhabitants [20]. By relying on various artificial intelligence techniques, AmI promises the successful interpretation of the wealth of contextual information obtained from such embedded sensors and it will adapt the environment to the user needs in a transparent and anticipatory manner.

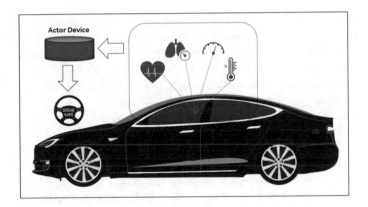

Fig. 1. Proposed system architecture.

4   Proposed System

In this work, we use fuzzy logic to implement the proposed system. Fuzzy sets and fuzzy logic have been developed to manage vagueness and uncertainty in a reasoning process of an intelligent system such as a knowledge based system, an expert system or a logic control system [8,11,13,15–19,22,23]. In Fig. 1, we show the architecture of our proposed system.

4.1   Proposed Fuzzy-Based Simulation System

The proposed system called Fuzzy-based System for Driving Risk Measurement (FSDRM) is shown in Fig. 2. For the implementation of our system, we consider four input parameters: Heart Rate (HR), Respiratory Rate (RR), Vehicle

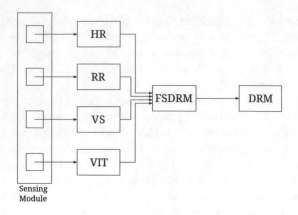

Fig. 2. Proposed system structure.

Speed (VS) and Vehicle's Inside Temperature (VIT) to determine the Driving Risk Measurement (DRM). These four input parameters are not correlated with each other, for this reason we use fuzzy system. The input parameters are fuzzified using the membership functions showed in Fig. 3(a), (b), (c) and (d). In Fig. 3(e) are shown the membership functions used for the output parameter. We use triangular and trapezoidal membership functions because they are suitable for real-time operation. The term sets for each linguistic parameter are shown in Table 1. We decided the number of term sets by carrying out many simulations. In Table 2, we show the Fuzzy Rule Base (FRB) of FSDRM, which consists of 81 rules. The control rules have the form: IF "conditions" THEN "control action". For instance, for Rule 1: "IF HR is S, RR is Sl, VS is Lo and VIT is L, THEN DRM is Hg" or for Rule 38: "IF HR is No, RR is Nm, VS is Lo and VIT is M, THEN DRM is Sf".

Table 1. Parameters and their term sets for FSDRM.

Parameters	Term sets
Heart Rate (HR)	Slow (S), Normal (No), Fast (F)
Respiratory Rate (RR)	Slow (Sl), Normal (Nm), Fast (Fa)
Vehicle Speed (VS)	Low (Lo), Moderate (Mo), High (Hi)
Vehicle's Inside Temperature (VIT)	Low (L), Medium (M), High (H)
Driving Risk Measurement (DRM)	Safe (Sf), Low (Lw), Moderate (Md), High (Hg), Very High (VH), Severe (Sv), Danger (D)

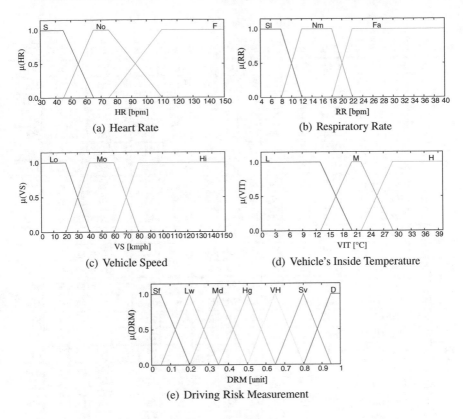

(a) Heart Rate

(b) Respiratory Rate

(c) Vehicle Speed

(d) Vehicle's Inside Temperature

(e) Driving Risk Measurement

Fig. 3. Membership functions.

4.2 Testbed Description

In order to evaluate the proposed system, we implemented a testbed and carried out experiments in a real scenario [4,7]. The FSDRM testbed is shown in Fig. 4. The testbed is composed of sensing and processing components. The sensing system consists of two parts. The first part is implemented in the Arduino Platform while the second one consists of a Microwave Sensor Module (MSM) called DC6M4JN3000. We set-up sensors on Arduino Uno to measure the vehicle's inside temperature and used the MSM to measure the driver's heart and respiratory rate. The vehicle speed is considered as a random value [6]. Then, we implemented a processing device to get the sensed data and to run our fuzzy system. The sensing components are connected to the processing device via USB cable. We used Arduino IDE and Processing language to get the sensed data from the first module, whereas the MSM generates the sensed data in the appropriate format itself. Then, we use FuzzyC to fuzzify these data and to determine the degree of risk which is the output of our proposed system. Based on the DRM an appropriate task can be performed.

Table 2. FRB of FSDRM.

Rule	HR	RR	VS	VIT	DRM	Rule	HR	RR	VS	VIT	DRM	Rule	HR	RR	VS	VIT	DRM
1	S	Sl	Lo	L	Hg	28	No	Sl	Lo	L	Md	55	F	Sl	Lo	L	VH
2	S	Sl	Lo	M	Md	29	No	Sl	Lo	M	Lw	56	F	Sl	Lo	M	Hg
3	S	Sl	Lo	H	Hg	30	No	Sl	Lo	H	Md	57	F	Sl	Lo	H	VH
4	S	Sl	Mo	L	VH	31	No	Sl	Mo	L	Hg	58	F	Sl	Mo	L	Sv
5	S	Sl	Mo	M	Hg	32	No	Sl	Mo	M	Md	59	F	Sl	Mo	M	Hg
6	S	Sl	Mo	H	VH	33	No	Sl	Mo	H	Hg	60	F	Sl	Mo	H	Sv
7	S	Sl	Hi	L	D	34	No	Sl	Hi	L	VH	61	F	Sl	Hi	L	D
8	S	Sl	Hi	M	Sv	35	No	Sl	Hi	M	Hg	62	F	Sl	Hi	M	Sv
9	S	Sl	Hi	H	D	36	No	Sl	Hi	H	VH	63	F	Sl	Hi	H	D
10	S	Nm	Lo	L	Md	37	No	Nm	Lo	L	Lw	64	F	Nm	Lo	L	Md
11	S	Nm	Lo	M	Lw	38	No	Nm	Lo	M	Sf	65	F	Nm	Lo	M	Lw
12	S	Nm	Lo	H	Md	39	No	Nm	Lo	H	Lw	66	F	Nm	Lo	H	Md
13	S	Nm	Mo	L	Hg	40	No	Nm	Mo	L	Lw	67	F	Nm	Mo	L	Hg
14	S	Nm	Mo	M	Md	41	No	Nm	Mo	M	Sf	68	F	Nm	Mo	M	Md
15	S	Nm	Mo	H	Hg	42	No	Nm	Mo	H	Lw	69	F	Nm	Mo	H	Hg
16	S	Nm	Hi	L	VH	43	No	Nm	Hi	L	Hg	70	F	Nm	Hi	L	Sv
17	S	Nm	Hi	M	Hg	44	No	Nm	Hi	M	Md	71	F	Nm	Hi	M	Hg
18	S	Nm	Hi	H	VH	45	No	Nm	Hi	H	Hg	72	F	Nm	Hi	H	Sv
19	S	Fa	Lo	L	VH	46	No	Fa	Lo	L	Md	73	F	Fa	Lo	L	VH
20	S	Fa	Lo	M	Hg	47	No	Fa	Lo	M	Lw	74	F	Fa	Lo	M	Hg
21	S	Fa	Lo	H	VH	48	No	Fa	Lo	H	Md	75	F	Fa	Lo	H	VH
22	S	Fa	Mo	L	Sv	49	No	Fa	Mo	L	Hg	76	F	Fa	Mo	L	Sv
23	S	Fa	Mo	M	Hg	50	No	Fa	Mo	M	Md	77	F	Fa	Mo	M	Hg
24	S	Fa	Mo	H	Sv	51	No	Fa	Mo	H	Hg	78	F	Fa	Mo	H	Sv
25	S	Fa	Hi	L	D	52	No	Fa	Hi	L	VH	79	F	Fa	Hi	L	D
26	S	Fa	Hi	M	Sv	53	No	Fa	Hi	M	Hg	80	F	Fa	Hi	M	Sv
27	S	Fa	Hi	H	D	54	No	Fa	Hi	H	VH	81	F	Fa	Hi	H	D

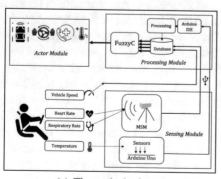

(a) The testbed scheme.

(b) Snapshot of testbed.

Fig. 4. FSDRM testbed.

5   Proposed System Evaluation

5.1   Simulation Results

In this subsection, we present the simulation results for our proposed system. The simulation results are presented in Figs. 5, 6 and 7. We consider the HR and RR as constant parameters. The VS values considered for simulations are from 10 to 100 kmph. We show the relation between DRM and VIT for different VS values. We vary the VIT parameter from 0 to 40 °C.

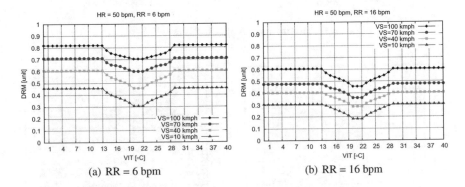

(a) RR = 6 bpm   (b) RR = 16 bpm

Fig. 5. Simulation results for HR = 50 bpm.

In Fig. 5, we consider the HR value 50 bpm and change the RR from 6 to 16 bpm. From Fig. 5(a) we can see that the DRM values are very high, especially when the vehicle speed is over 40 kmph and the vehicle's inside temperatures are not the best ones. In Fig. 5(b) is considered the same scenario but with a normal respiratory rate. It can be seen that the DRM is decreased compared with the first scenario. If the vehicle speed is under 40 kmph and the vehicle's inside temperature is between 18 and 24 °C, the risk level is under the moderate level.

In Fig. 6, we present the simulation results for HR 70 bpm. In Fig. 6(a) is considered the scenario with a normal respiratory rate. We can see that the DRM values are lower than all the other considered scenarios. This is due to the driver's vital signs, which indicate a very good status of the driver's body, and if the environment temperature is very comfortable, he could manage to drive safely even with a high speed. If the respiratory rate is increased (see Fig. 6(b)), it can be seen that the DRM is increased.

In Fig. 7, we increase the value of HR to 110 bpm. If the driver breathes normally we can see that there are some cases when the degree of risk is not high such as when the vehicle moves slowly or the ambient temperature is between 18 and 24 °C. On the other hand, when the driver breathes rapidly, the degree of risk is always above the high level.

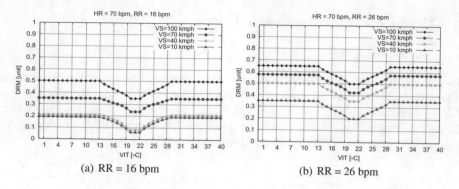

Fig. 6. Simulation results for HR = 70 bpm.

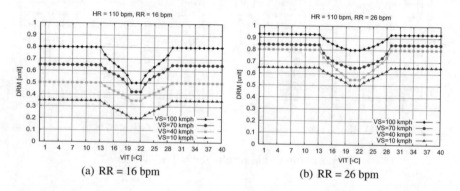

Fig. 7. Simulation results for HR = 110 bpm.

In the cases when the risk level is above the moderate level for a relatively long time, the system can perform a certain action. For example, when the DRM value is slightly above the moderate level the system may take an action to change the driver's mood, and when the DRM value is very high, the system could limit the vehicle's maximal speed to a speed that the risk level is decreased significantly.

5.2   Experimental Results

For the experiments, we considered the vehicle speed from 0 to 150 kmph. The experimental results are presented in Figs. 8, 9 and 10. In Fig. 8(a) are shown the results of DRM when HR and RR is "Slow". As we can see, there is not any DRM value that indicates a situation with a low risk. All the DRM values are above the moderate level. On the other hand, when the driver breathes normally, we can see a number of DRM values that are decided as situations with a low risk by our system (see Fig. 8(b)). These low DRM values are achieved when the driver is driving slowly and the temperature is within 18–24 °C range.

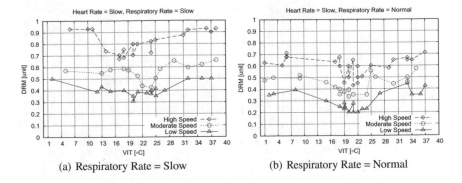

(a) Respiratory Rate = Slow

(b) Respiratory Rate = Normal

Fig. 8. Experimental results for slow heart rate.

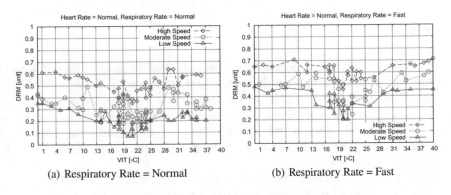

(a) Respiratory Rate = Normal

(b) Respiratory Rate = Fast

Fig. 9. Experimental results for normal heart rate.

The results of DRM for normal heart rate are presented in Fig. 9, with Fig. 9(a) and (b) presenting the experimental results for normal and fast respiratory rate, respectively. When the driver's heart beats normally, the risk is mostly under the moderate level. Moreover, when the vehicle is not moving at high speed, we can see several DRM values which are decided as safe and many other with low risk. From Fig. 9(b), we can see that even when the driver breathes rapidly, there are cases that the risk level is determined under the moderate level. These cases include situations when he drives slowly and in a comfortable environment temperature.

In Fig. 10 are shown the results of DRM for fast heart rate. The results are almost the same with that of Fig. 7 where the low values of DRM happen to be only when the driver's respiratory rate is normal, the vehicle moves slowly and the temperature is within 18–24 °C range. When he breathes rapidly, the degree of risk is decided to be even above the "Very High" level. In these situations, the driver should not drive fast as his situation is a potential risk for him and for other vehicles on the road. Therefore, the system decides to perform the appropriate action in order to provide the driving safety.

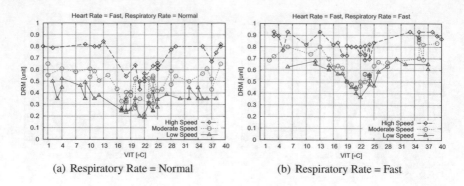

(a) Respiratory Rate = Normal   (b) Respiratory Rate = Fast

Fig. 10. Experimental results for fast heart rate.

6   Conclusions

In this paper, we proposed a new fuzzy-based system to decide the driving risk measurement. We took into consideration four parameters: driver's heart rate, driver's respiratory rate, vehicle speed and vehicle's inside temperatures. We evaluated the performance of proposed system by simulations and experiments. From the evaluation results, we conclude that driver's heart and respiratory rate, vehicle speed, and vehicle's inside temperature have different effects on the assessment of risk level.

In the future, we would like to make extensive simulations and experiments to evaluate the proposed system and compare the performance with other systems.

References

1. Aarts, E., De Ruyter, B.: New research perspectives on ambient intelligence. J. Ambient. Intell. Smart Environ. **1**(1), 5–14 (2009)
2. Aarts, E., Wichert, R.: Ambient intelligence. In: Technology Guide, pp. 244–249. Springer (2009)
3. Acampora, G., Cook, D.J., Rashidi, P., Vasilakos, A.V.: A survey on ambient intelligence in healthcare. Proc. IEEE **101**(12), 2470–2494 (2013)
4. Bylykbashi, K., Elmazi, D., Matsuo, K., Ikeda, M., Barolli, L.: Implementation of a fuzzy-based simulation system and a testbed for improving driving conditions in VANETs. In: International Conference on Complex, Intelligent, and Software Intensive Systems, pp. 3–12. Springer (2019)
5. Bylykbashi, K., Liu, Y., Ozera, K., Barolli, L., Takizawa, M.: A fuzzy-based system for safe driving information in VANETs. In: International Conference on Broadband and Wireless Computing, Communication and Applications, pp. 648–658. Springer (2018)
6. Bylykbashi, K., Qafzezi, E., Ikeda, M., Matsuo, K., Barolli, L.: A fuzzy-based system for driving risk measurement (FSDRM) in VANETs: a comparison study of simulation and experimental results. In: International Conference on P2P, Parallel, Grid, Cloud and Internet Computing, pp. 14–25. Springer (2019)

7. Bylykbashi, K., Qafzezi, E., Ikeda, M., Matsuo, K., Barolli, L.: Implementation of a fuzzy-based simulation system and a testbed for improving driving conditions in VANETs considering drivers's vital signs. In: International Conference on Network-Based Information Systems, pp. 37–48. Springer (2019)
8. Cuka, M., Elmazi, D., Ikeda, M., Matsuo, K., Barolli, L.: IoT node selection in Opportunistic Networks: Implementation of fuzzy-based simulation systems and testbed. Internet of Things **8**, 100105 (2019)
9. Gusikhin, O., Filev, D., Rychtyckyj, N.: Intelligent vehicle systems: applications and new trends. In: Informatics in Control Automation and Robotics, pp. 3–14. Springer (2008)
10. Hartenstein, H., Laberteaux, K.P.: A tutorial survey on vehicular ad hoc networks. IEEE Commun. Mag. **46**(6), 164–171 (2008)
11. Kandel, A.: Fuzzy Expert Systems. CRC Press Inc., Boca Raton (1991)
12. Karagiannis, G., Altintas, O., Ekici, E., Heijenk, G., Jarupan, B., Lin, K., Weil, T.: Vehicular networking: a survey and tutorial on requirements, architectures, challenges, standards and solutions. IEEE Commun. Surv. Tutor. **13**(4), 584–616 (2011)
13. Klir, G.J., Folger, T.A.: Fuzzy Sets, Uncertainty, and Information. Prentice Hall, Upper Saddle River (1988)
14. Lindwer, M., Marculescu, D., Basten, T., Zimmennann, R., Marculescu, R., Jung, S., Cantatore, E.: Ambient intelligence visions and achievements: linking abstract ideas to real-world concepts. In: 2003 Design, Automation and Test in Europe Conference and Exhibition, pp. 10–15 (2003)
15. Matsuo, K., Cuka, M., Inaba, T., Oda, T., Barolli, L., Barolli, A.: Performance analysis of two WMN architectures by WMN-GA simulation system considering different distributions and transmission rates. Int. J. Grid Util. Comput. **9**(1), 75 82 (2018)
16. McNeill, F.M., Thro, E.: Fuzzy Logic: A Practical Approach. Academic Press Professional, Inc., San Diego (1994)
17. Munakata, T., Jani, Y.: Fuzzy systems: an overview. Commun. ACM **37**(3), 69–77 (1994)
18. Ozera, K., Bylykbashi, K., Liu, Y., Barolli, L.: A fuzzy-based approach for cluster management in VANETs: performance evaluation for two fuzzy-based systems. Internet Things **3**, 120–133 (2018)
19. Ozera, K., Inaba, T., Bylykbashi, K., Sakamoto, S., Ikeda, M., Barolli, L.: A WLAN triage testbed based on fuzzy logic and its performance evaluation for different number of clients and throughput parameter. Int. J. Grid Util. Comput. **10**(2), 168–178 (2019)
20. Sadri, F.: Ambient intelligence: a survey. ACM Comput. Surv. (CSUR) **43**(4), 36 (2011)
21. Vasilakos, A., Pedrycz, W.: Ambient Intelligence, Wireless Networking, and Ubiquitous Computing. Artech House, Inc., Norwood (2006)
22. Zadeh, L.A., Kacprzyk, J.: Fuzzy Logic for the Management of Uncertainty. Wiley, New York (1992)
23. Zimmermann, H.J.: Fuzzy Set Theory and Its Applications. Springer, New York (1991)

Resource Management in SDN-VANETs: Coordination of Cloud-Fog-Edge Resources Using Fuzzy Logic

Ermioni Qafzezi[1]([⊠]), Kevin Bylykbashi[1], Tomoyuki Ishida[2], Keita Matsuo[2], Leonard Barolli[2], and Makoto Takizawa[3]

[1] Graduate School of Engineering, Fukuoka Institute of Technology (FIT), 3-30-1 Wajiro-Higashi, Higashi-Ku, Fukuoka 811–0295, Japan
eqafzezi@gmail.com, bylykbashi.kevin@gmail.com
[2] Department of Information and Communication Engineering, Fukuoka Institute of Technology (FIT), 3-30-1 Wajiro-Higashi, Higashi-Ku, Fukuoka 811-0295, Japan
{t-ishida,kt-matsuo,barolli}@fit.ac.jp
[3] Department of Advanced Sciences, Faculty of Science and Engineering, Hosei University, 3-7-2, Kajino-machi, Koganei-shi, Tokyo 184-8584, Japan
makoto.takizawa@computer.org

Abstract. In this work, we propose an intelligent system for coordination and management of the cloud-fog-edge resources in Vehicular Ad hoc Networks (VANETs) using Software Defined Networking (SDN) and Fuzzy Logic (FL) approaches. The proposed system called Fuzzy-based System for Resource Management (FSRM) determines the appropriate resources to be used by a vehicle to process different VANETs applications. The decision is made by prioritizing the application requirements: Time Sensitivity (TS) and Data Size (DS), and by considering the available connections of the vehicle i.e., Number of Neighboring Vehicles (NNV) and Vehicle Relative Speed with Neighboring Vehicles (VRSNV). We demonstrate in simulation the feasibility of FSRM to improve the management of the network resources.

1 Introduction

Traffic accidents, road congestion and environmental pollution are persistent problems faced by both developed and developing countries, which have made people live in difficult situations. Among these, the traffic incidents are the most serious ones because they result in huge loss of life and property. For decades, we have seen governments and car manufacturers struggle for safer roads and car accident prevention. The development in wireless communications has allowed companies, researchers and institutions to design communication systems that provide new solutions for these issues. Therefore, new types of networks, such as Vehicular Ad hoc Networks (VANETs) have been created. VANET consists of a network of vehicles in which vehicles are capable of communicating among themselves in order to deliver valuable information such as safety warnings and traffic information.

L. Barolli et al. (Eds.): EIDWT 2020, LNDECT 47, pp. 114–126, 2020.
https://doi.org/10.1007/978-3-030-39746-3_13

Nowadays, every car is likely to be equipped with various forms of smart sensors, wireless communication modules, storage and computational resources. The sensors will gather information about the road and environment conditions and share it with neighboring vehicles and adjacent roadside units (RSU) via vehicle-to-vehicle (V2V) or vehicle-to-infrastructure (V2I) communication.

As more and more smart sensors are installed on modern vehicles, massive amounts of data are generated from monitoring the on-road and in-board status. This exponential growth of generated vehicular data, together with the increasing data demands from in-vehicle users, has led to a tremendous amount of data in VANETs [20]. The traditional centralized VANET technology is no longer efficient in handling these massive amounts of traffic data, and in order to collect and process this amount of instantaneous traffic information, additional servers are required in distributed areas. Cloud computing was first considered as an appropriate solution for these types of situations.

Recently, fog computing extends cloud nearer to the user. This new architecture analyzes data close to devices for minimizing latency, decision making in real time and offloading massive traffic flow from the core networks. With edge computing, resources and services of computing, networking, storage and control capabilities are distributed anywhere along the continuum from the cloud to things [4].

By leveraging the fog/edge computing technology, a significant amount of computing power will be distributed near/to the vehicles. Therefore, most of data will be processed and stored at the fog/edge, which can minimize the latency and ensure better quality of service for connected vehicles [21].

Although the integration of cloud, fog and edge computing in VANETs is very promising to offer scalable access to storage, networking and computing resources, this network architecture lacks mechanisms needed for resources and connectivity management as it controls the network in a decentralized manner. The prospective solution to solve these problems is by augmenting Software Defined Networking (SDN) with this architecture. SDN provides flexibility, programmability, scalability and global knowledge by controlling the network in a centralized and programmable approach. In Fig. 1 we illustrate the topology of this novel VANET architecture which is composed of cloud computing data centers, fog servers with SDN Controllers (SDNCs), RSU Controllers (RSUCs), RSUs, Base Stations and vehicles. We also illustrate the infrastructure-to-infrastructure (I2I), V2I, and V2V communication links.

We are focused on the resource management in this new architecture and have proposed an intelligent approach based on fuzzy logic. We presented a cloud-fog-edge SDN-VANETs layered architecture which is coordinated by a fuzzy system implemented in the SDNC [2,16,17]. In this work, we propose a new Fuzzy-based System for Resource Management (FSRM) which decides the resources to be used by a particular vehicle based on its relative speed with the neighboring vehicles, the number of its neighbors, the time-sensitivity of the application data and the size of the data to be processed. We evaluate the performance of FSRM by computer simulations.

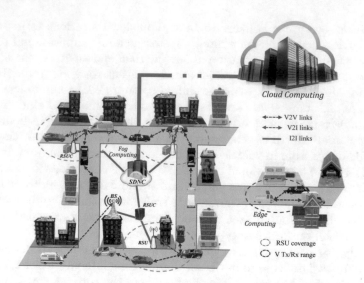

Fig. 1. Topology architecture of SDN-VANETs using Cloud, Fog and Edge computing.

The remainder of the paper is as follows. In Sect. 2, we present an overview of cloud-fog-edge computing, SDN and VANETs. In Sect. 3, we describe the proposed fuzzy-based system. In Sect. 4, we discuss the simulation results. Finally, conclusions and future work are given in Sect. 5.

2   Background Overview

In this section, we briefly introduce cloud-fog-edge computing and SDN as enabling technologies for full deployment and management of VANET applications and services. Moreover, we provide a short description of the VANET features, characteristics and communication issues, and present the content/resources distribution on the SDN-VANETs leveraging cloud-fog-edge computing architecture.

2.1   Cloud-Fog-Edge Computing

The notion of cloud computing started from the realization of the fact that instead of investing in infrastructure, businesses may find it useful to rent the infrastructure and sometimes the needed software to run their applications. The cloud computing is becoming a promising and prevalent service to replace traditional local systems due to its scalable access to computing resources.

The benefits of cloud computing have motivated many researchers to focus on shifting the conventional VANETs to Vehicular Cloud Computing (VCC) by merging VANET with cloud computing [5,12,13]. VCC is a very appealing technology due to its features and capabilities in supporting a series of novel, relevant, or sensitive applications. Additionally, VCCs are designed to initiate objectives that directly match everyday transportation needs, such as enabling computational services at low cost to authorized users, reducing traffic congestion, and implementing services to improve road safety [1].

However, there still are requirements such as low latency, high throughput, location awareness and mobility support that VCC can barely fulfill. Fog computing is proposed as a solution to overcome these issues between the vehicles and the conventional cloud [18]. Similar to cloud, fog computing provides data, compute, storage and application services at the proximity of the vehicular nodes.

The vehicular nodes are the edge of this layered architecture. With edge computing in VANETs, which means vehicles having resources and services of computing, networking, storage and control capabilities, a significant amount of data can be processed at/through the vehicles, consequently offloading massive traffic flow from the core networks.

2.2 Software Defined Networking

The core concept of SDN is the decoupling between the control plane and data plane, which provides dedicated mechanisms for resources and connectivity management. The first one is used for network traffic control and the latter for data forwarding. This separation will simplify network management that is extremely complicated when the number of nodes dramatically increases in such dynamic environment like VANET [19]. In addition, it will increase flexibility and programmability in the network by simplifying the development and deployment of new protocols and by bringing awareness into the system, so that it can adapt to changing conditions and requirements, i.e., emergency services [8]. This awareness allows SDN-VANET to make better decisions based on the combined information from multiple sources, not just individual perception from each node. Significant benefits of the incorporation of SDN are the reduction of interference, improvement of channels and wireless resources usage, as well as the routing of data in multi-hop and multi-path scenarios [8].

2.3 VANETs

VANETs are a special case of Mobile Adhoc Networks (MANETs) in which mobile nodes are vehicles. In VANETs, nodes (vehicles) have high mobility and tend to follow organized routes instead of moving at random. Moreover, vehicles offer attractive features such as higher computational capability and localization through GPS.

Despite the attractive features, VANETs are characterized by very large and dynamic topologies, variable capacity wireless links, bandwidth and hard delay constrains, and by short contact durations which are caused by the high mobility, high speed and low density of vehicles. In addition, limited transmission ranges, physical obstacles and interferences, make these networks characterized by disruptive and intermittent connectivity.

To make VANETs applications possible, it is necessary to design proper networking mechanisms that can overcome relevant problems that arise from vehicular environments.

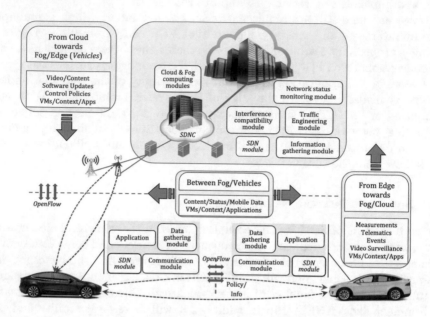

Fig. 2. Logical architecture of cloud-fog-edge SDN-VANET with content distribution.

While cloud, fog and edge computing in VANETs offer scalable access to storage, networking and computing resources, SDN provides higher flexibility, programmability, scalability and global knowledge. A detailed logical structure of this architecture with the content distribution is described in Fig. 2. The fog devices (such as fog servers and RSUs) are located between vehicles and the data centers of the main cloud environments. The safety applications data generated through on board and on road sensors are processed first in the vehicles as they require real-time processing. If more storing and computing resources are needed, the vehicle can request to use those of the other adjacent vehicles, assuming a connection can be established and maintained between them for a while. With the vehicles having created multiple virtual machines on other vehicles, the virtual machine migration must be achievable in order to provide continuity as one/some vehicle may move out of the range. However, to set-up virtual machines on the nearby vehicles, multiple requirements must be met and when these demands are not satisfied, the fog servers are used. Cloud servers are used as a repository for software updates, control policies and for the data that need long-term analytics and are not delay-sensitive. On the other side, SDN modules which offer flexibility and programmability, are used to simplify the network management by offering mechanisms that improve the network traffic control and provide mechanisms for coordination of resources. The implementation of this architecture promises to enable and improve the VANET applications: road and vehicle safety services, traffic optimization, video surveillance, telematics and commercial and entertainment applications.

3 Proposed Fuzzy-Based System

In this section, we present the layered cloud-fog-edge SDN-VANETs architecture which is coordinated by the global intelligence provided by SDNC. In this architecture, SDNC manages not only the computing and storage resources of fog, but also those of edge and cloud. It also controls the network behavior by selecting best routes and improving usage of channels and wireless resources. An illustration of this layered architecture is given in Fig. 3. Our proposed FSRM does not require the SDN-VANET to operate in a particular mode, i.e. central control mode, distributed control mode and hybrid control mode [8]. It supports either the central control and distributed control mode and consequently also the hybrid control mode. A vehicle that needs storage and computing resources for a particular application can use that of neighboring vehicles, fog servers or cloud data centers based on the application requirements and available connections. For instance, for a temporary application that needs real-time processing, the vehicle can use the resources of adjacent vehicles if the requirements to realize such operations are fulfilled. Otherwise, it will use the resources of fog servers, which offer low latency as well. Whereas real-time applications require the usage of edge and fog layer resources, for delay tolerant applications vehicles can use the cloud resources as these applications do not require low latency. FSRM is

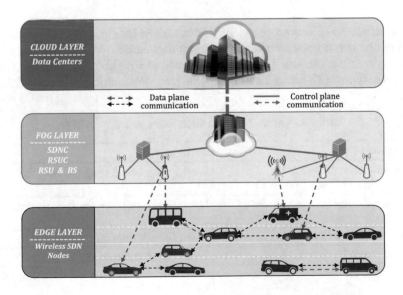

Fig. 3. Layered architecture of cloud-fog-edge SDN-VANETs.

implemented in the SDNC and in the vehicles which are equipped with SDN modules. If a vehicle does not have an SDN module, it sends the information to SDNC which sends back its decision. The FSRM uses the beacon messages received from the adjacent vehicles to extract information such as their current position, velocity, direction, and based on the application requirements, it decides the appropriate layer to run and process the application data. In the following we describe in detail our proposed FSRM.

3.1 Description of FLC

In this work, we use fuzzy logic to implement the proposed system. Fuzzy sets and fuzzy logic have been developed to manage vagueness and uncertainty in a reasoning process of an intelligent system such as a knowledge based system, an expert system or a logic control system [3,6,7,9–11,14,15,22,23].

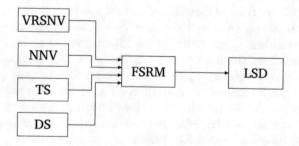

Fig. 4. FSRM structure.

The structure of the proposed FSRM is shown in Fig. 4. For the implementation of our system, we consider four input parameters: Vehicle Relative Speed with Neighboring Vehicles (VRSNV), Number of Neighboring Vehicles (NNV), Time Sensitivity (TS) and Data Size (DS) to determine the Layer Selection

Table 1. Parameters and their term sets for FSRM.

Parameters	Term sets
Vehicle Relative Speed with Neighboring Vehicles (VRSNV)	Slower (Sl), Same (Sa), Faster (Fa)
Number of Neighboring Vehicles (NNV)	Low (Lo), Moderate (Mo), High (Hi)
Time Sensitivity (TS)	Low (L), Middle (Mi), High (H)
Data Size (DS)	Small (S), Medium (M), Big (B)
Layer Selection Decision (LSD)	Decision Level 1 (DL1), DL2, DL3, DL4, DL5, DL6, DL7

Table 2. The fuzzy rule base of FSRM.

Rule	VRSNV	NNV	TS	DS	LSD	Rule	VRSNV	NNV	TS	DS	LSD
1	Sl	Lo	L	S	DL7	41	Sa	Mo	Mi	M	DL3
2	Sl	Lo	L	M	DL7	42	Sa	Mo	Mi	B	DL3
3	Sl	Lo	L	B	DL7	43	Sa	Mo	H	S	DL1
4	Sl	Lo	Mi	S	DL5	44	Sa	Mo	H	M	DL1
5	Sl	Lo	Mi	M	DL6	45	Sa	Mo	H	B	DL2
6	Sl	Lo	Mi	B	DL7	46	Sa	Hi	L	S	DL2
7	Sl	Lo	H	S	DL3	47	Sa	Hi	L	M	DL3
8	Sl	Lo	H	M	DL5	48	Sa	Hi	L	B	DL4
9	Sl	Lo	H	B	DL6	49	Sa	Hi	Mi	S	DL1
10	Sl	Mo	L	S	DL5	50	Sa	Hi	Mi	M	DL2
11	Sl	Mo	L	M	DL7	51	Sa	Hi	Mi	B	DL2
12	Sl	Mo	L	B	DL7	52	Sa	Hi	H	S	DL1
13	Sl	Mo	Mi	S	DL4	53	Sa	Hi	H	M	DL1
14	Sl	Mo	Mi	M	DL5	54	Sa	Hi	H	B	DL1
15	Sl	Mo	Mi	B	DL6	55	Fa	Lo	L	S	DL7
16	Sl	Mo	H	S	DL2	56	Fa	Lo	L	M	DL7
17	Sl	Mo	H	M	DL3	57	Fa	Lo	L	B	DL7
18	Sl	Mo	H	B	DL4	58	Fa	Lo	Mi	S	DL5
19	Sl	Hi	L	S	DL4	59	Fa	Lo	Mi	M	DL6
20	Sl	Hi	L	M	DL5	60	Fa	Lo	Mi	B	DL7
21	Sl	Hi	L	B	DL6	61	Fa	Lo	H	S	DL3
22	Sl	Hi	Mi	S	DL2	62	Fa	Lo	H	M	DL5
23	Sl	Hi	Mi	M	DL4	63	Fa	Lo	H	B	DL6
24	Sl	Hi	Mi	B	DL4	64	Fa	Mo	L	S	DL5
25	Sl	Hi	H	S	DL1	65	Fa	Mo	L	M	DL7
26	Sl	Hi	H	M	DL2	66	Fa	Mo	L	B	DL7
27	Sl	Hi	H	B	DL3	67	Fa	Mo	Mi	S	DL4
28	Sa	Lo	L	S	DL4	68	Fa	Mo	Mi	M	DL5
29	Sa	Lo	L	M	DL6	69	Fa	Mo	Mi	B	DL6
30	Sa	Lo	L	B	DL6	70	Fa	Mo	H	S	DL2
31	Sa	Lo	Mi	S	DL3	71	Fa	Mo	H	M	DL3
32	Sa	Lo	Mi	M	DL4	72	Fa	Mo	H	B	DL4
33	Sa	Lo	Mi	B	DL5	73	Fa	Hi	L	S	DL4
34	Sa	Lo	H	S	DL1	74	Fa	Hi	L	M	DL5
35	Sa	Lo	H	M	DL2	75	Fa	Hi	L	B	DL6
36	Sa	Lo	H	B	DL3	76	Fa	Hi	Mi	S	DL2
37	Sa	Mo	L	S	DL3	77	Fa	Hi	Mi	M	DL4
38	Sa	Mo	L	M	DL4	78	Fa	Hi	Mi	B	DL4
39	Sa	Mo	L	B	DL5	79	Fa	Hi	H	S	DL1
40	Sa	Mo	Mi	S	DL2	80	Fa	Hi	H	M	DL2
						81	Fa	Hi	H	B	DL3

Decision (LSD) value. These four input parameters are not correlated with each other, for this reason we use fuzzy sets. The input parameters are fuzzified using the membership functions showed in Fig. 5(a), (b), (c) and (d). In Fig. 5(e) are shown the membership functions used for the output parameter. We use triangular and trapezoidal membership functions because they are suitable for real-time operation. The term sets for each linguistic parameter are shown in Table 1. We decided the number of term sets by carrying out many simulations. In Table 2, we show the Fuzzy Rule Base (FRB) of FSRM, which consists of 81 rules. The control rules have the form: IF "conditions" THEN "control action". For instance, for Rule 1: "IF VRSNV is Sl, NNV is Lo, TS is L and DS is S, THEN LSD is DL7" or for Rule 54: "IF VRSNV is Sa, NNV is Hi, TS is H and DS is B, THEN LSD is DL1".

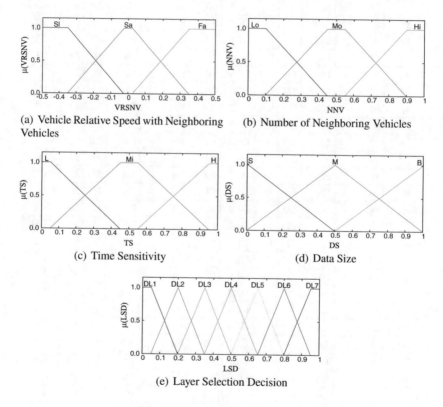

(a) Vehicle Relative Speed with Neighboring Vehicles

(b) Number of Neighboring Vehicles

(c) Time Sensitivity

(d) Data Size

(e) Layer Selection Decision

Fig. 5. Membership functions.

4 Simulation Results

In this section, we present the simulation results for our proposed system. The simulation results are presented in Figs. 6 and 7. We consider the VRSNV and NNV as a constant parameters. We change the TS value from 0.05 to 0.95 units.

In Fig. 6, we consider the VRSNV value −0.4 (0,4) in which we simulate the scenario where the vehicle is moving slower (faster) than the other vehicles in its vicinity. We change the value of NNV from 0.1 to 0.9. In Fig. 6(a), we show the results for VRSNV = −0.4 and NNV = 0.1. Having only a few vehicles within the communication range and moving much slower than them, makes it impossible for the vehicle to use the resources of its neighbors. The vehicle quickly moves out of the communication range of its neighbors, therefore, it is hard to establish and maintain a connection between them. Thus, from applications which need real-time computing, only the ones with the smallest data can be processed in the edge. However, even if the amount of application data is bigger, the vehicle can process these data at the fog servers which offer low latency as well. If the applications are delay tolerant, we can see that these applications will be processed at the cloud layer. If the number of the neighbors is increased (see Fig. 6(b) and (c)) the size of time-sensitive data that can be processed at the edge layer is also increased. This is due to the fact that the vehicle can distribute the data processing between many neighbors.

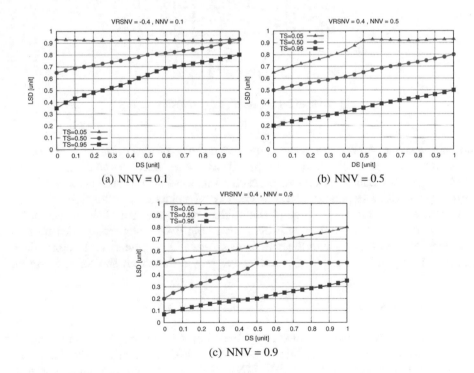

Fig. 6. Simulation results for VRSNV = −0.4 (0.4).

The scenario where the vehicle moves with the same speed as its neighbors is considered in Fig. 7. We can see that all real-time applications are processed at the edge, even when the size of data is big. The vehicle will have the same

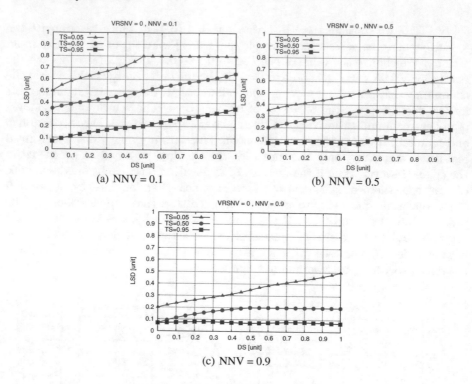

Fig. 7. Simulation results for VRSNV = 0.

adjacent vehicles for a while as they move with the same speed. Being in the range of each other for a time, creates the possibility of initiating virtual machines to these adjacent vehicles which can be used to process also a number of non real-time applications depending on the density of vehicles. If many vehicles are within the communication range of the vehicle which needs additional resources (see Fig. 7(c)), the possibility to process even big delay tolerant applications at the edge is increased. By avoiding the processing of big applications at cloud data centers, it is possible to offload an excessive traffic flow from the core networks.

5   Conclusions

In this paper, we proposed a new fuzzy-based system for resource management in a cloud-fog-edge layered architecture for SDN-VANET. FSRM takes into consideration four parameters: VRSNV, NNV, TS and DS. We evaluated the performance of FSRM by computer simulations. From the simulations results, we conclude as follows.

- Highly time-sensitive data are processed always in the edge and fog layer even if the vehicle does not have many neighbors within its communication range.

- If the number of neighboring vehicles is increased, the amount of data which can be processed in the edge layer is increased.
- If the vehicle has same relative speed with neighboring vehicles, in the edge layer will be processed not only the highly time-sensitive applications but also an amount data of delay tolerant applications depending on the number of these neighboring vehicles.
- If only a few vehicles are nearby and the vehicle moves relatively much slower/faster than them, the vehicle will always process all the delay tolerant applications in the cloud layer.

In the future, we would like to make extensive simulations to evaluate the proposed system and compare the performance with other systems.

References

1. Boukerche, A., Robson, E.: Vehicular cloud computing: architectures, applications, and mobility. Comput. Netw. **135**, 171–189 (2018)
2. Bylykbashi, K., Liu, Y., Matsuo, K., Ikeda, M., Barolli, L., Takizawa, M.: A fuzzy-based system for cloud-fog-edge selection in VANETs. In: International Conference on Emerging Internetworking, Data & Web Technologies, pp. 1–12. Springer (2019)
3. Cuka, M., Elmazi, D., Ikeda, M., Matsuo, K., Barolli, L.: IoT node selection in opportunistic networks: implementation of fuzzy-based simulation systems and testbed. Internet Things **8**, 100105 (2019)
4. Hu, Y.C., Patel, M., Sabella, D., Sprecher, N., Young, V.: Mobile edge computing-a key technology towards 5G. ETSI White Paper, vol. 11, no. 11, pp. 1–16 (2015)
5. Hussain, R., Son, J., Eun, H., Kim, S., Oh, H.: Rethinking vehicular communications: merging VANET with cloud computing. In: 4th IEEE International Conference on Cloud Computing Technology and Science Proceedings, pp. 606–609 (2012)
6. Kandel, A.: Fuzzy Expert Systems. CRC Press, Inc., Boca Raton (1992)
7. Klir, G.J., Folger, T.A.: Fuzzy Sets, Uncertainty, and Information. Prentice Hall, Upper Saddle River (1988)
8. Ku, I., Lu, Y., Gerla, M., Gomes, R.L., Ongaro, F., Cerqueira, E.: Towards software-defined VANET: architecture and services. In: 13th Annual Mediterranean Ad Hoc Networking Workshop (MED-HOC-NET), pp. 103–110 (2014)
9. Matsuo, K., Cuka, M., Inaba, T., Oda, T., Barolli, L., Barolli, A.: Performance analysis of two WMN architectures by WMN-GA simulation system considering different distributions and transmission rates. Int. J. Grid Util. Comput. **9**(1), 75–82 (2018)
10. McNeill, F.M., Thro, E.: Fuzzy Logic: A Practical Approach. Academic Press Professional, Inc., San Diego (1994)
11. Munakata, T., Jani, Y.: Fuzzy systems: an overview. Commun. ACM **37**(3), 69–77 (1994)
12. Olariu, S., Hristov, T., Yan, G.: The next paradigm shift: from vehicular networks to vehicular clouds. Mob. Ad Hoc Netw. Cut. Edge Dir. **56**(6), 645–700 (2013)
13. Olariu, S., Khalil, I., Abuelela, M.: Taking vanet to the clouds. Int. J. Pervasive Comput. Commun. **7**(1), 7–21 (2011)

14. Ozera, K., Bylykbashi, K., Liu, Y., Barolli, L.: A fuzzy-based approach for cluster management in vanets: performance evaluation for two fuzzy-based systems. Internet Things **3**, 120–133 (2018)
15. Ozera, K., Inaba, T., Bylykbashi, K., Sakamoto, S., Ikeda, M., Barolli, L.: A wlan triage testbed based on fuzzy logic and its performance evaluation for different number of clients and throughput parameter. Int. J. Grid Util. Comput. **10**(2), 168–178 (2019)
16. Qafzezi, E., Bylykbashi, K., Spaho, E., Barolli, L.: An intelligent approach for resource management in SDN-VANETs using fuzzy logic. In: International Conference on Broadband and Wireless Computing, Communication and Applications, pp. 747–756. Springer (2019)
17. Qafzezi, E., Bylykbashi, K., Spaho, E., Barolli, L.: A new fuzzy-based resource management system for SDN-VANETs. Int. J. Mob. Comput. Multimed. Commun. (IJMCMC) **10**(4), 1–12 (2019)
18. Stojmenovic, I., Wen, S., Huang, X., Luan, H.: An overview of fog computing and its security issues. Concurr. Comput. Pract. Exp. **28**(10), 2991–3005 (2016)
19. Truong, N.B., Lee, G.M., Ghamri-Doudane, Y.: Software defined networking-based vehicular adhoc network with fog computing. In: 2015 IFIP/IEEE International Symposium on Integrated Network Management (IM), pp. 1202–1207 (2015)
20. Xu, W., Zhou, H., Cheng, N., Lyu, F., Shi, W., Chen, J., Shen, X.: Internet of vehicles in big data era. IEEE/CAA J. Autom. Sin. **5**(1), 19–35 (2018)
21. Yuan, Q., Zhou, H., Li, J., Liu, Z., Yang, F., Shen, X.S.: Toward efficient content delivery for automated driving services: an edge computing solution. IEEE Netw. **32**(1), 80–86 (2018)
22. Zadeh, L.A., Kacprzyk, J.: Fuzzy Logic for the Management of Uncertainty. Wiley, New York (1992)
23. Zimmermann, H.J.: Fuzzy control. In: Fuzzy Set Theory and Its Applications, pp. 203–240. Springer (1996)

Performance Evaluation of WMNs Using WMN-PSOHC-DGA Considering Evolution Steps and Computation Time

Admir Barolli[1], Shinji Sakamoto[2(✉)], Seiji Ohara[3], Leonard Barolli[4], and Makoto Takizawa[5]

[1] Department of Information Technology, Aleksander Moisiu University of Durres, L.1, Rruga e Currilave, Durres, Albania
admir.barolli@gmail.com
[2] Department of Computer and Information Science, Seikei University, 3-3-1 Kichijoji-Kitamachi, Musashino-shi, Tokyo 180-8633, Japan
shinji.sakamoto@ieee.org
[3] Graduate School of Engineering, Fukuoka Institute of Technology, 3-30-1 Wajiro-Higashi, Higashi-Ku, Fukuoka 811-0295, Japan
seiji.ohara.19@gmail.com
[4] Department of Information and Communication Engineering, Fukuoka Institute of Technology, 3-30-1 Wajiro-Higashi, Higashi-Ku, Fukuoka 811-0295, Japan
barolli@fit.ac.jp
[5] Department of Advanced Sciences, Faculty of Science and Engineering, Hosei University, Kajino-Machi, Koganei-Shi, Tokyo 184-8584, Japan
makoto.takizawa@computer.org

Abstract. The Wireless Mesh Networks (WMNs) are becoming an important networking infrastructure because they have many advantages such as low cost and increased high speed wireless Internet connectivity. In our previous work, we implemented a Particle Swarm Optimization (PSO) and Hill Climbing (HC) based hybrid simulation system, called WMN-PSOHC, and a simulation system based on Genetic Algorithm (GA), called WMN-GA. Then, we implemented a hybrid simulation system based on PSOHC and distributed GA (DGA), called WMN-PSOHC-DGA. In this paper, we analyze the performance of WMNs using WMN-PSOHC-DGA simulation system considering the number of evolution steps and computation time. Simulation results show that the 8 evolution steps were enough and computation time was linearly increased with increasing of the evolution steps.

1 Introduction

The wireless networks and devices are becoming increasingly popular and they provide users access to information and communication anytime and anywhere [1,3,11,12,14,18,21,22]. Wireless Mesh Networks (WMNs) are gaining a lot of attention because of their low cost nature that makes them attractive for providing wireless Internet connectivity. A WMN is dynamically self-organized and self-configured, with the nodes in the network automatically establishing and maintaining mesh connectivity among them-selves (creating, in effect, an ad hoc network). This feature brings many advantages to WMNs such as low up-front cost, easy network maintenance, robustness and reliable service coverage [2]. Moreover, such infrastructure can be used to deploy community networks, metropolitan area networks, municipal and corporative networks, and to support applications for urban areas, medical, transport and surveillance systems.

Mesh node placement in WMN can be seen as a family of problems, which are shown (through graph theoretic approaches or placement problems, e.g. [9,16]) to be computationally hard to solve for most of the formulations [34].

We consider the version of the mesh router nodes placement problem in which we are given a grid area where to deploy a number of mesh router nodes and a number of mesh client nodes of fixed positions (of an arbitrary distribution) in the grid area. The objective is to find a location assignment for the mesh routers to the cells of the grid area that maximizes the network connectivity and client coverage. Network connectivity is measured by Size of Giant Component (SGC) of the resulting WMN graph, while the user coverage is simply the number of mesh client nodes that fall within the radio coverage of at least one mesh router node and is measured by Number of Covered Mesh Clients (NCMC). Node placement problems are known to be computationally hard to solve [13,35]. In previous works, some intelligent algorithms have been investigated for node placement problem [4,10,17,19,20,26,27,36].

In [25], we implemented a Particle Swarm Optimization (PSO) and Hill Climbing (HC) based simulation system, called WMN-PSOHC. Also, we implemented another simulation system based on Genetic Algorithm (GA), called WMN-GA [15], for solving node placement problem in WMNs. Then, we designed a Hybrid Intelligent System Based on PSO, HC and distributed GA (DGA), called WMN-PSOHC-DGA [23].

In this paper, we evaluate the performance of WMNs using WMN-PSOHC-DGA simulation system considering the number of evolution steps and computation time.

The rest of the paper is organized as follows. We present our designed and implemented hybrid simulation system in Sect. 2. The simulation results are given in Sect. 3. Finally, we give conclusions and future work in Sect. 4.

2   Proposed and Implemented Simulation System

2.1   WMN-PSOHC-DGA Hybrid Simulation System

Distributed Genetic Algorithm (DGA) has been focused from various fields of science. DGA has shown their usefulness for the resolution of many computationally hard combinatorial optimization problems. Also, Particle Swarm Optimization (PSO) has been investigated for solving NP-hard problem.

Velocities and Positions of Particles
WMN-PSOHC-DGA decide the velocity of particles by a random process considering the area size. For instance, when the area size is $W \times H$, the velocity

is decided randomly from $-\sqrt{W^2 + H^2}$ to $\sqrt{W^2 + H^2}$. Each particle's velocities are updated by simple rule.

For HC mechanism, next positions of each particle are used for neighbor solution s'. The fitness function f gives points to the current solution s. If $f(s')$ is better than $f(s)$, the s is updated to s'. However, if $f(s')$ is not better than $f(s)$, the s is not updated. It should be noted that the positions are not updated but the velocities are updated even if the $f(s)$ is better than $f(s')$.

Routers Replacement Methods

A mesh router has x, y positions and velocity. Mesh routers are moved based on velocities. There are many router replacement methods [24]. There are many moving methods in PSO field, such as:

Constriction Method (CM)
   CM is a method which PSO parameters are set to a week stable region ($\omega = 0.729$, $C_1 = C2 = 1.4955$) based on analysis of PSO by Clerc et al. [5,8,30].
Random Inertia Weight Method (RIWM)
   In RIWM, the ω parameter is changing randomly from 0.5 to 1.0. The C_1 and C_2 are kept 2.0. The ω can be estimated by the week stable region. The average of ω is 0.75 [7,28,32].
Linearly Decreasing Inertia Weight Method (LDIWM)
   In LDIWM, C_1 and C_2 are set to 2.0, constantly. On the other hand, the ω parameter is changed linearly from unstable region ($\omega = 0.9$) to stable region ($\omega = 0.4$) with increasing of iterations of computations [6,33].
Linearly Decreasing Vmax Method (LDVM)
   In LDVM, PSO parameters are set to unstable region ($\omega = 0.9$, $C_1 = C_2 = 2.0$). A value of V_{max} which is maximum velocity of particles is considered. With increasing of iteration of computations, the V_{max} is kept decreasing linearly [7,29,31].

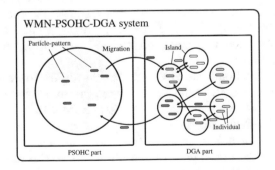

Fig. 1. Model of WMN-PSOHC-DGA migration.

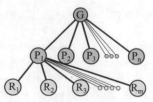

G: Global Solution
P: Particle-pattern
R: Mesh Router
n: Number of Particle-patterns
m: Number of Mesh Routers

Fig. 2. Relationship among global solution, particle-patterns and mesh routers in PSOHC part.

DGA Operations

Population of individuals: Unlike local search techniques that construct a path in the solution space jumping from one solution to another one through local perturbations, DGA use a population of individuals giving thus the search a larger scope and chances to find better solutions. This feature is also known as "exploration" process in difference to "exploitation" process of local search methods.

Selection: The selection of individuals to be crossed is another important aspect in DGA as it impacts on the convergence of the algorithm. Several selection schemes have been proposed in the literature for selection operators trying to cope with premature convergence of DGA. There are many selection methods in GA. In our system, we implement 2 selection methods: Random method and Roulette wheel method.

Crossover operators: Use of crossover operators is one of the most important characteristics. Crossover operator is the means of DGA to transmit best genetic features of parents to offsprings during generations of the evolution process. Many methods for crossover operators have been proposed such as Blend Crossover (BLX-α), Unimodal Normal Distribution Crossover (UNDX), Simplex Crossover (SPX).

Mutation operators: These operators intend to improve the individuals of a population by small local perturbations. They aim to provide a component of randomness in the neighborhood of the individuals of the population. In our system, we implemented two mutation methods: uniformly random mutation and boundary mutation.

Escaping from local optimal: GA itself has the ability to avoid falling prematurely into local optimal and can eventually escape from them during the search process. DGA has one more mechanism to escape from local optimal by considering some islands. Each island computes GA for optimizing and they migrate its gene to provide the ability to avoid from local optimal.

Convergence: The convergence of the algorithm is the mechanism of DGA to reach to good solutions. A premature convergence of the algorithm would cause that all individuals of the population be similar in their genetic features and thus the search would result ineffective and the algorithm getting stuck into local optimal. Maintaining the diversity of the population is therefore very important to this family of evolutionary algorithms.

In following, we present our proposed and implemented simulation system called WMN-PSOHC-DGA. We show the fitness function, migration function, particle-pattern, gene coding and client distributions.

Fitness Function

The determination of an appropriate fitness function, together with the chromosome encoding are crucial to the performance. Therefore, one of most important thing is to decide the determination of an appropriate objective function and its encoding. In our case, each particle-pattern and gene has an own fitness value which is comparable and compares it with other fitness value in order to share information of global solution. The fitness function follows a hierarchical approach in which the main objective is to maximize the SGC in WMN. Thus, the fitness function of this scenario is defined as

$$\text{Fitness} = 0.7 \times \text{SGC}(\boldsymbol{x}_{ij}, \boldsymbol{y}_{ij}) + 0.3 \times \text{NCMC}(\boldsymbol{x}_{ij}, \boldsymbol{y}_{ij}).$$

Migration Function

Our implemented simulation system uses Migration function as shown in Fig. 1. The Migration function swaps solutions between PSOHC part and DGA part.

Particle-Pattern and Gene Coding

In order to swap solutions, we design particle-patterns and gene coding carefully. A particle is a mesh router. Each particle has position in the considered area and velocities. A fitness value of a particle-pattern is computed by combination of mesh routers and mesh clients positions. In other words, each particle-pattern is a solution as shown is Fig. 2.

Table 1. WMN-PSOHC-DGA parameters.

Parameters	Values
Clients distribution	Normal distribution
Area size	32.0×32.0
Number of mesh routers	16
Number of mesh clients	48
Number of migrations	200
Evolution steps	From 1 to 32
Number of GA islands	16
Radius of a mesh router	2.0–2.5
Selection method	Roulette wheel method
Crossover method	SPX
Mutation method	Boundary mutation
Crossover rate	0.8
Mutation rate	0.2
Replacement method	LDVM

A gene describes a WMN. Each individual has its own combination of mesh nodes. In other words, each individual has a fitness value. Therefore, the combination of mesh nodes is a solution.

3   Simulation Results

In this section, we show simulation results using WMN-PSOHC-DGA system. In this work, we analyze the performance of WMNs considering the number of evolution steps and computation time. The number of mesh routers is considered 16 and the number of mesh clients 48. We conducted simulations 10 times, in order to avoid the effect of randomness and create a general view of results. We show the parameter setting for WMN-PSOHC-DGA in Table 1.

We show simulation results in Figs. 3 and 4. We see that for the SGC, solutions converge and reach maximum (100%) even if the number of evolution steps is 1. On the other hand, for the NCMC, WMN-PSOHC-DGA can reach

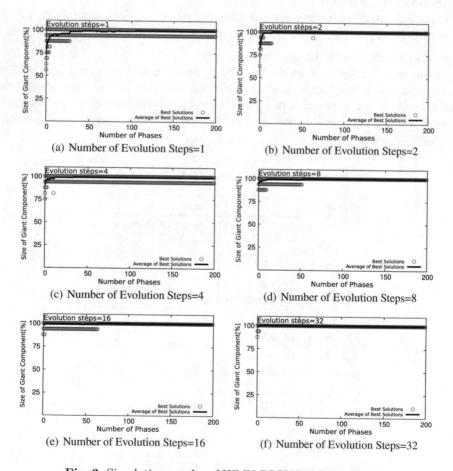

(a) Number of Evolution Steps=1

(b) Number of Evolution Steps=2

(c) Number of Evolution Steps=4

(d) Number of Evolution Steps=8

(e) Number of Evolution Steps=16

(f) Number of Evolution Steps=32

Fig. 3. Simulation results of WMN-PSOHC-DGA for SGC.

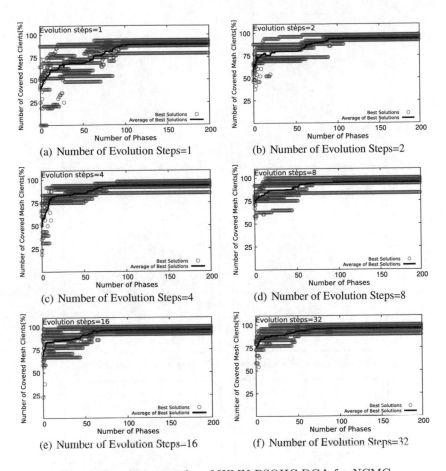

Fig. 4. Simulation results of WMN-PSOHC-DGA for NCMC.

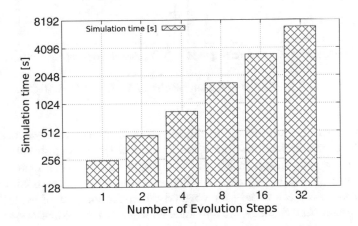

Fig. 5. Computation time.

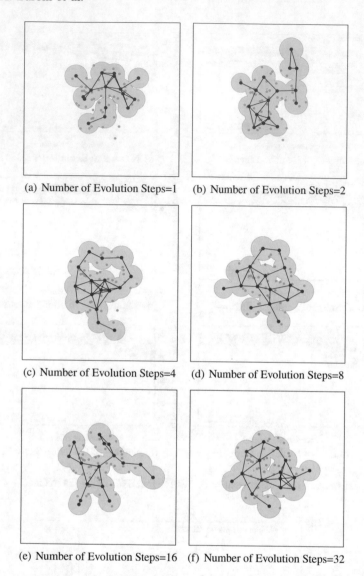

(a) Number of Evolution Steps=1   (b) Number of Evolution Steps=2

(c) Number of Evolution Steps=4   (d) Number of Evolution Steps=8

(e) Number of Evolution Steps=16   (f) Number of Evolution Steps=32

Fig. 6. Visualized simulation results of WMN-PSOHC-DGA.

maximum when the number of evolution steps is more or equal than 8 (see Fig. 4). The computation time increases linearly with the increment of number of evolution steps as shown in Fig. 5. We show the visualized image of the simulation results in Fig. 6. We can see that all mesh routers are connected with each other and all mesh clients are covered when the number of evolution steps is more or equal than 8. Thus, we conclude that the 8 evolution steps are enough for this considered scenario.

4   Conclusions

In this work, we evaluated the performance of WMNs using a hybrid simulation system based on PSOHC and DGA (called WMN-PSOHC-DGA) considering the number of evolution steps and computation time. Simulation results show that the 8 evolution steps were enough and computaion time was linearly increased by increasing the evolution steps.

In our future work, we would like to evaluate the performance of the proposed system for different parameters and patterns. Moreover, we would like to compare with other algorithms based systems.

References

1. Ahmed, S., Khan, M.A., Ishtiaq, A., Khan, Z.A., Ali, M.T.: Energy harvesting techniques for routing issues in wireless sensor networks. Int. J. Grid Util. Comput. **10**(1), 10–21 (2019)
2. Akyildiz, I.F., Wang, X., Wang, W.: Wireless mesh networks: a survey. Comput. Netw. **47**(4), 445–487 (2005)
3. Barolli, A., Sakamoto, S., Barolli, L., Takizawa, M.: A hybrid simulation system based on particle swarm optimization and distributed genetic algorithm for WMNs: performance evaluation considering normal and uniform distribution of mesh clients. In: International Conference on Network-Based Information Systems, pp. 42–55. Springer (2018)
4. Barolli, A., Sakamoto, S., Ozera, K., Barolli, L., Kulla, E., Takizawa, M.: Design and implementation of a hybrid intelligent system based on particle swarm optimization and distributed genetic algorithm. In: International Conference on Emerging Internetworking, Data & Web Technologies, pp. 79–93. Springer (2018)
5. Barolli, A., Sakamoto, S., Durresi, H., Ohara, S., Barolli, L., Takizawa, M.: A comparison study of constriction and linearly decreasing Vmax replacement methods for wireless mesh networks by WMN-PSOHC-DGA simulation system. In: International Conference on P2P, Parallel, Grid, Cloud and Internet Computing, pp. 26–34. Springer (2019)
6. Barolli, A., Sakamoto, S., Ohara, S., Barolli, L., Takizawa, M.: Performance analysis of WMNs by WMN-PSOHC-DGA simulation system considering linearly decreasing inertia weight and linearly decreasing Vmax replacement methods. In: International Conference on Intelligent Networking and Collaborative Systems, pp. 14–23. Springer (2019)
7. Barolli, A., Sakamoto, S., Ohara, S., Barolli, L., Takizawa, M.: Performance analysis of WMNs by WMN-PSOHC-DGA simulation system considering random inertia weight and linearly decreasing Vmax router replacement methods. In: Conference on Complex, Intelligent, and Software Intensive Systems, pp. 13–21. Springer (2019)
8. Clerc, M., Kennedy, J.: The particle swarm-explosion, stability, and convergence in a multidimensional complex space. IEEE Trans. Evol. Comput. **6**(1), 58–73 (2002)
9. Franklin, A.A., Murthy, C.S.R.: Node placement algorithm for deployment of two-tier wireless mesh networks. In: Proceedings of Global Telecommunications Conference, pp. 4823–4827 (2007)

10. Girgis, M.R., Mahmoud, T.M., Abdullatif, B.A., Rabie, A.M.: Solving the wireless mesh network design problem using genetic algorithm and simulated annealing optimization methods. Int. J. Comput. Appl. **96**(11), 1–10 (2014)
11. Gorrepotu, R., Korivi, N.S., Chandu, K., Deb, S.: Sub-1GHz miniature wireless sensor node for IoT applications. Internet Things **1**, 27–39 (2018)
12. Islam, M.M., Funabiki, N., Sudibyo, R.W., Munene, K.I., Kao, W.C.: A dynamic access-point transmission power minimization method using PI feedback control in elastic WLAN system for IoT applications. Internet Things **8**(100), 089 (2019)
13. Maolin, T., et al.: Gateways placement in backbone wireless mesh networks. Int. J. Commun. Netw. Syst. Sci. **2**(1), 44 (2009)
14. Marques, B., Coelho, I.M., Sena, A.D.C., Castro, M.C.: A network coding protocol for wireless sensor fog computing. Int. J. Grid Util. Comput. **10**(3), 224–234 (2019)
15. Matsuo, K., Sakamoto, S., Oda, T., Barolli, A., Ikeda, M., Barolli, L.: Performance analysis of WMNs by WMN-GA simulation system for two WMN architectures and different TCP congestion-avoidance algorithms and client distributions. Int. J. Commun. Netw. Distrib. Syst. **20**(3), 335–351 (2018)
16. Muthaiah, S.N., Rosenberg, C.P.: Single gateway placement in wireless mesh networks. In: Proceedings of 8th International IEEE Symposium on Computer Networks, pp. 4754–4759 (2008)
17. Naka, S., Genji, T., Yura, T., Fukuyama, Y.: A hybrid particle swarm optimization for distribution state estimation. IEEE Trans. Power Syst. **18**(1), 60–68 (2003)
18. Ohara, S., Barolli, A., Sakamoto, S., Barolli, L.: Performance analysis of WMNs by WMN-PSODGA simulation system considering load balancing and client uniform distribution. In: International Conference on Innovative Mobile and Internet Services in Ubiquitous Computing, pp. 25–38. Springer (2019)
19. Ozera, K., Sakamoto, S., Elmazi, D., Bylykbashi, K., Ikeda, M., Barolli, L.: A fuzzy approach for clustering in MANETs: performance evaluation for different parameters. Int. J. Space Based Situated Comput. **7**(3), 166–176 (2017)
20. Ozera, K., Bylykbashi, K., Liu, Y., Barolli, L.: A fuzzy-based approach for cluster management in VANETs: performance evaluation for two fuzzy-based systems. Internet Things **3**, 120–133 (2018)
21. Ozera, K., Inaba, T., Bylykbashi, K., Sakamoto, S., Ikeda, M., Barolli, L.: A WLAN triage testbed based on fuzzy logic and its performance evaluation for different number of clients and throughput parameter. Int. J. Grid Util. Comput. **10**(2), 168–178 (2019)
22. Petrakis, E.G., Sotiriadis, S., Soultanopoulos, T., Renta, P.T., Buyya, R., Bessis, N.: Internet of Things as a Service (iTaaS): challenges and solutions for management of sensor data on the cloud and the fog. Internet Things **3**, 156–174 (2018)
23. Sakamoto, S., Barolli, A., Barolli, L., Takizawa, M.: Design and implementation of a hybrid intelligent system based on particle swarm optimization, hill climbing and distributed genetic algorithm for node placement problem in WMNs: a comparison study. In: The 32nd IEEE International Conference on Advanced Information Networking and Applications (AINA 2018), pp. 678–685. IEEE (2018)
24. Sakamoto, S., Ozera, K., Barolli, A., Barolli, L., Kolici, V., Takizawa, M.: Performance evaluation of WMN-PSOSA considering four different replacement methods. In: International Conference on Emerging Internetworking, Data & Web Technologies, pp. 51–64. Springer (2018)
25. Sakamoto, S., Ozera, K., Ikeda, M., Barolli, L.: Implementation of intelligent hybrid systems for node placement problem in WMNs considering particle swarm optimization, hill climbing and simulated annealing. Mob. Netw. Appl. **23**(1), 27–33 (2018)

26. Sakamoto, S., Barolli, A., Barolli, L., Okamoto, S.: Implementation of a web interface for hybrid intelligent systems. Int. J. Web Inf. Syst. **15**(4), 420–431 (2019)
27. Sakamoto, S., Barolli, L., Okamoto, S.: WMN-PSOSA: an intelligent hybrid simulation system for WMNs and its performance evaluations. Int. J. Web Grid Serv. **15**(4), 353–366 (2019)
28. Sakamoto, S., Ohara, S., Barolli, L., Okamoto, S.: Performance evaluation of WMNs by WMN-PSOHC system considering random inertia weight and linearly decreasing inertia weight replacement methods. In: International Conference on Innovative Mobile and Internet Services in Ubiquitous Computing, pp. 39–48. Springer (2019)
29. Sakamoto, S., Ohara, S., Barolli, L., Okamoto, S.: Performance evaluation of WMNs by WMN-PSOHC system considering random inertia weight and linearly decreasing Vmax replacement methods. In: International Conference on Network-Based Information Systems, pp. 27–36. Springer (2019)
30. Sakamoto, S., Ohara, S., Barolli, L., Okamoto, S.: Performance evaluation of WMNs WMN-PSOHC system considering constriction and linearly decreasing inertia weight replacement methods. In: International Conference on Broadband and Wireless Computing, Communication and Applications, pp. 22–31. Springer (2019)
31. Schutte, J.F., Groenwold, A.A.: A study of global optimization using particle swarms. J. Glob. Optim. **31**(1), 93–108 (2005)
32. Shi, Y.: Particle swarm optimization. IEEE Connect. **2**(1), 8–13 (2004)
33. Shi, Y., Eberhart, R.C.: Parameter selection in particle swarm optimization. In: Evolutionary Programming VII, pp. 591–600 (1998)
34. Vanhatupa, T., Hannikainen, M., Hamalainen, T.: Genetic algorithm to optimize node placement and configuration for WLAN planning. In: The 4th IEEE International Symposium on Wireless Communication Systems, pp. 612–616 (2007)
35. Wang, J., Xie, B., Cai, K., Agrawal, D.P.: Efficient mesh router placement in wireless mesh networks. In: Proceedings of IEEE International Conference on Mobile Adhoc and Sensor Systems (MASS 2007), pp. 1–9 (2007)
36. Yaghoobirafi, K., Nazemi, E.: An autonomic mechanism based on ant colony pattern for detecting the source of incidents in complex enterprise systems. Int. J. Grid Util. Comput. **10**(5), 497–511 (2019)

Performance Evaluation of WMNs Using an Hybrid Intelligent System Based on Particle Swarm Optimization and Hill Climbing Considering Different Number of Iterations

Shinji Sakamoto[1(✉)], Seiji Ohara[2], Leonard Barolli[3], and Shusuke Okamoto[1]

[1] Department of Computer and Information Science, Seikei University,
3-3-1 Kichijoji-Kitamachi, Musashino-shi, Tokyo 180-8633, Japan
shinji.sakamoto@ieee.org, okam@st.seikei.ac.jp
[2] Graduate School of Engineering, Fukuoka Institute of Technology,
3-30-1 Wajiro-Higashi, Higashi-Ku, Fukuoka 811-0295, Japan
seiji.ohara.19@gmail.com
[3] Department of Information and Communication Engineering, Fukuoka Institute of Technology, 3-30-1 Wajiro-Higashi, Higashi-Ku, Fukuoka 811-0295, Japan
barolli@fit.ac.jp

Abstract. In our previous work, we implemented a Particle Swarm Optimization (PSO) based simulation system for node placement in WMNs, called WMN-PSO. Also, we implemented a simulation system based on Hill Climbing (HC), called WMN-HC. Then, we implemented a hybrid simulation system based on PSO and HC, called WMN-PSOHC. In this paper, we evaluate the performance of WMNs by using WMN-PSOHC considering different number of iterations. Simulation results show that, the simulation time increases with increasing the number of iterations. When the number of iterations increase twice, the simulation time increases more than twice. Thus, we conclude that the calculation time and quality of solution is a trade-off relation. In this considered scenario, 400 iterations are enough.

1 Introduction

The wireless networks and devices are becoming increasingly popular and they provide users access to information and communication anytime and anywhere [1,3,10,11,13,15,17,18]. Wireless Mesh Networks (WMNs) are gaining a lot of attention because of their low cost nature that makes them attractive for providing wireless Internet connectivity. A WMN is dynamically self-organized and self-configured, with the nodes in the network automatically establishing and maintaining mesh connectivity among them-selves (creating, in effect, an ad hoc network). This feature brings many advantages to WMNs such as low up-front cost, easy network maintenance, robustness and reliable service coverage [2]. Moreover, such infrastructure can be used to deploy community networks,

© Springer Nature Switzerland AG 2020
L. Barolli et al. (Eds.): EIDWT 2020, LNDECT 47, pp. 138–149, 2020.
https://doi.org/10.1007/978-3-030-39746-3_15

metropolitan area networks, municipal and corporative networks, and to support applications for urban areas, medical, transport and surveillance systems.

We consider the version of the mesh router nodes placement problem in which we are given a grid area where to deploy a number of mesh router nodes and a number of mesh client nodes of fixed positions (of an arbitrary distribution) in the grid area. The objective is to find a location assignment for the mesh routers to the cells of the grid area that maximizes the network connectivity and client coverage. Network connectivity is measured by Size of Giant Component (SGC) of the resulting WMN graph, while the user coverage is simply the number of mesh client nodes that fall within the radio coverage of at least one mesh router node and is measured by Number of Covered Mesh Clients (NCMC). Node placement problems are known to be computationally hard to solve [12, 28]. In some previous works, intelligent algorithms have been recently investigated [9, 14, 16, 22, 23, 30].

In our previous work [20, 21], we implemented a hybrid simulation system based on PSO and HC. We called this system WMN-PSOHC. In this paper, we analyze the performance of hybrid WMN-PSOHC system considering the number of iterations for an entire simulation.

The rest of the paper is organized as follows. We present our designed and implemented hybrid simulation system in Sect. 2. We introduce the WMN-PSOHC Web GUI tool in Sect. 3. The simulation results are given in Sect. 4. Finally, we give conclusions and future work in Sect. 5.

For this problem, we have a grid area arranged in cells we want to find where to distribute a number of mesh router nodes and a number of mesh client nodes of fixed positions (of an arbitrary distribution) in the considered area. The objective is to find a location assignment for the mesh routers to the area that maximizes the network connectivity and client coverage.

2   Proposed and Implemented Simulation System

2.1   Particle Swarm Optimization

In Particle Swarm Optimization (PSO) algorithm, a number of simple entities (the particles) are placed in the search space of some problem or function and each evaluates the objective function at its current location. The objective function is often minimized and the exploration of the search space is not through evolution [19]. However, following a widespread practice of borrowing from the evolutionary computation field, in this work, we consider the bi-objective function and fitness function interchangeably. Each particle then determines its movement through the search space by combining some aspect of the history of its own current and best (best-fitness) locations with those of one or more members of the swarm, with some random perturbations. The next iteration takes place after all particles have been moved. Eventually the swarm as a whole, like a flock of birds collectively foraging for food, is likely to move close to an optimum of the fitness function.

Each individual in the particle swarm is composed of three $\mathcal{D}$-dimensional vectors, where $\mathcal{D}$ is the dimensionality of the search space. These are the current position $\vec{x}_i$, the previous best position $\vec{p}_i$ and the velocity $\vec{v}_i$.

The particle swarm is more than just a collection of particles. A particle by itself has almost no power to solve any problem; progress occurs only when the particles interact. Problem solving is a population-wide phenomenon, emerging from the individual behaviors of the particles through their interactions. In any case, populations are organized according to some sort of communication structure or topology, often thought of as a social network. The topology typically consists of bidirectional edges connecting pairs of particles, so that if j is in i's neighborhood, i is also in j's. Each particle communicates with some other particles and is affected by the best point found by any member of its topological neighborhood. This is just the vector $\vec{p}_i$ for that best neighbor, which we will denote with $\vec{p}_g$. The potential kinds of population "social networks" are hugely varied, but in practice certain types have been used more frequently.

In the PSO process, the velocity of each particle is iteratively adjusted so that the particle stochastically oscillates around $\vec{p}_i$ and $\vec{p}_g$ locations.

2.2   Hill Climbing

Hill Climbing (HC) algorithm is a heuristic algorithm. The idea of HC is simple. In HC, the solution s' is accepted as the new current solution if $\delta \leq 0$ holds, where $\delta = f(s') - f(s)$. Here, the function f is called the fitness function. The fitness function gives points to a solution so that the system can evaluate the next solution s' and the current solution s.

The most important factor in HC is to define the neighbor solution, effectively. The definition of the neighbor solution affects HC performance directly. In our WMN-PSOHC system, we use the next step of particle-pattern positions as the neighbor solutions for the HC part.

2.3   WMN-PSOHC System Description

In following, we present the initialization, particle-pattern, fitness function and router replacement methods.

Initialization
Our proposed system starts by generating an initial solution randomly, by *ad hoc* methods [29]. We decide the velocity of particles by a random process considering the area size. For instance, when the area size is $W \times H$, the velocity is decided randomly from $-\sqrt{W^2 + H^2}$ to $\sqrt{W^2 + H^2}$.

Particle-Pattern
A particle is a mesh router. A fitness value of a particle-pattern is computed by combination of mesh routers and mesh clients positions. In other words, each particle-pattern is a solution as shown is Fig. 1. Therefore, the number of particle-patterns is a number of solutions.

Fitness Function
One of most important thing is to decide the determination of an appropriate objective function and its encoding. In our case, each particle-pattern has an own fitness value and compares other particle-patterns fitness value in order to share information of global solution. The fitness function follows a hierarchical approach in which the main objective is to maximize the SGC in WMN. Thus, we use α and β weight-coefficients for the fitness function and the fitness function of this scenario is defined as:

$$\text{Fitness} = \alpha \times \text{SGC}(\boldsymbol{x}_{ij}, \boldsymbol{y}_{ij}) + \beta \times \text{NCMC}(\boldsymbol{x}_{ij}, \boldsymbol{y}_{ij}).$$

Router Replacement Methods
A mesh router has x, y positions and velocity. Mesh routers are moved based on velocities. There are many router replacement methods in PSO field. In this paper, we use LDIWM to replace mesh router nodes.

Constriction Method (CM)
   CM is a method which PSO parameters are set to a week stable region ($\omega = 0.729$, $C_1 = C2 = 1.4955$) based on analysis of PSO by Clerc et al. [4,5,8].
Random Inertia Weight Method (RIWM)
   In RIWM, the ω parameter is changing randomly from 0.5 to 1.0. The C_1 and C_2 are kept 2.0. The ω can be estimated by the week stable region. The average of ω is 0.75 [24,26].
Linearly Decreasing Inertia Weight Method (LDIWM)
   In LDIWM, C_1 and C_2 are set to 2.0, constantly. On the other hand, the ω parameter is changed linearly from unstable region ($\omega = 0.9$) to stable region ($\omega = 0.4$) with increasing of iterations of computations [6,27].
Linearly Decreasing Vmax Method (LDVM)
   In LDVM, PSO parameters are set to unstable region ($\omega = 0.9$, $C_1 = C_2 = 2.0$). A value of V_{max} which is maximum velocity of particles is considered. With increasing of iteration of computations, the V_{max} is kept decreasing linearly [7,25].

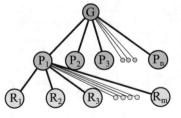

G: Global Solution
P: Particle-pattern
R: Mesh Router
n: Number of Particle-patterns
m: Number of Mesh Routers

Fig. 1. Relationship among global solution, particle-patterns and mesh routers.

Rational Decrement of Vmax Method (RDVM)

In RDVM, PSO parameters are set to unstable region ($\omega = 0.9$, $C_1 = C_2 = 2.0$). The V_{max} is kept decreasing with the increasing of iterations as

$$V_{max}(x) = \sqrt{W^2 + H^2} \times \frac{T - x}{x}.$$

Where, W and H are the width and the height of the considered area, respectively. Also, T and x are the total number of iterations and a current number of iteration, respectively.

3   WMN-PSOHC Web GUI Tool

The Web application follows a standard Client-Server architecture and is implemented using LAMP (Linux + Apache + MySQL + PHP) technology (see Fig. 2). We show the WMN-PSOHC Web GUI tool in Fig. 3. Remote users (clients) submit their requests by completing first the parameter setting. The parameter values to be provided by the user are classified into three groups, as follows.

- Parameters related to the problem instance: These include parameter values that determine a problem instance to be solved and consist of number of router nodes, number of mesh client nodes, client mesh distribution, radio coverage interval and size of the deployment area.
- Parameters of the resolution method: Each method has its own parameters.
- Execution parameters: These parameters are used for stopping condition of the resolution methods and include number of iterations and number of independent runs. The former is provided as a total number of iterations and depending on the method is also divided per phase (e.g., number of iterations in a exploration). The later is used to run the same configuration for the same problem instance and parameter configuration a certain number of times.

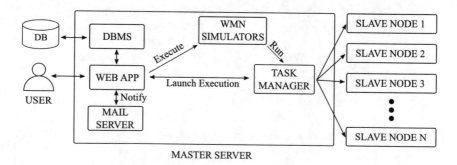

Fig. 2. System structure for web interface.

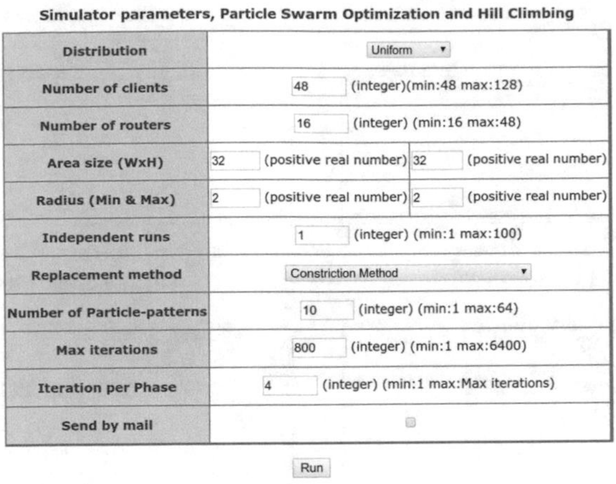

Simulator parameters, Particle Swarm Optimization and Hill Climbing

Distribution	Uniform ▾
Number of clients	48 (integer)(min:48 max:128)
Number of routers	16 (integer) (min:16 max:48)
Area size (WxH)	32 (positive real number) 32 (positive real number)
Radius (Min & Max)	2 (positive real number) 2 (positive real number)
Independent runs	1 (integer) (min:1 max:100)
Replacement method	Constriction Method ▾
Number of Particle-patterns	10 (integer) (min:1 max:64)
Max iterations	800 (integer) (min:1 max:6400)
Iteration per Phase	4 (integer) (min:1 max:Max iterations)
Send by mail	☐

Run

Fig. 3. WMN-PSOHC Web GUI tool.

4   Simulation Results

In this section, we show simulation results using WMN-PSOHC system. In this scenario, we consider Normal distributions of mesh clients. The number of mesh routers is considered 16 and the number of mesh clients 48. The total number of iterations is considered 200, 400, 800, 1600 and 3200, also the iterations per phase is considered 1, 2, 4, 8 and 16, respectively. We consider the number of particle-patterns 9. We conducted simulations 10 times, in order to avoid the effect of randomness and create a general view of results. We show the parameter setting for WMN-PSOHC in Table 1.

We show the simulation results in Figs. 4 and 5. For SGC, we see that the SGC reaches the maximum (100%) even if the number of iterations is 200 and the

Table 1. Parameter settings.

Parameters	Values
Clients distribution	Normal distribution
Area size	32.0×32.0
Number of mesh routers	16
Number of mesh clients	48
Total iterations	200, 400, 800, 1600, 3200
Iteration per phase	1, 2, 4, 8, 16
Number of particle-patterns	9
Radius of a mesh router	2.0
Fitness function weight-coefficients (α, β)	0.7, 0.3
Replacement method	LDIWM

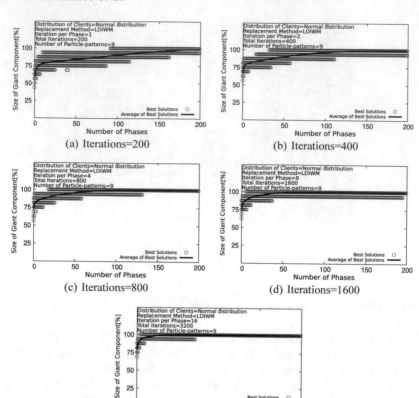

Fig. 4. Simulation results of WMN-PSOHC for SGC.

number of iterations per phase is 1. Also, we see that the convergence becomes faster in the entire simulation with increasing the number of iterations. For the NCMC, we can see that the NCMC reaches the maximum (100%) when the number of iteration is more or equal than 400 in this scenario. As shown in Fig. 6, we see that the simulation time increases with increasing the number of iterations. When the number of iterations increase twice, the simulation time increases more than twice. Thus, we conclude that the calculation time and quality of solution is a trade-off relation. In this scenario, 400 iterations are enough. We show the visualized results in Fig. 7. We see that all mesh router nodes are connected and all mesh client nodes are covered except the case of iteration is 200.

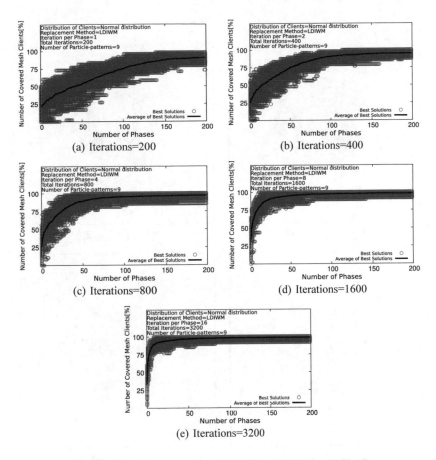

Fig. 5. Simulation results of WMN-PSOHC for NCMC.

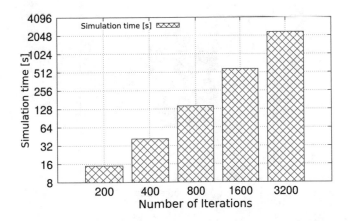

Fig. 6. Simulation time for different number of iterations.

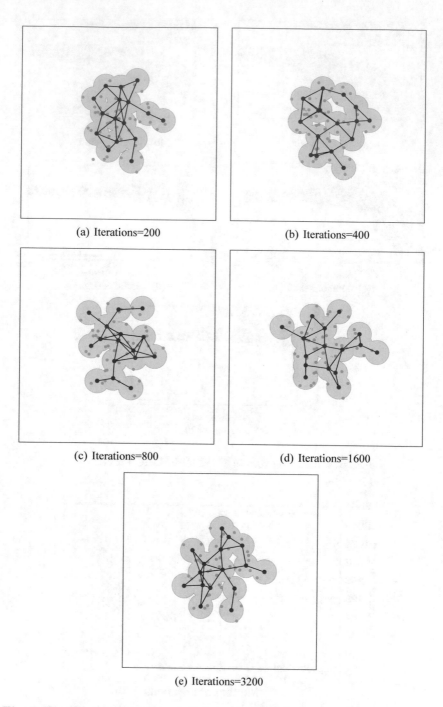

(a) Iterations=200

(b) Iterations=400

(c) Iterations=800

(d) Iterations=1600

(e) Iterations=3200

Fig. 7. Visualized image of simulation results for different number of iterations.

5   Conclusions

In this work, we evaluated the performance of a hybrid simulation system based on PSO and HC (called WMN-PSOHC) considering the number of iterations. Simulation results show that, the simulation time increases with increasing the number of iterations. When the number of iterations increase twice, the simulation time increases more than twice. Thus, we conclude that the calculation time and quality of solution is a trade-off relation. In this scenario, 400 iterations were enough.

In our future work, we would like to evaluate the performance of the proposed system for different parameters and scenarios.

References

1. Ahmed, S., Khan, M.A., Ishtiaq, A., Khan, Z.A., Ali, M.T.: Energy harvesting techniques for routing issues in wireless sensor networks. Int. J. Grid Util. Comput. **10**(1), 10–21 (2019)
2. Akyildiz, I.F., Wang, X., Wang, W.: Wireless mesh networks: a survey. Comput. Netw. **47**(4), 445–487 (2005)
3. Barolli, A., Sakamoto, S., Barolli, L., Takizawa, M.: A hybrid simulation system based on particle swarm optimization and distributed genetic algorithm for WMNs: performance evaluation considering normal and uniform distribution of mesh clients. In: International Conference on Network-Based Information Systems, pp. 42–55. Springer (2018)
4. Barolli, A., Sakamoto, S., Barolli, L., Takizawa, M.: Performance evaluation of WMN-PSODGA system for node placement problem in WMNs considering four different crossover methods. In: The 32nd IEEE International Conference on Advanced Information Networking and Applications (AINA 2018), pp. 850–857. IEEE (2018)
5. Barolli, A., Sakamoto, S., Durresi, H., Ohara, S., Barolli, L., Takizawa, M.: A comparison study of constriction and linearly decreasing Vmax replacement methods for wireless mesh networks by WMN-PSOHC-DGA simulation system. In: International Conference on P2P, Parallel, Grid, Cloud and Internet Computing, pp. 26–34. Springer (2019)
6. Barolli, A., Sakamoto, S., Ohara, S., Barolli, L., Takizawa, M.: Performance analysis of WMNs by WMN-PSOHC-DGA simulation system considering linearly decreasing inertia weight and linearly decreasing Vmax replacement methods. In: International Conference on Intelligent Networking and Collaborative Systems, pp. 14–23. Springer (2019)
7. Barolli, A., Sakamoto, S., Ohara, S., Barolli, L., Takizawa, M.: Performance analysis of WMNs by WMN-PSOHC-DGA simulation system considering random inertia weight and linearly decreasing Vmax router replacement methods. In: Conference on Complex, Intelligent, and Software Intensive Systems, pp. 13–21. Springer (2019)
8. Clerc, M., Kennedy, J.: The particle swarm-explosion, stability, and convergence in a multidimensional complex space. IEEE Trans. Evol. Comput. **6**(1), 58–73 (2002)
9. Girgis, M.R., Mahmoud, T.M., Abdullatif, B.A., Rabie, A.M.: Solving the wireless mesh network design problem using genetic algorithm and simulated annealing optimization methods. Int. J. Comput. Appl. **96**(11), 1–10 (2014)

10. Gorrepotu, R., Korivi, N.S., Chandu, K., Deb, S.: Sub-1GHz miniature wireless sensor node for IoT applications. Internet Things **1**, 27–39 (2018)
11. Islam, M.M., Funabiki, N., Sudibyo, R.W., Munene, K.I., Kao, W.C.: A dynamic access-point transmission power minimization method using PI feedback control in elastic WLAN system for IoT applications. Internet Things **8**, 100,089 (2019)
12. Maolin, T., et al.: Gateways placement in backbone wireless mesh networks. Int. J. Commun. Netw. Syst. Sci. **2**(1), 44 (2009)
13. Marques, B., Coelho, I.M., Sena, A.D.C., Castro, M.C.: A network coding protocol for wireless sensor fog computing. Int. J. Grid Util. Comput. **10**(3), 224–234 (2019)
14. Naka, S., Genji, T., Yura, T., Fukuyama, Y.: A hybrid particle swarm optimization for distribution state estimation. IEEE Trans. Power Syst. **18**(1), 60–68 (2003)
15. Ohara, S., Barolli, A., Sakamoto, S., Barolli, L.: Performance analysis of WMNs by WMN-PSODGA simulation system considering load balancing and client uniform distribution. In: International Conference on Innovative Mobile and Internet Services in Ubiquitous Computing, pp. 25–38. Springer (2019)
16. Ozera, K., Bylykbashi, K., Liu, Y., Barolli, L.: A fuzzy-based approach for cluster management in VANETs: performance evaluation for two fuzzy-based systems. Internet Things **3**, 120–133 (2018)
17. Ozera, K., Inaba, T., Bylykbashi, K., Sakamoto, S., Ikeda, M., Barolli, L.: A WLAN triage testbed based on fuzzy logic and its performance evaluation for different number of clients and throughput parameter. Int. J. Grid Util. Comput. **10**(2), 168–178 (2019)
18. Petrakis, E.G., Sotiriadis, S., Soultanopoulos, T., Renta, P.T., Buyya, R., Bessis, N.: Internet of Things as a Service (iTaaS): challenges and solutions for management of sensor data on the cloud and the fog. Internet Things **3**, 156–174 (2018)
19. Poli, R., Kennedy, J., Blackwell, T.: Particle swarm optimization. Swarm Intell. **1**(1), 33–57 (2007)
20. Sakamoto, S., Oda, T., Ikeda, M., Barolli, L., Xhafa, F.: Implementation and evaluation of a simulation system based on particle swarm optimisation for node placement problem in wireless mesh networks. Int. J. Commun. Netw. Distrib. Syst. **17**(1), 1–13 (2016)
21. Sakamoto, S., Ozera, K., Ikeda, M., Barolli, L.: Implementation of intelligent hybrid systems for node placement problem in WMNs considering particle swarm optimization, hill climbing and simulated annealing. Mob. Netw. Appl. **23**(1), 27–33 (2018)
22. Sakamoto, S., Barolli, A., Barolli, L., Okamoto, S.: Implementation of a web interface for hybrid intelligent systems. Int. J. Web Inf. Syst. **15**(4), 420–431 (2019)
23. Sakamoto, S., Barolli, L., Okamoto, S.: WMN-PSOSA: an intelligent hybrid simulation system for WMNs and its performance evaluations. Int. J. Web Grid Serv. **15**(4), 353–366 (2019)
24. Sakamoto, S., Ohara, S., Barolli, L., Okamoto, S.: Performance evaluation of WMNs by WMN-PSOHC system considering random inertia weight and linearly decreasing inertia weight replacement methods. In: International Conference on Innovative Mobile and Internet Services in Ubiquitous Computing, pp. 39–48. Springer (2019)
25. Schutte, J.F., Groenwold, A.A.: A study of global optimization using particle swarms. J. Global Optim. **31**(1), 93–108 (2005)
26. Shi, Y.: Particle swarm optimization. IEEE Connect. **2**(1), 8–13 (2004)
27. Shi, Y., Eberhart, R.C.: Parameter selection in particle swarm optimization. In: Porto, V.W., Saravanan, N., Waagen, D., Eiben, A.E. (eds.) Evolutionary Programming VII, pp. 591–600. Springer, Heidelberg (1998)

28. Wang, J., Xie, B., Cai, K., Agrawal, D.P.: Efficient mesh router placement in wireless mesh networks. In: Proceedings of IEEE International Conference on Mobile Adhoc and Sensor Systems (MASS 2007), pp. 1–9 (2007)
29. Xhafa, F., Sanchez, C., Barolli, L.: Ad hoc and neighborhood search methods for placement of mesh routers in wireless mesh networks. In: Proceedings of 29th IEEE International Conference on Distributed Computing Systems Workshops (ICDCS 2009), pp. 400–405 (2009)
30. Yaghoobirafi, K., Nazemi, E.: An autonomic mechanism based on ant colony pattern for detecting the source of incidents in complex enterprise systems. Int. J. Grid Util. Comput. **10**(5), 497–511 (2019)

A Software-Oriented Approach to Energy-Efficiently Unicasting Messages in Wireless Ad-Hoc Networks

Ryota Sakai[✉], Takumi Saito, Ryuji Oma, Shigenari Nakamura, Tomoya Enokido, and Makoto Takizawa

Hosei University, Tokyo, Japan
{ryota.sakai.7i,takumi.saito.3j,ryuji.oma.6r}@stu.hosei.ac.jp,
nakamura.shigenari@gmail.com, makoto.takizawa@computer.org

Abstract. Wireless ad-hoc networks are getting more important in various applications like (vehicle-to-vehicle) V2V networks and opportunistic networks. Here, it is critical to reduce the electric energy consumption of nodes in networks. Each node consumes more energy to perform software processes to send and receive messages than the wireless communication devices. The farther node p_j sends messages to a node p_i, the more number of messages the node p_i loses due to noise. Then, the node p_j retransmits a lost message to the node p_i. This means, the more number of times the node p_i transmits a message, the more energy the node p_i consumes. The energy to be consumed by a node more depends on the number of transmissions of each message than the strength of radio emission. We have to decrease the number of transmissions of a message of each node in order to reduce the energy consumption of the node. In this paper, we first define a model to give the energy to be consumed by each node to deliver a message to another node. Then, we propose two types of EEU (Energy-Efficient Unicast) protocols, EEUE and EEUP to find an energy-efficient route from a source node to a destination node.

Keywords: Wireless ad-hoc networks · Energy-Efficient Unicast (EEU) protocol · EEUE protocol · EEUP protocol

1 Introduction

Infrastructure-less networks like wireless ad-hoc networks [7] are important in various types of applications like V2V (vehicle-to-vehicle) networks and opportunistic networks [1,2]. Here, it is critical to reduce the electric energy consumption of nodes to deliver messages from a source node to a destination node in networks. In energy-efficient broadcast protocols [6,11] and unicast protocols [7,8], energy consumed by a sender node to send a message is assumed to be proportioned to a square of distance with a destination node [12].

A node consumes electric energy not only by communication devices but also communication software, i.e. processes to send and receive messages by taking usage of

© Springer Nature Switzerland AG 2020
L. Barolli et al. (Eds.): EIDWT 2020, LNDECT 47, pp. 150–157, 2020.
https://doi.org/10.1007/978-3-030-39746-3_16

using the communication devices. We implement communication processes in a Raspberry PI3 model B [10] node and measure the energy consumption of the node to send and receive messages. Through the measurement [3–5,8], we make clear each node mainly consumes energy to perform communication processes to send and receive messages compared with the strength of radio emission of the communication devices. Even if a node p_i transmits a message to a neighbor node p_j in the wireless communication range, the node p_j may fail to receive the message due to noise. Then, the node p_i retransmits the message to the node p_j. The message loss ratio monotonically, even exponentially increases as the distance between a pair of the neighboring nodes p_i and p_j gets longer. The energy consumed by a node p_i to deliver a message to a node p_j depends on how many times the node p_i transmits a message to the node p_j. Messages sent by a source node are forwarded by nodes in a route to a destination node. The total energy consumed by nodes in a route depends on the total number of transmissions of the nodes.

Based on this observation, we newly propose a model to estimate the energy consumption of nodes to deliver messages from a node to a neighbor node in wireless networks. Then, we propose a pair of EEU (Energy-Efficient Unicast) protocols, EEUE and EEUP protocols to make a route from a source node to a destination node so as to decrease the number of transmissions of nodes in the route to reduce the total energy consumption of the nodes. Here, a shortest route from a source node to a destination node is first detected in a flooding algorithm of the AODV protocol [9]. Then, a more efficient route is obtained by changing a preceding node of each node in the route by backtracking the route from the destination node. In the EEUE protocol, a route from a source node to a destination node is obtained where the total energy consumption of the nodes is minimized. That is, a route whose total number of transmissions of the nodes is minimum is obtained. In another EEUP algorithm, a route is detected so that the probability that messages are delivered to nodes is maximized.

In Sect. 2, we present a system model and propose an energy consumption model of a node. In Sect. 3, we propose the routing protocols, EEUE and EEUP protocols.

2 System Model

A network N is composed of nodes p_1, ..., p_n ($n \geq 1$) which are interconnected in wireless communication links. Let d_{ij} be distance between a pair of nodes p_i and p_j. Each node p_i sends a message m in a wireless link. We assume $d_{ij} = d_{ji}$ for every pair of nodes p_i and p_j. Here, a node p_j in the communication range of the node p_i can receive the message m. Other nodes out of the communication range cannot receive any message sent by the node p_i. Let $maxd_i$ show the largest communication range of a node p_i. If $d_{ij} \leq maxd_i$, a node p_j can receive a message sent by a node p_i. Otherwise, the node p_j can receive no message from the node p_i. A peer p_j is referred to as *first-neighbor* node of a node p_i ($p_i \leftrightarrow p_j$) iff (if and only if) $d_{ij} \leq maxd_i$. Let $FNC(p_i)$ be a set of first-neighbor nodes of a node p_i, i.e. $\{p_j | p_i \leftrightarrow p_j$, i.e. $d_{ij} \leq maxd_i\}$. In this paper, we assume the first-neighbor relation " $\leftrightarrow$ " is symmetric, i.e. $p_i \leftrightarrow p_j$ if $p_j \leftrightarrow p_i$ for every pair of nodes p_i and p_j. This means, $maxd_i = maxd_j$ for every pair of nodes p_i and p_j.

Even if a node p_j is a first-neighbor node of a node p_i ($p_j \leftrightarrow p_i$), the node p_j cannot necessarily receive a message sent by the node p_i, i.e. the node p_j may lose some message sent by the node p_i due to noise. The message loss ratio $l_{ij}(d_{ij})$ of the node p_j depends on the distance d_{ij} between a pair of the nodes p_i and p_j. The farther a node p_j is from a node p_i, the larger message loss ratio $l_{ij}(d_{ij})$, i.e. $l_{ij}(x) \geq l_{ij}(y)$ if $x \geq y$. In this paper, the message loss ratio $l_{ij}(d)$ for distance d between a pair of nodes p_i and p_j is given as follows:

$$l_{ij}(d) = \begin{cases} 0 & (d < mind_i). \\ (d - mind_i)^2/(maxd_i - mind_i)^2 & (mind_i \leq d < maxd_i). \\ 1 & (d \geq maxd_i). \end{cases} \qquad (1)$$

Figure 1 shows the message loss ratio $l_{ij}(d)$ for distance ratio $d/maxd_i$ where d is distance d_{ij} between a pair of nodes p_i and p_j and $mind_i/maxd_i = 0.2$. If the node p_j is in the distance $mind_i$ of a node p_i, the node p_j receives every message sent by the node p_i without any message loss, i.e. $l_{ij}(d) = 0$ for $d \leq mind_i$. If the node p_j is at longer distance than $maxd_i$, a node p_j receives no message from the node p_i, i.e. p_j is not a first-neighbor node of p_i, $p_j \notin FNC(p_i)$. Here, $l_{ij}(d) = 1$ for $d \geq maxd_i$. If the node p_j is at distance d longer than $mind_i$ and shorter than $maxd_i$, i.e. $mind_i \leq d \leq maxd_i$, the message loss ratio $l_{ij}(d)$ is proportional to a square of the distance d between the nodes p_i and p_j, i.e. $(d - mind_i)^2$. The farther a node p_j is from a node p_i, the more number of messages sent by the node p_i are lost by the node p_j.

If a node p_j loses a message m sent by a node p_i, the node p_i retransmits the message m until the node p_j receives the message m. This means, the each node p_i retransmits a message m infinite times in the worst case. We calculate how many times a node p_i transmits a message m until a node p_j receives the message m. The expected number $NT_{ij}(d)$ of transmissions of a message m from a node p_i to a node p_j is given for distance d between the nodes p_i and p_j as follows:

$$NT_{ij}(d) = 1/(1 - l_{ij}(d)). \qquad (2)$$

Here, if $l_{ij}(d) = 0$, i.e. no message is lost, the node p_i retransmits no message to the node p_j. i.e. $NT_{ij}(d) = 1$. That is, the node p_i can deliver a message m to the node p_j by just once sending the message m. On the other hand, $NT_{ij}(d) = \infty$ if $l_{ij}(d) = 1$. This means, the node p_i retransmits a message m infinite times.

In reality, each node p_i does not retransmit a message infinite times. This means, maximum number mx of transmissions of a message is *a priori* given. A node p_i gives up to deliver a message to a node p_j if the node p_j does not receive the message after transmitting the message mx times. The expected number $NT_{ij}(d)(\leq mx)$ of transmissions of each message from a node p_i to a node p_j with maximum number mx of transmissions and distance d ($=d_{ij}$) between a pair of the nodes p_i and p_j is $(1 - l_{ij}(d))(1 + 2 \cdot l_{ij}(d) + 3 \cdot l_{ij}(d)^2 + \cdots + mx \cdot l_{ij}(d)^{mx-1}) + mx \cdot l_{ij}(d)^{mx}$. Hence, the expected number $NT_{ij}(d)$ of transmissions is given as follows:

$$NT_{ij}(d) = (1 - l_{ij}(d)^{mx})/(1 - l_{ij}(d)). \qquad (3)$$

Figure 2 shows the ratio $NT_{ij}(d)/mx$ of the expected transmission number $NT_{ij}(d)$ to the maximum transmission number mx for distance d ($=d_{ij}$) where $mx = 5, 10$.

Here, $mind_i/maxd_i = 0.2$. $NT_{ij}(d) = 1$ for $d/maxd_i \leq 0.2$, and $NT_{ij}(d) = mx$ for $d/maxd_i \geq 1$. For example, in order to deliver a message to a node p_j, the node p_i sends the message once for $d/maxd_i \leq 0.2$. The node p_i sends a message in average 1.16 times for $d/maxd_i = 0.5$ and 2.28 times for $d/maxd_i = 0.8$ where $mx = 10$.

The probability $LP_{ij}(d)$ that a node p_i fails to deliver a message to a node p_j is $l_{ij}(d)^{mx}$ for distance d $(=d_{ij})$. On the other hand, the probability $DP_{ij}(d)$ that a node p_i makes a success at delivering a message to a node p_j is $1 - LP_{ij}(d) = 1 - l_{ij}(d)^{mx}$.

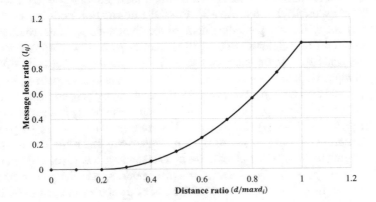

Fig. 1. Message loss ratio ($mind_i/maxd_i = 0.2$).

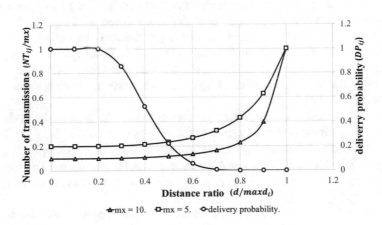

Fig. 2. Expected number of transmissions and delivery probability ($mx = 5$, $mx = 10$).

3 Routing Protocols

3.1 EEUE Protocol

We consider a network N which is composed of nodes $p_1, ..., p_n (n \geq 1)$ interconnected in wireless links. Let P be a set of nodes $p_i, \cdots, p_n$ $(n \geq 1)$ in the network N.

First, a shortest route from a source node p_s to a destination node p_d is obtained by using the flooding algorithm of the AODV protocol [9]. A route from a source node p_s to a destination node p_d is a totally ordered subset $\langle SP, \rightarrow \rangle$ where $SP \subseteq P$, $\{p_s p_d\} \subseteq SP$, and p_s is a top and p_d is a bottom, i.e. their are no nodes p_i and p_j in SP such that $p_i \rightarrow p_s$ and $p_d \rightarrow p_j$. Here, "$p_i \rightarrow p_j$" means that $p_i \leftrightarrow p_j$, and p_j is a succeeding node of p_i and in turn p_i is a preceding node of p_j in the route. The ordered relation $\rightarrow$ is not transitive, i.e. $p_i \rightarrow p_j$ means there is no node p_k such than $p_i \rightarrow p_k \rightarrow p_j$.

Starting from the shortest route R, we obtain a more energy-efficient route ER of the node set P by modifying the route R. First, the destination node p_d is taken as a *pivot* node. Here, p_j shows a preceding node of the pivot node p_d in the route R. We try to find a common first-neighbor node p_k of the nodes p_d and p_j. Suppose a common first-neighbor node p_k is found such that $p_k \leftrightarrow p_j$ and $p_k \leftrightarrow p_d$, i.e. $p_k \in FNC(p_i) \cap FNC(p_j)$. Suppose a path $p_j \rightarrow p_k \rightarrow p_d$ is better than the current path $p_j \rightarrow p_d$ from the node p_j to the pivot node p_d. For example, the total energy consumption of the nodes p_j and p_k in the path $p_j \rightarrow p_k \rightarrow p_d$ is smaller than the node p_j in the path $p_j \rightarrow p_d$. That is, $NT_{jk}(d_{jk}) + NT_{ki}(d_{ki}) < NT_{ji}(d_{ji})$. The route $p_s \rightarrow \cdots \rightarrow p_i \rightarrow p_j \rightarrow p_d$ is changed with $p_s \rightarrow \cdots \rightarrow p_i \rightarrow p_j \rightarrow p_k \rightarrow p_d$. Otherwise, the route R is not changed. Next, a preceding node p_k of the pivot node p_d is taken as a new pivot node in the route $p_s \rightarrow \cdots \rightarrow p_i \rightarrow p_k \rightarrow p_d$. For a pair of the pivot node p_k and the preceding node p_j of p_k, a common first-neighbor node f_h of the nodes p_j and p_k is tried to be found. If found, the path $p_k \rightarrow p_h \rightarrow p_i$ is compared with the current path $p_k \rightarrow p_i$. A better path is taken as presented here. This produce is iterated until the source node p_s is taken as a pivot node.

Suppose a node p_j is a preceding node of a pivot node p_j, i.e. $p_j \rightarrow p_i$ in a route R. Let p_k be a common first-neighbor node of the nodes p_i and p_j, i.e. $p_k \in (FNC(p_i) \cap FNC(p_j))$. We assume the maximum number mx_k of transmissions of every node p_k to deliver each message to another node is the same mx. If a node p_j forwards a message m to a succeeding node p_i ((1) of Fig. 3), the node p_j is expected to $NT_{ji}(d_{ji})$ ($\leq mx$) times transmit the message m to the node p_i. Here, the node p_i can receive the message m with probability $DP_{ji}(d_{ji})$.

On the other hand, suppose a node p_j forwards a message m to the node p_k and then the node p_k forwards the message m to the node p_i as shown in Fig. 3 (2). The total number $TNT_{jki}(d_{jk}, d_{ki})$ of message transmissions of the nodes p_j and p_k is $NT_{jk}(d_{jk}) + NT_{ki}(d_{ki})$ in case the node p_i receives a message sent by a node p_j. The probability $TDP_{jki}(d_{jk}, d_{ki})$ that the node p_j can receive the message m via the node p_k from the node p_i is $DP_{jk}(d_{jk}) \cdot DP_{ki}(d_{ki})$.

First, we consider how to reduce the total energy consumption of nodes in a route from a source node p_s to a destination node p_d. The total energy to be consumed by each node p_i to deliver a message to another nodes p_j is proportional to the expected number $NT_{ij}(d_{ij})$ of transmissions of each node. If the condition $TNT_{jki}(d_{jk}, d_{ki})$ ($= NT_{jk}(d_{jk}) + NT_{ki}(d_{ki})$) $< NT_{ji}(d_{ji})$ is satisfied, the total energy consumption of the nodes p_j and p_k in the path (2) is smaller than the energy consumption of the node p_j in the path (1). Hence, the current path (1) $p_j \rightarrow p_i$ is changed with a new path (2) $p_j \rightarrow p_k \rightarrow p_i$ in the route. Let TN_{ji} be a set $\{p_k | p_k \in (FNC(p_i) \cap FNC(p_j))$ and $TNT_{jki}(d_{jk}, d_{ki})$ $< NT_{ji}(d_{ji})\}$ of first-neighbor nodes of nodes p_j and p_i which satisfy the condition.

If $|TN_{ji}| > 1$, a node p_k in the set TN_{ji} is taken as a new preceding node of the pivot node p_i and a new succeeding node of the node p_j where $TNT_{jki}(d_{jk}, d_{ki})$ is minimum. Here, the node p_k becomes a new pivot node. Otherwise, the preceding node p_j of the current pivot node p_i becomes a new pivot node. For a new pivot node, the procedure is performed. If the source node p_s becomes a pivot node, the route from the source node p_s to the destination node p_d is obtained.

A shortest route R is first obtained by the AODV protocol [9] from a source node p_s to a destination node p_d, i.e. $R = p_s \rightarrow \cdots \rightarrow p_k \rightarrow p_i \rightarrow p_j \rightarrow \cdots p_d$. In the route R, let $next(p_i)$ and $prior\ (p_i)$ be succeeding and preceding nodes p_j and p_k of a node p_i such that $p_k \rightarrow p_i \rightarrow p_j$, respectively. An energy-efficient route ER from the source node p_s to the destination node p_d is obtained from the shortest route R by the EEUE protocol (Algorithm 1).

3.2   EEUP Protocol

Next, we consider another approach. Suppose a node p_j is a preceding node of a pivot node p_i ($p_j \rightarrow p_i$) in a route R. The node p_j delivers a message m to a neighbor node p_i with probability $DP_{ji}(d_{ji}) = (1 - l_{ji}(d_{ji})^{mx})$ as shown in Fig. 3 (1). Next, we consider case a node p_j sends a message m to a neighbor node p_k and then the node p_k forwards the message m to the node p_i in the way (2) shown in Fig. 3. The probability $TDP_{jki}(d_{jk}, d_{ki})$ that a message m is delivered to the node p_i from the node p_{ij} is $(1 - l_{jk}(d_{jk})^{mx}) \cdot (1 - l_{ki}(d_{ki})^{mx})$. Even if $TNT_{jki}(d_{jk}, d_{ki}) < NT_{ji}(d_{ji})$, the way (1) is taken, i.e. the node p_j directly sends a message to the node p_i if $TDP_{jki}(d_{jk}, d_{ki}) < DP_{ji}(d_{ji})$. Because the node p_i can receive a message from the node p_j with higher probability than the way (2).

In this paper, let $NP_{ji}(d_{ji})$ be $(1 - l_{ji}(d_{ji})^{NT_{ji}(d_{ji})})$ where $NT_{ji}(d_{ji})$ is the expected number of transmissions of each message. $NP_{ji}(d_{ji})$ shows the probability that a node p_i receives a message m from a node p_j by transmitting the message m at most $NT_{ji}(d)$ times. $TNP_{jki}(d_{jk}, d_{ki})$ is $NP_{jk}(d_{jk}) \cdot NP_{ki}(d_{ki})$. Let TP_{ji} be a set $\{p_k | p_k \in (FNC(p_i) \cap FNC(p_j))$ and $TNP_{jki}(d_{jk}, d_{ki}) > NP_{ji}(d_{ji})\}$ of neighbor nodes of a node p_j and p_i. If $|TP_{ji}| > 1$, a node p_k in TP_{ji} is taken as a predecessor of the pivot node p_i where $TNP_{jk}(d_{jk}, d_{ki})$ is maximum. Otherwise, the node p_j becomes a new pivot node.

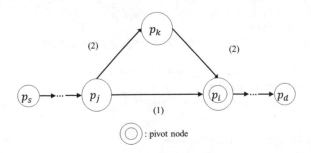

Fig. 3. Routes.

Algorithm 1: EEUE protocol

---

1 **input** : p_s = source node;
2   p_d = destination node;
3   R = shortest-route from p_s to p_d, i.e. AODV route;
4 **output** : ER = energy-efficient route from p_s to p_d;
5 $p_i = p_d$; /* p_i is a pivot node */
6 **while** $p_i \neq p_s$ **do**
7 | $p_j = prior\ (p_i)$; /*$p_j \rightarrow p_i$ */
8 | FN_{ij} = set of common first-neighbor nodes of p_i and p_j, i.e. $\{ p_k \in FNC(p_i) \cap FNC(p_j) \mid p_j \leftrightarrow p_k$ and $p_k \leftrightarrow p_i \}$;
9 | $TN_{ji} = \{p_k \in FN_{ji} \mid TNT_{jki}(d_{jk}, d_{ki}) < NT_{ji}(d_{ji})\}$;
10 | **if** $\mid TN_{ji} \mid\ \geq 1$, **then**
11 | | **select** a node p_k in TN_{ji} where $TNT_{jki}(d_{jk}, d_{ki})$ is minimum;
12 | | **if** p_k *is found,* **then**
13 | | | **connect** p_j **to** p_k and p_k **to** p_i in ER; /* $p_j \rightarrow p_k \rightarrow p_i$ */
14 | | | **disconnect** p_j and p_i $(p_j \nrightarrow p_i)$;
15 | | | $p_i = p_k$; /* p_k is a new pivot node */
16 | **else**
17 | | **connect** p_j to p_i in ER; /* $p_j \rightarrow p_i$ */
18 | | $p_i = p_j$; /* p_j is a new pivot node */
19 | **while end;**

Algorithm 2: EEUP protocol

---

1 **input** : p_s = source node;
2   p_d = destination node;
3   R = shortest-route from p_s to p_d, i.e. AODV route;
4 **output** ; RR = reliable route from p_s to p_d;
5 $p_i = p_d$; /* p_i is a pivot node */
6 **while** $p_i \neq p_s$ **do**
7 | $p_j = prior\ (p_i)$; /*$p_j \rightarrow p_i$ */
8 | FN_{ij} = set of common first-neighbor nodes of p_i and p_j i.e. $\{ p_k \in FNC(p_i) \cap FNC(p_j) \mid p_j \leftrightarrow p_k$ and $p_k \leftrightarrow p_i \}$;
9 | $TP_{ji} = \{p_k \in FN_{ij} \mid TNP_{jki}(d_{jk}, d_{ki}) > NP_{ji}(d_{ji})\}$;
10 | **if** $\mid TP_{ji} \mid\ \geq 1$, **then**
11 | | **select** a node p_k in TP_{ij} where $TNP_{jki}(d_{jk}, d_{ki})$ is maximum;
12 | | **if** p_k *is found,* **then**
13 | | | **connect** p_j **to** p_k and p_k **to** p_i in RR; /* $p_j \rightarrow p_k \rightarrow p_i$ */
14 | | | **disconnect** p_j and p_i $(p_j \nrightarrow p_i)$;
15 | | | $p_i = p_k$; /* p_k is a new pivot node */
16 | **else**
17 | | **connect** p_j to p_i in RR; /* $p_j \rightarrow p_i$ */
18 | | $p_i = p_j$; /* p_j is a new pivot node */
19 | **while end;**

Thus, a reliable route RR is obtained from the shortest AODV route R by the EEUP protocol which is shown in Algorithm 2.

4 Concluding Remarks

In wireless networks, a node consumes electric energy to send and receive messages. In our experiment, a node mainly consumes energy to perform a communication process to send and receive messages than communication devices to emit and receive radio. The more frequently a node retransmits a message, the more energy the node consumes. The energy consumption of a node depends on the number of transmissions of the node. In this paper, we proposed the model to estimate the energy consumption of a node to deliver a message to a neighbor node given the maximum number of retransmissions. Then, we proposed the EEUE and EEUP protocols to make energy-efficient and reliable routes from a source node to a destination node, respectively.

References

1. Spaho, E., Bylykbashi, K., Barolli, L., Takizawa, M.: An integrated system considering WLAN and DTN for improving network performance: evaluation for different scenarios and parameters. In: Proceedings of the 6th International Conference on Emerging Internet, Data Web Technologies (EIDWT 2018), pp. 339–348 (2018)
2. Dhurandher, S.K., Sharma, D.K., Woungang, I., Saini, A.: An energy-efficient history-based routing scheme for opportunistic networks. Int. J. Commun. Syst. **30**(7), 1–13 (2015)
3. Enokido, T., Aikebaier, A., Takizawa, M.: A model for reducing power consumption in peer-to-peer systems. IEEE Syst. J. **4**(2), 221–229 (2010)
4. Enokido, T., Aikebaier, A., Takizawa, M.: Process allocation algorithms for saving power consumption in peer-to-peer systems. IEEE Trans. Industr. Electron. **58**(6), 2097–2105 (2011)
5. Enokido, T., Aikebaier, A., Takizawa, M.: An extended simple power consumption model for selecting a server to perform computation type processes in digital ecosystems. IEEE Trans. Industr. Inf. **10**(2), 1627–1636 (2014)
6. Nakamura, S., Sugino, M., Takizawa, M.: Algorithms for energy-efficient broadcasting messages in wireless networks. J. High Speed Netw. (JHS) **24**(1), 1–15 (2018)
7. Ogawa, E., Nakamura, S., Enokido, T., Takizawa, M.: One-to-one routing protocols for wireless ad-hoc networks considering the electric energy consumption. In: Proceedings of the 12th International Conference on Innovative Mobile and Internet Services in Ubiquitous Computing (IMIS 2018), pp. 105–115 (2018)
8. Oma, R., Nakamura, S., Duolikun, D., Enokido, T., Takizawa, M.: Evaluation of data and subprocess transmission strategies in the tree-based fog computing model. In: Proceedings of the 22nd International Conference on Network-Based Information Systems (NBiS-2019), pp. 15–26 (2019)
9. Perkins, E.C., Royer, M.E., Das, R.S.: Adhoc On-demand Distance Vector (AODV) Routing (1997). https://www.ietf.org/rfc/rfc3561.txt
10. Raspberry Pi 3 Model B. https://www.raspberrypi.org/products/raspberry-pi-3-model-b/
11. Sugino, M., Nakamura, S., Enokido, T., Takizawa, M.: Energy-efficient broadcast protocols in wireless networks. In: Proceedings of the 18th International Conference on Network-Based Information Systems (NBiS 2015), pp. 357–364 (2015)
12. Zhao, F., Guibas, L.: Wireless Sensor Networks: An Information Processing Approach. Morgan Kauhmann Publishers, Burlington (2004)

Evaluation of Parallel Data Transmission in the Mobile Fog Computing Model

Kosuke Gima[1]($\boxtimes$), Takumi Saito[1], Ryuji Oma[1], Shigenari Nakamura[1],
Tomoya Enokido[2], and Makoto Takizawa[1]

[1] Hosei University, Tokyo, Japan
{kosuke.gima.3r,takumi.saito.3j,ryuji.oma.6r}@stu.hosei.ac.jp,
nakamura.shigenari@gmail.com, makoto.takizawa@computer.org
[2] Rissho University, Tokyo, Japan
eno@ris.ac.jp

Abstract. In this paper, we discuss the mobile fog computing (MFC) model including mobile fog nodes which communicate with other fog nodes in wireless networks. Here, each fog node supports some process by which output data is obtained by processing input data received from other fog nodes and sensor nodes and sends the output data to target fog nodes which supports a process to handle the output data. Here, even if there are multiple target nodes in the communication range, a source node sends data to one of the target nodes in the MFC algorithm proposed in our previous studies. In this paper, we newly propose a pair of parallel data transmission (PDT) algorithms, PDT1 and PDT2 where a source node divides the output data to segments and sends each segment to a target node in the communication range. In the PDT1 algorithm, each segment has the same size. In the PDT2 algorithm, source node sends segments to target nodes so that each target node can process the same size of data. In the evaluation, the PDT1 algorithm and the PDT2 algorithm, the energy consumption and execution time of source and target fog nodes can be more reduced.

Keywords: IoT · Fog computing (FC) model · Mobile fog computing (MFC) model · Parallel data transmission (PDT) algorithms · Energy consumption

1 Introduction

The IoT (Internet of Things) includes millions of sensors and actuators which are interconnected with clouds of servers in networks [9]. In the fog computing (FC) models [11,16] proposed to efficiently realize the IoT, the network and server traffic are reduced by distributing processes to handle sensor data on not only servers but also fog nodes. Sensor data is processed by a process of a fog node p_i and the processed data is sent to another fog node p_j which supports a process to further process the data. Here, the node p_j is a *target* node of the

© Springer Nature Switzerland AG 2020
L. Barolli et al. (Eds.): EIDWT 2020, LNDECT 47, pp. 158–170, 2020.
https://doi.org/10.1007/978-3-030-39746-3_17

source node p_i. The TBFC (Tree-Based FC) model of the IoT is proposed in our previous studies [12,15], where fog nodes are hierarchically structured in a tree to reduce the energy consumption of the fog nodes. Here, a parent node is a target node of each child node. In addition to improving the performance and reducing the energy consumption of the IoT, the FC model is required to be tolerant of faults of fog nodes. In order to make the TBFC model fault-tolerant, the FTBFC (Fault-tolerant TBFC) model is proposed by selecting alternate nodes which support processes of faulty nodes [10,14].

In the MFC (Mobile FC) model [1,5,6], mobile fog nodes like vehicles are moving and communicate with one another in wireless communication networks to exchange not only output data but also processes. Here, a source node sends the output data to one target node in the MFC algorithm [5]. In this paper, we consider the parallel data transmission (PDT) model [6], where each fog node sends output data to one or more than one target node in the communication range. Here, a source fog node divides the output data to multiple segments and sends each segment to one target node. Thus, data segments are in parallel processed by multiple target fog nodes in order to reduce the total energy consumption and execution time of the target nodes. We propose a pair of PDT1 and PDT2 algorithms to select target fog nodes in the communication range to which a source node sends segments of output data so that the total energy consumption and execution time of the source node and target fog nodes can be reduced. In the PDT1 algorithm, a source node sends a segment of same size to each target node selected. In the PDT2 algorithm, a source node sends a segment of different size to each target node selected so that each target node has the same total size of its own input data and the segment that is each target node processes, i.e. the amount of data. In the evaluation, we show the total energy consumption and execution time of a source node and target nodes can be reduced in the PDT2 algorithm than the PDT1 and MFC algorithms.

In Sect. 2, we present a system model. In Sect. 3, we discuss the power consumption and computation models of a fog node of the FC model. In Sect. 4, we propose the PDT1 and PDT2 algorithms to select multiple target nodes to which segments of the output data are transmitted. In Sect. 5, we evaluate the PDT1 and PDT2 algorithms.

2   System Model

A fog computing (FC) model is composed of nodes, i.e. server nodes, fog nodes, and device nodes with sensors and actuators [16]. In this paper, we consider the MFC (Mobile FC) model [5–7] which is composed of mobile fog nodes interconnected in wireless networks. Mobile fog nodes can communicate with one another only if the fog nodes are in the communication range of each other.

In the TBFC model [7,8,13,15], a fog node f_i supports a process $p(f_i)$ to process input data from sensors and other fog nodes. Here, the node f_i is a host node of the process $p(f_i)$. A node which supports a process to handle data is a *target* node of the data. The sensor data d is received and processed by a target node f_i. The output data d_i is obtained by the process $p(f_i)$ through processing the input data d. The output data d_i has to be sent to a target fog node f_j which supports a process $p(f_j)$ to process the output data d_i.

A process F_i *precedes* a process F_j (or F_j *follows* F_i) ($F_i \rightarrow F_j$) if and only if (iff) the output data of the process F_i is processed by the process F_j [5]. Here, F_j is a target process of F_i (Fig. 1). Output data obtained by a process $p(f_i) = F_i$ in a fog node f_i is sent to a fog node f_j with a process $p(f_j)$ ($= F_j$) such that $p(f_i) \rightarrow p(f_j)$. Thus, sensor data is processed by a sequence of processes. A node f_i *precedes* a node f_j ($f_i \rightarrow f_j$) iff $p(f_i) \rightarrow p(f_j)$. Here, f_j is a *target* node of the node f_i. On receipt of data d, a fog node f_i stores the data d in the memory M_i. Let $maxM_i$ be the maximum size of the memory storage M_i of a fog node f_i. Output data obtained by processing the input data is also stored in the memory M_i.

Let FF_i be a set of fog nodes which are in the communication range of a fog node f_i. FN_i ($\subseteq FF_i$) is a set of target fog nodes of a fog node f_i in the communication range, i.e. $FN_i = \{f_j \mid f_j \in FF_i \text{ and } f_i \rightarrow f_j\}$. In the MFC model [6], each source node f_i selects one target node f_j in the set FN_i. In this paper, we consider a way where one or more than one target node is selected in the set FN_i and a source node sends the output data to the target nodes selected.

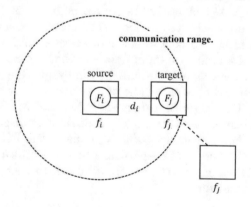

Fig. 1. Source and target nodes.

3   Power Consumption and Computation Models of a Fog Node

A fog node f_i receives and processes input data D_i and sends a target node output data d_i obtained by processing the input data D_i. Let $TT_i(x)$ and $RT_i(x)$ be a pair of time [sec] for a node f_i to transmit and receive data of size x to and from another fog node, respectively. The transmission time $TT_i(x)$ and receiving time $RT_i(x)$ are proportional to the size x of the data where tc_i, tt_i, rc_i, and rt_i are constants. [5,6]:

$$TT_i(x) = tc_i + tt_i \cdot x. \tag{1}$$

$$RT_i(x) = rc_i + rt_i \cdot x. \tag{2}$$

Suppose there are multiple target nodes $f_{i1}, \cdots, f_{i,s_i}$ $(s_i \geq 1)$ in the communication range of a node f_i, i.e. $FN_i = \{f_{i1}, \cdots, f_{i,s_i}\}$. The output data d_i of a node f_i is divided into segments $d_{i1}, \cdots, d_{is_i}$ $(s_i \geq 1)$. Then, the node f_i sends each segment d_{ij} to a target node f_{ij} in the set FN_i $(j = 1, \cdots, s_i)$. Let x be the size of the output data d_i and x_j be the size $|d_{ij}|$ of each segment d_{ij}. Here, $x = x_1 + \cdots + x_{s_i}$. It takes time $TTT_i(x)$ [sec] for the node f_i to send the segments $d_{i1}, \cdots, d_{i,s_i}$ to the target nodes $f_{i1}, \cdots, f_{i,s_i}$, respectively:

$$TTT_i(x) = TT_i(x_1) + \cdots + TT_i(x_{s_i}) \tag{3}$$
$$= tc_i \cdot s_i + tt_i \cdot x.$$

It takes time $CT_i(x)$ [sec] for a fog node f_i to process input data D_i of size x. The execution time $CT_i(x)$ is assumed to be proportional to x and x^2 depending on the computational complexity of the process $p(f_i)$:

$$CT_i(x) = ct_i \cdot C_i(x). \tag{4}$$

Here, $C_i(x) = x$ or x^2 and ct_i is a constant. In a Raspberry Pi 3 Model B node f_i [1], $tt_i = 0.00001$, $rt_i = 0.00009$, $tc_i = 0.00007$, $rc_i = 0$, and $ct_i = 0.00009$. The process $p(f_i)$ creates output data d_i of size $\rho_i \cdot x$ by processing input data D_i of size x. Here, ρ_i is the output ratio of the node f_i. For example, if the process $p(f_i)$ obtains an average value d_i of a collection of ten values the output ratio, ρ_i is 0.1. A fog node f_i is assumed to follow the SPC (Simple Power Consumption) model [2–4]. Here, a node f_i consumes the maximum electric power $maxE_i$ [W] if at least one process is performed. Otherwise, i.e. no process is performed, the node f_i consumes the minimum power $minE_i$ [W].

A fog node f_i consumes the electric power TP_i and RP_i [W] to transmit and receive data. TP_i and RP_i are given as follows (Fig. 2);

$$TP_i = te_i \cdot maxE_i. \tag{5}$$

$$RP_i = re_i \cdot maxE_i. \tag{6}$$

Here, te_i and re_i are constants. $te_i \leq 1$ and $re_i \leq 1$. In a Raspberry Pi 3 Model B node f_i, $maxE_i = 3.7$ [W], $minE_i = 2.1$ [W], $te_i = 0.676$, and $re_i = 0.729$ according to our experiment [1]. In a PC node f_i with an Intel Core i7-6700K CPU, $maxE_i$ is 89.5 [W] and $minE_i = 41.3$ [W].

Since a fog node f_i consumes the power $TP_i(x) = te_i \cdot maxE_i$ [W] for time $RT_i(x)$ [sec], the node f_i consumes the energy $RE_i(x)$ [J] $= RP_i$ [W] $\cdot RT_i(x)$ [sec] to receive input data of size x. A node f_i consumes the energy $TE_i(\rho_i \cdot x)$

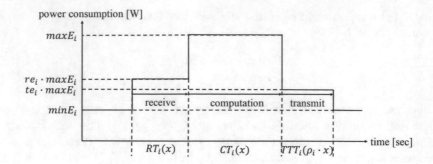

Fig. 2. Power consumption of a node f_i.

$[J] = TP_i$ [W] $\cdot TTT_i(\rho_i \cdot x)$ [sec] to send output data d_i of size $\rho_i \cdot x$ for $TTT_i(\rho_i \cdot x)$ [sec].

$$RE_i(x) = RP_i \cdot RT_i(x) = re_i \cdot (rc_i + rt_i \cdot x) \cdot maxE_i. \tag{7}$$

$$TE_i(\rho_i \cdot x) = TP_i \cdot TT_i(\rho_i \cdot x) = te_i \cdot (tc_i + tt_i \cdot \rho_i \cdot x) \cdot maxE_i. \tag{8}$$

If a node f_i sends the segments $d_{i1}, \cdots, d_{i,s_i}$ of the output data d_i of size x to target nodes $f_{i1}, \cdots, f_{i,s_i}$, respectively, the node f_i consumes the energy $TTE_i(x)$ [J] where $x = x_1 + \cdots + x_{s_i}$;

$$TTE_i(x) = TP_i \cdot TTE_i(x) = te_i \cdot (tc_i \cdot s_i + tt_i \cdot x) \cdot maxE_i. \tag{9}$$

A fog node f_i consumes the maximum power $maxE_i$ [W] to process input data of size x for time $CT_i(x)$ [sec]. Hence, the energy $CE_i(x)$ [J] $= maxE_i$ [W] $\cdot CT_i(x)$ [sec] is consumed by the fog node f_i to process data of size x:

$$CE_i(x) = CT_i(x) \cdot maxE_i = ct_i \cdot C_i(x) \cdot maxE_i. \tag{10}$$

A fog node f_i receives input data D_i of size x and generates output d_i of size $\rho_i \cdot x$. The fog node f_i sends segments $d_{i1}, \cdots, d_{i,s_i}$ of the output data d_i to s_i (≥ 1) target nodes $f_{i1}, \cdots, f_{i,s_i}$. Here, y_j shows the size of each segment d_{ij}. Here, $\rho_i \cdot x = x_1 + \cdots + x_{s_i}$. Here, the fog node f_i totally consumes the energy $EE_i(x)$:

$$EE_i(x) = RE_i(x) + CE_i(x) + TTE_i(\rho_i \cdot x) \tag{11}$$
$$= RE_i(x) + CE_i(x) + TE_1(y_1) + \cdots + TE_{s_i}(y_{s_i})$$
$$= \{re_i \cdot (rc_i + rt_i \cdot x) + ct_i \cdot C_i(x) + te_i \cdot (s_i \cdot tc_i + tt_i \cdot \rho_i \cdot x)\} \cdot maxE_i.$$

If $y_j = y_k$ for every pair of segments d_{ij} and d_{ik}, $y_j = y_k = \rho_i \cdot x/s_i$. The node f_i consumes the energy $EE_i(x)$:

$$EE_i(x) = RE_i(x) + CE_i(x) + s_i \cdot TE_i(\rho_i \cdot x/s_i). \tag{12}$$

4   Selection Algorithms in the Parallel Data Transmission (PDT) Model

4.1   PDT Model

Each fog node f_i moves and communicates with other fog nodes in wireless networks. A fog node f_i sends output data d_i in the memory M_i to a target node f_j in FN_i. We discuss how fog nodes exchange data with one another. In our previous studies [5], the data transmission (DT), process transmission (PT), and data exchange (DE) algorithms are proposed for a fog node to communicate with another node in the communication range [6]. In the DT algorithm, a fog node f_i sends the output data d_i to a target node f_j. In the PT algorithm, a source node f_i does not send output data d_i to a target node. A target node f_j sends a process $p(f_j)$ to a node f_i. Then, the node f_i processes the output data d_i by using the process $p(f_j)$. In the DE algorithm, a node f_i communicates with a node f_j in the communication range but the node f_j is not a target node. Here, the nodes f_i and f_j exchange the output data so that the usage ratios of the memories get the same.

In this paper, we discuss the parallel data transmission (PDT) model [6] so that the output data d_i can be more efficiently processed by multiple target nodes. Suppose there are multiple target nodes $f_{i1}, \cdots, f_{i,s_i}$ of a node f_i which are in the communication range of a node f_i, i.e. $FN_i = \{f_{i1}, \cdots, f_{i,s_i}\}$. Let x_i be the size $|d_i|$ of the output data d_i of the node f_i. Suppose the execution time $CT_j(x)$ of the target process $p(f_j)$ is $O(x_i^2)$, i.e. $ct_j \cdot x_i^2$ for the size x of the output data d_i. If the node f_i sends the half of the data d_i to a target node f_{ij} and the other half to another target node f_{ik}, the total execution time of the nodes f_{ij} and f_{ik} is $(ct_{ij} + ct_{lk}) \cdot (x_i/2)^2$. If $ct_{ij} = ct_{ik} = c$, the total execution time of the target nodes f_{ij} and f_{ik} is $c \cdot (x_i/2)^2 + c \cdot (x_i/2)^2 = c \cdot x_i^2/2$. If the node f_i sends the output data d_i to only the node f_{ij}, the execution time of the process $p(f_j)$ on the target node f_{ij} is $c \cdot x_i^2$. Since, $c \cdot x_i^2/2$ is smaller than $c \cdot x_i^2$, the total execution time and total energy consumption of target fog nodes can be reduced by dividing output data to segments and in parallel processing the segments in different target nodes (Fig. 3).

4.2   Algorithms to Select Target Nodes

We discuss how each fog node f_i selects target fog nodes $f_{i1}, \cdots, f_{ik}$ in the communication range to which segments $d_{i1}, \cdots, d_{ik}$ of output data d_i of the node f_i are sent. In this paper, we assume every fog node is homogeneous, i.e. $ct_i = ct_j$ for every pair of nodes f_i and f_j. Let FF_i be a set of fog nodes which are in the communication range of a fog node f_i. Let FN_i be a subset of target nodes of a node f_i, i.e. $FN_i = \{f_j \in FF_i \mid f_i \to f_j\}\ (\subseteq FF_i)$. In the set FN_i, fog nodes are sorted with respected to the available size of memory. Let a_i be the size of available memory in a node f_i. FN_i is an ordered set of target nodes $f_{i1}, \cdots, f_{il}\ (l \le s_i)$. Here, if $a_{ij} > a_{ih}$ for a pair of nodes f_{ij} and f_{ih} in FF_i, f_{ij}

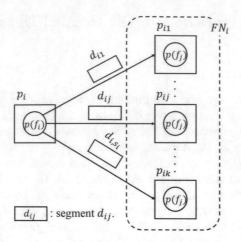

Fig. 3. Parallel data transmission (PDT).

precedes f_{ih}, i.e. $j < h$. Let x be the size $|d_i|$ of output data d_i of a source fog node f_i.

Suppose a fog node f_i divides the output data d_i of size $x_i = |d_i|$ into k ($\geq$ 1) segments $d_{i1}, \cdots, d_{ik}$ where $x_{ih} = |d_{ih}|$ for $h = 1, \cdots, k$. Here, $x_i = x_{i1} + \cdots + x_{ik}$. The fog node f_i sends a segment d_{ih} to a target fog node f_{ih} in the set FN_i in the communication range ($h = 1, \cdots, k$). Let y_{ih} show the size of input data dd_{ih} which a fog node f_{ih} holds before receiving the segment d_{ij}. In each fog node f_{ih}, a segment d_{ih} has to be processed in addition to its own input data dd_{ih}. Hence, the target node f_{ih} consumes the energy $RE_{ih}(x_{ih})$ to receive the segment d_{ih} and the energy $CE_{ih}(y_{ih} + x_{ih})$ to process the input data d_{ih} and dd_{ih}, i.e. $RE_{ih}(x_{ih}) + CE_{ih}(y_{ih} + x_{ih})$. The source node f_i consumes the energy $TE_i(x_{ih})$ to send the segment d_{ih} to each target node f_{ih} ($h = 1, \cdots, k$). The source node f_i and the k target nodes $f_{i1}, \cdots, f_{ik}$ totally consume the energy EE_k:

$$EE_k = \sum_{h=1}^{k}[TE_i(x_{ih}) + RE_{ih}(x_{ih}) + CE_{ih}(y_{ih} + x_{ih})]. \qquad (13)$$

First, suppose a fog node f_i sends segments $d_{i1}, \cdots, d_{ik}$ ($k \leq s_i$) where the size x_{ih} of each segment d_{ih} is the same x_i/k. A first node f_{i1} is selected in the ordered set FN_i. If the output data d_{i1} ($= d_i$) of size x_{i1} ($= x_i$) is sent to the selected fog node f_{i1}, the node f_i consumes the energy $TE_i(x)$ to send data of size x to the node f_{i1}. The fog node f_{i1} processes not only the data d_{i1} of size x sent by the fog node f_i but also its own data dd_{i1} whose size is y_{i1}. Hence, the target node f_{i1} consumes the energy $RE_{i1}(x_i) + CE_{i1}(y_{i1} + x_i)$ to receive and process the segment d_{i1} of size x. Totally, $EE_1 = TE_i(x_i) + RE_{i1}(x_i) + CE_{i1}(y_{i1} + x_i)$.

Next, a pair of fog nodes f_{i1} and f_{i2} are selected in the set FN_i. Here, the output data d_i is divided to a pair of segments d_{i1} and d_{i2} whose sizes are x_{i1} and x_{i2}, respectively. The fog node f_i sends segments d_{i1} and d_{i2} to each of the target fog nodes f_{i1} and f_{i2}, respectively. The fog node f_i and the selected target nodes f_{i1} and f_{i2} totally consume the energy $EE_2 = TE_i(x_{i1}) + TE_i(x_{i2}) + (RE_{i1}(x_{i1}) + CE_{i1}(y_{i1} + x_{i1})) + (RE_{i2}(x_{i2}) + CE_{i2}(y_{i2} + x_{i2}))$.

Thus, the k nodes $f_{i1}, \cdots, f_{ik}$ $(1 \le k \le s_i)$ are taken in the ordered set FN_i Fig. 4. The total energy consumption EE_k of the node f_i and the k target nodes $f_{i1}, \cdots, f_{ik}$ is given in formula (13). If the size of each segment d_{ih} is the same x/k $(h = 1, \cdots, k)$, the total energy consumption EE_k is given as follow:

$$EE_k = k \cdot TE_i(x_i/k) + \sum_{h=1}^{k} [RE_{ih}(x_i/k) + CE_{ih}(y_{ih} + x_i/k)]. \qquad (14)$$

We select the number k $(\le s_i)$ of target fog nodes so that the total energy consumption EE_k of the source node f_i and the target nodes $f_{i1}, \cdots, f_{ik}$ is minimum in the PDT1 algorithm (Algorithm 1). First, EE_h is obtained for $h = 1$. While $EE_h < EE_{h-1}$, h is incremented by one. If $EE_h > EE_{h-1}$, $k = h - 1$. The node f_i divides the output data d_i into k segments $d_{i1}, \cdots, d_{ik}$ of same size x_i/k. The node f_i sends a segment d_{ij} to each node f_{ih} $(h = 1, \cdots, k)$ in the set FN_i (Fig. 4).

Algorithm 1: PDT1 selection algorithms

---

1 **input** : $FN_i = \langle f_{i1}, \cdots, f_{i,s_i} \rangle$ = ordered set of target nodes in the communication range;

2   x = size of output data d_i;

3   y_{ij} = size of data dd_{ij} which f_{ij} holds $(j = 1, \cdots, s_i)$;

4 **output** : k = number of target fog nodes to which f_i sends segments $d_{i1}, \cdots,$ d_{ik};

5 $E_i = \infty$;

6 **for** $l = 1, \cdots, s_i$ **do**

7   **if** $y_{il} + x/l < maxBF_{il}$, **then**

8         $E = l \cdot TE_i(x/l) + \sum_{j=1}^{l} [RE_{ij}(x/l) + CE_{ij}(y_{ij} + x/l)]$;

9   **if** $E < EE$, **then**

10             $EE = E$; $k = l$;

11   **if end**;

12   **else**

13   **break**;

14   **for end**;

Next, a fog node f_i sends segments $d_{i1}, \cdots, d_{ik}$ to target fog nodes $f_{i1}, \cdots,$ f_{ik}, respectively, so that the total amount $y_{ih} + x_{ih}$ of data in each target fog

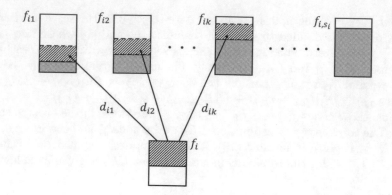

Fig. 4. PDT1 algorithm.

node f_{ih} is the same z (Fig. 5), i.e. $z = y_{ih} + x_{ih}$. This means, the processing load of each node f_{ij} ($j = 1, \cdots, k$) is the same. After sending the segments, each selected target node f_{ih} has the same amount z of the input data dd_{ih} and the segment d_{ih}. By using the PDT2 algorithm (Algorithm 2), k (≥ 1) target nodes $f_{i1}, \cdots, f_{ik}$ are selected in the ordered set FN_i and a segment d_{ih} of size x_{ih} is sent to each node f_{ih}. First, the total size b of the input data $dd_{i1}, \cdots, dd_{ik}$ of the target nodes $f_{i1}, \cdots, f_{ik}$ and the output data d_i of the node f_i is calculated, i.e. $b = y_{i1} + \cdots + y_{ik} + x_i$. Here, each target node f_{ih} has to totally process the data of size $z = b/k$. The node f_i sends a segment d_{ih} of size $z - y_{ih}$ to each target node f_{ih}. In formula (13), $y_{ih} + x_{ih}$ is the same z for every target fog node f_{ih}. The more amount of input data dd_{ih} a node f_{ih} holds, the smaller segment d_{ih} is sent to the node f_{ih}. The PDT2 algorithm gives the number k and the energy E_k as EE.

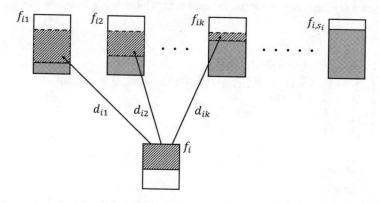

Fig. 5. PDT2 algorithm.

Algorithm 2: PDT2 selection algorithms

---

1 **input** : $FN_i = \langle f_{i1}, \cdots, f_{i,s_i} \rangle$ = ordered set of target nodes in the communication range;

2 $\quad\quad y_{ij}$ = size of data dd_{ij} which f_{ij} holds $(j = 1, \cdots, s_i)$;

3 $\quad\quad x$ = size of output data d_i;

4 **output** : k = number of target fog nodes to which f_i sends segments $d_{i1}, \cdots,$ d_{ik};

5 $\quad\quad x_{ih}$ = size of each segment d_{ih} $(h = 1, \cdots, k)$;

6 $EE = \infty;$ $\quad k = -1;$

7 **for** $l = 1, \cdots, s_i$ **do**

8 $\quad\quad z = (x + y_{i1} + \cdots + y_{il}) \,/\, l;$

9 $\quad\quad$ **if** $z < maxBF_{ij}$ $(j = 1, \cdots, l)$ **and** $z > y_{il}$, **then**

10 $\quad\quad\quad\quad E = \sum_{h=1}^{l} [TE_i(z - y_{ih}) + RE_{ih}(z - y_{ih}) + CE_{ih}(z)];$

11 $\quad\quad\quad\quad$ **if** $E < EE$, **then**

12 $\quad\quad\quad\quad\quad\quad EE = E;$ $\quad k = l;$

13 $\quad\quad\quad\quad\quad\quad$ **if end**;

14 $\quad\quad$ **else**

15 $\quad\quad\quad\quad$ **break**;

16 $\quad\quad$ **for end**;

---

5   Evaluation

We evaluate the PDT1 and PDT2 algorithms in terms of energy consumption and execution time of fog nodes. We consider a fog node f_i and a set FN_i of target fog nodes $f_{i1}, \cdots, f_{i,s_i}$ $(s_i > 1)$ in the communication range of the node f_i. The set FN_i is ordered as $f_{i1}, \cdots, f_{i,s_i}$ in terms of the available size of memory as presented in the preceding section. Every fog node is assumed to be homogeneous in this evaluation. Each target node f_{ih} $(h \le s_i)$ holds input data dd_{ih} of size y_{ih}. The number s_i of target node is 10. In this evaluation, $y_{ij} = 100 \cdot j$ [KB] $(j = 1, \cdots, 10)$. The source node f_i holds output data d_i of size x_i. The size x_i of the output data d_i of the fog node f_i. The maximum memory size $maxM_{ij}$ of each target node f_{ij} is 2 [MB]. In each of the PDT1 and PDT2 algorithms, the number k of target fog nodes $f_{i1}, \cdots f_{ik}$ is decided in the set FN_i.

Figure 6 shows the energy consumption for number k of target nodes. The energy consumption of the PDT2 algorithm is slightly smaller than the PDT1. Figure 7 shows the longest execution time of the k target nodes, i.e. response time. As shown in the figure, the response time of the PDT2 algorithm is one third shorter than the PDT1 algorithm.

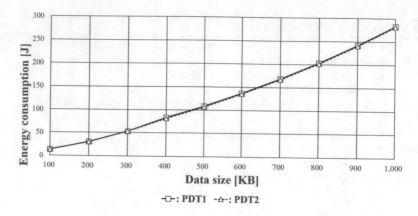

Fig. 6. Energy consumption of $O(x^2)$.

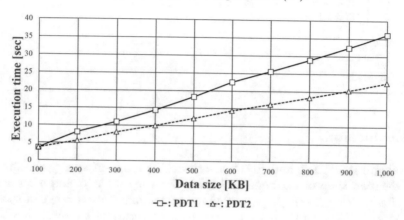

Fig. 7. Execution time of $O(x^2)$.

6　Concluding Remarks

The MFC (Mobile Fog Computing) model is composed of mobile fog nodes which communicate with one another in wireless networks. A fog node receives data from other nodes and processes the data by a process supported by the fog node. In this paper, we newly proposed the parallel data transmission (PDT) algorithms, PDT1 and PDT2 in the MFC model to efficiently realize the IoT. Here, a fog node divides the output data to segments and sends in parallel the segments to target nodes. In the PDT1 algorithm, a source node sends a segment of same size to each target node. In the PDT2 algorithm, a source node sends a segment of different size to each target node so that the total size of the segment and the input data is the same. In the evaluation, we showed the energy consumption and execution time of the target nodes can be reduced in the PDT1 and PDT2 algorithms.

References

1. Raspberry pi 3 model b. https://www.raspberrypi.org/products/raspberry-pi-3-model-b/
2. Enokido, T., Ailixier, A., Takizawa, M.: A model for reducing power consumption in peer-to-peer systems. IEEE Syst. J. **4**(2), 221–229 (2010)
3. Enokido, T., Ailixier, A., Takizawa, M.: Process allocation algorithms for saving power consumption in peer-to-peer systems. IEEE Trans. Industr. Electron. **58**(6), 2097–2105 (2011)
4. Enokido, T., Ailixier, A., Takizawa, M.: An extended simple power consumption model for selecting a server to perform computation type processes in digital ecosystems. IEEE Trans. Industr. Inf. **10**(2), 1627–1636 (2014)
5. Gima, K., Oma, R., Nakamura, S., Enokido, T., Takizawa, M.: A model for mobile fog computing in the IoT (accepted). In: Proceedings of the 22nd International Conference on Network-Based Information Systems (NBiS 2019), pp. 447–458 (2019)
6. Gima, K., Oma, R., Nakamura, S., Enokido, T., Takizawa, M.: Parallel data transmission protocols in the mobile fog computing model. In: Proceedings of the 14th International Conference on Broad-Band Wireless Computing, Communication and Applications (BWCCA 2019), pp. 494–503 (2019)
7. Guo, Y., Oma, R., Nakamura, S., Duolikun, D., Enokido, T., Takizawa, M.: Data and subprocess transmission on the edge node of TWTBFC model. In: Proceedings of the 11th International Conference on Intelligent Networking and Collaborative Systems (INCoS 2019), pp. 80–90 (2019)
8. Guo, Y., Oma, R., Nakamura, S., Duolikun, D., Enokido, T., Takizawa, M.: Evaluation of a two-way tree-based fog computing (TWTBFC) model. In: Proceedings of the 13th International Conference on Innovative Mobile and Internet Services in Ubiquitous Computing (IMIS 2019), pp. 72–81 (2019)
9. Hanes, D., Salgueiro, G., Grossetete, P., Barton, R., Henry, J.: IoT Fundamentals: Networking Technologies, Protocols, and Use Cases for the Internet of Things. Cisco Press, Indianapolis (2018)
10. Oma, R., Nakamura, S., Duolikun, D., Enokido, T., Takizawa, M.: Evaluation of data and subprocess transmission strategies in the tree-based fog computing model (accepted). In: Proceedings of the 22nd International Conference on Network-Based Information Systems (NBiS 2019), pp. 15–26 (2019)
11. Oma, R., Nakamura, S., Duolikun, D., Enokido, T., Takizawa, M.: An energy-efficient model for fog computing in the Internet of Things (IoT). Internet Things **1–2**, 14–26 (2018)
12. Oma, R., Nakamura, S., Duolikun, D., Enokido, T., Takizawa, M.: Evaluation of an energy-efficient tree-based model of fog computing. In: Proceedings of the 21st International Conference on Network-Based Information Systems (NBiS 2018), pp. 99–109 (2018)
13. Oma, R., Nakamura, S., Duolikun, D., Enokido, T., Takizawa, M.: Energy-efficient recovery algorithm in the fault-tolerant tree-based fog computing (FTBFC) model. In: Proceedings of the 33rd International Conference on Advanced Information Networking and Applications (AINA 2019), pp. 132–143 (2019)
14. Oma, R., Nakamura, S., Duolikun, D., Enokido, T., Takizawa, M.: A fault-tolerant tree-based fog computing model (accepted). Int. J. Web Grid Serv. (IJWGS) **15**(3), 219–239 (2019)

15. Oma, R., Nakamura, S., Enokido, T., Takizawa, M.: A tree-based model of energy-efficient fog computing systems in IoT. In: Proceedings of the 12th International Conference on Complex, Intelligent, and Software Intensive Systems (CISIS 2018), pp. 991–1001 (2018)
16. Rahmani, A.M., Liljeberg, P., Preden, J.S., Jantsch, A.: Fog Computing in the Internet of Things. Springer, Cham (2018)

Performance Evaluation of VegeCare Tool for Insect Pest Classification with Different Life Cycles

Natwadee Ruedeeniraman[1], Makoto Ikeda[2(✉)], and Leonard Barolli[2]

[1] Graduate School of Engineering, Fukuoka Institute of Technology, 3-30-1 Wajiro-higashi, Higashi-ku, Fukuoka 811-0295, Japan
mgm19108@bene.fit.ac.jp
[2] Department of Information and Communication Engineering, Fukuoka Institute of Technology, 3-30-1 Wajiro-higashi, Higashi-ku, Fukuoka 811-0295, Japan
makoto.ikd@acm.org, barolli@fit.ac.jp

Abstract. In this paper, we present the performance evaluation of VegeCare tool for insect pest classification. We collect the main pests of rice and corn crops for different life cycles. The dataset belongs to 4 categories, which look very similar, but they are different species. Moreover, they have different appearances depending on their life cycle. In some stages of their life, they resemble each other. For this reason, the conventional recognition systems could not detect them precisely. To improve the performance of the insect pest classification, we classify each insect pest by considering every stage of the life cycle. Experimental results show that the proposed tool achieve a good accuracy. We found that the tool can detect corn borer and armyworm which are ambiguous insects. The accuracy is more than 70%.

Keywords: VegeCare · Insect pest classification · Corn · Rice · Life cycle

1 Introduction

Smart agriculture systems based on deep learning can deliver high quality and safe food, which is very important for people's healthy and happy life [16]. With the spread of Low Power Wide Area (LPWA) services, we can develop a system with agricultural sensed data to collect from a long distance. Therefore, the establishment of advanced data processing technology for handling the agriculture big data is required, and the realization of the new generation AI framework is expected [9].

Deep Neural Networks (DNNs) have been adopted in intelligent systems for various application and they can be applied to extremely difficult problems for humans [10]. Fixed and mobile cameras and sensors are scattered in our life. In some application areas, intelligent algorithm has made it possible to make decisions even better than humans [2, 8, 11, 12, 17, 18].

In [15], we proposed a vegetable detection and classification system considering TensorFlow framework. Also, we evaluated the performance of the proposed VegeCare

© Springer Nature Switzerland AG 2020
L. Barolli et al. (Eds.): EIDWT 2020, LNDECT 47, pp. 171–180, 2020.
https://doi.org/10.1007/978-3-030-39746-3_18

tool for tomato disease classification for managing the plant growth [14]. The growth management of plants is important for increasing the productivity of vegetables.

In this paper, we present the performance evaluation of VegeCare tool for rice and corn insect pests classification. We used 4 kinds of insect pest. The insect classification is one of the functions of the proposed VegeCare tool, which helps the growth of vegetables for farmers. For evaluation, we use as metrics the learning accuracy, loss and insect pest classification.

The structure of the paper is as follows. In Sect. 2, we give the related work. In Sect. 3, we describe the Neural Networks. In Sect. 4, we describe the proposed system. In Sect. 5, we provide the description of the evaluation system. In Sect. 6, we provide the evaluation results. Finally, conclusions and future work are given in Sect. 7.

2   Related Work

In recent years, new systems are used for rice cultivation, which automatically can measure the water level and water temperature of the paddy field. Also, a cultivation management system was proposed to collect the knowledge and skill of farmers. However, there are many problems in the field of biological information sensing such as the growing conditions and the physiological conditions.

Hokkaido agricultural research center is aiming for developing the unmanned system of agricultural machines (Rice planter, tractor, etc.) equipped with remote monitoring [4]. The system is linked with the quasi-zenith satellite by applying the object detection and straight-line assist.

In [1], the authors presented a framework for classifier fusion, which can support the automatic recognition of fruits and vegetables in a supermarket environment. The authors show that the proposed framework yields better results than several related works found in the literature.

In [19], the authors proposed a target recognition method based on improved VGG19 for quickly and accurately detecting insects in image data.

In [13], the authors proposed a feature reuse residual network for insect pest recognition, but they did not divide the periods of insect growth.

3   Neural Networks

Deep Learning (DL) has a deep hierarchy that connects multiple internal layers for feature detection and representation learning. Representation learning is used to express the extracting essential information from observation data in the real world. Feature extraction uses trial and error by artificial operations. However, DL uses the pixel level of the image as an input value and acquires the characteristic that is most suitable to identify it [3, 6]. The simplest kind of Neural Network (NN) is a single-layer perceptron network, which consists of a single layer of output, the inputs are fed directly to the outputs. In this way, it can be considered the simplest kind of feed-forward network.

The learning method in a CNN uses the backpropagation model like a conventional multi-layer perceptron. To update the weighting filter and coupling coefficient, CNN uses stochastic gradient descent. In this way, CNN recognizes the optimized feature by using convolutional and pooling operations [5,7]. For the task of vegetable classification, Rectified Linear Unit (ReLU) is used in CNN to speed up training. CNN has been applied to object recognition.

4   Proposed System

The structure of our proposed system is shown in Fig. 1. Also, we designed VegeCare, which is a mobile application for managing the plant growth for the farmer.

The processing system runs on Ubuntu Linux 14.04. The CPU is equipped with Intel Core i3 3.3 GHz, NVIDIA GeForce GTX 1060 for GPU, and RAM for 16 GB. Based on insect pest classification characteristics and challenges, we consider the accuracy and loss, which is computed by the TensorFlow framework (version 1.13.1).

4.1   VegeCare Functions

The mobile application called VegeCare has three functions:

1. Insect pest classification,
2. Vegetable classification,
3. Plant disease classification.

The mobile application runs on the Android terminal.

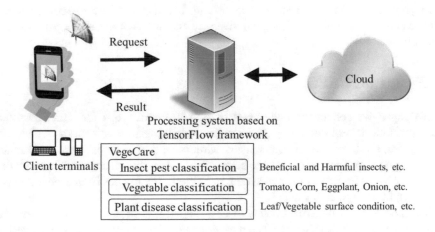

Fig. 1. System model.

4.2   Insect Pest Classification

4.2.1   Overview

We coexist with animals and crops on earth, and crops coexist with insects. Some predatory species of insects are the natural enemies of common pests. Beneficial insects increase the ecological diversity of fields and defend vegetables from harmful insects. Beneficial insects have positive effects on a field, they aid in pollination and some cases serve as natural pesticides. The proposed detection system considers insect classification such as beneficial insects and harmful insects. The system can warn about the problem of harmful insects. We assume to use a camera in a mobile device rather than fixed cameras. The user can take a picture of insects using the mobile application and can determine if the insects are harmful or harmless to vegetables and crops.

The insect pests can categorize complete metamorphosis or incomplete metamorphosis. For complete metamorphosis, there are big differences between larval and adult forms. Complete metamorphosis or holometaboly is almost completely specific to winged insects, that have four different forms such as egg, larva, pupa and adult.

On the other hand, incomplete metamorphosis or hemimetabolous development has only one stage that is anatomically and physiologically different. Thus, hemimetabolous insect has egg, young nymph, late nymph and adult stages. After egg stage, the form is very similar.

4.2.2   Diversity and Difficulty

The insect pests at different stages of the life cycle can damage agricultural products in different degrees. The existing recognition systems are not be able to detect them precisely. To improve the performance of the insect pest classification system, we classify each insect pest considering every stage of their life cycle. For the classification model, classifying them to the same category is difficult because it is hard to extract discriminant features. In addition to biological diversity, the imbalanced data distribution also cannot be ignored.

5   Evaluation Setting

In this paper, we collect the main pests of rice and corn crops for insect pest classification. The dataset contains 2,233 images belonging to 4 categories, which look very similar, but they are different species. Moreover, they have different appearances depending on their life cycle. In some stage of their life, they can resemble each other. We show a list of insect pests for each crop in Table 1. The images from each insect pest are shown from Figs. 2, 3, 4 and 5.

We show a list of hyperparameters of CNN in Table 2. In order to create a learning model, we used 200 epochs and batch size is 32. The image size has been resized to 256×256. The network is based on the sequential model, which consists of three convolutional layers, three pooling layers and four fully connected layers.

Table 1. List of insect pest classification for rice and corn crops.

Crop	Super-class	Class	Memo
Rice	1: Rice leaf roller	Egg Larva Pupa Adult	Harmful insect
	2: Rice leafhopper	Eggs Nymph Adult	Harmful insect
Corn	3: Corn borer	Eggs Larva Pupa Adult	Harmful insect
	4: Armyworm	Eggs Larva Pupa Adult	Harmful insect

(a)   (b)   (c)   (d)

Fig. 2. Different forms of rice leaf roller images containing (a) egg, (b) larva, (c) pupa and (d) adult belonging to the same super-class.

(a)   (b)   (c)

Fig. 3. Different forms of rice leafhopper images containing (a) egg, (b) nymph and (c) adult belonging to the same super-class.

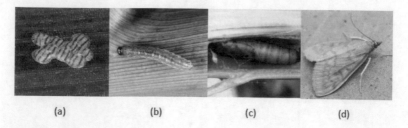

(a)   (b)   (c)   (d)

Fig. 4. Different forms of corn borer images containing (a) egg, (b) larva, (c) pupa and (d) adult belonging to the same super-class.

(a)   (b)   (c)   (d)

Fig. 5. Different forms of armyworm images containing (a) egg, (b) larva, (c) pupa and (d) adult belonging to the same super-class.

Table 2. List of hyperparameters.

Function	Values
Epoch	200
Batch size	32
Filter sizes for convolution layer	3×3
Activation function	ReLU
Loss function	Categorical cross-entropy
Optimizer	RMSprop
Dropout	0.5

6   Evaluation Results

6.1   Learning Results

The dataset contains 2,233 images belonging to 4 categories of insect pests. For the imbalanced data distribution, we split the dataset to 7:1:2. The training, validation and testing sets are split at the sub-class level. We used 1,565, 222 and 446 images respectively. Also, the dataset should be diverse to prevent over-learning and improve classification results. We use various images of insect pests, which are converted from original images by using random zooming and shear conversion.

The results of learning accuracy and loss are shown in Fig. 6. The results of accuracy are increased with the increase of epochs. We observed that the results of accuracy for training are more than 70% after 130 epochs and reached more than 75% at the end of learning. The training loss is less than 0.8. Also, we confirmed that validation accuracy and validation loss improve with the increase of epochs, but they have some oscillations. When classification produces random results, the number of correct random guesses is causing the fluctuation of the accuracy.

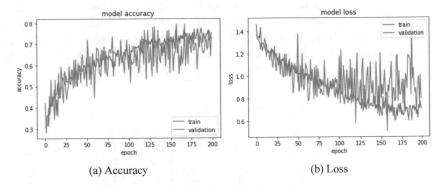

(a) Accuracy   (b) Loss

Fig. 6. Learning results.

6.2 Classification Results

We evaluated the insect pests classification of rice and corn insect pests based on the learning result. In Table 4 are shown the results of insect pest classification for every stage of life cycle. Experimental results show that the proposed tool achieve a good accuracy. We found that the tool can detect corn borer and armyworm which are ambiguous insects. The accuracy is more than 70%.

Rice leaf roller adults are grayish-brown weevils with a dark brown V-shaped mark on their back (see Fig. 2(d)). Even their shape looks similar to the adults' corn borer and armyworm (see Figs. 4(d) and 5(d)), the accuracy of rice leaf roller adults is more than 70%. For the rice leaf roller larva, they are transparent yellow-greenish while the army-worms are opaque yellow-greenish so they are mistaken sometimes (see Figs. 2(b) and 5(b)). The accuracy of rice leaf roller egg and pupa shows that we need more training data to precisely detect them.

The accuracy of rice leafhopper egg and nymph is lower than 34% because of the insufficient training data. For the rice leafhopper adult, they are bright green with vari-able black markings, wedge-shaped with a characteristic diagonal movement. The male insect has a black spot in the middle of the forewings that is absent in females (see Fig. 3(c)). Some of them can only be soft ochre. Mainly due to the biological diversity, the accuracy of rice leaf roller adult is about 44%.

For corn borer, the accuracy of every stage of the life cycle is more than 60%. Especially, the accuracy of the corn borer larva is reached 80%. The larvae tend to be light brown or pinkish-gray in color dorsally, with a brown to the blackhead capsule and a yellowish-brown thoracic plate (see Fig. 4(b)). So, they look different from the others for the same stage of the life cycle. For all of the insect pests at the egg stage, the corn borer's accuracy can be reached to 68.75% because the eggs are oval, flattened, and creamy white in color the same as the others but they usually with an iridescent appearance (see Fig. 4(a)).

The armyworm egg can not be detected because of insufficient training data. The identity of the armyworm egg is the group of laid eggs covered over with grey hairs in company with the others which are spherical and creamy in color as the same (see Fig. 5(a)). Likewise, the pupa is only dark brown which has not the identity of themselves in this way (see Fig. 5(c)), the tool mistakenly detected them as the corn borer pupa. We found that the stage of life cycle such as pupa, eggs and nymph need more training data to detect their identity for classification. However, the accuracy of armyworm adults is more than 75% even their shape looks similar (Table 3).

Table 3. Results for classification.

Super-class	Class	Quantity	Train.	Valid.	Test	Accuracy
1: Rice leaf roller	Eggs	27	19	3	5	40.00
	Larva	317	222	32	63	49.21
	Pupa	54	38	5	11	63.63
	Adult	140	98	14	28	71.43
2: Rice leafhopper	Eggs	13	9	1	3	33.33
	Nymph	34	24	3	7	28.57
	Adult	203	142	20	41	43.90
3: Corn borer	Eggs	80	56	8	16	68.75
	Larva	431	302	43	86	80.23
	Pupa	26	18	3	5	60.00
	Adult	264	185	26	53	64.15
4: Armyworm	Eggs	13	10	1	2	0.00
	Larva	451	316	45	90	77.78
	Pupa	14	10	1	3	33.33
	Adult	166	116	17	33	75.76
Total		2233	1565	222	446	

Table 4. Classification results for each super-class.

Super-class	Accuracy			
	1	2	3	4
1: Rice leaf roller	56.07	4.67	18.69	20.56
2: Rice leafhopper	5.88	39.22	23.53	31.37
3: Corn borer	5.00	3.75	73.13	18.13
4: Armyworm	3.13	1.56	20.31	75.00

7 Conclusions

In this paper, we proposed an insect pest classification tool using the TensorFlow framework. We evaluated the performance of learning accuracy, loss and classification of our proposed tool for insect pests. From the evaluation results, we found that our proposed classification tool can detect corn borer and armyworm which are ambiguous insects. The accuracy is more than 70%.

In future work, we would like to develop an extensional tool to improve the accuracy that can help farmers to prevent damage that occurs with their agricultural products. Moreover, we will consider different structures of CNN.

References

1. Faria, F.A., dos Santos, J.A., Rocha, A., da Silva Torres, R.: Automatic classifier fusion for produce recognition. In: Proceedings of the 25th International Conference on Graphics, Patterns and Images (SIBGRAPI-2012), pp. 252–259 (2012)
2. Gentile, A., Santangelo, A., Sorce, S., Vitabile, S.: Human-to-human interfaces: emerging trendsz and challenges. Int. J. Space Based Situated Comput. 1(1), 3–17 (2011)
3. Hinton, G.E., Osindero, S., Teh, Y.W.: A fast learning algorithm for deep belief nets. Neural Comput. 18(7), 1527–1554 (2006)
4. NARO Hokkaido Agricultural Research Center: HARC brochure. http://www.naro.affrc.go.jp/publicity_report/publication/files/2017NARO_english_1.pdf (2017)
5. Kang, L., Kumar, J., Ye, P., Li, Y., Doermann, D.: Convolutional neural networks for document image classification. In: Proceedings of 22nd International Conference on Pattern Recognition 2014 (ICPR-2014), pp. 3168–3172, August 2014
6. Le, Q.V.: Building high-level features using large scale unsupervised learning. In: Proceedings of IEEE International Conference on Acoustics, Speech and Signal Processing 2013 (ICASSP-2013), pp. 8595–8598, May 2013
7. Lee, H., Grosse, R., Ranganath, R., Ng, A.Y.: Convolutional deep belief networks for scalable unsupervised learning of hierarchical representations. In: Proceedings of the 26th Annual International Conference on Machine Learning. pp. 609–616, June 2009
8. Mahesha, P., Vinod, D.: Support vector machine-based stuttering dysfluency classification using gmm supervectors. Int. J. Grid Utility Comput. 6(3/4), 143–149 (2015)
9. Mattihalli, C., Gedefaye, E., Endalamaw, F., Necho, A.: Plant leaf diseases detection and auto-medicine. Internet of Things 1–2, 67–73 (2018)
10. Mnih, V., Kavukcuoglu, K., Silver, D., Rusu, A.A., Veness, J., Bellemare, M.G., Graves, A., Riedmiller, M., Fidjeland, A.K., Ostrovski, G., Petersen, S., Beattie, C., Sadik, A., Antonoglou, I., King, H., Kumaran, D., Wierstra, D., Legg, S., Hassabis, D.: Human-level control through deep reinforcement learning. Nature 518, 529–533 (2015)
11. Okamoto, K., Yanai, K.: Real-time eating action recognition system on a smartphone. In: Proceedings of the IEEE International Conference on Multimedia and Expo Workshops (ICMEW-2014), pp. 1–6 (2014)
12. Petrakis, E.G.M., Sotiriadis, S., Soultanopoulos, T., Renta, P.T., Buyya, R., Bessis, N.: Internet of Things as a service (iTaaS): challenges and solutions for management of sensor data on the cloud and the fog. Internet of Things 3–4, 156–174 (2018)
13. Ren, F., Liu, W., Wu, G.: Feature reuse residual networks for insect pest recognition. IEEE Access 7, 122758–122768 (2019)

14. Ruedeeniraman, N., Ikeda, M., Barolli, L.: Performance evaluation of vegecare tool for tomato disease classification. In: Proceedings of the 22nd International Conference on Network-Based Information Systems (NBiS-2019), pp. 595–603, September 2019
15. Ruedeeniraman, N., Ikeda, M., Barolli, L.: Tensorflow: a vegetable classification system and its performance evaluation. In: Proceedings of the 13th International Conference on Innovative Mobile and Internet Services in Ubiquitous Computing (IMIS-2019), July 2019
16. Sardogan, M., Tuncer, A., Ozen, Y.: Plant leaf disease detection and classification based on CNN with LVQ algorithm. In: Proceedings of the 3rd International Conference on Computer Science and Engineering (UBMK-2018), pp. 382–385, September 2018
17. Silver, D., Huang, A., Maddison, C.J., Guez, A., Sifre, L., van den Driessche, G., Schrittwieser, J., Antonoglou, I., Panneershelvam, V., Lanctot, M., Dieleman, S., Grewe, D., Nham, J., Kalchbrenner, N., Sutskever, I., Lillicrap, T., Leach, M., Kavukcuoglu, K., Graepel, T., Hassabis, D.: Mastering the game of Go with deep neural networks and tree search. Nature **529**, 484–489 (2016)
18. Silver, D., Schrittwieser, J., Simonyan, K., Antonoglou, I., Huang, A., Guez, A., Hubert, T., Baker, L., Lai, M., Bolton, A., Chen, Y., Lillicrap, T., Hui, F., Sifre, L., van den Driessche, G., Graepel, T., Hassabis, D.: Mastering the game of Go without human knowledge. Nature **550**, 354–359 (2017)
19. Xia, D., Chen, P., Wang, B., Zhang, J., Xie, C.: Insect detection and classification based on an improved convolutional neural network. Sensors **18**(12), 4169 (2018). https://www.mdpi.com/1424-8220/18/12/4169

The Improved Redundant Energy Consumption Laxity-Based Algorithm with Differentiating Starting Time of Process Replicas

Tomoya Enokido[1]($\boxtimes$) and Makoto Takizawa[2]

[1] Faculty of Business Administration, Rissho University, Tokyo, Japan
eno@ris.ac.jp
[2] Department of Advanced Sciences, Faculty of Science and Engineering,
Hosei University, Tokyo, Japan
makoto.takizawa@computer.org

Abstract. Multiple replicas of each application process can be redundantly performed on multiple virtual machines in a server cluster system to realize reliable application services. However, a large amount of electric energy is consumed in a server cluster since multiple replicas of each application process are performed on multiple virtual machines. In this paper, the IRECLB-DST (improved redundant energy consumption laxity-based algorithm with differentiating starting time of process replicas) algorithm is newly proposed to reduce the total electric energy consumption of a server cluster by differentiating starting time of replicas. We evaluate the IRECLB-DST algorithm in terms of the total electric energy consumption of a server cluster and the average response time of each process compared with the RECLB (redundant energy consumption laxity based) algorithm previously proposed.

Keywords: Virtual machines · Green computing · Fault-tolerant · Meaningless replicas · Process replication

1 Introduction

In current information systems, various kinds of distributed applications like vehicle network services [1] are implemented on cloud computing systems [2] since cloud computing systems can provide scalable and fault-tolerant computing systems. A cloud computing system is implemented by using a server cluster system [3–5] composed of a large number of physical servers. Computation resources of each physical server have to be more efficiently utilized to provide scalable and fault-tolerant computing systems in a server cluster system. Virtual machine technologies [5,6] are widely used to more efficiently utilize the computation resources of each physical server in a server cluster system. Some servers might stop by faults [7] in a server cluster system. Then, virtual machines performed on the servers also stop and application processes allocated to the virtual

L. Barolli et al. (Eds.): EIDWT 2020, LNDECT 47, pp. 181–188, 2020.
https://doi.org/10.1007/978-3-030-39746-3_19

machines cannot be successfully terminated. Hence, process replication [8–10] is widely used to reliably perform application processes. However, a server cluster system consumes a large amount of electric energy to perform multiple replicas of each application process on multiple virtual machines. The total electric energy consumption of a server cluster to redundantly perform each application process on multiple virtual machine has to be reduced as discussed in Green computing [2,5,8,11].

In our previous studies, the *RECLB* (*redundant energy consumption laxity based*) algorithm [9] is proposed to reduce the total electric energy consumption of a server cluster for redundantly performing each application process. In the RECLB algorithm, each time a load balancer receives a request process, the load balancer selects a set of virtual machines to redundantly perform the request process so that the total electric energy to redundantly perform the request process can be reduced. Then, the load balancer broadcasts the request process to the selected virtual machines and replicas of the request process is performed on the virtual machines almost at the same time. Here, if a replica of the request process successfully terminates on a virtual machine, other replicas still being performed on other virtual machines are *meaningless*. Here, meaningless replicas are not required to be performed since the request process can commit without performing the meaningless replicas.

In this paper, *IRECLB-DST* (*improved redundant energy consumption laxity-based algorithm with differentiating starting time of process replicas*) algorithm is proposed to reduce the total electric energy consumption of a server cluster by reducing the execution of meaningless replicas. In the IRECLB-DST algorithm, a load balancer serially forwards a request process to each virtual machine every δ time unit. Then, the execution of meaningless replicas can be reduced in the RECLB-DST algorithm. As a result, the total electric energy consumption of a server cluster can be reduced. The evaluation result shows the total electric energy consumption of a server cluster can be more reduced in the IRECLB-DST algorithm than the RECLB algorithm.

In Sect. 2, we discuss the system model. In Sect. 3, we propose the IRECLB-DST algorithm. In Sect. 4, we evaluate the IRECLB-DST algorithm compared with the RECLB algorithm.

2   System Model

2.1   Server Cluster

Multiple servers s_1, ..., s_n ($n \geq 1$) construct a server cluster S. A notation nc_t ($nc_t \geq 1$) denotes the total number of cores in a server s_t. A notation C_t denotes a set of cores c_{1t}, ..., $c_{nc_t t}$ in a server s_t. A notation ct_t ($ct_t \geq 1$) denotes the total number of threads on each core c_{ht} in a server s_t. The total number nt_t ($nt_t \geq 1$) of threads in a server s_t is $nc_t \cdot ct_t$. Each server s_t holds a set TH_t of threads th_{1t}, ..., $th_{nt_t t}$ and threads $th_{(h-1) \cdot ct_t+1}$, ..., $th_{h \cdot ct_t}$ ($1 \leq h \leq nc_t$) are bounded to a core c_{ht}. A set V_t of virtual machines VM_{1t}, ..., $VM_{nt_t t}$ are installed in a server s_t and each virtual machine VM_{kt} is exclusively performed on one thread th_{kt}

in a server s_t. We consider *computation type application processes* (*computation processes*) which mainly consume CPU resources of a virtual machine VM_{kt}. A term *process* stands for a computation process in this paper. Each time a load balancer K receives a request process p^i from a client cl^i, the load balancer K selects a subset VMS^i of virtual machines in the server cluster S and forwards the process p^i to every virtual machines VM_{kt} in the subset VMS^i. On receipt of a request process p^i, a virtual machine VM_{kt} creates and performs a replica p^i_{kt} of a process p^i. On termination of a replica p^i_{kt}, the virtual machine VM_{kt} sends a reply r^i_{kt} to the load balancer K. The load balancer K takes only the first reply r^i_{kt} and ignores every other reply. A notation NF denotes the maximum number of servers which concurrently stop by fault in the cluster S. A notation rd^i denotes the *redundancy* of a process p^i, i.e. $rd^i = |VMS^i|$. We assume $NF + 1 \leq rd^i \leq n$. Hence, a load balancer K can receive at least one reply r^i_{kt} from a virtual machine VM_{kt} even if NF servers stop by fault in a server cluster S. A virtual machine VM_{kt} is *active* iff (if and only if) at least one replica is performed on the virtual machine VM_{kt}. A virtual machine VM_{kt} is *idle* iff the virtual machine VM_{kt} is not active. A core c_{ht} is *active* iff at least one virtual machine VM_{kt} is active on a thread th_{kt} in the core c_{ht}. A core c_{ht} is *idle* if the core c_{ht} is not active.

2.2 Computation Model of a Virtual Machine

A notation $CP_{kt}(\tau)$ denotes a set of replicas being performed on a virtual machine VM_{kt} at time τ. The total number $NC_{kt}(\tau)$ of replicas being performed on a virtual machine VM_{kt} at time τ is $|CP_{kt}(\tau)|$. A notation $minT^i_{kt}$ denotes the minimum computation time of a replica p^i_{kt} where the replica p^i_{kt} is exclusively performed on a virtual machine VM_{kt} and the other virtual machines are idle in a server s_t. The maximum computation rate of every virtual machine VM_{kt} in the server s_t is the same, i.e. $minT^i_{1t} = \cdots = minT^i_{n_t t}$. $minT^i = minT^i_{kt}$ on the fastest server s_t. We assume one virtual computation step [vs] is performed for one time unit [tu] on a virtual machine VM_{kt} in the fastest server s_t. The maximum computation rate $Maxf_{kt}$ of the fastest virtual machine VM_{kt} is 1 [vs/msec]. $Maxf_{ku} \leq Maxf_{kt}$ on a slower server s_u. $Maxf = max(Maxf_{k1}, ..., Maxf_{kn})$. A replica p^i_{kt} is considered to be composed of VS^i_{kt} virtual computation steps. $VS^i_{kt} = minT^i_{kt} \cdot Maxf = minT^i_{kt}$ [vs].

The computation rate $f^i_{kt}(\tau)$ of a replica p^i_{kt} performed on a virtual machine VM_{kt} at time τ is given as follows [9]:

$$f^i_{kt}(\tau) = \alpha_{kt}(\tau) \cdot VS^i / (minT^i_{kt} \cdot NC_{kt}(\tau)) \cdot \beta_{kt}(nv_{kt}(\tau)). \tag{1}$$

Here, $\alpha_{kt}(\tau)$ is the *computation degradation ratio* of a virtual machine VM_{kt} at time τ ($0 \leq \alpha_{kt}(\tau) \leq 1$). $\alpha_{kt}(\tau_1) \leq \alpha_{kt}(\tau_2) \leq 1$ if $NC_{kt}(\tau_1) \geq NC_{kt}(\tau_2)$. $\alpha_{kt}(\tau) = 1$ if $NC_{kt}(\tau) = 1$. $\alpha_{kt}(\tau)$ is assumed to be $\varepsilon_{kt}^{NC_{kt}(\tau)-1}$ where $0 \leq \varepsilon_{kt} \leq 1$. $nv_{kt}(\tau)$ is the number of active virtual machines on a core which performs a virtual machine VM_{kt} at time τ. A notation $\beta_{kt}(nv_{kt}(\tau))$ denotes the *performance degradation ratio* of a virtual machine VM_{kt} at time τ ($0 \leq \beta_{kt}(nv_{kt}(\tau)) \leq 1$) where multiple virtual machines are active on the same core. $\beta_{kt}(nv_{kt}(\tau)) = 1$ if $nv_{kt}(\tau) = 1$. $\beta_{kt}(nv_{kt}(\tau_1)) \leq \beta_{kt}(nv_{kt}(\tau_2))$ if $nv_{kt}(\tau_1) \geq nv_{kt}(\tau_2)$.

Suppose that a replica p_{kt}^i starts and terminates on a virtual machine VM_{kt} at time st_{kt}^i and et_{kt}^i, respectively. Here, the total computation time T_{kt}^i [msec] of a replica p_{kt}^i is et_{kt}^i - st_{kt}^i and $\sum_{\tau=st_{kt}^i}^{et_{kt}^i} f_{kt}^i(\tau) = VS^i$ [vs]. The computation laxity $lc_{kt}^i(\tau)$ [vs] of a replica p_{kt}^i at time τ is VS^i - $\sum_{x=st_{kt}^i}^{\tau} f_{kt}^i(x)$.

2.3 Power Consumption Model of a Server

A notation $E_t(\tau)$ denotes the electric power [W] of a server s_t at time τ. Notations $maxE_t$ and $minE_t$ denote the maximum and minimum electric power [W] of a server s_t, respectively. A notation $ac_t(\tau)$ denotes the number of active cores in a server s_t at time τ. A notation $minC_t$ denotes the electric power [W] where at least one core c_{ht} is active on a server s_t. A notation cE_t denotes the electric power [W] consumed by a server s_t to make one core active.

The electric power $E_t(\tau)$ [W] of a server s_t to perform processes on virtual machines at time τ is given as follows [9]:

$$E_t(\tau) = minE_t + \sigma_t(\tau) \cdot (minC_t + ac_t(\tau) \cdot cE_t). \qquad (2)$$

Here, $\sigma_t(\tau) = 1$ if at least one core c_{ht} is active on a server s_t at time τ. Otherwise, $\sigma_t(\tau) = 0$.

The total electric energy $TE_t(\tau_1, \tau_2)$ [J] of a server s_t from time τ_1 to τ_2 is $\sum_{\tau=\tau_1}^{\tau_2} E_t(\tau)$. The processing power $PE_t(\tau)$ [W] of a server s_t at time τ is $E_t(\tau)$ - $minE_t$. The total processing electric energy $TPE_t(\tau_1, \tau_2)$ [J] of a server s_t from time τ_1 to τ_2 is $\sum_{\tau=\tau_1}^{\tau_2} PE_t(\tau)$. The total processing electric energy laxity $tpel_t(\tau)$ [J] shows how much electric energy a server s_t has to consume to perform every replica being performed on every active virtual machine in the server s_t at time τ. In our previous studies, the ELaxity algorithm [9] is proposed to obtain the total processing electric energy laxity $tpel_t(\tau)$ of a server s_t at time τ. Suppose a load balancer K receives a new request process p^i from a client cl^i and would like to allocate a replica p_{kt}^i to a virtual machine VM_{kt} performed on a server s_t at time τ. Then, the load balancer K can obtain the total processing electric energy laxity $tpel_t(\tau)$ of the server s_t at time τ where the replica p_{kt}^i is allocated to the virtual machine VM_{kt} at time τ by using the ELaxity algorithm.

3 Replication Algorithm

3.1 Meaningless Replicas

The RECLB (redundant energy consumption laxity based) algorithm [9] is proposed to select multiple virtual machines for redundantly performing each process so that the total processing electric energy consumption of a server cluster can be reduced. In the RECLB algorithm, a subset VMS^i of virtual machines in a server cluster S, whose total amount of processing electric energy laxity $\sum_{VM_{kt} \in VMS^i} tpel_t(\tau)$ is the minimum, is selected for each request process p^i at time τ. Then, a load balancer K broadcasts the request process p^i to every virtual machine VM_{kt} in the subset VMS^i. Here, every replica p^i_{kt} of a process p^i is started on each virtual machine VM_{kt} in the subset VMS^i almost at the same time. The process p^i commits when the load balancer K receives the first reply r^i_{kt} from a virtual machine VM_{kt} in the subset VMS^i.

Suppose a replica p^i_{kt} of a process p^i successfully terminates on a virtual machine VM_{kt} but another replica p^i_{ku} is still being performed on another virtual machine VM_{ku}. Here, the replica p^i_{ku} is *meaningless* since the replica p^i_{kt} already successfully terminates and the process p^i can commit without performing the replica p^i_{ku}.

[Definition]. A replica p^i_{ku} is *meaningless* iff the replica p^i_{ku} is to be performed after another replica p^i_{kt} successfully terminates.

3.2   The IRECLB-DST Algorithm

The *IRECLB-DST (improved redundant energy consumption laxity-based algorithm with differentiating starting time of process replicas)* algorithm is proposed to reduce the total electric energy consumption of a server cluster by reducing the execution of meaningless replicas. A notation δ^i denotes the inter-request time [msec] to forward a request process p^i to each virtual machine VM_{kt} in the subset VMS^i. In the IRECLB-DST algorithm, a load balancer K serially forwards a request process p^i to each virtual machine VM_{kt} in the subset VMS^i every δ^i time unit in order to reduce the execution of meaningless replicas.

Suppose a load balancer K receives a request process p^i and the redundancy rd^i of the process p^i is three ($rd^i = 3$). Suppose a load balancer K receives a first reply r^i_{kt} from a virtual machine VM_{kt} while forwarding the request process p^i to virtual machines in a subset $VMS^i (= \{VM_{kt}, VM_{ku}, VM_{kv}\})$ as shown in Fig. 1. Here, a replica p^i_{ku} performed on the virtual machine VM_{ku} is meaningless since a replica p^i_{kt} has correctly terminated on the virtual machine VM_{kt} and the load balancer K has received a reply r^i_{kt}. On the other hand, the load balancer K stops forwarding the third request to perform a replica p^i_{kv} to a virtual machine VM_{kv} since the load balancer K has received a reply r^i_{kt} from the virtual machine VM_{kt} before time $\tau_3 (= \tau_2 + \delta^i)$. In the IRECLB-DST algorithm, the number of replicas for a process p^i might be smaller than the redundancy rd^i while rd^i replicas are performed for every process p^i in the RECLB algorithm. As a result, the total electric energy consumption of a server cluster can be reduced by differentiating the starting time of each replica since the execution of meaningless replicas can be reduced.

Suppose a pair of replicas p^i_{kt} and p^i_{ku} are performed on a pair of virtual machines VM_{kt} and VM_{ku}, respectively, as shown in Fig. 2. A notation $d_{K,kt}$ denotes the delay time between a load balancer K and a virtual machine VM_{kt}. Suppose the virtual machine VM_{kt} stops by fault while performing the replica p^i_{kt}. The load balancer K can receive a reply r^i_{ku} from the virtual machine VM_{ku}. Suppose the virtual machine VM_{ku} is the q-th virtual machine ($1 \leq q \leq rd^i$)

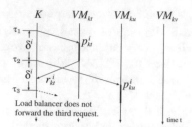

Fig. 1. The stop forwarding.

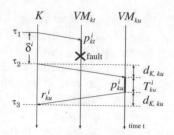

Fig. 2. The response time.

where the load balancer K forwards a request process p^i. Then, the response time of a process p^i is $(q-1) \cdot \delta^i + 2 \cdot d_{K,ku} + T_{ku}^i$. In Fig. 2, the response time of a process p^i is $\delta^i + 2 \cdot d_{K,ku} + T_{ku}^i$.

4   Evaluation

We evaluate the IRECLB-DST algorithm in terms of the total processing electric energy consumption of a server cluster and the average response time of each process compared with the RECLB algorithm. A homogeneous cluster S is composed of three servers s_1, s_2, s_3 ($n = 3$). Every server s_t ($1 \leq t \leq 3$) follows the same computation model and the same power consumption model. Every server s_t is equipped with a dual-core CPU ($nc_t = 2$). Two threads are bounded for each core in a server s_t, i.e. $ct_t = 2$. The number nt_t of threads in each server s_t is four, i.e. $nt_t = nc_t \cdot ct_t = 2 \cdot 2 = 4$. $minE_t = 14.8$ [W], $minC_t = 6.3$ [W], $cE_t = 3.9$ [W], and $maxE_t = 33.8$ [W]. Each virtual machine VM_{kt} is bounded to one thread th_{kt} in a server s_t ($k = 1, ..., 4$ and $t = 1, 2, 3$). There are twelve virtual machines in the server cluster S. Every virtual machine VM_{kt} follows the same computation model. The maximum computation rate $Maxf_{kt}$ is 1 [vs/msec]. The parameter ε_{kt} in the computation degradation ratio $\alpha_{kt}(\tau)$ is 1. The performance degradation ratios $\beta_{kt}(1) = 1$ and $\beta_{kt}(2) = 0.5$. We assume the fault probability fr_t for every server s_t is the same $fr = 0.1$.

The number m of processes $p^1, ..., p^m$ ($0 \leq m \leq 100,000$) are issued in the simulation. The starting time of each process p^i is randomly selected in a unit of one millisecond between 1 and 60 [sec]. The minimum computation time $minT^i$ of every process p^i is assumed to be 1 [msec]. We assume the redundancy rd^i for each process p^i is the same $rd = 3$ and $NF = rd - 1 = 2$. The delay time $d_{K,kt}$ of every pair of a load balancer K and every virtual machine VM_{kt} is 1 [msec] in the server cluster S. The minimum response time of every process p^i is $2d_{K,kt} + minT_{kt}^i = 2 \cdot 1 + 1 = 3$ [msec].

Figure 3 shows the average total processing electric energy consumption of the server cluster S to perform the total number m of processes in the RECLB and IRECLB-DST algorithms. In Fig. 3, IRECLB-DST(δ) stands for the total processing electric energy consumption of the server cluster S in the IRECLB-DST algorithm with inter-request time δ ($= \{1, 2, 3, 4, 5\}$). In the RECLB

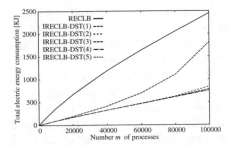 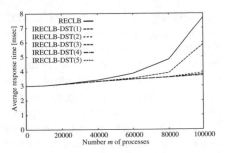

Fig. 3. The electric energy consumption.   **Fig. 4.** The response time of each process.

and IRECLB-DST(δ) algorithms, the average total processing electric energy consumption increases as the number m of processes increases. For $1 \leq \delta \leq 5$, the average total processing electric energy consumption can be more reduced in the IRECLB-DST(δ) algorithm than the RECLB algorithm since the execution of meaningless replicas can be reduced by differentiating starting time of replicas in the IRECLB-DST(δ) algorithm. The larger inter-request time δ, the smaller total processing electric energy is consumed in the server cluster S in the IRECLB-DST(δ) algorithm.

Figure 4 shows the average response time of each process in the RECLB and IRECLB-DST algorithms. In Fig. 4, IRECLB-DST(δ) stands for the average response time of each process in the IRECLB-DST algorithm with inter-request time δ ($= \{1, 2, 3, 4, 5\}$). The average response time of each process increases as the number m of processes increases in the RECLB and IRECLB(δ) algorithms. The execution of meaningless replicas can be reduced by differentiating starting time of replicas in the IRECLB-DST(δ) algorithm. Then, the computation resources of each virtual machine to perform meaningless replicas can be used to perform other replicas. As a result, for $1 \leq \delta \leq 5$, the average response time of each process can be more reduced in the IRECLB-DST(δ) algorithm than the RECLB algorithm. The larger inter-request time δ, the shorter average response time each process takes in the IRECLB-DST(δ) algorithm.

From the evaluation, the average total processing electric energy consumption of a homogeneous server cluster S to redundantly perform processes on virtual machines is shown to be more reduced in the IRECLB-DST algorithm than the RECLB algorithm for $rd = 3$ and $1 \leq \delta \leq 5$. In addition, the average response time of each process is shown to be more reduced in the IRECLB-DST algorithm than the RECLB algorithm for $rd = 3$ and $1 \leq \delta \leq 5$. Following the evaluation, we conclude the IRECLB-DST algorithm is more useful in a homogeneous server cluster than the RECLB algorithm.

5   Concluding Remarks

In this paper, we proposed the IRECLB-DST algorithm to reduce the total electric energy consumption of a server cluster by differentiating starting time of process replicas. In the IRECLB-DST algorithm, the average total processing electric energy consumption of a server cluster and the average response time of each process can be furthermore reduced than the RECLB algorithm since the execution of meaningless replicas can be reduced by differentiating starting time of each replica. We evaluated the IRECLB-DST algorithm in terms of the total processing electric energy consumption of a server cluster and the average response time of each process compared with the RECLB algorithm. The average total processing electric energy consumption of the server cluster and the average response time of each process are shown to be more reduced in the IRECLB-DST algorithm than the RECLB algorithm.

References

1. Bylykbashi, K., Spaho, E., Barolli, L., Xhafa, F.: Routing in a many-to-one communication scenario in a realistic VDTN. J. High Speed Netw. **24**(2), 107–118 (2018)
2. Natural Resources Defense Council (NRDS): Data center efficiency assessment - scaling up energy efficiency across the data center Industry: Evaluating key drivers and barriers (2014). http://www.nrdc.org/energy/files/data-center-efficiency-assessment-IP.pdf
3. Enokido, T., Takizawa, M.: Integrated power consumption model for distributed systems. IEEE Trans. Ind. Electron. **60**(2), 824–836 (2013)
4. Enokido, T., Aikebaier, A., Takizawa, M.: An extended simple power consumption model for selecting a server to perform computation type processes in digital ecosystems. IEEE Trans. Industr. Inf. **10**(2), 1627–1636 (2014)
5. Enokido, T., Duolikun, D., Takizawa, M.: The energy consumption laxity-based algorithm to perform computation processes in virtual machine environments. Int. J. Grid Utility Comput. (IJGUC) **10**(5), 545–555 (2019)
6. Red Hat OpenShift Online: Kernel Virtual Machine (2018). http://www.linux-kvm.org/page/Main_Page
7. Lamport, R., Shostak, R., Pease, M.: The byzantine generals problems. ACM Trans. Prog. Lang. Syst. **4**(3), 382–401 (1982)
8. Enokido, T., Duolikun, D., Takizawa, M.: Energy consumption laxity-based quorum selection for distributed object-based systems. Evol. Intel. (2018). https://doi.org/10.1007/s12065-018-0157-1
9. Enokido, T., Takizawa, M.: An energy-efficient process replication algorithm in virtual machine environments. In: Proceedings of the 11th International Conference on Broadband and Wireless Computing, Communication and Applications (BWCCA-2016), pp. 105–114 (2016)
10. Schneider, F.B.: Replication management using the state-machine approach. In: Distributed Systems, 2nd edn. ACM Press, Vancouver (1993)
11. Khan, S., Kolodziej, J., Li, J., Zomaya, A.Y.: Evolutionary Based Solutions for Green Computing. Springer, Heidelberg (2013)

Implementations on Static Body Detections by Locational Sensors on Mobile Phone for Disaster Information System

Noriki Uchida[1]([✉]), Misaki Fukumoto[1], Tomoyuki Ishida[1], and Yoshitaka Shibata[2]

[1] Fukuoka Institute of Technology, 3-30-1 Wajirohigashi,
Fukuoka Higashi-ku, Fukuoka 811-0214, Japan
{n-uchida,t-ishida}@fit.ac.jp, s16b1042@bene.fit.ac.jp
[2] Iwate Prefectural University, 152-52 Sugo, Takizawa, Iwate 020-0693, Japan
Shibata@iwate-pu.ac.jp

Abstract. It is considered that various disaster information services on smartphones have widely spread over the world, but that some significant subjects for the usages of the injured evacuators or information weakness still exist. Therefore, this paper introduces the life safety information system for smartphones with the detections of the abnormal body static conditions by the sensors on smartphones. In the proposed methods, the locational sensor on the smartphone firstly observes the movements of evacuators after the detection of the earthquake by the gyro sensors on the smartphone. Secondly, the observed values over the locational and time threshold are checked by the gyro sensors and the battery conditions for the detection of the abnormal static body conditions. At last, the calculated results are used for the autonomous emergent messages and the DTN (Delay Tolerant Networks) routing based on the Data Triage Method in the disaster information system. Then, the paper reports the implementation of the prototype system for the proposed methods, and the future works including the additional algorithm based on the Markov Chain Model are discussed.

1 Introduction

It is considered that the DIS (Disaster Information System) such as the life safety information system or the disaster alert system has spread over the world after the East Japan Great Earthquake in 2011. Especially, the previous paper [1] reports the communication means just after the disasters, and it is supposed that the applications on smartphones are significantly useful methods for emergent communication means. In fact, the life safety information services such as the Google Person Finder [2] are well known for the disaster information services, and various applications are currently provided for the smartphones.

However, it is considered that there are some significant subjects for the current systems on smartphones. One of the subjects is the usages of the injured evacuators. If the evacuators get injured by the disasters, the users would be hard to operate the smartphones in order to send the emergent messages. Secondly, the information weakness who are not good at the operations of the smartphone would be hard to send the messages. Especially,

© Springer Nature Switzerland AG 2020
L. Barolli et al. (Eds.): EIDWT 2020, LNDECT 47, pp. 189–196, 2020.
https://doi.org/10.1007/978-3-030-39746-3_20

older people are likely to get injured by the disaster, and those people might be hard to use the applications.

Therefore, this paper introduces the life safety information system for smartphones with the detections of the abnormal body static conditions [3] by the sensors on smartphones. In the proposed methods, the locational sensor on the smartphone firstly is observed for the movements of evacuators after the detection of the earthquake by the gyro sensors. Secondly, the observed values over the locational and time threshold are checked for the detection of the abnormal static body conditions. Then, the gyro sensors and the battery conditions are checked for the static body conditions. At last, the calculated results are used for the autonomous emergent messages and the DTN (Delay Tolerant Networks) routing [4] based on the Data Triage Method [5] in the disaster information system.

Then, this paper reports the way of the implementations for the prototype system of the proposed methods, and the experimental results are presented for the future works of this studies. It is also discussed about the additional anomaly detection algorithm based on Markov Chain Model [6] for the future works.

In the following, section II introduces the system configuration and major functions of the assumed DIS, and section III explains the proposed static body detection methods on smartphone and the anomaly detection algorithm based on Markov Chain Model. Then, the implementations of the prototype system are discussed in section V, and the conclusion and future study are discussed in section VI.

2   Proposed Disaster Information System

The assumed DIS networks consist of the DTN networks and the IP networks as shown in Fig. 1. If the communication networks were damaged by the disaster, the assumed DIS would communicate the emergent messages such as life safety or the rescue by the DTN routing. Then, the message from the evacuators are duplicated until the gateway smartphone that is located in the IP network available area. Then, the emergent messages are transmitted to the DIS servers that is previously configured.

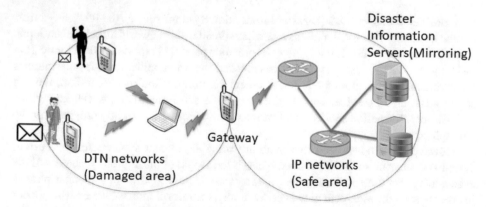

Fig. 1. The assumed network configurations in the DIS.

In details, in the DTN networks, the evacuators can firstly operate the application of the life safety information in order to tell their safety information for their relatives as usual. However, in the proposed system, if the evacuators were not able to operate the applications because of the injury by the disaster, the application automatically start to observe their locational data in order to sense the abnormal static body conditions. Then, when the application detects the abnormal static body conditions, the emergent messages are automatically created and transmitted with their locations and the priority levels of the emergency for the other transmittable smartphone by the DTN routing. The static body conditions are continuously detected by certain periods, and the priority level is increased during the smartphone senses the abnormal static body conditions in the application.

Next, the priority level is used for the DTN routing with the Data Triage Method, and so the high priority messages are delivered to the DIS servers in the IP networks as shown in Fig. 2.

FIFO(Epidemic)   Priority Order(Data Triage)

Fig. 2. The message duplications by the Epidemic Method and the Data Triage Methods in the DTN routing.

The Data Triage Method [5] is one of the queue restored typed DTN routing such as the Maxprop [8], and the stored messages in the queue on the smartphone rearranged by the priority levels of the messages unlike the Maxprop. Although the previous papers such as [9, 10] pointed out the lower delivery rates and latency of the epidemic typed DTN routing, the papers such as [7] discussed the effectivity of the DTN in the robust network conditions.

Then, the evacuator's messages are duplicated to another smartphone by the GSM/LTE/5G or WiFi D2D (Device-to-Device) in the damaged areas when the other smartphone are located in the radio transmittable range. Here, the global IP address of the DIS servers are previously configured in the application, and so the smartphones automatically merge the evacuator's messages to the DIS server when the smartphones reach within the IP network areas [7].

3 Static Body Detections

In the proposed DIS, the priority levels are changed by the static body detections [3] on smartphones as previously mentioned. Currently, the merchants of the smartphones usually equip various sensors such as the GPS, gyro, proximity, and so on. Here, the

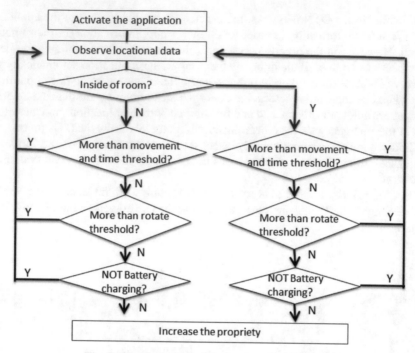

Fig. 3. The flow chart of the static body detections.

proposed detection method continuously observes the locational data, the gyro sensor, and the battery conditions, and Fig. 3 shows the flow chart of the method.

At first, after the activation of the application by the certain gyro sensor's pattern of the earthquake, the locational data on smartphones are continuously observed with the local time on smartphones. Then, by using the noise of the observed locational data, the smartphones are detected whether inside of the room or not. This is because the locational data has higher noises when the smartphone is located inside of the room. Although most of the current smartphones use the Wi-Fi and Bluetooth locational data other than the GPS data in order to improve the accuracy of the locations, there are still significant differences between the inside and the outside of rooms.

Next, the proposed body detection methods check whether the observed locational data on the smartphones is more than the movement and time threshold values or not as shown in Fig. 4. In this paper, the movement and time threshold values are previously determined by the field experiments of the prototype system.

As shown in Fig. 4, the abnormal static body detects if the locational data is not moving during certain periods. Then, if the abnormal static body is detected, the rotational data on the smartphones by the gyro sensor is checked. According to field experiments of the prototype system, the high values of the locational error sometimes occurs on the smartphones, and so it is considered that the additional error check function is necessary for the assumptions of the actual usages. At last, the battery charging is checked for the detection of the abnormal static body condition in the system.

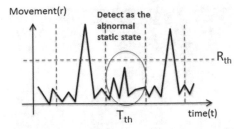

Fig. 4. Abnormal static body detections by the movement and time threshold.

Additionally, the autonomous calculations by the movement and time threshold values are planning for the future works. Since this paper applies the fixed threshold values in the case of inside and outside of rooms from the field experiments, it is necessary to consider various radio conditions in the actual fields. The anomaly detection algorithm based on Markov Chain Model is one of the considerable algorithms for the future works. In detail, the authors are implementing the MCMS functions for the calculations of the movement and time threshold values. In the MCMS calculations [4], it approximates the value function $V(s)$ and the motivation function $Q(s, a)$ by the reward r_i as the formula (1) and formula (2), where α is the learning rate $(0 < \alpha < 1)$.

$$V(s) \leftarrow V(s) + \alpha[R_i - V(s)] \tag{1}$$

$$Q(s, a) \leftarrow Q(s, a) + \alpha[R_i - Q(s, a)] \tag{2}$$

Here, R_i is also defined as the formula (3).

$$R_l - r_{i\,|\,1} + \gamma r_{i+2} + \gamma^2 r_{i+3} + \ldots = \sum_{r=0}^{\infty} \gamma^r r_{i+1+\gamma} \tag{3}$$

The MCMS based calculations are held for the setting of the movement threshold (Rth) and the time threshold (Tth) in the proposed methods for considering the different environments in the injured victims.

4 Prototype System

The implementations are confirmed by Google Pixel3a (4 GB MEM, 64 GB Storage, Android OS 9.0, IEEE802.11a/b/g/n/ac), Java 8.0, and Android Studio 3.5. Also, the SQLite is introduced for the prototype system, and the LocationManager in the Android API is mainly used for the implementations.

Figure 5 shows the captured window menu of the prototype system. The captured window shows the observed locational data and the rotational data on the smartphone, and the inside of rooms and the battery charging is also detecting in the application.

Fig. 5. Captured windows of the prototype system

Then, the beep alert and pop-up messages are shown if the abnormal static body condition is detected, and the priority level is attached to the transmitted messages in the prototype system. Figures 6 and 7 shows the results of the locational data and the rotational data in the field experiments that the smartphone is located inside of rooms. Here, the threshold values of the movements are set to ±0.0003, and the rotations are ±0.003 by the previous field experiments.

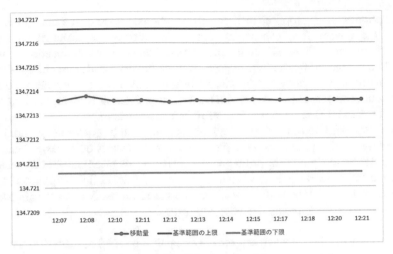

Fig. 6. Experimental results of the locational data on smartphones

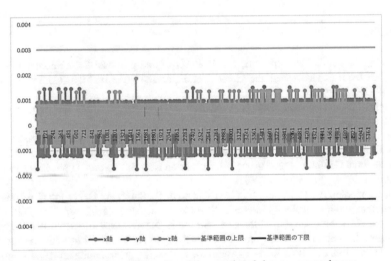

Fig. 7. Experimental results of the rotational data on smartphones

According to the results, it is obvious that the observed values of the locational and gyro sensors are within the higher and the lower threshold values, and so the evaluator is sensed as the abnormal body static conditions in this field experiments.

5   The Conclusion and Future Study

It is considered that various DIS services on smartphones have widely spread over the world, but it is necessary to consider the injured evacuators or the information weakness for the DIS. Therefore, this paper introduces the life safety information system for smartphones with the detections of the abnormal body static conditions by the sensors on smartphones. In the proposed methods, the locational sensor on the smartphone firstly observes the movements of evacuators after the detection of the earthquake by the gyro sensors on the smartphone. Secondly, the observed values over the locational and time threshold are checked by the gyro sensors and the battery conditions for the detection of the abnormal static body conditions. At last, the calculated results are used for the autonomous emergent messages and the DTN (Delay Tolerant Networks) routing based on the Data Triage Method in the disaster information system.

Then, the paper introduced the implementations of the prototype system, and the results of the field experiments were presented. As the results of the experiments, it is considered the effectivity of the prototype system for the abnormal static body detections.

For the future works, we now planning for the additional field experiments by the prototype system, and the additional algorithm based on the Markov Chain Model is implementing for the anomaly detection algorithm.

Acknowledgment. This work was supported by supported by JSPS KAKENHI Grant Number 19K04972.

References

1. Shibata, Y., Uchida, N., Shiratori, N.: Analysis of and proposal for a disaster information network from experience of the Great East Japan Earthquake. IEEE Commun. Mag. **52**, 44–50 (2014)
2. Google: Google Person Finder. https://www.google.org/personfinder/global/home.html. Accessed Jan 2018
3. Uchida, N., Shingai, T., Shigetome, T., Ishida, T., Shibata, Y.: Proposal of static body object detection methods with the DTN routing for life safety information systems. In: The 32nd International Conference on Advanced Information Networking and Applications Workshops (WAINA 2018), pp. 112–117 (2018)
4. Fall, K., Hooke, A., Torgerson, L., Cerf, V., Durst, B., Scott, K.: Delay-tolerant networking: an approach to interplanetary Internet. IEEE Commun. Mag. **41**(6), 128–136 (2003)
5. Uchida, N., Kawamura, N., Shibata, Y., Shiratori, N.: Proposal of data triage methods for disaster information network system based on delay tolerant networking. In: The 7th International Conference on Broadband and Wireless Computing, Communication and Applications (BWCCA 2013), pp. 15–21 (2013)
6. Andrieu, C., Freitas, N.D., Doucet, A., Jordan, M.I.: An introduction to MCMC for machine learning. Mach. Learn. **50**(2), 5–43 (2003)
7. Uchida, N., Kawamura, N., Sato, G., Shibata, Y.: Delay tolerant networking with data triage method based on emergent user policies for disaster information networks. Mob. Inf. Syst. **10**(4), 347–359 (2014)
8. Burgess, J., Gallagher, B., Jensen, D., Levine, B.: MaxProp: routing for vehicle-based disruption-tolerant networks. In: Proceedings INFOCOM 2006. 25th IEEE International Conference on Computer Communications, pp. 1–11 (2006)
9. Spyropoulos, T., Psounis, K., Raghavendra, C.S.: Spray and wait: an efficient routing scheme for intermittently connected mobile networks. In: WDTN 2005 Proceedings of the 2005 ACM SIGCOMM Workshop on Delay-Tolerant Networking, pp. 252–259 (2005)
10. Lindgren, A., Doria, A., Scheln, O.: Probabilistic routing in intermittently connected networks. ACM SIGMOBILE Mob. Comput. Commun. Rev. **7**(3), 19–20 (2003)

A DQN Based Mobile Actor Node Control in WSAN: Simulation Results of Different Distributions of Events Considering Three-Dimensional Environment

Kyohei Toyoshima[1], Tetsuya Oda[1(✉)], Masaharu Hirota[2], Kengo Katayama[1], and Leonard Barolli[3]

[1] Department of Information and Computer Engineering,
Okayama University of Science (OUS),
1-1 Ridaicho, Kita-ku, Okayama 700–0005, Japan
t18j056tk@ous.jp, {oda,katayama}@ice.ous.ac.jp
[2] Department of Information Science, Okayama University of Science (OUS),
1-1 Ridaicho, Kita-ku, Okayama 700–0005, Japan
hirota@mis.ous.ac.jp
[3] Department of Information and Communication Engineering,
Fukuoka Institute of Technology (FIT),
3-30-1 Wajiro-Higashi, Higashi-Ku, Fukuoka 811–0295, Japan
barolli@fit.ac.jp

Abstract. Wireless Sensor Actor Networks (WSANs) consist of wireless network nodes, with the ability to sense events (sensors) and to perform actuations (actors) based on the sensing data collected by all sensors. This paper describes a design of a simulation system based on Deep Q-Network (DQN) for actor node mobility control in WSANs. DQN is a deep neural network structure used for estimation of Q value of the Q-learning technique. The proposed simulation system is implemented in Rust programming language. We evaluate the performance of the proposed system for different distributions of event placement considering three-dimensional environment. For this scenario, the simulation results show that for normal distribution of events actor nodes are connected in the best case.

1 Introduction

Wireless Sensor Actor Networks (WSANs) consist of wireless network nodes, with the ability to sense events (sensors) and to perform actuations (actors) based on the sensing data collected by all sensors [1,2]. WSANs include data processing, mobile controls, and functionalities of the different types of sensors and actuations. Actor nodes can be integrated with sensor and actor nodes which include both sensing and actuation functions. The integrated sensor and actor nodes are resource rich nodes that are equipped with better processing power,

© Springer Nature Switzerland AG 2020
L. Barolli et al. (Eds.): EIDWT 2020, LNDECT 47, pp. 197–209, 2020.
https://doi.org/10.1007/978-3-030-39746-3_21

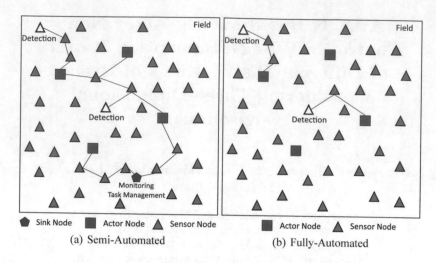

Fig. 1. WSAN architectures.

higher transmission power, more energy resources and may contain additional capabilities such as mobility. An application of such node is the swarm robot. In such applications, the robots have both sensing and actuation capabilities, function as integrated sensor and actor nodes.

One advantage of mobile actor node is that more effective actions can be taken, since the actor node can get as close as possible to the event location. Additionally, in a dense network the presence of a mobile actor eases the problem of overloaded forwarding nodes by balancing the load among all the nodes in the network. In a sparse network, a mobile actor node can connect between groups of isolated nodes. However, there is no guarantee that the existing mobility models will work efficiently for a mobile actor node in a sparse network. An example application is waterfront monitoring where sensor nodes may be deployed to monitor water levels or meteorological conditions, wave characteristics and water temperature. When an event occurs in this environment, a possible action taken by the actor node could be of collecting a water sample for further analysis.

One of the main advantages of WSANs is the ability to exploit node mobility for various purposes [3]. Several performance metrics including connectivity, accuracy, dependability, coverage and energy can be improved by moving and relocating various nodes in these networks [4,5].

WSANs requires actor node connectivity to perform collaborative tasks. Mobility improves the reliability of these networks in various ways. For example, by continuously carrying spare nodes, the failed nodes can be replaced with another spare nodes. Likewise, if the network is partitioned, mobility can restore connectivity by moving one or more nodes to the selected locations. Recently, the above dependency issues has been extensively studied from the context of WSANs [6,7]. These tasks were focused on handling individual failures and restoring connectivity of actor nodes or connectivity of WSN with the sink nodes.

On the other hand, there are many critical issues in WSANs. In the specific application, different objectives can be taken into account such as energy consumption, network performance, coverage, mobility control, etc [8,9]. This paper proposes a simulation system based on Deep Q-Network (DQN) as a controller for actor node mobility which enables different tests on connectivity and coverage problems in WSAN. The performance of proposed system is evaluated for different distributions of events considering three-dimensional environments [10–13].

The paper is structured as follows. Section 2 describes the basics of WSANs including architectures and research challenges. Section 3 presents a brief introduction of DQN. Section 4 gives the description and design of the simulation system. In Sect. 5, we show our simulation results. Finally, conclusions and future work are given in Sect. 6.

2 WSAN

2.1 WSAN Architecture

The development of Internet of Things (IoT) devices have contributed in a way to the recent expansion of WSANs [14]. WSANs are normally composed of a large number of nodes, which are low power devices generally equipped with one or more sensing units (composed of sensors and, eventually, analog to digital converters), an actuator, a processing unit that includes the power supply, a radio unit, and a limited memory.

The main functionality of the WSANs is to allow the actor node to perform the appropriate action in the environment based on the data sensed by the sensor node or another actor nodes. If it is necessary to transmit important data (an event occurred), the sensor nodes may transmit data back to the sink node, which controls the tasks of the actor node from the distance or transmit their data to the actor nodes (These actor nodes can operate independently of the sink node). Here, the former is called Semi-Automated Architecture (as seen in Fig. 1(a)), the latter is called Fully-Automated Architecture (Fig. 1(b)). Obviously, both architectures can be used in different applications. This paper emerges the need to develop new sophisticated algorithms for Fully-Automated Architecture to realize appropriate coordination between WSAN nodes. On the other hand, it has advantages, such as *higher local position accuracy, low energy consumption, higher reliability, low latency, long network lifetime* and so on.

2.2 WSAN Challenge

Some of the key challenges in WSAN are related to the participation of actors and their functionalities [15].

Mobility: In WSANs, sensor nodes and actor nodes can be mobile [15]. For instance, robots used in flying drones over a disaster recovery area or industrial monitoring sites. Thus, protocols developed for WSANs must support the mobility of sensor nodes and actor nodes, where dynamic topology changes, unstable routes and network isolations are present.

Placement and Positioning: WSAN are heterogeneous networks, where actor nodes and sensor nodes have different mobility abilities, processing powers and functionalities [16]. Therefore, node placement algorithms should consider to optimize the number of sensor nodes and actor nodes and their initial locations based on application [17,18].

Node Heterogeneity: WSAN is composed of many sensor nodes and actors nodes. Sensor nodes are cheap, small devices with limited wireless communication, computation and sensing capabilities [19]. But, since actuation is more complicated and energy consuming activity than sensing, actor nodes are resource-rich nodes equipped with longer battery life, stronger transmission powers and better processing capabilities [20].

Power Management: Same as energy constrained WSNs, in WSANs sensor nodes have limited power supplies, which limits the network lifetime [21]. Actor nodes have more powerful power supplies but their functionalities are more sophisticated, so they spend more energy when completing complex tasks. Therefore, WSAN protocols must be designed with minimized energy consumption for both sensor nodes and actor nodes [22].

Self Healing: One of the serious problems in mobile Self Organizing Networks is the frequent node isolations when the network is running. Faults in the actor node can cause network isolation and further hinder the fulfillment of application requirements. A lot of research have been done to recover connectivity by utilizing the function of the actor nodes that does not require so much cost for movement. The actor node may also be specialized to perform extra energy supply to charge sensor nodes or other actor nodes in the network [23].

Scalability: Smart Cities are emerging fast and WSAN, with its practical functions of simultaneous sensing and actuating will play an important role. The heterogeneity is not limited and most of the systems will continue to grow together with cities. In order to keep the functionality of WSAN applicable, scalability should be considered when designing WSAN protocols and algorithms. Data replication, clustering and so on, can be used in order to support growing networks [24].

Algorithm 1. Deep Q-learning with Experience Replay

1: Initialize replay memory D to capacity N
2: Initialize action-value function Q with random weights
3: **for** $episode = 1, M$ **do**
4:     Initialise sequence $m_1 = \{v_1\}$ and preprocessed sequenced $\phi_1 = \phi(m_1)$
5:     **for** $t = 1, T$ **do**
6:         With probability ε select a random action a_t
7:         otherwise select at $= \max_a Q^*(\phi(m_t), a; \theta)$
8:         Execute action at in emulator and observe reward r_t and image v_{t+1}
9:         Set $m_{t+1} = m_t, a_t, v_{t+1}$ and preprocess $\phi_{t+1} = \phi(m_{t+1})$
10:       Store transition $(\phi_t, a_t, r_t, \phi_{t+1})$ in D
11:       Sample random minibatch of transitions $(\phi_j, a_j, r_j, \phi_{j+1})$ from D
12:       Set $y_j = \begin{cases} r_j & \text{for terminal } \phi_{j+1} \\ r_j + \max_a(\phi_{j+1}, a_j; \theta) & \text{for non-terminal } \phi_{j+1} \end{cases}$
13:       Perform a gradient descent step on $(y_j - Q(\phi_j, a'_j; \theta))^2$ according to equation 3
14:     **end for**
15: **end for**

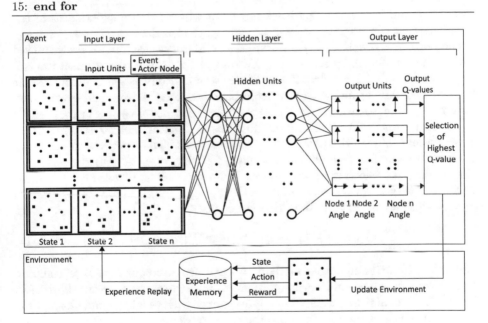

Fig. 2. The structure of DQN based mobile actor node control simulation system.

3 DQN

The algorithm for training Deep Q-learning is presented in Algorithm 1. Deep reinforcement learning is a function approximation method using deep neural network for value function and policy function in reinforcement learning. Deep Q-Network (DQN) using a Convolution Neural Network (CNN) as a function approximation of Q-leaning is a Deep Reinforcement Learning method proposed by Mnih et al. [25,26]. In paper [25], the authors present the implementation and performance evaluation of DQN for different Atari 2600 games. For experimental results, DQN can use game screen input without feature value design, and better than conventional reinforcement learning method with feature value design using linear function [27]. DQN combines the methods of neural fitted Q iteration [28], experience replay [29], sharing the hidden layer of action value function in each behavior pattern, and learning can be stabilized even with a nonlinear function such as CNN [30,31].

In this work, we use the Deep Belief Network (DBN), where computational complexity is smaller than CNN for DNN part in DQN. The environment is set as v_i. At each step, the agent selects an action a_t from the action sets of the mobile actor nodes and observes a communication and sensing coverage v_t from the current state. The change of the mobile actor node score r_t was regarded as the reward for the action. For a reinforcement learning, we can complete all of these mobile actor nodes sequences m_t as Markov decision process [32] directly, where sequences of observations and actions $m_t = v_1, a_1, v_2, \ldots, a_{t-1}, v_t$. Likewise, it uses a method known as experience replay in which it store experiences of the agent at each timestep, $e_t = (m_t, a_t, r_t, m_{t+1})$ in a dataset $D = e_1, \ldots, e_N$, cached over many episodes into a Experience Memory. Defining the discounted reward for the future by a factor γ, the sum of the future reward until the end would be $R_t = \sum_{t'=t}^{T} \gamma^{t'-t} r_{t'}$. T means the termination time-step of the mobile actor nodes. After running experience replay, the agent selects and executes an action according to an ε-greedy strategy. Since using histories of arbitrary length as inputs to a neural network can be difficult, Q function instead works on fixed length format of histories produced by a function ϕ. The target was to maximize the action value function $Q^*(m, a) = \max_\pi E[R_t | m_t = m, a_t = a, \pi]$, where π is the strategy for selecting best action. From the Bellman equation, it is equal to maximize the expected value of $r + \gamma Q^*(m', a')$, if the optimal value $Q^*(m', a')$ of the sequence at the next time step is known.

$$Q^*(m', a') = E_{m' \sim \xi}[r + \gamma \max_{a'} Q^*(m', a') | m, a] \tag{1}$$

Not using iterative updating method to optimize the equation, it is common to estimate the equation by using a function approximator. Q-network in DQN was such a neural network function approximator with weights θ and $Q(s, a; \theta) \approx Q^*(m, a)$. The loss function to train the Q-network is:

$$L_i(\theta_i) = E_{s, a \sim \rho(.)}[(y_i - Q(s, a; \theta_i))^2]. \tag{2}$$

The y_i is the target, which is calculated by the previous iteration result θ_{i-1}. $\rho(m, a)$ is the probability distribution of sequences m and a. The gradient of the loss function is shown in Eq. (3):

$$\nabla_{\theta_i} L_i(\theta_i) = E_{m, a \sim \rho(.); s' \sim \xi}[(y_i - Q(m, a; \theta_i))\nabla_{\theta_i} Q(m, a; \theta_i)]. \tag{3}$$

4 Design and Implementation of Proposed Simulation System

In this section, we present the design and implementation of proposed simulation system based on DQN for actor node mobility control in WSANs. The actor node can choose its networking, moving direction, actuation and sensing policies to maximize the number of sensing events and the number of connected integrated sensor and actor nodes. The simulation system structure is shown in Fig. 2. The proposed simulating system is implemented by Rust programming language [33, 34]. Rust is a system programming language focused on three goals: safety, speed, and concurrency [35]. Rust supports a mixture of programming styles: imperative procedural, concurrent actor, object-oriented and functional.

We consider tasks in which an agent interacts with an environment. In this case, the actor node moves step by step in a sequence of observations, actions and rewards. We took in consideration the connectivity, sensing and mobility of actor nodes.

For an actor node are considered 7 mobile patterns (up, down, back, forward, right, left, stop). The actor nodes have networking, sensing, mobility and actuation mechanisms. In order to decide the reward function, we considered Number of Sensing Events (NSE) and Number of Connected Actor Nodes (NCAN) parameters. The reward function r is defined as follows:

$$r = \begin{cases} -5 \times (NSE + NCAN) \ (if\ NCAN = 0) \\ 5 \times (NSE + NCAN) \quad (if\ NCAN \geq 1). \end{cases} \tag{4}$$

The initial weights values are assigned as Normal Initialization [36]. The input layer is using actor nodes and the position of events, total reward values in Experience Memory and mobile actor node patterns. The hidden layer is connected with 256 rectifier units in Rectified Linear Units (ReLU) [37]. The output Q values are actor node movement patterns.

5 Simulation Results

The simulation parameters are shown in Tables 1 and 2. The simulations are done for 24 events for normal, uniform, exponential and gamma distributions and 4 actor nodes considering three-dimensional environment.

In Table 3, the simulation results of total reward of episodes are summarized. Only the worst, median and best episodes are shown. The total reward value means that the DQN performance increases. For normal distribution of events, the total reward is higher than other distributions of events.

In Fig. 3 are shown the simulation results for reward vs. number of iterations for normal, uniform, exponential and gamma distributions. In Fig. 3(a) are shown the simulation results for normal distribution of events. From Fig. 3(a), we can see that with higher reward values, the actor node can move and keep connection with other actor nodes. In Fig. 3(b) are shown the simulation results for uniform distribution of events. As we can see from the results, the actor node do not keep

Table 1. Simulation parameters of DQN.

Parameters	Values
Number of episode	30000
Number of iteration	200
Number of hidden layers	3
Number of hidden units	15
Initial weight value	Normal Initialization
Activation function	ReLU
Action selection probability (ε)	$0.999 - (t/\text{Number of episode})$ $(t = 0, 1, 2, \ldots, \text{Number of episode})$
Learning rate (α)	0.04
Discount rate (γ)	0.9
Experience memory size	300×100
Batch size	32
Number of events	24
Number of actor nodes	4

Table 2. Simulation parameters of WSAN.

Parameters	Values
Area size	$32 \times 32 \times 32$
Types of actor node movement	Forward, back, left, right, up, down, stop
Positions of events	Normal, uniform, exponential and gamma distributions
Initial positions of actor nodes	Center
Number of events	24
Number of actor nodes	4

Table 3. Simulation results of total reward.

Episode	Total reward			
	Normal dist.	Uniform dist.	Exponential dist.	Gamma dist.
Best episode	10600	3590	2920	4000
Median episode	4890	1775	1710	1880
Worst episode	−2440	−935	−315	−1010

connection with other actor nodes because the exponential distribution of events is scattered. In Fig. 3(c) and (d), we show exponential and gamma distributions. The performance of exponential and gamma distributions of events is almost the same as uniform distribution of events.

In Fig. 4 is shown visualization interface of implemented simulation system for normal, uniform, exponential and gamma distributions considering three-dimensional environment. As we can see from the figures, in the normal distribution (see Fig. 4(a)), the events are located in a part of the area and are not scattered as in the case of uniform, exponential and gamma distributions (see Fig. 4(b), (c) and (d)) so the actor nodes can cover more events.

From all simulation results, we can see that for different distributions of events, the best performance is for normal distribution of events.

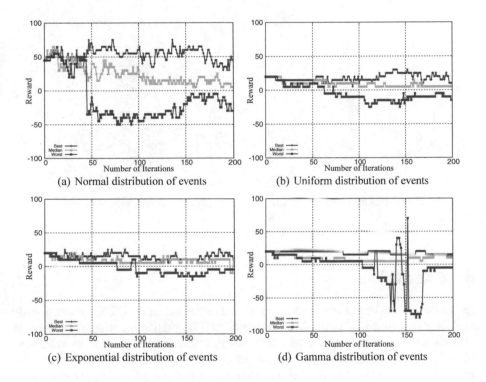

(a) Normal distribution of events

(b) Uniform distribution of events

(c) Exponential distribution of events

(d) Gamma distribution of events

Fig. 3. Simulation results of reward.

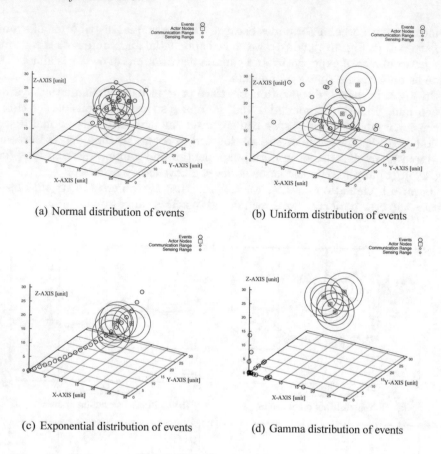

(a) Normal distribution of events

(b) Uniform distribution of events

(c) Exponential distribution of events

(d) Gamma distribution of events

Fig. 4. Visualization interface.

6 Conclusions

In this work, we implemented a simulation system based on DQN for actor node mobility control in WSANs. We proposed and implemented simulation system and showed also the interface in a simulation scenario for different distributions of events considering three-dimensional environments. For this scenario, the simulation results show that for normal distribution of events and the best episode all actor nodes are connected. However other distributions of events, the actor node do not keep connection with other actor nodes.

In the future, we would like to make extensive simulations for different simulation scenarios and improve the simulation system.

References

1. Akyildiz, I.F., Kasimoglu, I.H.: Wireless sensor and actor networks: research challenges. Ad Hoc Netw. **2**(4), 351–367 (2004)

2. Krishnakumar, S.S., Abler, R.T.: Intelligent actor mobility in wireless sensor and actor networks. In: IFIP WG 6.8 1-st International Conference on Wireless Sensor and Actor Networks (WSAN 2007), pp. 13–22 (2007)

3. Sir, M.Y., Senturk, I.F., Sisikoglu, E., Akkaya, K.: An optimization-based approach for connecting partitioned mobile sensor/actuator networks. In: IEEE Conference on Computer Communications Workshops, pp. 525–530 (2011)

4. Younis, M., Akkaya, K.: Strategies and techniques for node placement in wireless sensor networks: a survey. Ad-Hoc Netw. **6**(4), 621–655 (2008)

5. Liu, H., Chu, X., Leung, Y.-W., Du, R.: Simple movement control algorithm for bi-connectivity in robotic sensor networks. IEEE J. Sel. Areas Commun. **28**(7), 994–1005 (2010)

6. Abbasi, A., Younis, M., Akkaya, K.: Movement-assisted connectivity restoration in wireless sensor and actor networks. IEEE Trans. Parallel Distrib. Syst. **20**(9), 1366–1379 (2009)

7. Akkaya, K., Senel, F., Thimmapuram, A., Uludag, S.: Distributed recovery from network partitioning in movable sensor/actor networks via controlled mobility. IEEE Trans. Comput. **59**(2), 258–271 (2010)

8. Costanzo, C., Loscri, V., Natalizio, E., Razafindralambo, T.: Nodes self-deployment for coverage maximization in mobile robot networks using an evolving neural network. Comput. Commun. **35**(9), 1047–1055 (2012)

9. Li, Y., Li, H., Wang, Y.: Neural-based control of a mobile robot: a test model for merging biological intelligence into mechanical system. In: IEEE 7-th Joint International Information Technology and Artificial Intelligence Conference (ITAIC 2014), pp. 186–190 (2014)

10. Oda, T., Obukata, R., Ikeda, M., Barolli, L., Takizawa, M.: Design and implementation of a simulation system based on deep Q-network for mobile actor node control in wireless sensor and actor networks. In: The 31-th IEEE International Conference on Advanced Information Networking and Applications Workshops (IEEE WAINA 2017) (2017)

11. Oda, T., Kulla, E., Cuka, M., Elmazi, D., Ikeda, M., Barolli, L.: Performance evaluation of a deep Q-network based simulation system for actor node mobility control in wireless sensor and actor networks considering different distributions of events. In: The 11-th International Conference on Innovative Mobile and Internet Services in Ubiquitous Computing (IMIS 2017), pp. 36–49 (2017)

12. Oda, T., Elmazi, D., Cuka, M., Kulla, E., Ikeda, M., Barolli, L.: Performance evaluation of a deep Q-network based simulation system for actor node mobility control in wireless sensor and actor networks considering three-dimensional environment. In: The 9-th International Conference on Intelligent Networking and Collaborative Systems (INCoS 2017), pp. 41–52 (2017)

13. Oda, T., Kulla, E., Katayama, K., Ikeda, M., Barolli, L.: A deep Q-network based simulation system for actor node mobility control in WSANs considering three-dimensional environment: a comparison study for normal and uniform distributions. In: The 12-th International Conference on Complex, Intelligent, and Software Intensive Systems (CISIS 2018), pp. 842–852 (2018)

14. Llaria, A., Terrasson, G., Curea, O., Jiménez, J.: Application of wireless sensor and actuator networks to achieve intelligent microgrids: a promising approach towards a global smart grid deployment. Appl. Sci. **6**(3), 61–71 (2016)

15. Kulla, E., Oda, T., Ikeda, M., Barolli, L.: SAMI: a sensor actor network Matlab implementation. In: The 18-th International Conference on Network-Based Information Systems (NBiS 2015), pp. 554–560 (2015)

16. Kruger, J., Polajnar, D., Polajnar, J.: An open simulator architecture for heterogeneous self-organizing networks. In: Canadian Conference on Electrical and Computer Engineering, (CCECE 2006), pp. 754–757 (2006)
17. Akbas, M., Turgut, D.: APAWSAN: actor positioning for aerial wireless sensor and actor networks. In: IEEE 36-th Conference on Local Computer Networks (LCN 2011), pp. 563–570 (2011)
18. Akbas, M., Brust, M., Turgut, D.: Local positioning for environmental monitoring in wireless sensor and actor networks. In: IEEE 35-th Conference on Local Computer Networks (LCN 2010), pp. 806–813 (2010)
19. Melodia, T., Pompili, D., Gungor, V., Akyildiz, I.: Communication and coordination in wireless sensor and actor networks. IEEE Trans. Mob. Comput. 6(10), 1126–1129 (2007)
20. Gungor, V., Akan, O., Akyildiz, I.: A real-time and reliable transport (RT2) protocol for wireless sensor and actor networks. IEEE/ACM Trans. Netw. 16(2), 359–370 (2008)
21. Mo, L., Xu, B.: Node coordination mechanism based on distributed estimation and control in wireless sensor and actuator networks. J. Control Theory Appl. 11(4), 570–578 (2013)
22. Selvaradjou, K., Handigol, N., Franklin, A., Murthy, C.: Energy-efficient directional routing between partitioned actors in wireless sensor and actor networks. IET Commun. 4(1), 102–115 (2010)
23. Kantaros, Y., Zavlanos, M.M.: Communication-aware coverage control for robotic sensor networks. In: IEEE Conference on Decision and Control, pp. 6863–6868 (2014)
24. Melodia, T., Pompili, D., Akyldiz, I.: Handling mobility in wireless sensor and actor networks. IEEE Trans. Mob. Comput. 9(2), 160–173 (2010)
25. Mnih, V., Kavukcuoglu, K., Silver, D., Rusu, A.A., Veness, J., Bellemare, M.G., Graves, A., Riedmiller, M., Fidjeland, A.K., Ostrovski, G., Petersen, S., Beattie, C., Sadik, A., Antonoglou, I., King, H., Kumaran, D., Wierstra, D., Legg, S., Hassabis, D.: Human-level control through deep reinforcement learning. Nature 518, 529–533 (2015)
26. Mnih, V., Kavukcuoglu, K., Silver, D., Graves, A., Antonoglou, I., Wierstra, D., Riedmiller, M.: Playing Atari with deep reinforcement learning, pp. 1–9. arXiv:1312.5602v1 (2013)
27. Lei, T., Ming, L.: A robot exploration strategy based on Q-learning network. In: IEEE International Conference on Real-Time Computing and Robotics (RCAR 2016), pp. 57–62 (2016)
28. Riedmiller, M.: Neural fitted Q iteration - first experiences with a data efficient neural reinforcement learning method. In: The 16-th European Conference on Machine Learning (ECML 2005). Lecture Notes in Computer Science, vol. 3720, pp. 317–328 (2005)
29. Lin, L.J.: Reinforcement learning for robots using neural networks. Technical report, DTIC Document (1993)
30. Lange, S., Riedmiller, M.: Deep auto-encoder neural networks in reinforcement learning. In: The 2010 International Joint Conference on Neural Networks (IJCNN 2010), pp. 1–8 (2010)
31. Kaelbling, L.P., Littman, M.L., Cassandra, A.R.: Planning and acting in partially observable stochastic domains. Artif. Intell. 101(1–2), 99–134 (1998)
32. Kaelbling, L.P., Littman, M.L., Moore, A.W.: Reinforcement learning: a survey. J. Artif. Intell. Res. 4, 237–285 (1996)

33. The Rust Programming Language. https://www.rust-lang.org/. Accessed 14 Oct 2019
34. GitHub - rust-lang/rust: A safe, concurrent, practical language. https://github.com/rust-lang/. Accessed 14 Oct 2019
35. 'rust' tag wiki - Stack Overflow. http://stackoverflow.com/tags/rust/info/. Accessed 14 Oct 2019
36. Glorot, X., Bengio, Y.: Understanding the difficulty of training deep feedforward neural networks. In: The 13-th International Conference on Artificial Intelligence and Statistics (AISTATS 2010), pp. 249–256 (2010)
37. Glorot, X., Bordes, A., Bengio, Y.: Deep sparse rectifier neural networks. In: The 14-th International Conference on Artificial Intelligence and Statistics (AISTATS 2011), pp. 315–323 (2011)

Cognitive Approaches for Sensor Data Analysis in Transformative Computing

Marek R. Ogiela[1]([⊠]) and Lidia Ogiela[2]

[1] Cryptography and Cognitive Informatics Research Group,
AGH University of Science and Technology, 30 Mickiewicza Avenue, 30-059 Kraków, Poland
mogiela@agh.edu.pl
[2] Department of Cryptography and Cognitive Informatics, Pedagogical University of Krakow,
Podchorążych 2 Street, 30-084 Kraków, Poland
lidia.ogiela@gmail.com

Abstract. Recently a Transformative Computing paradigm has been defined. It is a new scientific approach, which allow to join global wireless communication with distributed sensor networks and AI. In this paper we'll try to describe the possible extensions of such computing technologies towards application of advanced cognitive systems for data analysis, allowing semantic interpretation of context data, in which computation task is performed. This allow to define a new approach called cognitive transformative computation oriented on specific user working in particular environment.

Keywords: Transformative computing · Cognitive cryptography · Security protocols

1 Introduction

During last several years we can observe development of few important computational approaches connected with development of global IT structures and computation paradigms. Among them we can find Big Data technologies, Cloud Computing, Ubiquitous Computing, Smart Computation, IoT, 5G technologies, and also Transformative Computing [1].

Transformative computing is a new approach, which allow to combine three elements like sensor networks, global communication infrastructure, as well as Artificial Intelligence techniques [2, 3]. Everything shows that this technology will play an important role in future computer systems, and internet technologies. Among the most important computational approaches we can also find cognitive informatics oriented on application of brain reasoning models for solving complex problems, and also for semantic evaluation of analyzed patterns [4–7].

In this paper we will try show cognitive systems can be adopted for semantic data analysis performed in Transformative Computing.

© Springer Nature Switzerland AG 2020
L. Barolli et al. (Eds.): EIDWT 2020, LNDECT 47, pp. 210–214, 2020.
https://doi.org/10.1007/978-3-030-39746-3_22

2 Application of Cognitive Systems in Transformative Computing

Transformative computing join sensor devices with wireless communication technologies, allowing to collect various sensor data in one place, where such information can be next analyzed using selected AI approaches. Depending on the data structure and collected information AI technologies allow to solve different problems on collected data. For such analysis it is possible to apply different approaches oriented on solving particular analytical tasks or numerical problems. Such analysis can be also connected with deeper semantic evaluation of collected data what can be performed using cognitive information systems previously defined in [3, 8, 9]. Such possibilities of making semantic interpretation of acquired information implies that Transformative Computing can be enriched toward application of cognitive systems instead of well-known AI techniques. Further we'll try to describe how we can use such systems for deep and semantic analysis of data, and signals registered by sensors and collected in processing unit.

Possibilities of application of cognitive systems for data analysis is very important because such systems allow to focus analysis processes on semantic meaning evaluation, and understanding the context, in which data are processed and analyzed. Registering external signals using sensors networks, we can obtain a great amount of information, which characterize external world, in which computing systems are placed or have to solve particular problem. Transformative computing module can then consider different external features during analysis, and continuously monitor changes of such features over the time. It also allow to adopt optimal solutions for features registered at particular time or place, and consider changes in external environment.

When during analysis we can obtain data or parameters connected with particular person or group of persons, further solutions and computation strategy we can also adjust for such participants of protocol. Having such possibilities we can see that cognitive systems offer much deeper analysis of collected data that simple AI techniques. It also allow to focus analysis on particular persons or object, mostly important for final results. In next section will be presented examples of application of cognitive systems for analysis, in which we can consider external features in security protocols [10, 11].

3 Context Data Analysis Using Cognitive Systems

As was described in previous section cognitive systems can considerably extend transformative computing capabilities towards using signals from external world or personal sensors. Cognitive systems allow to apply more advance semantic extraction algorithms for evaluation of external conditions, personal features, and context data. Such parameters can be next used in creation of security protocols, oriented for particular user, users groups or environment, in which this new protocols will be used.

As an example of such procedure we can define visual verification codes for user authentication. Such codes can be oriented for particular user or user groups by using, very special and expertise information from particular areas connected with authorized user or group of persons [12, 13].

These areas can be connected with technology, sport, music, arts, medicine etc., and allow perform user authentication based on his knowledge. In this manner we can define user-oriented protocols dedicated for group of persons representing selected area of expertise. Additionally we can consider external parameters describing the context, in which user is located or external environment features. Sensors allow to measure parameters like actual position, time, environmental features, external infrastructure, user movements, motion direction, velocity etc. Considering such features allow to perform particular procedure, when some requirements will be satisfied i.e. only in particular places, time, devices etc.

In Fig. 1 is presented an example of such context dependent protocol.

Fig. 1. Security protocol can consider external parameters like position, direction of motion, velocity etc. People in blue area (siting in the restaurant, walking on the left) can perform selected procedure, which consider not only theirs personal parameters but also theirs location, time, speed etc. People in yellow rectangle on the right side, can be unavailable from lunching procedure until they changed direction and location.

Example presented in Fig. 1 shows how external sensor signals can be considered in authentication protocols. Such procedure can work only in particular external situation. We can use smart wearable sensors to prevent or allowing lunching procedures in particular location, time weather conditions etc. This is very important feature of computing technologies, which can be oriented or focused on particular user, users groups, and also particular environmental context or external conditions, monitored and measured by sensors devices. All such external signals can be analyzed not only by AI procedures, but also by cognitive systems, which allow describing semantic context, and the meaning of situation in which computing tasks are performed.

4 Conclusions

In this paper, it was presented a new approach for sensor data analysis oriented for transformative computing technologies. In particular cognitive system application for semantic data evaluation was described. Transformative computing join AI technologies with sensor signals and wireless communication. In this paper we described the way of extension of AI application by cognitive systems, which allow to imitate the brain functions, and allow to evaluate the semantic content for analyzed patterns, or situations. Application of cognitive systems allow to compile and understand signals from external sensors in the same manner as it can be interpreted by human users.

It was also presented the way of creation of new security protocols, oriented for particular user, which allow to use personal features, and sensor information from external or wearable devices. Such authentication approach can consider external parameters like position, time, motion, and personal features e.g. beat rate, temperature, saturation etc. Such approach allow to create a very useful protocols, which can be oriented for particular person, but also dependent from context parameters or external configuration. Presented approach considerably extend traditional security procedures, and cryptographic protocols, which are not dependent from personal characteristic and external features [14–16].

Acknowledgments. This work has been supported by the AGH University of Science and Technology research Grant No 16.16.120.773. This work has been supported by the National Science Centre, Poland, under project number DEC-2016/23/B/HS4/00616.

References

1. Youssef, M., Kawsar, F.: Transformative computing and communication. Computer **52**(7), 12–14 (2019)
2. Grossberg, S.: Adaptive resonance theory: how a brain learns to consciously attend, learn, and recognize a changing world. Neural Netw. **37**, 1–47 (2012)
3. Ogiela, M.R., Ogiela, L.: On using cognitive models in cryptography. In: IEEE AINA 2016 - The IEEE 30th International Conference on Advanced Information Networking and Applications, Crans-Montana, Switzerland, 23–25 March, pp. 1055–1058 (2016)
4. Ogiela, M.R., Ogiela, L.: Cognitive keys in personalized cryptography. In: IEEE AINA 2017 - The 31st IEEE International Conference on Advanced Information Networking and Applications, Taipei, Taiwan, 27–29 March, pp. 1050–1054 (2017)
5. Ogiela, M.R., Ogiela, U., Ogiela, L.: Secure information sharing using personal biometric characteristics. In: Kim, T.-H., et al. (eds.) Computer Applications for Bio-technology, Multimedia and Ubiquitous City. CCIS, vol. 353, pp. 369–373. Springer, Heidelberg (2012)
6. Ogiela, M.R., Ogiela, L., Ogiela, U.: Biometric methods for advanced strategic data sharing protocols. In: The Ninth International Conference on Innovative Mobile and Internet Services in Ubiquitous Computing (IMIS 2015), Blumenau, Brazil, 8–10 July, pp. 179–183 (2015)
7. Ogiela, L., Ogiela, M.R.: Bio-inspired cryptographic techniques in information management applications. In: IEEE AINA 2016 - The IEEE 30th International Conference on Advanced Information Networking and Applications, Crans-Montana, Switzerland, 23–25 March, pp. 1059–1063 (2016)

8. Ogiela, U., Ogiela, L.: Linguistic techniques for cryptographic data sharing algorithms. Concurr. Comput. Pract. E. **30**(3), e4275 (2018). https://doi.org/10.1002/cpe.4275
9. Ogiela, L., Ogiela, M.R.: Insider threats and cryptographic techniques in secure information management. IEEE Syst. J. **11**, 405–414 (2017)
10. Ogiela, M.R., Ogiela, U.: Secure information management in hierarchical structures. In: Kim, T.-H., et al. (eds.) AST 2011. CCIS, vol. 195, pp. 31–35. Springer, Heidelberg (2011)
11. Ogiela, L., Ogiela, M.R., Ogiela, U.: Efficiency of strategic data sharing and management protocols. In: The 10th International Conference on Innovative Mobile and Internet Services in Ubiquitous Computing (IMIS 2016), Fukuoka, Japan, 6–8 July, pp. 198–201 (2016)
12. Ogiela, L.: Advanced techniques for knowledge management and access to strategic information. Int. J. Inf. Manage. **35**(2), 154–159 (2015)
13. Meiappane, A., Premanand, V.: CAPTCHA as Graphical Passwords - A New Security Primitive: Based on Hard AI Problems. Scholars' Press (2015)
14. Osadchy, M., Hernandez-Castro, J., Gibson, S., Dunkelman, O., Perez-Cabo, D.: No bot expects the DeepCAPTCHA! Introducing immutable adversarial examples, with applications to CAPTCHA generation. IEEE Trans. Inf. Forensics Secur. **12**(11), 2640–2653 (2017)
15. Easttom, Ch.: Modern Cryptography: Applied Mathematics for Encryption and Information Security. McGraw-Hill Education, New York (2015)
16. Schneier, B.: Applied Cryptography. Wiley, Indianapolis (2015)

A Study on Access Control Scheme Based on ABE Using Searchable Encryption in Cloud Environment

Yong-Woon Hwang[1], Im-Yeong Lee[1(✉)], and Kangbin Yim[2]

[1] Department of Computer Science and Engineering, Soonchunhyang University,
Asan, South Korea
{hyw0123,imylee}@sch.ac.kr
[2] Department of Information Security Engineering, Soonchunhyang University,
Asan, South Korea
yim@sch.ac.kr

Abstract. Recently, with the development of cloud computing technology, data can be stored and shared in the cloud. However, because the cloud involves a network, various security threats can occur. Therefore, it is important to encrypt and store data on cloud servers, and CP-ABE (Ciphertext-Polly Attribute Based Encryption) based access control technology is used to access the encrypted data. However, some of the previously studied CP-ABE schemes are inefficient and vulnerable to security threats. This paper is a data access control technique using searchable encryption and attribute-based encryption in cloud environment, and aims to secure and efficient data sharing system in cloud environment. Searchable encryption system technology allows you to efficiently search for the desired data among the many encrypted data stored in the cloud environment. In addition, CP-ABE is used to access encrypted data, so that the user can access the encrypted data securely as a user attribute and decrypt the data.

1 Introduction

In recent years, owing to the development of cloud computing, the cloud computing environment has been used widely. Instead of renting storage that is traditionally costly to the cloud, the cloud can store or share data in cloud environments as required. First, the service provider cannot be trusted fully. If you are a user of the cloud, the data stored in the cloud can be safe from external threats, but the provider providing the cloud knows your data at will. Other examples include data leakage or loss owing to malicious users. An attacker can intercept data during storage, thereby leaking stored data. Therefore, it is important to encrypt and store data in a cloud environment, and technology is required to access stored encryption data. However, search efficiency is degraded by having to decrypt all of the cloud's stored encryption data to obtain the desired encrypted data. Searchable encryption technology enables multiple owners to efficiently obtain the desired data, even if they are encrypted and stored in a cloud environment. Attribute-based encryption is a suitable cryptographic technology for accessing encrypted data by multiple users; it performs encryption and replication with various user attributes and

© Springer Nature Switzerland AG 2020
L. Barolli et al. (Eds.): EIDWT 2020, LNDECT 47, pp. 215–221, 2020.
https://doi.org/10.1007/978-3-030-39746-3_23

is often used as an access control technology to access data, In particular, among the attribute-based encryption, CP-ABE (Ciphertext-Polly Attribute Based Encryption) is widely used in the cloud environment. However, some of the previously studied CP-ABE schemes are inefficient and vulnerable to security threats [1]. In particular, in the previously studied CP-ABE scheme, the use of cloud storage space can be inefficient because the size of the ciphertext increases with the number of attributes. In addition, users' computations that are being decoded are not proportional to the number of attributes [2]. In this paper is a data access control technique using searchable encryption and attribute-based encryption in cloud environment, and aims to secure and efficient data sharing system in cloud environment. By using searchable encryption system technology, it is possible to efficiently search for desired data among many encrypted data stored in cloud environment, and it is safe because it accesses searched data based on user's attribute. This paper is suitable to be applied to the environment where data can be securely shared in the cloud that is shared by many users in the company, between companies. Each session of this paper is as follows. In Sect. 2, we will introduce CP-ABE and Searchable Encryption as a related works, and in Sect. 3, we will describe the proposed scheme. In Sect. 4 will analyze the security and efficiency of the proposed system, and Sect. 5 concludes with a conclusion.

2   Related Works

This sections describes CP-ABE, and search encryption.

2.1   CP-ABE

Attribute-based encryption is an encryption algorithm that performs encryption and decryption according to each user's attribute set and the access policy created by a given attribute set. CP-ABE is attribute-based encryption that is used in various fields as an access control technique for accessing data in a cloud environment. The CP-ABE scheme encrypts data by creating an access structure with the attributes of the user who wants to access the data when the data owner generates the ciphertext. The ciphertext is then sent to the recipient, who then decrypts it according to his attribute set. If the recipient who wishes to access the data has [Hospital A] and [Doctor] attribute, the data owner creates and encrypts the access structure with [Hospital A, Doctor] attributes. As a result, only recipients who satisfy the access structure can decrypt the ciphertext [1].

2.2   Searchable Encryption

Searchable encryption technology makes it easy for users to navigate and obtain the data they need without having to decrypt all encrypted data, even if the data is stored encrypted in the cloud environment. The earliest version of searchable encryption technology was proposed by Song, Wagner, and Perrig in 2000. Hidden search was designed to search plain text without leaking information. However, as it is an initial method, no specific safety is defined. In 2003, Goh designed the first searchable cryptographic system using the Bloom filter. However, because the Bloom filter is used, more errors may be included in the search results. To reduce the probability of this error, a larger Bloom filter should be used; however, if the Bloom filter used is extremely large, the search becomes inefficient. Since then, searchable encryption using symmetric and public keys have been continuously studied. Currently, the searchable cryptography technology is applied to the CP-ABE scheme to provide secure data access techniques in the cloud environment. In this study, by adding a searchable encryption scheme to the CP-ABE scheme, we provide a function to efficiently search encrypted data in the encrypted state without performing decryption [3–5].

3 Proposed Scheme

This sections propose scheme that introduces searchable ciphers to provide a system that enables multiple data owners and users to securely share data in the cloud environment. The scenario of the proposed method is shown in Fig. 1. The proposed scheme is safe against various security threats such as collusion attack and camouflage attack, and the password size can be expressed in a specific size to efficiently use the space of the existing cloud storage. In addition, compared with the existing CP-ABE scheme, the computational efficiency is increased by reducing the amount of computation that users decrypt through a trusted server that supports outsourcing. The detailed descriptions of the proposed protocols are as follows.

3.1 System Parameters

The system parameters used in the proposed scheme are as follows (Fig. 1).

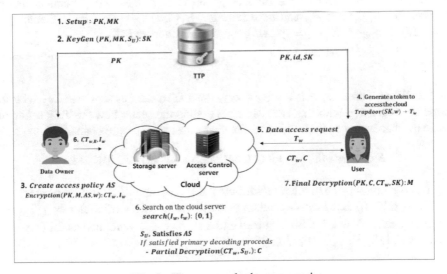

Fig. 1. The proposed scheme scenario

- TTP: manages user attributes with a trusted third party
- Storage: Servers that manage data
- AC: User access control management
- PK, MK: Public parameter, master key
- SK: User security key (decryption key)
- ID_{user}: User Identifier
- S_U, S: User attribute data, a set of attribute data
- AS: Access policy or Access structure
- T_w: Tokens with access to the cloud
- w: keyword
- I_w: Keyword index
- CT_w: Encrypted data
- T: Timestamp

3.2 Initial Phase

Step 1. Initially, the user sends a registration message to the TTP to request registration. The TTP adds the user information to a user list when user registration is completed. Subsequently, the PK and MK are created through the setup process and KeyGen process.

- *Setup steps*: Generate pk, mk
  Each attribute has multiple values $V_i = [v_{i,1}, v_{i,2}, v_{i,ni},]$, v_i, n_i and can be represented as att_i as a set. W = $[W_1, W_2, W_3, W_n]$ is an access policy, where $W_i \subset V_i$. When the prime order of the bilinear group G is p, the TTP generates random generators α, β, $t_i \in Z_p$ and hash function $H_1\{0, 1\}^* \rightarrow$ G. After calculating Y = e (,) , h = $g^\beta \in G_0$, $T_i = g^{t_i}$, we generate a PK and an MK.
- $PK < e, G, g, \{T_i\}_{i \in [1,n]}, Y, h >$
- $MK < \alpha, t_i >$

Step 2. When a user requests a private key, the TTP generates a private key with the properties registered with the user. After generating the secret key, the TTP sends the PK to the data owner and PK, SK, and TN to the user via a secure channel.

$$KeyGen(\text{MK}, S_U, \text{PK}) : \text{Generate SK via set of } S, \text{ MK, PK}$$

- $j \in S$ (j represents the number of attributes, and S represents a set of attributes)
- $r_1 \ldots r_j \in Z_p$ (r_j represents a random value given for each attribute), r = $\sum_{j=1}^{n} r_n$
  Subsequently, $D' = g^{\alpha-r}$ is calculated and a secret key is generated as follows.
- $SK < S, \{D_{i,1}\}_{i \in [1,n]}, D', H_1(v_i)^\beta >$

3.3 Data Encryption Phase

Step 3. Data owners create access structures based on attributes that allow their data to be accessed within the cloud environment. Subsequently, multiple values of the attributes specified in the access structure are calculated according to the condition, and ciphertext CT is generated through this. In addition, the user generates an index value I_w through the keyword specified in the ciphertext and transmits it to the cloud with the ciphertext CT_w.

$Encrypt(PK, M, AS, w) = CT_w, I_w$: Data encryption step with disclosure parameter PK, access structure AS, and keyword w

- $C_0 = Y^s = M \cdot e(g, g)^{\alpha s}$, $C_1 = g^s$, $C_2 : h^s$, $\tilde{C}_1 = e\left(h, g^{ws'}\right)$, $\tilde{C}_2 = g^{ss'}$

  If $v_{i,1} \in W_i$, computes $C_i = g^{t_n s}$

  If $v_{i,1} \notin W_i$, computes $C_i = g^{t_{n+1} s}$

- $CT_w < AS, C_0, C_1, (C_i)_{i \in N} >$
- $C' = \left(h \cdot \prod_{i \in AS} C_i\right)^s = \left(h \cdot \prod_{i \in AS} g^{t_i}\right)^s$, $g^t = \prod_{i=1}^{n} g^{t_i}$
- $CT_w < AS, C_0, C_1, C_2, C' >$, $I_w < \tilde{C}_1, \tilde{C}_2, (C_i)_{i \in N} >$

3.4 User Data Access and Data Decryption Phase

Step 4. The user requests the access by generating the token T_w through the trapdoor process with the SK received from the TTP and the keyword set w, and then transmits it to the cloud.

- Trapdoor(SK, w) $= T_w$, $T_w = e\left(\prod_{i=1}^{n} H_1(x_i)^\beta, g^w\right)$

The cloud server retrieves the ciphertext stored in the server through T_w received from the user and retrieves the ciphertext that matches I_w specified in the ciphertext. The search result is shown as $\{0, 1\}$ and provides the user with the desired ciphertext.

- Search$(I_w, T_w) : e\left(\tilde{C}_2, T_w\right) = e\left(\prod_{i=1}^{n} C_{1,i}, \tilde{C}_1\right)$

Step 5. When the user requests a ciphertext, the AC server performs a partial decoding by calculating and comparing the computed attribute value specified in the ciphertext with the user attribute value. Subsequently, partial decryption is performed, and C and ciphertext CT are transmitted to the user.

$Partial\ decrypt(CT, S_{U*}) = C$: When partial decryption is performed with ciphertext and user attributes.

- $C = \dfrac{e(g^r, C')}{e\left(\check{C}, \left(\prod_{j \in S} g^{t_i}\right)^s\right)} = e(g, g)^{rs}$

Step 6. The user performs a final decryption of the CT received from the AC server with the SK, C, and PK.

$Final\ decrypt(PK, C, CT, SK) = M$: The user proceeds with the final decoding to acquire the message M.

- $M = \dfrac{\check{C}}{e\left(\check{C}, D'\right) \cdot C} = \dfrac{M \cdot e(g,g)^{\alpha s}}{e(g^s, g^{\alpha-r}) \cdot e(g,g)^{rs}}$.

4   Analysis of Proposed Scheme

- User collusion: Because the proposed scheme uses nonce in addition to the attribute value in the TTP when generating the secret key, even if the user infers the attribute through a collusion attack, the secret key cannot be generated. Furthermore, when partial decryption is performed by comparing the attributes of the user with those of the access policy specified in the ciphertext, the message M cannot be viewed because a secret key of the user does not exist. Therefore, collusion attacks between users or between users and service providers are prevented.
- Access control for unauthorized users: In this proposal, only the creator of the token can access the cloud through the secret key received from the TTP. In addition, only users who satisfy the access policy specified by the data owner can access the data to access the data stored in the cloud. In AC, the user's access is blocked first, and if the conditions above are satisfied, then the AC is partially decrypted and the ciphertext and partial decryption results are transmitted to the user. Therefore, the proposed scheme blocks the access of unauthorized users, thereby ensuring the integrity and confidentiality of the data.
- Efficiency of cloud storage space and decryption calculations: In the CP-ABE scheme, when the ciphertext is generated, it increases proportionally to the number of attributes specified in the access policy, thus wasting space in the stored storage. Li scheme [6], for the ciphertext CT $<\tilde{C}, C_0, C_0^*, \{\{C_{i,1}\}, \left\{C_{i,t,2}, C_{i,t}^*\right\}_{t\in[1,n_i]}\}_{i\in[1,n]}>$, the size increases with the number of attributes. In this proposal, the number of attributes specified in the access policy is calculated separately, $C' = \left(h \cdot \prod_{i\in AS} C_i\right)^s = \left(h \cdot \prod_{i\in AS} g^{t_i}\right)^s$ to represent a single number, which results in a fixed-size ciphertext. Thus, the proposed scheme can efficiently use the cloud storage space compared with the existing CP-ABE method. In addition, partial decoding is performed by placing an AC server that supports outsourcing techniques. As the user receives the result C and the ciphertext CT that have been partially decoded from the AC and only performs the final decryption, the user can obtain the message M. Therefore, the amount of ciphertextc decryption of the user is reduced compared to the existing CP-ABE

Table 1. Comparison with existing searchable CP-ABE scheme with proposed scheme

Scheme items	Helil scheme [6]	Wu scheme [7]	The proposed scheme
User collusion	Safe	Safe	Safe
Camouflage attack	Safe	Safe	Safe
Ciphertext length	It is proportional to the number of Attribute		Constant-size ciphertext
Keyword search	×	○	○
Partial decryption (server)	$2(n+1)c_e + 2nE$	–	$2c_e + nM + 2E$
Final decryption (user)	$1c_e + 2M + 2E$	$(2n+1)c_e + 2E$	$1c_e + 2M$
Outsourcing	○	×	○

c_e: *Pairing operation; M: Multiplication operation; n: Number of attributes; E: Exponentiation operation*

method. As shown in Table 1, the amount of user decryption operations is greatly reduced compared to the Wu scheme, and the decryption operation amount is similar to the Helil scheme using partial decryption. However, the Helil scheme did not consider the requirement of searching for encrypted data in the cloud and we were satisfied with this proposed scheme.

5 Conclusions

In this paper, we propose a data access scheme using searchable encryption and attribute-based encryption in cloud environment for secure and efficient data sharing system in cloud environment. The proposed scheme is safe for various security threats such as collusion attacks and camouflage attacks. Moreover, because the ciphertext size was constant-size output regardless of the number of attributes, the cloud storage could be used efficiently. In addition, part of the decoding operation was outsourced in the user side during decoding in the AC; therefore, computational efficiency was improved for the user who was burdened with the decoding process owing to the lack of computing resources. In the proposed scheme a searchable cipher could be introduced to efficiently search for the desired data among a large number of data encrypted in the existing cloud environment. This can be effectively applied to a system that securely shares data in all cloud environments such as medical, military, and enterprise. In future research, it is necessary to study the privacy protection of users accessing data in the cloud environment and to track the users leaked through the key when the user leaks the decryption key.

Acknowledgments. This research was supported by Basic Science Research Program through the National Research Foundation of Korea (NRF) funded by the Ministry of Education (NRF-2019R1A2C1085718) and the MSIT (Ministry of Science and ICT), Korea, under the ITRC (Information Technology Research Center) support program (IITP-2019-0-00403) supervised by the IITP (Institute for Information & communications Technology Planning & Evaluation).

References

1. Bethencourt, J., Sahai, A., Waters, B.: Ciphertext-policy attribute-based encryption. In: 2007 IEEE Symposium on Security and Privacy (SP 2007) (2007)
2. Hahn, C., Hur, J.: Constant-size ciphertext-policy attribute-based data access and outsourceable decryption scheme. J. KIISE **43**(8), 933–945 (2016)
3. Cui, J., Zhou, H., Zhong, H., Xu, Y.: AKSER: attribute-based keyword search with efficient revocation in cloud computing. Inf. Sci. **423**, 343–352 (2018)
4. Cao, L., Zhang, J., Dong, X., Xi, C., Wang, Y., Zhang, Y., Guo, X., Feng, T.: A based on blinded CP-ABE searchable encryption cloud storage service scheme. Int. J. Commun. Syst **31**(10), e3566 (2018)
5. Wang, H., Dong, X., Cao, Z., Li, D.: Secure and efficient attribute-based encryption with keyword search. Comput. J. **61**(8), 1133–1142 (2018)
6. Helil, N., Rahman, K.: CP-ABE access control scheme for sensitive data set constraint with hidden access policy and constraint policy. Secur. Commun. Netw. (2017)
7. Wu, A., Zheng, D., Zhang, Y., Yang, M.: Hidden policy attribute-based data sharing with direct revocation and keyword search in cloud computing. Sensors **18**(7), 2158–2175 (2018)

Transformative Computing in Knowledge Extraction and Service Management Processes

Lidia Ogiela[1]([✉]), Makoto Takizawa[2], and Urszula Ogiela[3]

[1] Department of Cryptography and Cognitive Informatics, Pedagogical University of Krakow, Podchorążych 2 Street, 30-084 Kraków, Poland
lidia.ogiela@gmail.com
[2] Department of Advanced Sciences, Hosei University, 3-7-2, Kajino-cho, Koganei-shi, Tokyo 184-8584, Japan
makoto.takizawa@computer.org
[3] Pedagogical University of Krakow, Podchorążych 2 Street, 30-084 Kraków, Poland
uogiela@gmail.com

Abstract. In this paper will be described a new methodology of service management processes. This issue will be classified by cognitive and transformative computing. One of the most important direction of proposed methods will be oriented for the Cloud services. New methodology based on signals from external sensors can be used at different levels of data analysis processes in distributed infrastructure. The impact of service management processes will be analysed and presented in research results.

Keywords: Transformative computing · Knowledge extraction · Service management processes

1 Introduction

Knowledge extraction processes, which are dedicated to the tasks of signal acquisition, and at the same time collecting extensive data and information play an increasingly important role in data management areas [8, 9]. These processes presently are expanded with new stages of data analysis and interpretation, performing semantic description and content evaluation, that can be implemented on the basis of cognitive analysis of interpreted data sets. The essence of this approach is to point to the most important semantic elements that can be described by the semantic content present in the evaluated data sets, which determine the knowledge in relation to the analyzed information [3, 5–7].

The processes of semantic analysis have been described in literature i.e. in papers [4–9], where their influence and significance in the processes of semantic evaluation of data, which can be performed on the basis of application of mathematical linguistics methods. Until now, knowledge extraction processes have been described in the context of semantic analysis conducted by cognitive systems [4–9]. Transformative computing [1, 2, 10], which is the main subject of this work, is a new direction of their development, and area of their possible application.

© Springer Nature Switzerland AG 2020
L. Barolli et al. (Eds.): EIDWT 2020, LNDECT 47, pp. 222–225, 2020.
https://doi.org/10.1007/978-3-030-39746-3_24

2 Transformative Computing as a New Direction of Knowledge Extraction Methodology

In the processes of knowledge extraction performed for the needs of a deeper, and more accurate description of the data, it is necessary to define the characteristics of the analyzed data sets, and determine their importance in the information processing process. Knowledge extraction can therefore occur at various levels of data collection, knowledge acquisition and information processing. The knowledge gathering process may also take place at various semantic levels, taking into account selected or all available knowledge levels, both basic and expert.

A new approach in the field of knowledge extraction processes is the application of transformative computing methods. This technique allows obtaining data from various external sensors, which are then collected in one place, where properly processed, allow to obtain optimal results of the conducted cognitive analysis. Therefore, they provide the opportunity to acquire knowledge from extensive and independent of each other external sources, which are a kind of external network subject to modification depending on the changes taking place in the environment. The data obtained from external sensors allow to extract knowledge which, being subject to the processing process, can be used in data management support processes.

Schematic views of transformative computing are shown in Fig. 1.

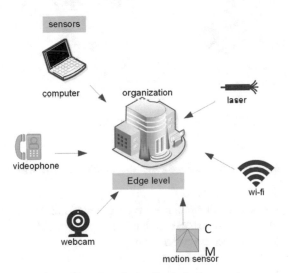

Fig. 1. Scheme of transformative computing.

3 Transformative Computing for Service Management Processes

An innovative approach to issues of transformative computing as a method that allows obtaining signals from external sensors is the possibility of its application in service

management processes. These processes as a special type of data management procedures can be implemented from various levels of the entity, which is the foundation
of their implementation. Initially, therefore, they are conducted at the level of a given
structure (Edge Computing). Enriched with the possibility of obtaining data from external sensors, they enable the implementation of management processes depending on the
information obtained, determining their degree of priority (priority of the service), and
the number of external sensors from which data is obtained.

Acquisition of data from external sensors (external sources) allows the collection of
independent information about processed data (relating to sets of services performed),
coming from various sources and complementary to each other.

Service management processes are not limited only to the level of the entity in
which the process is carried out. Service management can also take place at other levels
(external to the basic structure), which include fog levels and clouds (Fig. 2).

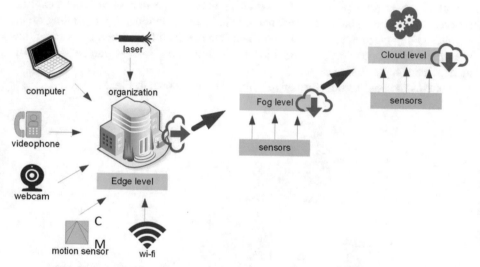

Fig. 2. Process of transformative computing at different management levels.

Service management implemented from the fog and cloud level allows for optimization of management costs, easy access to data, ability to quickly modify and update it
with new information, and secure service management. In terms of the three-level structure of service management, transformative computing methods that allow obtaining
information from various external sources, given the opportunity to take into account
the state of the environment, the impact of individual information obtained from various
(independent) sources on the entire process of service management.

4   Conclusions

An innovative approach to the issues of services management is the possibility of application transformative computing methods to data acquisition tasks, and based on them

also the knowledge extraction from the analyzed data sets. An example of extremely extensive data sets are sets of services provided by a specific entity, taking into account constant changes caused by the necessity of market changes. The possibility of obtaining data and information based on the signals recorded by external sensors allows you to constantly monitor and take into account emerging changes. Immediate reaction to changes taking place from the entity's environment, improves service management processes by reducing unnecessary stages of this process.

Support for service management processes based on transformative computing methods can be implemented at various levels of distributed structures.

Acknowledgment. This work has been supported by the National Science Centre, Poland, under project number DEC-2016/23/B/HS4/00616.

This work was partially supported by JSPS KAKENHI grant number 15H0295.

References

1. Benlian, A., et al.: The transformative value of cloud computing: a decoupling, platformization, and recombination theoretical framework. J. Manag. Inf. Syst. **35**(3), 719–739 (2018)
2. Gil, S.S., et al.: Transformative effects of IoT, blockchain and artificial intelligence on cloud computing: evolution, vision, trends and open challenges. Internet Things **8**, 100118 (2019)
3. Nakamura, S., Ogiela, L., Enokido, T., Takizawa, M.: Flexible synchronization protocol to prevent illegal information flow in peer-to-peer publish/subscribe systems. In: Barolli, L., Terzo, O. (eds.) Complex, Intelligent, and Software Intensive Systems. Advances in Intelligent Systems and Computing, vol. 611, pp. 82–93 (2018)
4. Ogiela, L.: Intelligent techniques for secure financial management in cloud computing. Electron. Commer. Res. Appl. **14**(6), 456–464 (2015)
5. Ogiela, L.: Advanced techniques for knowledge management and access to strategic information. Int. J. Inf. Manag. **35**(2), 154–159 (2015)
6. Ogiela, L.: Cryptographic techniques of strategic data splitting and secure information management. Pervasive Mob. Comput. **29**, 130–141 (2016)
7. Ogiela, L., Ogiela, M.R.: Insider threats and cryptographic techniques in secure information management. IEEE Syst. J. **11**(2), 405–414 (2017)
8. Ogiela, M.R., Ogiela, U.: Secure information management in hierarchical structures. In: Advanced Computer Science and Information Technology. Communications in Computer and Information Science, vol. 195, pp. 31–35 (2011)
9. Ogiela, U., Takizawa, M., Ogiela, L.: Classification of cognitive service management systems in cloud computing. In: Barolli, L., Xhafa, F., Conesa, J. (eds.) Advances on Broad-Band Wireless Computing, Communication and Applications, BWCCA 2017. Lecture Notes on Data Engineering and Communications Technologies, vol. 12, pp. 309–313. Springer, Cham (2018). https://doi.org/10.1007/978-3-319-69811-3_28
10. Wei, Y., Blake, M.B.: Service-oriented computing and cloud computing: challenges and opportunities. IEEE Internet Comput. **14**(6), 72–75 (2010)

Input Amplitude Dependency of Duplexer with Dispersive and Nonlinear Dielectric in 2-D Photonic Crystal Waveguide

Naoki Higashinaka[1] and Hiroshi Maeda[2(⊠)]

[1] Graduate School of Communication and Information Networking,
Fukuoka Institute of Technology, Fukuoka, Japan
mgm18104@bene.fit.ac.jp
[2] Department of Information and Communication Engineering,
Fukuoka Institute of Technology, Fukuoka, Japan
hiroshi@fit.ac.jp

Abstract. In two-dimensional photonic crystal waveguide with passive switching function based on the input light wavelength, the input amplitude characteristic is simulated by using frequency dependent FDTD method. For the input amplitude $E_0 = 0.1, 0.5, 1.0, 1.5[V/m]$, distribution ratio is calculated from the electric field profile of output ports. In addition, cross-correlation coefficient and normalized power spectra are used as other considerations. As a result, signal switch based on the wavelength of the input signal is confirmed. However, as the input amplitude exceeded $E_0 = 0.1[V/m]$, it is noted that the optical field penetrates into the photonic crystal.

1 Introduction

In recent years, Internet traffic has increased by the development of optical fiber communication and wireless communication. Along with that, information processing of electronic integrated circuits is increasing, and there are problems such as heat generation and increase in power consumption. There is a possibility that these problems can be solved by using optical technology. Therefore, optical integrated circuits have been researching in various research institutions [1–3].

Photonic crystal has attracted attention for the realization of optical integrated circuits. The photonic crystal is artificial structure that blocks light of a specific wavelength depending on the material, the periodic lattice and the thickness of the pillar etc. The wavelength band is called the photonic band gap (PBG). In addition, the photonic crystal has one, two, and three-dimensional structures [4].

In this research, linear dispersion and nonlinear dielectric material is installed as a resonator in branch point of the two-dimensional photonic crystal waveguide. In previous study, optical switching was confirmed by using the structure [5]. In this paper, distribution characteristic according to the change of the amplitude of the input signal is examined by frequency dependent FDTD (Finite Difference Time Domain) method. The method is used to calculate the electromagnetic field in the waveguide, and the

© Springer Nature Switzerland AG 2020
L. Barolli et al. (Eds.): EIDWT 2020, LNDECT 47, pp. 226–236, 2020.
https://doi.org/10.1007/978-3-030-39746-3_25

method can be applied to linear dispersion and nonlinear dielectric materials [6]. In addition, normalized power spectra and cross-correlation coefficient are used as other considerations.

2 Frequency Dependent FDTD Method

According to Yee's FDTD scheme [7], the two dimensional $\left(i.e. \frac{\partial}{\partial y} = 0\right)$, discretized Maxwell's equation for transverse electric (TE) mode, which propagates to x and z axis, are described as follows;

$$
\begin{aligned}
H_x^{n+\frac{1}{2}}\left(i + \frac{1}{2}, k\right) = {} & H_x^{n-\frac{1}{2}}\left(i + \frac{1}{2}, k\right) \\
& + \frac{\Delta t}{\mu_0 \Delta z}\left\{E_y^n\left(i + \frac{1}{2}, k + \frac{1}{2}\right) - E_y^n\left(i + \frac{1}{2}, k - \frac{1}{2}\right)\right\}
\end{aligned}
\tag{1}
$$

$$
\begin{aligned}
H_z^{n+\frac{1}{2}}\left(i, k + \frac{1}{2}\right) = {} & H_z^{n-\frac{1}{2}}\left(i, k + \frac{1}{2}\right) \\
& - \frac{\Delta t}{\mu_0 \Delta x}\left\{E_y^n\left(i + \frac{1}{2}, k + \frac{1}{2}\right) - E_y^n\left(i - \frac{1}{2}, k - \frac{1}{2}\right)\right\}
\end{aligned}
\tag{2}
$$

$$
\begin{aligned}
D_y^{n+\frac{1}{2}}\left(i + \frac{1}{2}, k + \frac{1}{2}\right) = {} & D_y^n\left(i + \frac{1}{2}, k + \frac{1}{2}\right) \\
& + \frac{\Delta t}{\Delta z}\left\{H_x^{n+\frac{1}{2}}\left(i + \frac{1}{2}, k + 1\right) - H_x^{n+\frac{1}{2}}\left(i + \frac{1}{2}, k\right)\right\} \\
& - \frac{\Delta t}{\Delta x}\left\{H_z^{n+\frac{1}{2}}\left(i + 1, k + \frac{1}{2}\right) - H_z^{n+\frac{1}{2}}\left(i, k + \frac{1}{2}\right)\right\}
\end{aligned}
\tag{3}
$$

where E and H are electric and magnetic field and D is electric flux intensity. Space and time discretization are $\Delta x, \Delta z, \Delta t$. Number of corresponding grids are i, k, n. Giving a set of suitable initial conditions and a constitutive equation between D and E to above equations, the latest field is calculated successively as the increase of time step number.

Let us consider composite space of free space, linear dielectric material and dispersive linear and nonlinear dielectric material.

$$
E_y^n = \frac{D_y^n}{\varepsilon_0}
\tag{4}
$$

where ε_0 is permittivity of free space.

The electric field in linear dielectric material is described as follows;

$$
E_y^n = \frac{D_y^n}{\varepsilon_0 \varepsilon_r}
\tag{5}
$$

where ε_r is relative permittivity of linear dielectric material.

The electric field in dispersive linear and nonlinear dielectric material is described as follows [6];

$$E_y^n = \frac{\frac{1}{\varepsilon_0} D_y^n - S_L^{n-1} + 2\chi_0^{(3)}\alpha \left(E_y^{n-1}\right)^3}{\varepsilon_\infty + \chi_0^{(3)}(1-\alpha)S_R^{n-1} + 3\chi_0^{(3)}\alpha \left(E_y^{n-1}\right)^2} \tag{6}$$

$$S_L(z) = 2e^{-\alpha_L \Delta t}\cos(\beta_L \Delta t) \cdot z^{-1}S_L(z) - e^{-2\alpha_L \Delta t} \cdot z^{-2}S_L(z)$$
$$+ \gamma_L \Delta t e^{-\alpha_L \Delta t}\sin(\beta_L \Delta t) \cdot E(z) \tag{7}$$

$$\alpha_L = 2\pi f_L \cdot \delta_L \tag{8}$$

$$\beta_L = 2\pi f_L \sqrt{1 - \delta_L^2} \tag{9}$$

$$\gamma_L = \frac{2\pi f_L \cdot (\varepsilon_s - \varepsilon_\infty)}{\sqrt{1 - \delta_L^2}} \tag{10}$$

$$S_R^n = 2e^{-\alpha_R \Delta t}\cos(\beta_R \Delta t) \cdot S_R^{n-1} - e^{-2\alpha_R \Delta t} \cdot S_R^{n-2}$$
$$+ \gamma_R \Delta t e^{-\alpha_R \Delta t}\sin(\beta_R \Delta t) \cdot \left(E^n\right)^2 \tag{11}$$

$$\alpha_R = 2\pi f_{NL} \cdot \delta_{NL} \tag{12}$$

$$\beta_R = 2\pi f_{NL}\sqrt{1 - \delta_{NL}^2} \tag{13}$$

$$\gamma_R = \frac{2\pi f_{NL} \cdot \chi_0^{(3)} \cdot (1-\alpha)}{\sqrt{1 - \delta_{NL}^2}} \tag{14}$$

where ε_∞ and ε_s are relative permittivity of dispersive linear and nonlinear dielectric material, f_L is linear relaxation frequency, f_{NL} is nonlinear relaxation frequency, δ_L is linear decaying factor and δ_{NL} is nonlinear decaying factor.

Following equation is developed from the above equation.

$$E_y^n = \frac{\frac{1}{\varepsilon_0} D_y^n - S_L^{n-1} + 2\chi_0^{(3)}\alpha (E_y^{n-1})^3}{\varepsilon_\infty + \chi_0^{(3)}(1-\alpha)S_R^{n-1} + 3\chi_0^{(3)}\alpha (E_y^{n-1})^2} \tag{15}$$

The electric field in linearly dispersive and nonlinear dielectrics is calculated from this equation.

3 Settings for Simulation

Optical signal propagating in two-dimensional photonic crystal waveguide with a duplexer with three ports is analyzed. Furthermore, in order to catch the optical signal leaked into the photonic crystal, the output port is added in the photonic crystal.

Figure 1 is shown that two-dimensional photonic crystal waveguide used in simulation. In Fig. 1, background of the waveguide is free space and the refractive index $n_0 = 1.0$. Black circles are linear dielectric material and the refractive index $n_1 = 3.6$ as is in Ref. [8]. In addition, green circles are dispersive linear and nonlinear dielectric material. The refractive index $n_2 = 1.5$, relative permittivity of dispersive linear and nonlinear dielectric material $\varepsilon_\infty = 2.25$, $\varepsilon_s = 5.25$, linear relaxation frequency $f_L = 63.7[THz]$, linear decaying factor $\delta_L = 2.5 \times 10^{-4}$, nonlinear relaxation frequency $f_{NL} = 14.8[THz]$, nonlinear decaying factor $\delta_{NL} = 3.36 \times 10^{-1}$, nonlinear susceptibility $\chi_0^{(3)} = 0.07$, and weight factor for Kerr effect $\alpha = 0.7$ as is in Ref. [6]. The lattice period $L = 551.8[nm]$, and the radius of dielectric material $R = 0.2L$ are used. The FDTD discretization are $\Delta x = \Delta z = 9.85[nm]$ and $\Delta t = 0.0209[fs]$, respectively. The analysis area is $W = 9.69[\mu m]$ in the propagation axis and $H = 9.69[\mu m]$ in the transverse axis. The wavelength $\lambda = 1.350 - 1.600[m]$. Berenger's perfectly matched layer [9] is installed as absorbing boundary condition.

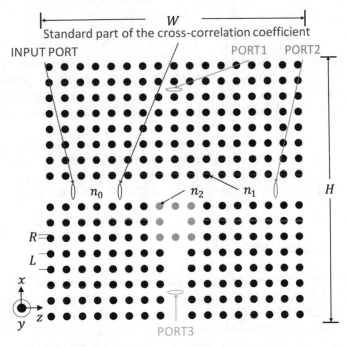

Fig. 1. Top of view 2-D photonic crystal waveguide (duplexer is composed of linear dispersive and nonlinear material at the center of the waveguide)

The electric field input to port 1 is Gaussian beam and given by

$$E_y(x, t) = E_0 \exp\left\{ -\left(\frac{x}{w_0}\right)^2 \right\} \sin\left(\frac{2\pi ct}{\lambda}\right) \qquad (16)$$

where the amplitude $E_0 = 0.1, 0.5, 1.0$ and $1.5[V/m]$, the beam spot $w_0 = 28[nm]$, optical speed in vacuum $c = 2.998 \times 10^8[m/s]$.

Optical signal propagating in two-dimensional photonic crystal waveguide with a duplexer with three ports is analyzed. Furthermore, in order to catch the optical signal leaked into the photonic crystal, the output port is added in the photonic crystal. Figure 1 is shown that two-dimensional photonic crystal waveguide used in simulation. The light intensity $P_i (i = 1, 2, 3)$ is calculated from the integral of the square of the output electric field of each port. Additionally, definition of distribution ratio S_i is described as follows;

$$S_i = \frac{\int_{PORTi} P_i dw}{\int_{PORT1} P_1 dw + \int_{PORT2} P_2 dw + \int_{PORT3} P_3 dw} \quad (i = 1, 2, 3) \quad (17)$$

where w is the direction traversing the waveguide.

Cross-correlation coefficient C_i is calculated from the normalized electric field mode output from each port. C_i is described as follows;

$$C_i = \frac{\frac{1}{n}\sum_{j=1}^{n}\left(E_{y|CC}(j) - \overline{E_{y|CC}}\right)\left(E_{y|porti}(j) - \overline{E_{y|porti}}\right)}{\sqrt{\frac{1}{n}\sum_{j=1}^{n}\left(E_{y|CC}(j) - \overline{E_{y|CC}}\right)^2}\sqrt{\frac{1}{n}\sum_{j=1}^{n}\left(E_{y|porti}(j) - \overline{E_{y|porti}}\right)^2}} (i = 2, 3)$$

$$(18)$$

where, $E_{y|CC}(j)$ is the standard part, $E_{y|porti}(j)$ is the output mode at each port, $\overline{E_{y|CC}}$ and $\overline{E_{y|porti}}$ are the time average value of each output mode, and is the maximum cell number of the waveguide. Also, the mode changes between plus and minus with time changes. In this calculation, the positive maximum mode in the steady state of the electric field is used.

Finally, in order to examine the wavelength component included in the output signal, wavelength analysis is performed by the Fast Fourier Transform (FFT) using the time waveform of the electric field output to each port. Parameters used for the FFT are sampling time 0.0209[fs], sampling number 65536, time window length 1.3697[ps] and sampling frequency 0.3650 [THz].

4   Simulation Results

Distribution ratio and cross-correlation coefficient were calculated by the converged electric field while the input amplitude is changed. Input amplitude E_0 was changed in 0.1, 0.5, 1.0, 1.5[V/m]. At first, Fig. 2 shows distribution ratio of port 1, port 2 and port 3 and Cross-correlation coefficient of port 2 and port 3 in $E_0 = 0.1$[V/m]. It was confirmed that the distribution ratio to the port 2 and port 3 is changed along with change of the input signal wavelength. The optical signal does not reach port 1 at any wavelength because it is closed by linear dielectric pillars.

In addition, the cross-correlation coefficient exceeds 0.9 in most of the wavelength band, and high correlation is confirmed between the reference part and the mode of each port. However, at the wavelength $\lambda = 1.388$[μm], the cross-correlation coefficient of port 3 is very small. In other words, it is shown that there is no correlation between the mode of the reference part and the mode of port 3. At the wavelength $\lambda = 1.388$[μm] and 1.516[μm], the electric field distribution is shown in Fig. 3. It is shown that the optical signal does not reach port 3 at the wavelength $\lambda = 1.388$[μm] from Fig. 3.

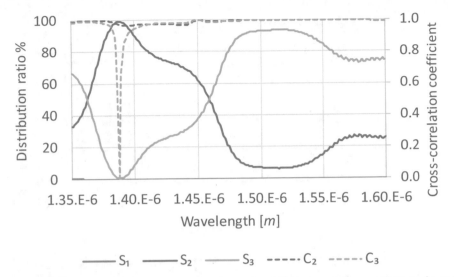

Fig. 2. Distribution ratio of port 1, port 2 and port 3 and Cross-correlation coefficient of port 2 and port 3 ($E_0 = 0.1[V/m]$)

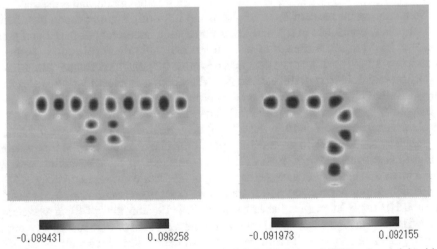

Fig. 3. Electric field distribution at $E_0 = 0.1[V/m]$ (left side is $\lambda = 1.388[\mu m]$ and right side is $\lambda = 1.516[\mu m]$)

Next, Fig. 4 shows distribution ratio of port 1, port 2 and port 3 and Cross-correlation coefficient of port 2 and port 3 in $E_0 = 0.5[V/m]$. Distribution ratio and cross-correlation coefficient similar to $E_0 = 0.1[V/m]$ are confirmed. However, there was small change in the distribution ratio and cross-correlation coefficient around $\lambda = 1.500[\mu m]$. At the wavelength $\lambda = 1.389[\mu m]$ and $1.513[\mu m]$, the electric field distribution is shown in Fig. 5. It was confirmed that optical signal leaks into the photonic crystal at $\lambda = 1.513[\mu m]$.

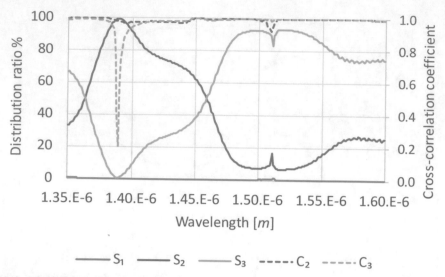

Fig. 4. Distribution ratio of port 1, port 2 and port 3 and Cross-correlation coefficient of port 2 and port 3 ($E_0 = 0.5[V/m]$)

Next, confirm the result of $E_0 = 1.0[V/m]$ and $1.5[V/m]$. Figure 6 shows distribution ratio of port 1, port 2 and port 3 and Cross-correlation coefficient of port 2 and port 3 in $E_0 = 1.0[V/m]$. The change of distribution rate and cross-correlation coefficient around $\lambda = 1.500[\mu m]$ is larger than that of $E_0 = 0.5[V/m]$. At the wavelength $\lambda = 1.389[\mu m]$ and $1.518[\mu m]$, the electric field distribution is shown in Fig. 7. Figure 8 shows distribution ratio of port 1, port 2 and port 3 and Cross-correlation coefficient of port 2 and port 3 in $E_0 = 1.5[V/m]$. The change of distribution rate and cross-correlation coefficient around $\lambda = 1.500[\mu m]$ is larger than that of $E_0 = 1.0[V/m]$. At the wavelength $\lambda = 1.392[\mu m]$ and $1.516[\mu m]$, the electric field distribution is shown in Fig. 9. From these results, it was found that changes in the distribution ratio and cross-correlation coefficient increase as the input amplitude increases.

Finally, output normalized power spectra of input wavelength $\lambda = 1.516[\mu m]$ at $E_0 = 0.1[V/m]$ is shown on the left side of Fig. 10, and that of input wavelength $\lambda = 1.516[\mu m]$ at $E_0 = 1.5[V/m]$ is shown on the right side of Fig. 10. Let us compare the two graphs. At $E_0 = 0.1[V/m]$, only the power spectrum of the wavelength of the input optical signal is output. However, at $E_0 = 1.5[V/m]$, the power spectrum of the wavelength of the input optical signal and the power spectrum of the short wavelength are output. Therefore, the optical signals leaking into the photonic crystal waveguide are those wavelengths.

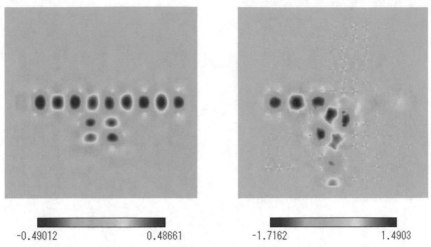

Fig. 5. Electric field distribution at $E_0 = 0.5$[V/m] (left side is $\lambda = 1.388$[μm] and right side is $\lambda = 1.516$[μm])

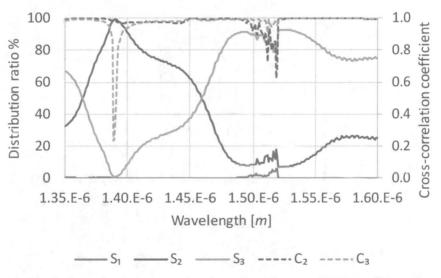

Fig. 6. Distribution ratio of port 1, port 2 and port 3 and Cross-correlation coefficient of port 2 and port 3 ($E_0 = 1.0$[V/m])

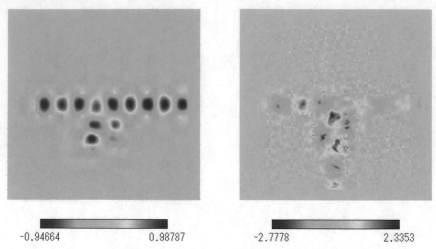

-0.94664   0.98787   -2.7778   2.3353

Fig. 7. Electric field distribution at $E_0 = 1.0$[V/m] (left side is $\lambda = 1.389$[μm] and right side is $\lambda = 1.518$[μm])

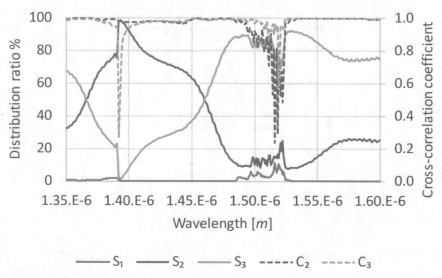

Fig. 8. Distribution ratio of port 1, port 2 and port 3 and Cross-correlation coefficient of port 2 and port 3 ($E_0 = 1.5$[V/m])

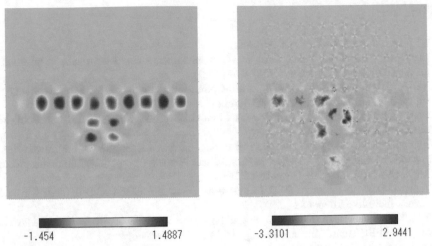

Fig. 9. Electric field distribution at $E_0 = 1.5$[V/m] (left side is $\lambda = 1.392$[μm] and right side is $\lambda = 1.516$[μm])

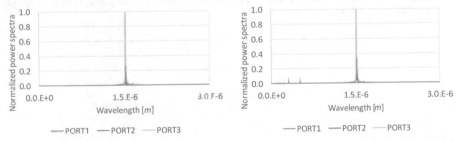

Fig. 10. Normalized power spectra (left side is $E_0 = 0.1$[V/m] and $\lambda = 1.516$[μm], and right side is $E_0 = 1.5$[V/m] and $\lambda = 1.516$[μm])

5 Conclusion

In two-dimensional photonic crystal waveguide with a linear dispersive and nonlinear medium duplexer, it was found that the distribution ratio and cross-correlation coefficient are changed depending on the input amplitude $E_0 = 0.1, 0.5, 1.0, 1.5$[V/m]. As the input amplitude increased, the fluctuation of the distribution ratio and cross-correlation coefficient increased around $\lambda = 1.500$[μm], and the optical signal penetrated into the photonic crystal. From this simulation, it is found that $E_0 = 0.5, 1.0, 1.5$[V/m] is not suitable for optical signal switching, and the most suitable input amplitude for optical signal switching is $E_0 = 0.1$[V/m]. Therefore, it is necessary to consider the amplitude level of the input optical signal for optical switching. As a future work, we would like to switch multiple optical signals in two-dimensional photonic crystal waveguide with multiple resonators.

References

1. NTT Tech. J. **22**(5) (2010). The Telecommunications Association
2. Takahashi, Y., Inui, Y., Chihara, M., et al.: A micrometre-scale Raman silicon laser with a microwatt threshold. Nature **498**, 470–474 (2013)
3. Al Islam, P., Sultan, N., Nayeem, S., et al.: Optimization of photonic crystal waveguide based optical filter. In: The IEEE Region 10 Symposium, pp. 162–167 (2014)
4. Noda, S., Baba, T. (eds.): Roadmap on Photonic Crystals. Kluwer Academic Publishers, Dordrecht (2003)
5. Higashinaka, N., Maeda, H.: Routing of optical baseband signal depending on wavelength in periodic structure. In: Advance on Broad-Band Wireless Computing, Communication and Applications, pp. 621–629, November 2019
6. Sullivan, D.M.: Nonlinear FDTD formulations using Z transforms. IEEE Trans. Microw. Theory Tech. **43**(3), 676–682 (1995)
7. Yee, K.S.: Numerical solution of initial boundary value problems involving Maxwell's equation. IEEE Trans. Antennas Propag. **14**(3), 302–307 (1966)
8. Li, X., Maeda, H.: Numerical analysis of photonic crystal directional coupler with kerr-type nonlinearity. In: Proceedings of 5th Asia-Pacific Engineering Research Forum on Microwaves and Electromagnetic Theory, pp. 165–168 (2004)
9. Berenger, J.: A perfectly matched layer for the absorption of electromagnetic waves. J. Comput. Phys. **114**(2), 185–200 (1994)

A Beam Power Allocating Method for Ka-band Multi-beam Broadcasting Satellite Based on Meteorological Data

Takumi Iwamoto[1] and Kiyotaka Fujisaki[2(✉)]

[1] Graduate School of Engineering, Fukuoka Institute of Technology,
3-30-1 Wajiro-higashi, Higashi-ku, Fukuoka 811-0295, Japan
mgm18101@bene.fit.ac.jp
[2] Department of Information and Communication Engineering,
Fukuoka Institute of Technology, 3-30-1 Wajiro-higashi, Higashi-ku,
Fukuoka 811-0295, Japan
fujisaki@fit.ac.jp

Abstract. The Ka-band such as 21 GHz-band can be used to achieve future broadcasting satellite services. However, the rain attenuation above Ka-band is significant problem to be solved. Therefore, we consider the compensation technique using multi-beam satellite covering service area with several beams. In the proposed method, the transmission power of each beam is adaptively controlled based on weather condition in each service area. In this paper, we consider the effectiveness of some methods to allocate the transmission power of each beam.

1 Introduction

Since allocation of 12 GHz-band (Ku-band) for broadcasting satellite services (BSS) in Japan, the BSS continues to be developed and 4 K/8 K ultra high definition television (UHDTV) started from the end of 2018. With the development of BSS, it is considered to make use of higher frequency band such as 21 GHz-band (Ka-band) in addition to current 12 GHz-band for the future BSS. The 21 GHz-band was allocated to some regions including Japan for BSS at WARC-92 [1]. Toward the realization of 21 GHz-band BSS, many research has been made on onboard antenna, transponder, and receiving antenna [2–4]. One of the significant problem to be solved is the rain attenuation toward the realization of the 21 GHz BSS. The rain attenuation dramatically increases in accordance with the frequency. According to the recommendation [5] by ITU-R, the rain attenuation of the 21 GHz-band at Tokyo is approximately four times bigger than that of the 12 GHz-band in decibels. One of conventional measures in 12 GHz-band BSS is to give the rain fade margin on the link based on the statistics of the rain attenuation. But, this measure is not considered to giving 21 GHz-band satellite which has large amount of the rain attenuation compared with that of 12 GHz-band because satellite has the strict limitation of the electricity. Thus, it is essential to realize a technique to control link parameters such as transmission

© Springer Nature Switzerland AG 2020
L. Barolli et al. (Eds.): EIDWT 2020, LNDECT 47, pp. 237–246, 2020.
https://doi.org/10.1007/978-3-030-39746-3_26

power based on weather condition to compensate the rain attenuation with the limited electricity. One of such techniques is adaptive control using multi-beam satellite which covers service area with several beams [6–13].

In references [8–13], we considered the adaptive control method using the 21 GHz-band multi-beam satellite as the compensation technique for the rain attenuation. In the proposed method, the transmission power of some beams is increased based on the prediction of the link quality with the precipitation data of each service area. We showed the method to predict the link quality of next 1 h [8,9] and effectiveness of the precipitation data to predict the link quality of next 1 h [10,11]. Furthermore, in the simulation, we made use of short time precipitation data such as 10-minutes precipitation data and considered the power density in the service area based on directivity of onboard antenna to realize a situation more similar to the reality [12,13].

In previous works, there is large difference of the link quality of each service area. Thus, we attempt to achieve uniform link quality throughout the service area. Also, we consider the power allocation based on the rain effect because it is likely that the necessary power is different from each beam.

The structure of this paper is as follows. In Sect. 2, we introduce the adaptive control method and its evaluation method. Then, simulation results have been shown to consider the effectiveness of the proposed method in Sect. 3. Finally, we conclude this paper.

2   Adaptive Control Method and Evaluation Method

We propose the compensation technique for the rain attenuation of the 21 GHz-band BSS using multi-beam satellite. As a multi-beam satellite, we focus on Wideband InterNetworking engineering and Demonstration Satellite (WINDS) developed by National Institute of Information and Communication Technology (NICT) and Japan Aerospace Exploration Agency (JAXA). Because WINDS has a Ka-band multi-beam antenna using a multi-port amplifier, it can realize flexible power allocation according to the weather condition [14].

2.1   Overview of Simulation

As seen in Fig. 1, nine beams are placed so that they cover all area of Japan based on WINDS. Following simulation is carried out to evaluate the effectiveness of the proposed method. First, the link quality of next 1 h in each beam is predicted by using predicted precipitation data. Next, several beams whose link quality is expected to be degraded due to the rain are selected based on the prediction. Then, the transmission power toward these selected beams is increased in addition to normal power. We call these selected beams "enhanced-beam" hereinafter in this paper. Finally, the validity of the adaptive control method proposed by us is evaluated by using measured precipitation data.

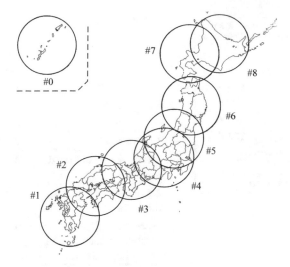

Fig. 1. Beam placement in simulation.

2.2 Adaptive Control Method

We propose following compensation technique for the rain attenuation.

- **Precipitation data for prediction of link quality**
  Radar-AMcDAS precipitation data by Japan Meteorological Agency (JMA) is used to predict the link quality of next 1 h [10,11]. Radar-AMeDAS is obtained by weather radar and AMeDAS (Automated Meteorological Data Acquisition System) at 1 h intervals. As shown in Fig. 2, Radar-AMeDAS is image data consisting of 640×640 pixels. In the simulation, a pixel of the data is assumed to be a ground receiving station of radio wave from a broadcasting satellite.

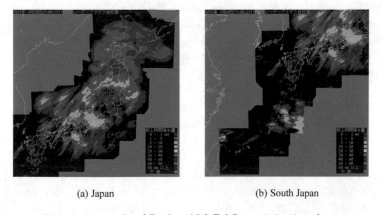

(a) Japan   (b) South Japan

Fig. 2. Example of Radar-AMeDAS precipitation data.

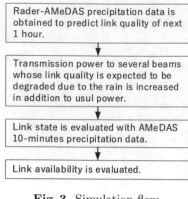

Fig. 3. Simulation flow.

Table 1. Parameters in the simulation.

Usual power	100 W
	(11.1 W/1 beam)
Compensation power	100 W
Satellite orbit	143° East
	(Geostationary)
Frequency	21 GHz
Diameter of onboard antenna	2.4 m
Diameter of receiving antenna	0.6 m
Aperture efficiency of onboard antenna and receiving antenna	0.7

- **Beam selection method**
  In order to select the enhanced-beam, we consider a method based on the number of the receiving stations where the link quality is expected to be degraded due to the rain [8]. Specifically, reference precipitation is defined to evaluate the rain effect on the link, and the enhanced-beam is selected in descending order of the number of the receiving stations where the precipitation is expected to be the reference precipitation in each service area.

2.3   Evaluation Method

Figure 3 shows simulation flow. Link parameters of this simulation are shown in Table 1 and following conditions are considered to evaluate the validity of the proposed method.

- **Precipitation data for evaluation of link quality**
  It is necessary to make use of short time precipitation data because rainfall tends to change in a short time [12]. Thus, we make use of AMeDAS 10-minutes precipitation data for the evaluation of the link quality.
- **Threshold precipitation**
  Since the precipitation data is used for the evaluation of the link quality, the precipitation needs to be converted into the rain attenuation. According to the recommendation [15] by ITU-R, specific rain attenuation γ [dB/km] is given by:

$$\gamma = kR^{\alpha}, \tag{1}$$

where R is rainfall intensity [mm/h]. The k and α are frequency dependent coefficients. In the simulation, $k = 0.0814$ and $\alpha = 1.0754$ are based on the 20 GHz-band [6]. A product of γ and the propagation distance of the rain area gives the amount of rain attenuation. Reference [16] claims that mean

$0\,°C$ isotherm height H_0 which is critical rain height depends on season and latitude L $(24 < L < 45°N)$ as follows:

$$H_0 = \begin{cases} 2.18 + 0.220L - 0.00409L^2 \text{ (June to September)} \\ 3.66 + 0.140L - 0.00342L^2 \text{ (otherwise)}. \end{cases} \tag{2}$$

By using Eq. (2), the propagation distance of the rain area D_r [km] is given by:

$$D_r = \frac{H_0}{\sin \phi}, \tag{3}$$

where ϕ is an elevation angle at a receiving station. Threshold precipitation that can maintain the link quality is given by Eqs. (1), (3) and rain fade margin. As seen in Table 1, the total of usual and compensation power are set to $100\,W$ respectively in the proposed method. Specifically, $11.1\,W$ ($= 100\,W/9$ beams) is allocated to each beam as the usual power. On the other hand, the compensation power is allocated to the enhanced-beam in addition to the usual power. That is, the allocated power to an enhanced-beam is up to $111.1\,W$ ($= 100\,W + 11.1\,W$). Table 2 shows relation between the allocated power to a beam and the threshold precipitation in case of the parameters shown in Table 1.

- **Power density distribution in service area**
The threshold precipitation depends on a location of a receiving station because of beam directivity of onboard antenna as well as the $0\,°C$ isotherm height. The beam directivity $G(\theta)$ of an aperture antenna is given by:

$$G(\theta) = \eta \left(\frac{\pi D}{\lambda} \right)^2 \left| \frac{2J_1(\xi(\theta))}{\xi(\theta)} \right|, \tag{4}$$

where η is aperture efficiency, D is physical diameter [m] of the antenna and J_1 is Bessel function of the first kind. The first term in Eq. (4) shows antenna gain of a parabolic antenna. $\xi(\theta)$ is expressed by:

$$\xi(\theta) = \frac{\pi D}{\lambda} \sin \theta, \tag{5}$$

where θ is angle difference between the center of a beam shown in Fig. 1 and a receiving station. In the simulation, $\theta = 0$ is defined at the center of each beam. That is, if a receiving station is located at the center of a beam, the threshold precipitation is the value shown in Table 2, and it gets smaller as a receiving station separates farther from the center of a beam.

- **Service availability**
Link state is evaluated at 10-minute intervals whether precipitation at the evaluation time exceeds the threshold precipitation or not as follows:

$$\begin{cases} \text{precipitation} > \text{threshold precipitation : incommunicable state} \\ \text{precipitation} \leq \text{threshold precipitation : communicable state}. \end{cases} \tag{6}$$

Table 2. Allocated power and threshold precipitation.

Allocated power [W]		11.1	21.1	31.1	41.1	51.1	61.1	71.1	81.1	91.1	101.1	111.1
Threshold precipitation [mm/h]	Annual season	16.1	20.4	22.9	24.7	26.2	27.3	28.3	29.1	29.9	30.6	31.2
	Rainy season*	15.5	19.6	22.1	23.9	25.2	26.4	27.3	28.1	28.8	29.5	30.1

* June to September

Eventually, service availability is defined as following equation.

$$\text{service availability} = \frac{\text{total communicable time}}{\text{total evaluation time}} \tag{7}$$

Because recommendation [17] by ITU-R suggests the service availability of BSS should be better than 99.86% of an average year, we adopt it as the target value.

3   Simulation Results

Table 3 shows the service availability without compensation from 2014 to 2015. From the results shown in Table 3, there is large difference of the service availability of each beam. In particular, the service availability of northern beams 6 to 8 achieves the target value even for the case of no compensation. Thus, these beams 6 to 8 are excluded from the selection of the enhanced-beam so that other southern beams are selected as the enhanced-beam, and we attempt to achieve the uniform service availability throughout the service area.

Table 3. Service availability [%] without compensation.

Beam No.	0	1	2	3	4	5	6	7	8	Avg.
Service availability	99.444	99.400	99.536	99.545	99.697	99.791	99.878	99.891	99.904	99.690

3.1   Achieving Uniform Service Availability

Table 4 shows the parameters for the adaptive control. We consider the number of enhanced-beam is 1 to 4 with the northern beams 6 to 8 excluded from the selection of enhanced-beam. Also, the reference precipitation to select the enhanced-beam is set to 15 mm/h based on reference [8]. Table 5 shows the number of enhanced-beam and the allocated power to each enhanced-beam.

Table 6 shows the service availability by the adaptive control. "Excl." and "Not-excl." in Table 6 mean that beams 6 to 8 are excluded from the selection of the enhanced-beam and all beams are the target of the beam selection

Table 4. Parameters for adaptive control.

The number of enhanced-beam	1–4
Excluded beams from enhanced-beam	Beams 6–8
Reference precipitation for selection of enhanced-beam	15 mm/h

Table 5. Allocated power to enhanced-beam.

The number of enhanced-beam	Allocated power [W]
1	100.0 + 11.1
2	50.0 + 11.1
3	33.3 + 11.1
4	25.0 + 11.1

respectively. From the results, we can see that the service availability of southern beams 0 to 5 improves by 0.003–0.013% by excluding beams 6 to 8 from the selection of the enhanced-beam. In terms of the average service availability, it is degraded by 0.004–0.010% by the exclusion. However, the difference of the service availability of each beam gets smaller by the exclusion. Table 7 shows standard deviation (SD) of the service availability shown in Table 6. From this table, SD in all cases is smaller by the exclusion.

When the number of enhanced-beam is 1, the service availability of southern beam 0 is the highest compared with other cases of the number of enhanced-beam although that of other beams is the lowest and SD is the highest in this case. On the other hand, when the number of enhanced-beam is more than 2, the service availability of each beam except beam 0 is basically improved as the number of enhanced-beam increases. Specifically, the service availability of beams 4 and 5 in addition to northern beams 6 to 8 achieves the target value when the number of enhanced-beam is 3 and 4 while that of these beams couldn't achieve it when the number of enhanced-beam is 1 and 2.

3.2 Power Allocation to Enhanced-Beam

As mentioned above, the enhanced-beam is selected based on the descending order of the number of receiving stations suffered from the rain effect, and the same compensation power is allocated to each enhanced-beam regardless of the selection order as shown in Table 5. However, it is likely that the necessary power is different from each enhanced-beam. Thus, we consider the allocation of the compensation power based on the selection order of each enhanced-beam. Table 8 shows some cases of the allocation of the compensation power when the number of the enhanced-beam is 4. The even case in Table 8 is the conventional allocation shown in Table 5.

Table 9 shows the service availability of each power allocation in the condition that beams 6 to 8 are excluded from the beam selection. It showed that the service availability of each beam in addition to the average service availability of case A and case B is higher than that of the even power allocation. In particular,

Table 6. Service availability [%].

Beam No.	N = 1		N = 2		N = 3		N = 4	
	Excl.	Not-excl.	Excl.	Not-excl.	Excl.	Not-excl.	Excl	Not-excl
0	99.744	99.739	99.739	99.735	99.729	99.722	99.713	99.710
1	99.690	99.680	99.754	99.744	99.761	99.756	99.747	99.742
2	99.649	99.643	99.786	99.773	99.817	99.806	99.818	99.811
3	99.708	99.701	99.777	99.769	99.820	99.808	99.833	99.821
4	99.780	99.776	99.848	99.838	99.865	99.857	99.871	99.864
5	99.830	99.820	99.868	99.859	99.892	99.884	99.900	99.894
6	99.878	99.911	99.878	99.924	99.878	99.933	99.878	99.938
7	99.891	99.923	99.891	99.944	99.891	99.947	99.891	99.949
8	99.904	99.936	99.904	99.950	99.904	99.953	99.904	99.954
Avg.	99.780	99.784	99.834	99.839	99.852	99.860	99.855	99.865

N: the number of enhanced-beam

Table 7. Standard deviation of service availability [%].

	N = 1	N = 2	N = 3	N = 4
Excl.	0.089	0.060	0.058	0.065
Not-excl.	0.105	0.082	0.080	0.084

Table 8. Allocation of compensation power to each enhanced-beam (Total allocated power is sum of the compensation power and 11.1 W of the usual power).

Selection order	Even case	Case A	Case B
1 st	25 W	40 W	30 W
2 nd	25 W	30 W	30 W
3 rd	25 W	20 W	20 W
4 th	25 W	10 W	20 W

Table 9. Service availability [%] of each power allocation (The number of enhanced-beam: 4, Excluded beams: 6 to 8).

Beam No.	Even case	Case A	Case B
0	99.713	99.747	99.726
1	99.747	99.773	99.761
2	99.818	99.827	99.827
3	99.833	99.841	99.839
4	99.871	99.876	99.875
5	99.900	99.899	99.901
6	99.878	99.878	99.878
7	99.891	99.891	99.891
8	99.904	99.904	99.904
Avg.	99.855	99.861	99.860

the service availability of southern beams is effectively improved. As mentioned in previous section, although the service availability of southern beam 0 is the highest when the number of enhanced-beam is 1, the average service availability couldn't achieve the target value. On the other hand, compared with the service availability, that of case A is improved largely, and the average service availability achieves the target value. Nevertheless, since the service availability of beams 0 to 3 couldn't achieve the target value, we need additional measure such as enlarging the size of receiving antenna.

4 Conclusions

In this paper, we proposed a compensation technique for the rain attenuation of the 21 GHz-band BSS by using the multi-beam satellite. The transmission power of each beam was controlled to reduce the rain effect based on the weather condition in each service area. In this paper, we considered some improvement of the proposed method and following results were shown by the simulation.

- It was considered to exclude some northern beams from the selection of enhanced-beam because the link quality of these beams achieves the target value even for the case of no compensation. As a result, the service availability of southern beams was improved, and the difference of the service availability of each beam was reduced by the exclusion.
- We considered the power allocation to the enhanced-beam because there was possibility that the necessary power was different from each enhanced-beam. As a result, the service availability of southern beams was effectively improved by changing the transmission power based on the selection order of the enhanced-beam.

For the future study, we will consider the optimization of the power allocation and enlarging the size of the receiving antenna at the southern area so that the service availability of these areas achieves the target value.

References

1. Radio Regulation ITU-R Resolution 525 (1992)
2. Nagasaka, M., Nakazawa, S., Kamei, M., Tanaka, S., Ito, Y.: Designing an array-fed imaging reflector antenna engineering model for 21-GHz band satellite broadcasting system. In: 2013 IEEE Antennas and Propagation Society International Symposium, pp. 300–301 (2013)
3. Kamei, M., Matsusaki, Y., Nagasaka, M., Nakazawa, S., Tanaka, S., Ikeda, T.: Development of on-board output filter for wideband and high output power transponder of 21GHz-band broadcasting satellite. In: 32nd AIAA International Communications Satellite Systems Conference (2014)
4. Nagasaka, M., Kojima, M., Sujikai, H., Hirokawa, J.: 12- and 21-GHz dual-band dual-circularly polarized offset parabolic reflector antenna fed by microstrip antenna arrays for satellite broadcasting reception. IEICE Trans. Commun. **E102–B**(7), 1323–1333 (2019)

5. Recommendation ITU-R BO.1659, Mitigation techniques for rain attenuation for broadcasting-satellite service systems in frequency bands between 17.3 GHz and 42.5 GHz (2012)
6. Yoshino, T., Ito, S.: A selection method of beams compensated using AMeDAS data for the multi-beam satellite broadcasting. IEICE Trans. Commun. **J82–B**(1), 64–70 (1999). (in Japanese)
7. Chodkaveekityada, P., Fukuchi, H.: Evaluation of adaptive satellite power control method using rain radar data. IEICE Trans. Commun. **E99–B**(1), 2450–2457 (2016)
8. Iwamoto, T., Fujisaki, K.: Study of the improvement of the link quality using AMeDAS rainfall data for multibeam satellite broadcasting system. IEICE Tech. Rep. **118**(176), 85–90 (2018). (in Japanese)
9. Fujisaki, K.: Study on improvement of link quality for Ka-band multibeam satellite system I - evaluation of the link availability under the most suitable beam control. IEICE Tech. Rep. **115**(448), 65–70 (2016). (in Japanese)
10. Iwamoto, T., Fujisaki, K.: Study of the beam control method of the Ka-band multi-beam broadcasting satellite system using meteorological data - evaluation of the effectiveness of radar AMeDAS precipitation data. ITE Tech. Rep. **43**(2), 71–74 (2019). (in Japanese)
11. Iwamoto, T., Fujisaki, K.: Study of beam power control of Ka-band multi-beam broadcasting satellite using meteorological data. In: Proceedings of AINA Workshops (2019)
12. Iwamoto, T., Fujisaki, K.: Study of beam power control methods of Ka-band multi-beam broadcasting satellite system using meteorological data - evaluation of the link quality using AMeDAS 10-minutes precipitation data. IEICE Tech. Rep. **119**(175), 7–12 (2019). (in Japanese)
13. Iwamoto, T., Fujisaki, K.: Study of beam power control methods of Ka-band multi-beam broadcasting satellite system using meteorological data - evaluation of link quality in consideration of beam-directivity. IEICE Tech. Rep. **119**(220), 93–98 (2019)
14. Special issue on wideband internetworking engineering test and demonstration satellite (WINDS). J. NICT **54**(4), 3–10 (2007)
15. Recommendation ITU-R P.618-13, Propagation data and prediction methods required for the design of Earth-space telecommunication systems (2017)
16. Satoh, K.: Studies on spatial correlation of rain rate and raindrop layer height. IEICE Trans. Commun. **J66–B**(4), 493–500 (1983). (in Japanese)
17. Recommendation ITU-R BO.1696, Methodologies for determining the availability performance for digital multi-programme BSS systems, and their associated feeder links operating in the planned bands (2005)

3-Party Adversarial Cryptography

Ishak Meraouche[1](✉), Sabyasachi Dutta[2], and Kouichi Sakurai[1]

[1] Kyushu University, Fukuoka, Japan
meraouche.ishak.768@s.kyushu-u.ac.jp, sakurai@inf.kyushu-u.ac.jp
[2] University of Calgary, Calgary, Canada
saby.math@gmail.com

Abstract. The domain of Artificial Intelligence (AI) has seen an outstanding growth during the last two decades. It has proven its efficiency in handling complex domains including speech recognition, image recognition and many more. One interesting and evolving branch that was put forward years ago but have seen a good growth only during the past few years is encryption using AI. After Google announced that it has succeeded teaching neural networks encryption in the presence of Eavesdroppers, research in this particular area has seen a rapid spread of interest among different researchers all over the world to develop new Neural Networks capable of operating different cryptographic tasks. In this paper, we take initial steps to achieve secure communication among more than two parties using neural network based encryption. We forward the idea of two party symmetric encryption scheme of Google to a multi party Encryption scheme. In this paper we will focus on a 3-Party case.

Keywords: Deep learning · Neural networks · Cryptography · Symmetric key encryption

1 Introduction

Digital communication world heavily relies on cryptographic techniques for resolving security threats. Provably secure cryptographic protocols have become basic building blocks for communication engineering where the channels are most often insecure. However, provable security incurs a huge overhead in the communication. Several faster cryptographic algorithms are proposed but in most cases they are not provably secure and often they are broken. Nowadays researchers are trying to develop cryptographic techniques based on artificial neural networks. There are quite a few works proposing basic encryption schemes based on neural networks (NN) [9,13,15] and also attacks [11].

One very interesting proposal was put forward by Google Brain Researchers [1] in 2016. In their proposal, two neural networks were able to learn (on their own) how to communicate securely in the presence of an eavesdropper. This work initiated a flow of research along this direction and even got improved in [3]. The idea of these works is a follow up of the work by Goodfellow et al. [5,6]. Recently,

© Springer Nature Switzerland AG 2020
L. Barolli et al. (Eds.): EIDWT 2020, LNDECT 47, pp. 247–258, 2020.
https://doi.org/10.1007/978-3-030-39746-3_27

Zhou et al. [16] initiated the study of security analysis of neural network based encryption schemes (in particular [1]) by using several statistical tests e.g. χ^2 test, Kolmogorov-Smirnov test.

The work done by Google Brain researchers is considered a big step in the world of cryptography as it ensures secure communication between two parties. The security is in the fact that the two neural networks are not taught any specific encryption method, all relies on the synchronisation of the two neural networks and their structure. It has been proved in [1] that even with a neural network with a similar structure, it is difficult to decrypt the cipher text by cryptanalysis.

Our aim in this paper is doing a similar work but among three parties instead of just two. As synchronising multiple parties might be a bit challenging, we study a single case and try to synchronize three neural networks so they communicate securely in different scenarios. In other words, we synchronize three neural networks to communicate securely as in [1] in the presence of an eavesdropper with the same neural network structure while trying to improve the accuracy and preserving the training time.

2 Related Works

2.1 Classical Symmetric-Key Cryptography

Symmetric cryptography is widely used in the world of cryptography. Usually, there are two parties Alice and Bob who are willing to communicate with eachother. These two parties share in advance, a key K and a pre-fixed encryption/decryption algorithm. Alice uses the plain text P and the key K as inputs to the encryption algorithm, process it and output the cipher text C. Once C is sent to Bob, he will process it with the decryption algorithm (which also depends on K) to output the original plain text P. Many existing encryption algorithms are widely used in networking security. A good and widely used algorithm is the Rijndael Algorithm (The advanced encryption standard) Approved by the NIST as the Advanced Encryption standard [8]. Other very efficient algorithms are Serpent [2] and Blowfish [12]. Symmetric key encryption schemes are very fast and incur a low communication overhead.

2.2 The Two Party Neural Network Based Encryption by Google Brain [1]

The Training Model
In this model, Alice and Bob are both neural networks sharing a secret key K willing to communicate securely. They have the same neural network structure but will be initialized with random parameters. In order to train the neural networks, a third neural network (called Eve) is added to the scenario and she plays the role of an eavesdropper. Eve has also the same neural network structure but does not know the key K. Figure 1 shows the schematic of the model.

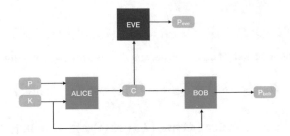

Fig. 1. Two party encryption using neural networks

Alice has as an input the plaintext P and the secret key K. They are used as inputs to Alice's neural network. After being processed, the neural network outputs the cipher text C.

C is sent to Bob who uses it as an input to his neural network along the secret key K and output P_{Bob}.

Eve will intercept the ciphertext C and use it as an input to her neural network but without the secret key K.

In order to train the neural networks Alice, Bob and Eve are trained with the following parameters respectively: $\theta_A, \theta_B, \theta_E$. The function $E_A(\theta_A, P, K)$ represents the encryption function for Alice with output as $\mathbf{C}$ and $D_B(\theta_B, C, K)$ represents the decryption algorithm of Bob with output as $\mathbf{P_{Bob}}$. Regarding Eve, she has the following decryption function: $D_E(\theta_E, C)$ with output as P_E.

In order to calculate the distance between the original plaintext and each deciphered text the L_1 metric is used:

$$d(P, P') = \sum_{i=0}^{N} \mid P_i - P_i' \mid$$

The loss function for Eve can be defined as follow:

$$L_E(\theta_A, \theta_E, P, K) = d(P, D_E(\theta_E, E_A(\theta_A, P, K)))$$

As we can see, L_E measures the error in the decryption process of Eve. By taking an expected value, a loss function over the distribution of the plain texts and keys can be defined as follow:

$$L_E(\theta_A, \theta_E, P, K) = E_{P,K}(d(P, D_E(\theta_E, E_A(\theta_A, P, K))))$$

The optimal Eve is obtained by minimizing the loss:

$$O(E)(\theta_A) = argmin_{\theta_E}(L_E(\theta_A, \theta_E, P, K))$$

Similarly, the definition of the loss function for Bob is as follow:

$$L_B(\theta_A, \theta_B, P, K) = d(P, D_B(\theta_B, E_A(\theta_A, P, K), K))$$

The expected value over a distribution of plain texts and keys gives:

$$L_B(\theta_A, \theta_B, P, K) = E_{P,K}(d(P, D_E(\theta_E, E_A(\theta_A, P, K)), K))$$

With the above, the optimal values for Alice and Bob are as follow:

$$L_{AB}(\theta_A, \theta_B) = L_B(\theta_A, \theta_B) - L_E(\theta_A, O_E(\theta_A))$$

Training Process

Using the loss function defined above ($L_{AB}(\theta_A, \theta_B)$), first initialize neural networks with random parameters ($\theta_A, \theta_B, \theta_C$) and then update them after every iteration till minimization is done.

Once the minimization is done, Alice and Bob are said to be synchronized and can communicate securely even in the presence of eavesdroppers as their final parameters are optimal and can guarantee high decryption error for eavesdroppers and low decryption error for members of the same party.

How Is the Data Actually Being Encrypted?

The way data is being encrypted might not be so clear from the perspective of the calculation of the minimum distance.

Concretely, each neural network's input is the concatenation of the plain text with the secret key. After this, the bit stream will go through every layer of the neural network in the following order: First a fully connected layer that will output a bit stream of the same size as the input. Next, we will have a serie of Convolutional layers that will mix the bits of the input vector and output a cipher text. You will notice that the Mix & Transform method is being applied to the input vector in order to output a cipher text. What is happening inside the neural network when it receives the input vector is a kind of black-box and cannot concretely be analysed.

2.3   Steganography Based on Adversarial Neural Networks

Besides cryptography, Steganography is also an interesting field. It aims to hide an information inside another one. Usually we hide a text message or a small image inside another one using a secret key and make it impossible to extract the hidden information without the key.

Recent works show methods to do secure steganography based on adversarial neural networks. We have for example the work done in [10] that shows a way to train the cover image to generate the secret information without modifying the original image. Another method for steganography is also shown in [7].

A very recent paper [14] also features a method to do steganography using neural networks. In their work, there are two neural networks Alice and Bob sharing a key K and willing to communicate in the presence of the eavesdropper Eve. Alice trains to hide a message inside a cover image so that only Bob can extract it.

2.4 Over the Air Communication

Another promising work has been done in [4] where a full implementation of a communication over the air (radio frequencies) has been done based on deep learning.

Our Contribution

In traditional symmetric key encryption, if any number of parties hold the same key then they are able to communicate among themselves securely. The order in which the parties communicate does not matter. We ask the same question in case of NN based symmetric key encryption when all the parties share the common secret state/key. We consider the case of three neural networks who are to be trained to achieve a symmetric encryption rule. We consider three possible scenarios to train the neural networks. The challenging issue is to get a low error rate for decryption while and after training. We demonstrate that it is possible to include a third neural network to the system of Alice and Bob, called Charlie and that they will be able to synchronize and communicate securely in the presence of eavesdroppers. In the first scenario, there is no hindrance in the communication among the parties. In the last two scenarios, Alice has to act as an intermediary for secure communication to happen.

3 Results

Our aim is to synchronize three neural networks. Alice, Bob and Charlie are the parties who want to communicate securely even in the presence of eavesdroppers. Alice, Bob and Charlie share the same secret key K and have the same neural network structure. As before, Eve has access to any ciphertext communicated among the three parties but do not have access to the secret key K.

As in the two-party model and in order to synchronize our neural networks we need a loss function.

As we added Charlie to the group, we need to add his loss function too, which is defined as follows:

$$L_C(\theta_A, \theta_C, P, K) = d(P, D_C(\theta_C, E_A(\theta_A, P, K), K))$$

The expectation over a distribution on plain texts and keys is defined as follows:

$$L_C(\theta_A, \theta_C, P, K) = E_{P,K}(d(P, D_C(\theta_C, E_A(\theta_A, P, K)), K))$$

The final loss function is defined as follows:

$$L_{ABC}(\theta_A, \theta_B, \theta_C) = L_B(\theta_A, \theta_B, \theta_C) - L_E(\theta_A, O_E(\theta_A))$$

3.1   First Scenario

Training Process

In this scenario we suppose that one Neural network is sending messages to the two other neural networks in the party. We choose Alice as the sender and Bob, Charlie as receivers.

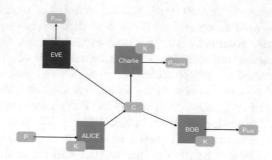

Fig. 2. Setup of experiment 1: In this setup Alice acts as a leader and sends messages to both Bob & Charlie so that the parameters of the NNs are optimized.

Experiment

We make Alice generate a plain text P of 32 bits, encrypting it and sending it to both Bob and Charlie.

We make Bob and Charlie decrypt the Cipher text and output $P_{Bob}, P_{charlie}$ respectively. Eve intercepts and tries to decrypt the cipher text C and outputs P_{eve}.

In every iteration, we take calculate the minimum distance between the original plaintext and decrypted ones respectively: $P_{Bob}, P_{charlie}$ and P_{eve}. Figure 2 shows the plot of the decryption error made during each iteration (We train by a set of epochs consisting of 2000 iterations each.) (Fig. 3).

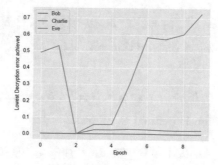

Fig. 3. Test results of experiment 1

Analysis

In this experiment, we can see that Bob and Charlie start with almost 100% accuracy and about 50% accuracy for Eve. It all goes good but Eve quickly learns how to decrypt the messages made between Alice and Bob. Alice starting from the second epoch starts making her encryption method harder and as a result, Eve's decryption error goes up to 70% by the end of the epoch. The change made by Alice in the encryption process results in decryption error of both Bob and Charlie. The accuracy of decryption is lowered to about 98%.

3.2   64-Bit Message Length

We noticed that rising the message and key length to 64 bits gives much more accuracy to Alice and Bob while training. However and most importantly, Eve's accuracy never goes beyond 50%. The training results are shown in Fig. 4.

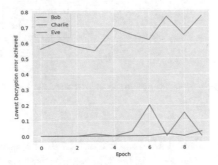

Fig. 4. Test results of experiment 1 with 64 bits key and message length

3.3   Second Scenario

Training Process

The Second Scenario consists into making two pairs of neural networks. Concretely, Alice will communicate with Charlie and Charlie will communicate with Bob. There is no direct communication between Alice and Bob.

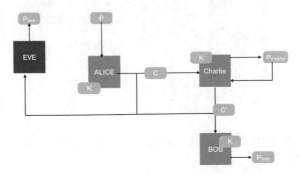

Fig. 5. Setup of the second scenario

Experiment

In Fig. 5, Alice generates a ciphertext and sends it to Charlie. Charlie deciphers the ciphertext, encrypts it again with the same method and send the new ciphertext to Bob. Bob decrypts the new ciphertext to output P_{Bob}. Eve has access to both of the communications. We train and calculate the error between the original ciphertext and the three outputs: $P_{Bob}, P_{charlie}$ and P_{eve}. We plot the decryption error made by every Charlie, Bob and Eve as shown in Fig. 6.

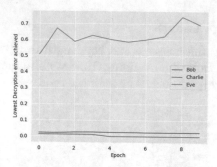

Fig. 6. Second scenario experimental results

Analysis

We see that Charlie has almost 100% accuracy in decryption after 4 epochs. Bob though has about 97.5% accuracy after the decryption process.

After synchronizing Alice with Charlie and Charlie with Bob, we tried to make Alice communicate directly with Bob with the synchronized parameters but unfortunately the communication was not accurate and therefore this is not possible. Figure 7 shows the decryption error achieved which is around 50%.

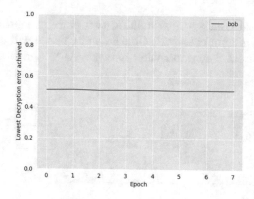

Fig. 7. Error made by Bob when Alice communicates directly with him without going through Charlie

3.4 Third Scenario

Training Process

The third and last scenario is to make Alice synchronize with Bob and Charlie separately. Concretely, Alice will synchronize with a set of parameters with Charlie and another one to communicate with Bob. In this scenario, if Bob wants to communicate with Charlie, he will need to use Alice as a bridge. Figure 8 shows the model of the third scenario.

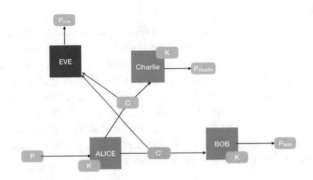

Fig. 8. Setup of the third scenario

Experiment

We make Alice generate two set of parameters for her neural network. When Alice generates a plain text P, when sending this plain text to Bob, she will use the first set of parameters. When sending a message to Charlie, she will use the second set of parameters. The decryption errors are recorded and plotted in Figs. 9 and 10.

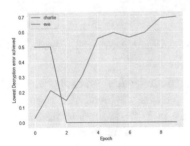

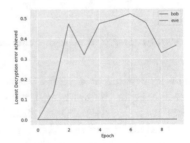

Fig. 9. Decryption error achieved when Alice communicates with Charlie

Fig. 10. Decryption error achieved when Alice communicates with Bob

Analysis

We can see that Charlie starts with a decryption error of 50% but quickly synchronizes with Alice (after 2000 iterations) and gets a nearly perfect accuracy.

Eve has at the start of the training a very good accuracy but it starts to go down quickly after about 2000 iteration.

In this training experiment, Bob had a perfect accuracy from the beginning. All Alice had to do while training is to rise the decryption error for Eve which is done after about 4 epochs.

As in the second scenario, Bob cannot talk properly to Charlie without a proper synchronization. As we can see in Fig. 11 Bob's error is around 45%.

3.5   Discussions

Despite having a different way to communicate, during our experiments all the three scenarios had pretty much the same training time and accuracy. The usage of cases depend on scenario. If one entity talks to two other entities at the same time, then the first scenario is appropriate. Second and Third scenario are best for talking separately. Third scenario differs from the second scenario by letting Alice use one single key per neural network which will allow to "hide" the messages sent by Alice to Bob from Charlie and vice-versa. In a multi-party perspective, The best usage cases will be the same, First Scenario for a group talk; Second and third scenario to talk separately.

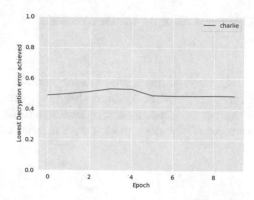

Fig. 11. Decryption error achieved when Bob communicates with Charlie

The extension to a multi-party scheme is theoretically possible, the only inconvenience is the training time – the more parties we have, the more time it takes to train them. One possible way to do it is to train the neural networks in advance and save the parameters in a trusted server or party; then, the trusted party will serve as much parameters as required for the entities willing to communicate.

4   Conclusion and Further Work

We presented our method to make a 3-party adversarial network based on Google's 2-party model. We proposed three different scenarios where we proved in each of them very good accuracy for the receiving parties.

As an interesting future work, one may consider extending three party neural networks towards achieving resistance against more concrete adversaries (e.g. chosen plaintext attack) and extending to multiple-party cases.

Acknowledgement

Ishak Meraouche is supported by the Ministry of Education, Culture, Sports, Science and Technology (MEXT), Japan for his studies at Kyushu University.

Sabyasachi Dutta was financially supported by the National Institute of Information and Communications Technology (NICT), Japan, under the NICT International Invitation Program during his stay at Kyushu University where the initial phase of the research work was carried out.

References

1. Abadi, M., Andersen, D.G.: Learning to protect communications with adversarial neural cryptography. CoRR **abs/1610.06918** (2016)
2. Anderson, R.J., Biham, E., Knudsen, L.R.: The case for serpent. In: The Third Advanced Encryption Standard Candidate Conference, 13-14 April 2000, New York, New York, USA, pp. 349–354 (2000)
3. Coutinho, M., de Oliveira Albuquerque, R., Borges, F., García-Villalba, L.J., Kim, T.: Learning perfectly secure cryptography to protect communications with adversarial neural cryptography. Sensors **18**(5), 1306 (2018)
4. Dörner, S., Cammerer, S., Hoydis, J., ten Brink, S.: Deep learning-based communication over the air. IEEE J. Sel. Top. Sign. Proces. **12**(1), 132–143 (2017)
5. Goodfellow, I.J., Pouget-Abadie, J., Mirza, M., Xu, B., Warde-Farley, D., Ozair, S., Courville, A.C., Bengio, Y.: Generative adversarial nets. In: Advances in Neural Information Processing Systems 27: Annual Conference on Neural Information Processing Systems 2014, 8-13 December 2014, Montreal, Quebec, Canada, pp. 2672–2680 (2014)
6. Goodfellow, I.J., Shlens, J., Szegedy, C.: Explaining and harnessing adversarial examples. In: 3rd International Conference on Learning Representations, ICLR 2015, Conference Track Proceedings, San Diego, CA, USA, 7–9 May 2015 (2015)
7. Hayes, J., Danezis, G.: Generating steganographic images via adversarial training. In: Advances in Neural Information Processing Systems 30: Annual Conference on Neural Information Processing Systems 2017, 4-9 December 2017, Long Beach, CA, USA, pp. 1954–1963 (2017)
8. Jamil, T.: The rijndael algorithm. IEEE Potentials **23**(2), 36–38 (2004)
9. Kanter, I., Kinzel, W., Kanter, E.: Secure exchange of information by synchronization of neural networks. EPL (Europhys. Lett.) **57**(1), 141 (2002)
10. Ke, Y., Zhang, M., Liu, J., Su, T.: Generative steganography with kerckhoffs' principle based on generative adversarial networks. CoRR **abs/1711.04916** (2017)
11. Klimov, A., Mityagin, A., Shamir, A.: Analysis of neural cryptography. In: Advances in Cryptology - ASIACRYPT 2002, 8th International Conference on the Theory and Application of Cryptology and Information Security, Proceedings, Queenstown, New Zealand, 1-5 December 2002, pp. 288–298 (2002)
12. Schneier, B.: Description of a new variable-length key, 64-bit block cipher (blowfish). In: Fast Software Encryption, Cambridge Security Workshop, Proceedings, Cambridge, UK, 9-11 December 1993, pp. 191–204 (1993)

258   I. Meraouche et al.

13. Wang, X.Y., Yang, L., Liu, R., Kadir, A.: A chaotic image encryption algorithm based on perceptron model. Nonlinear Dyn. **62**(3), 615–621 (2010)
14. Yedroudj, M., Comby, F., Chaumont, M.: Steganography using a 3 player game. CoRR **abs/1907.06956** (2019)
15. Yu, W., Cao, J.: Cryptography based on delayed chaotic neural networks. Phys. Lett. A **356**(4), 333–338 (2006)
16. Zhou, L., Chen, J., Zhang, Y., Su, C., James, M.A.: Security analysis and new models on the intelligent symmetric key encryption. Comput. Secur. **80**, 14–24 (2019)

A Framework for Automatically Generating IoT Security Quizzes in 360VR Images/Videos Based on Linked Data

Wei Shi[1]([✉]), Tianhao Gao[2], Srishti Kulshrestha[3], Ranjan Bose[4],
Akira Haga[1], and Yoshihiro Okada[1]

[1] The Innovation Center for Educational Resources, Kyushu University Library,
Kyushu University, Fukuoka, Japan
{shi.wei.243,akira.haga.879}@m.kyushu-u.ac.jp
[2] The Graduate School of ISEE, Kyushu University, Fukuoka, Japan
koutengou@yahoo.co.jp
[3] Department of Electrical Engineering, Indian Institute of Technology Delhi,
New Delhi, India
srish.kul@gmail.com
[4] Indraprastha Institute of Information Technology Delhi, New Delhi, India
bose@iiitd.ac.in

Abstract. Internet of Things (IoT) technology has greatly developed in the past ten years. Many smart buildings and smart homes are constructed and used. Because of the popularity of smart buildings and smart homes, the security problem becomes more and more important. How to effectively enhance people's IoT security knowledge is discussed frequently. In the past, we have already developed a framework that can automatically generate the IoT quizzes based on Linked Data in a virtual 3D environment. In this paper, we discuss how to extend it to support the quiz generation using the images and/or videos captured by a 360-degree camera. In this paper, we will apply the object recognition technology to automatically find out the IoT devices from these images and videos. According to the recognizing result, our framework will generate related questions for test or train learners.

1 Introduction

Internet of Things (IoT) is an important concept in the following era. Applying IoT technology, many smart buildings and smart homes are constructed. With the development of IoT technology, many researchers start to focus the IoT security problems. How to effectively provide people the IoT security knowledge are also wildly discussed. In a smart building/home, there are many different types of IoT devices. For education purposes, many e-textbooks and e-learning materials are developed and provided to users. Quizzes are also a type of effective method for helping learners to remember knowledge and to test learners. As a kind of e-learning materials, on-line quizzes are widely used. Comparing to traditional paper quizzes, on-line quizzes are more flexible and have richer question formats. Some quizzes are developed to provide different questions to different learners, according to learners' characters (such as

© Springer Nature Switzerland AG 2020
L. Barolli et al. (Eds.): EIDWT 2020, LNDECT 47, pp. 259–267, 2020.
https://doi.org/10.1007/978-3-030-39746-3_28

knowledge levels). Furthermore, comparing to the traditional paper quizzes, it is possible for on-line quizzes to record users answering action patterns for further analytics.

In our previous research [1], we have already introduced our framework which supports users to create their customized quizzes. We also introduced how to apply our framework to create IoT security quizzes in a virtual 3D environment [2]. In this paper, we will furtherly extend our framework for supporting the generation of an IoT quiz using the 360-degree images and/or videos. Such kind of environment is more closers to a real environment, so learners can be more effectively trained or tested. In such kind of quizzes, our framework will automatically recognize the IoT devices in learners' view, and according to different devices to provide corresponding questions.

This paper is organized as follow: In Sect. 2, we introduce the related works; Sect. 3 explains the details of how to use our framework to create quizzes; Sect. 4 shows how to extend our framework for creating quizzes using 360-degree images/videos; the last section concludes our framework.

2 Related Work

In past years, many researchers discussed how to manage data, define ontologies and then realize the automatic generation of quizzes. They research this topic from different aspects. Some researchers are considering how to generate smooth question and choice sentences. They apply the natural language processing technology to the data for this purpose. Edison [3] and Akhil [4] discussed this problem in their papers. Both are focusing the algorithm researches. Then, other researchers focus on how to use the machine learning technology to control the difficulty levels of quiz questions and to provide good questions, such as the framework introduced in [5]. These research aspects are very important but out of the scope of this paper. We also plan to follow their steps to improve our framework in the future.

Another important research aspect is to discuss how to obtain data from a Linked Data set according to the quiz creators' requests. Oscar showed us the research results on this topic in two papers[6, 7]. They discussed how to specify the data, which is retrieved from DBpedia and used for generating quizzes, in two aspects, based on the domain specification [6] or the educational standards [7]. Because Linked Data has a more flexible structure, it is more difficult to use than the relational database, which structures data in a tabular structure. These papers have not provided any generic solution to this problem. Comparing to these researches, our framework provides a more intuitive schema, which is dynamically extracted from the Link Data set, and is represented as a directed graph. This method will reduce the difficulty to use the Linked Data for quiz creators.

As well, some researchers have already developed their systems which support the creations of quizzes. Foulonneau [8] developed a streamline which can and only can use one triple from the Linked Data set such as DBpedia to generate quizzes. This method ignored the linkage between triples which is the most important feature of the Linked Data. Liu and Lin [9] also developed a system, named "Sherlock". It can use the data from DBpedia and BBC to generate quizzes and can realize the control of the difficulty level of quizzes. However, the templates provided for creating the quizzes are still too basic and may be difficult to improve. The linkages among RDF triples are not effectively used as well. There are also some other successful quiz generation systems in specific fields. For example, Hazriani [10] proposed a quiz generation schema for learning "movie based context-aware", and Sung [11] discussed the generation of English quiz. They are powerful and useful. But the common insufficiency of these systems is that their methods are not generic. It is difficult to create quizzes in other fields by reusing these research results.

3   IoT Security Quiz Generation Framework Based on Linked Data

We have already developed a quiz generation framework basked on Linked Data [1]. Then, we have already extended our framework for supporting quizzes in a virtual 3D environment [2]. In this section, I will show the details of our framework.

This framework includes two tools. One is an RDF data schema extraction and representation tool. This tool can extract the schema of the RDF data and represent the extracted schema as a graph. By using this tool, instead of knowing the schema of the RDF data which is used to create the quiz, the schema will be visualized and then users can visually manipulate it to define quizzes. The other tool is an authoring tool. This authoring tool provides a set of HTML components for composing quiz page templates. By manually assigning the connections between HTML components and nodes of the visualized RDF schema, the system will automatically generate a set of SPARQL queries. Using these queries, our system can dynamically retrieve the necessary data from our RDF dataset and provide the values to the corresponding HTML components. This will lead to the dynamic generation of the quiz pages. Figure 1 shows an overview of using our framework to create an IoT security quiz. In each question page, there are two questions. For answering the first question, quiz users need to select the name of the IoT device shown in the pictures. The second question is an extension of the first question. Quiz users need to answer what attack may be performed on the selected device in the first question. In this section, we will explain the details of our framework by introducing how to use our framework to create the IoT Security quiz shown in Fig. 1. To simplify the description of the URI in our database, we use the "IoTSecurity:" to replace "http://.......IoTSecurity/", which is the prefix of the URIs of the RDF data in our database. We also use another two prefixes of the URIs in this paper, the "rdfs:" to replace "http://www.w3.org/2000/01/rdf-schema#" and use the "rdf:" to replace "http://www.w3.org/1999/02/22-rdf-syntax-ns#", which are defined by W3C.

To define this quiz, we first extract the RDF data schema. The left lower part of Fig. 1 shows the schema of the RDF data for creating our quiz application. In this framework, the schema is simplified to be composed of edges and two kinds of nodes. One kind of nodes is the Class node, and the other kind is the Property node. The edges in such a graph may be labeled or not. We suppose there is an RDF dataset R. As we introduced, the data item t in R is identified by a unique URI. We suppose t is an instance of class T. To generate the schema of R, we first extract all the classes in this dataset. A class T_c will be represented as a circular Class node (e.g. the node

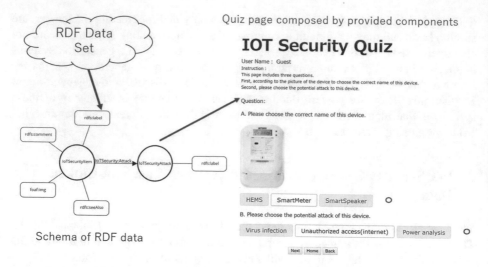

Fig. 1. An IoT security quiz generated by our framework

"IoTSecurityItem" in Fig. 1). Next, we extract all the instances of T_c and all the tuples whose subject is these instances. Then, we extract the objects of these tuples. If the value of an object is a fixed value, such as number or text, we add a rectangular Property node to the schema. The label of this node is the predict (e.g., the node "rdfs:comment" in Fig. 1). The edge from the Class node T_c to this property node will be unlabeled. If the value of an object is an instance of another Class T_a, an edge from the Class nodes T_a to T_c will be drawn. The label of this edge is the triple's predict.

Next, we use the authoring tool for supporting quiz makers to create quiz games as web applications. This authoring tool is an extension of C. Ma's research [10]. In our framework, we use such an authoring tool to define the templates of the quiz pages. To simplify users' manipulations, we provide a set of template components. These components are text components, image components, video components sound components, navigation button components, checkbox components, and choice components. Besides these provided components, users can add the necessary HTML elements by directly programming. These components are provided in the form of HTML + JavaScript codes. In the page template definition, users also need to specify the class nodes which are used as the question, the right choice, and the wrong choice. Then users can associate the component with the Property nodes in the RDF schema. When a quiz page is generated, the values of corresponding properties will be provided to the HTML components and be displayed on the web page. In the first question of our example quiz, we defined the Node "IoTSecurityItem" is used as the class of the question and its choices, the question image is related to the Node "ofaf:img" to the question's picture, the text of each choice is related to its Property Node "rdfs.label".

After composing a quiz page and associating the components with the nodes of the RDF schema, our system will automatically generate SPARQL queries for retrieving necessary data for generating the quiz pages. The SPARQL queries are generated in two steps. First, users need to specify how to generate quiz questions and their right answers. Supposing we specify a class A as the question and its linked class B as the right answer, the edge between A and B are labeled as L, and the properties "A.P1" and "B.P2" are associated with the HTML components in the quiz page, the system will generate a query as follows:

```
SELECT ?v1 ?v2 ?v2 ?v3
    WHERE {
            ?v1 rdf:type "A".
            ?v2 L ?v1.
            ?v1 P1 ?v3.
            ?v3 P2 ?v4.
    }
    LIMIT 1
    OFFSET randomNumber.
```

In this query, "randomNumber" is a random number generated by the system to retrieve a random item from the database as the question and the right answer. To generate the first question of our example, the SPARQL query should be

```
SELECT    ?v1 ?v2 ?v3
WHERE {
?v1 rdf:type "IoTSecurityItem
?v1 rdfs:label ?v2.
?v1 pt:IMG ?v3.
}
LIMIT 1
OFFSET randomNumber.
```

Next step, users need to specify how to generate the wrong choices. As a question's wrong choices, they should have one or more common features with the right choice. Users need to define the constraints to specify these common features through the RDF schema, and then our system can translate such specifications into the SPARQL query. Normally, the wrong choices and the right choices should belong to the same class, but users can specify another class for the wrong choices. To define a new constraint, we suppose that the URI of the right answer is Ur, the wrong choices are Uw, the Ur's property Pcr, and the Uw's property Pcw are used to define the constraint between the wrong answers and the right answer, and the constraint between these two properties is f. The constraint is defined as:

$$Uw.Pcw = f(Ur.Pcr).$$

Currently, our framework supports nine kinds of relationships between the right answer and wrong answers. They are "equal", "large than", "less than", "unequal", "union", "and", "minus", "contain", and "not contain". In the future, we will add necessary computations according to the requests from the quiz makers.

In the first question of our quiz example, we only specify the default constraints between the right answer and the wrong answers: the right answer and wrong answers belong to the same class "IoTSecurityItem". Then, we suppose the URI of the right answer is U and we need n wrong answers, then the SPARQL query for generating one wrong choice is shown as follows:

```
SELECT    ?v1 ?v2 ?v3 ?v4 ?v5
WHERE {
     U rdf:type ?v1.
     ?v2 rdf:type ?v1.
     ?v2 rdfs:label ?v3.
}
LIMIT 1
OFFSET randomNumber
```

This query will be performed n times to retrieve n wrong answers for this quiz.

4 IoT Security Quiz Generation in 360-Degree Images/Videos

In this paper, we extend our framework for supporting the generation of quizzes in 360-degree images/videos. Such an environment is much close to the real situation and can improve the learners' abilities to deal with the real problem. The quizzes generated by this extend framework can show a 360-degree first. We take such images/videos of smart homes and smart buildings using Ricoh R. In these images/videos, there will be many different types of IoT devices. These devices will be automatically recognized by our framework. When a device is detected and recognized, it will be used to trigger a question related to this device. The questions and choices still are provided using the text format.

This extended framework works as shown in Fig. 2. In this paper, we use the Cascade Classifier provided by OpenCV [12] to realize the object detection and recognition. We use Google to search the images of the IoT devices registered in our quiz resources database. The searched results will be used as the training data set. Next, we need to pre-process these images. After realizing the learning process, this network will be used to realize the IoT devices in a picture.

To process the 360-degree images, we first cut the images according to the learners' views. Then, our system will detect the devices in this image. If there is only one device, its name will be used as the keyword to search our quiz resources database to generate the questions and choices. If there is more than one device is detected, the device who is closest to the center of the view will be used as the keyword to generate quizzes. To process the 360-degree videos, our system will obtain n images from the videos per second. n is determined according to the performance of the user's device. The obtained images will be processed in the same way explained before.

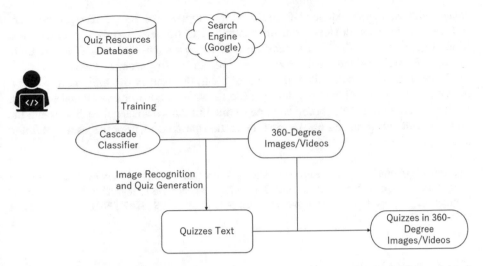

Fig. 2. The workflow of using our framework to generate quizzes in 360-degree images/videos

5 Discussion

We will discuss how to evaluate our framework in this section in the three aspects. First, we will evaluate the F value (2Recall*Precision/(Recall + Precision)) of the IoT device recognition method in our framework. In this part, we want to evaluate if our framework is possible to correctly recognize the IoT devices when providing quiz questions. We also plan to improve the precision of this method. In our framework, we hope to realize the high precision even if we have to reduce the recall.

Next, we will evaluate the educational effects of the quizzes generated by our framework. We will ask some students from the Graduate School of Information Science and Electrical Engineering, Kyushu University to help us. In this evaluation, we will divide these students into two groups. They separately use paper quizzes and the quizzes generated by our framework. In this experiment, students first should use the quizzes to practice. Then, they will be asked to answer quizzes for obtaining their final scores. We will compare their scores in these two groups.

Last, we will questionnaire these students for collecting the feedback on these two quizzes as well. The questionnairing result will be used to improve the generated quizzes and our framework.

6 Conclusion

In this research, we have introduced our framework for automatically generating quizzes based on the linked data. In our framework, it includes an RDF data schema extraction and visualization tool, and a question page authoring tool. With the support of these two tools, users can easily define quizzes as web applications. Furthermore, we extend our framework for supporting users to create the quizzes in 360-degree images/videos. In the extended framework, we applied the image recognition technology to extract the devices from images/videos to trigger the quiz questions. Such kind of quizzes is much closer to real life. We suppose such kind of quizzes are more effective for training and testing users especially in the IoT security field.

In the future, we will collect the data of using our framework and analyze users' activities. The analyzation result will be used to evaluate our framework and to update the generated quiz. Furthermore, we hope to find out the weakness of each user and the common difficult point during the IoT education process to improve users' learning activities.

Acknowledgments. The is research was mainly supported by Strategic International Research Cooperative Program, Japan Science and Technology Agency (JST) regarding "Security in the Internet of Things Space", and partially supported by JSPS KAKENHI Grant Number JP16H02923 and JP17H00773.

References

1. Shi, W., Kaneko, K., Ma, C., Okada, Y.: A framework for automatically generating quiz-type serious games based on linked data. Int. J. Inf. Educ. Technol. **9**(4), 250–256 (2019)
2. Shi, W., Ma, C., Kulshrestha, S., Bose, R.: Okada, Y: A framework for automatically generating IoT security quizzes in a virtual 3D environment based on linked data. In: Barolli, L., Xhafa, F., Khan, Z., Odhabi, H. (eds.) Advances in Internet, Data and Web Technologies, EIDWT 2019. Lecture Notes on Data Engineering and Communications Technologies, vol. 29, pp. 103–113. Springer, Cham (2019)
3. Marrese-Taylor, E., Nakajima, A., Matsuo, Y., Ono, Y.: Learning to automatically generate fill-in-the-blank quizzes. In: Proceedings of the 5th Workshop on Natural Language Processing Techniques for Educational Applications, pp. 152–156 (2018)
4. Killawala, A., Khokhlov, I., Reznik, L.: Computational intelligence framework for automatic quiz question generation. In: 2018 IEEE International Conference on Fuzzy Systems (FUZZ-IEEE), pp. 1–8 (2018)
5. Amria, A., Ewais, A., Hodrob R.: A framework for automatic exam generation based on intended learning outcomes. In: Proceedings of the 10th International Conference on Computer Supported Education - Volume 1: CSEDU, pp. 474–480 (2018)
6. Rocha, O.R., Zucker, C.F., Giboin, A.: Extraction of relevant resources and questions from DBpedia to automatically generate quizzes on specific domains. In: International Conference on Intelligent Tutoring Systems 2018, Jun 2018, Montreal, Canada
7. Rocha, O.R., Zucker, C.F.: Automatic generation of quizzes from DBpedia according to educational standards. In: The 3rd Educational Knowledge Management Workshop (EKM 2018), Lyon, France, April 2018
8. Foulonneau, M.: Generating educational assessment items from linked open data: the case of DBpedia. In: The Semantic Web: ESWC 2011 Workshops: ESWC 2011 Workshops, pp. 16–27 Springer, Heidelberg (2012)
9. Liu, D. Lin, C.: Sherlock: a semi-automatic quiz generation system using linked data. In: ISWC-PD 2014 Proceedings of the 2014 International Conference on Posters & Demonstrations Track, vol. 1272, pp. 9–12. CEUR-WS.org Aachen, Germany (2014)

10. Nakanishi, T.H., Fukuda, A.: Architecture, textual context description, and quiz generation scheme for the movie based context-aware learning system. In: 2016 IEEE Region 10 Conference (TENCON), pp. 2410–2413 (2016)
11. Sung, L., Lin, Y., Chen, M.C.: An automatic quiz generation system for English text. In: Seventh IEEE International Conference on Advanced Learning Technologies (ICALT 2007), pp. 196–197 (2007)
12. Huamán, A.: Cascade Classifier. https://docs.opencv.org/master/db/d28/tutorial_cascade_classifier.html

Development and Evaluation of Road State Information Platform Based on Various Environmental Sensors in Snow Countries

Yoshitaka Shibata[1](✉), Yoshikazu Arai[2], Yoshiya Saito[2],
and Jun Hakura[2]

[1] Regional Corporate Research Center, Iwate Prefectural University,
152-89 Sugo, Takizawa, Iwate, Japan
shibata@iwate-pu.ac.jp
[2] Faculty of Software and Information, Iwate Prefectural University,
152-52 Sugo, Takizawa, Iwate, Japan
{arai,y-saito,hakura}@iwate-pu.ac.jp

Abstract. In this research, various road state information including dry, wet, slush, snowy, icy states on the road are determined in realtime using various environmental sensors. These state information is not only exchanged directly between drivers through V2X, but also collected into cloud servers on Internet and organized as wide area road state information platform for ordinal users to know road state using smartphone and tablet terminals. In this paper, the basic system, architecture, crowd sensing, V2X communication system, prototype system and GIS system of Road State Information Platform are precisely discussed.

1 Introduction

As advent of IoT, sensor and mobility technologies, automotive driving cars have emerged and developed in well developed countries and running on the highways and big road as mainly at level 2. In several countries, such as the U.S. and China, test driving of autonomous driving at level 3 and 4 are proceeding on public road. In Japan, autonomous driving cars and public buses service by connected environment using 5G are expected around several areas at Tokyo Olympic Game on July in 2020.

In order to realize the safer and reliable automobile driving environment, not only car conditions, road condition, current running positions, the surrounding terrain have to be timely identified and feedbacked to vehicles. However, current automotive driving is objective for more ideal road and environmental conditions such as well paved and flat roads, good weather with no obstacles, and clear center line with limited car and environmental sensor and 3D roadmap.

However, on the other hand, the road infrastructures in many developing countries are not well maintained compared with developed countries and so bad and dangerous due to luck of regular road maintenance, falling objects from other vehicles, overloaded trucks or buses. Therefore, in order to maintain safe and reliable auto driving, the vehicles have to detect those obstacles and road conditions in advance and avoid when they pass away.

Second, in the cold or snow countries, such as Japan and Northern countries, most of the road surfaces are occupied with heavy snow and iced surface in winter and many slip accidents occurred even though the vehicles attach snow specific tires. In fact, almost more 90% of traffic accidents in northern part of Japan is caused from slipping car on snowy or iced roads. Thus, both cases, traffic accidents are rapidly increased. Therefore, safer and more reliable road monitoring and warning system which can transmit the road condition information to drivers and new self-driving method before passing through the dangerous road area is indispensable. Furthermore, the information and communication environment in local areas, is not well developed and their mobile and wireless communication facilities are unstable along the roads compared with urban area. Thus, once a traffic accidents or disaster occurred, information collection, transmission and sharing are delayed or even cannot be made. Eventually the resident's safe and reliable daily lives cannot be maintained. More robust and resilient information infrastructure and proper and quick information services with road environmental conditions are indispensable.

In order to resolve those problems, we introduce a new generation wide area road surface state information platform based on crowd sensing technology. In crowd sensing, many data from many vehicles with various environmental sensors including accelerator, gyro sensor, infrared temperature sensor, quasi electrical static sensor, camera and GPS are integrated to precisely detect the various road surface states and identify the dangerous locations on GIS. This road information is transmitted to the neighbor vehicles and roadside server in realtime using V2X communication network and shared in realtime. This road information is also widely collected in wide areas as bigdata to the cloud server on Internet through the roadside servers and analyzed to predict the future road conditions and undedicated local roads by combining with weather data, 3D terrain data along streets. Thus, wide area road state GIS information can be attained.

In the following, the related works with the road state information state sensing and analysis systems are introduced in section two. Then general system and architecture of Road Surface State Information Platform are explained in section three. Next, the crowed sensing system and its functions with various environmental sensors are precisely shown in section four. After that, the V2X communication system and its function are explained in section five. Then, a prototype system to evaluate function and performance of the proposed system is explained in section six. Finally conclusion and future works are summarized in section seven.

2 Related Works

With the road state sensing, there are several related works so far. In the paper [1], the road surface temperature model by taking accounting the effects of surrounding road environment to facilitate proper snow and ice control operations is introduced. In his research the fixed sensor system along road is used to observe the precise temperature and build road surface temperature model using heat balance method. In this paper [2], road ice surface state is simply estimated using the predicted temperature and amount of meshed fallen snow and evaluated the predicted. Although the predicted accuracy is high, the difference between the ice and snow was not resolved. In the paper [3], road state data collection system of roughness of urban area roads is introduced. In this system, mobile profilometer by use of conventional accelerometers to measure realtime roughness and road state GIS. This system provides general and wide area road state monitoring facility for urban area, but snow and icy states are note considered. In the paper [5], the system which collects the information with road surface condition and delivers it while moving by telephone is discussed. In the paper [6], a system which extracts, collects and processes sensor information from each vehicle and provides it to users using Internet is introduced. In the paper [7], a system which provides information with frozen road condition to the car navigation system of taxi using browser function. In the paper [8], Road surface state analysis method is introduced for automobile tire sensing by using quasi-electric field technology.

In all of the above systems, construction of communication infrastructure is essential and those systems cannot work out at challenged network environment in at inter-mountain areas. In the followings, a new road state information platform is proposed to overcome those problem.

3   Wide Road Surface State Information Platform

In order to resolve those problems in previous session, we introduce a new generation wide area road surface state information platform based on crowd sensing and V2X technologies as shown in Fig. 1. The wide area road surface state information platform mainly consists of multiple roadside wireless nodes, namely Smart Relay Shelters (SRS), Gateways, and mobile nodes, namely Smart Mobile Box (SMB). Each SRS or SMB is furthermore organized by a sensor information part and communication network part.

The vehicle has sensor information part includes various sensor devices such as semi-electrostatic field sensor, an acceleration sensor, gyro sensor, temperature sensor, humidity sensor, infrared sensor and sensor server. Using those sensor devices, various road surface states such as dry, rough, wet, snowy and icy roads can be quantitatively decided.

In our system, SRS and SMB organize a large scale information infrastructure without conventional wired network such as Internet. The SMB on the car collects various sensor data including acceleration, temperature, humidity and frozen sensor data as well as GPS data and carries and exchanges to other smart node as message ferry while moving from one end to another along the roads.

On the other hand, SRS not only collects and stores sensor data from its own sensors in its database server but exchanges the sensor data from SMB in vehicle nodes when it passes through the SRS in roadside wireless node by V2X communication protocol. Therefore, both sensor data at SRS and SMB are periodically uploaded to cloud system through the Gateway and synchronized. Thus, SMB performs as mobile communication means even through the communication infrastructure is challenged environment or not prepared.

This network not only performs various road sensor data collection and transmission functions, but also performs Internet access network function to transmit the various data, such as sightseeing information, disaster prevention information and shopping and so on as ordinal public wide area network for residents. Therefore, many applications and services can be realized.

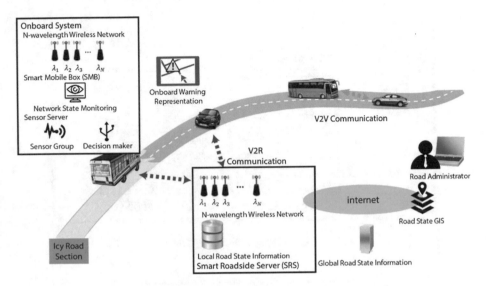

Fig. 1. Wide road surface state information platform

4   Sensing System with Various Sensors

In order to detect the precise road surface conditions, such as dry, wet, dumpy, showy, frozen roads, various sensing devices including accelerator, gyro sensor, infrared temperature sensor, humidity sensor, quasi electrical static sensor, camera and GPS are integrated to precisely and quantitatively detect the various road surface states and determine the dangerous locations on GIS in sensor server as shown in Fig. 2. The 9 axis dynamic sensors including accelerator, gyro sensor and electromagnetic sensors can measure vertical amplitude of roughness, rut along the road. The infrared temperature sensor observes the road surface temperature without touching the road surface. The quasi electrical static sensor detects the snow and icy conditions by observing the quasi electrical static field intensity. Camera can detect the obstacles on the road. The far-infrared laser sensor precisely measures the friction rate of snow and icy states. The sensor server periodically samples those sensor signals and performs AD conversion and signal filtering in Receiver module, analyzes the sensor data in Analyzer module to quantitatively determine the road surface state and learning from the sensor data in AI module to classify the road surface state as shown in Fig. 3. As result, the correct road surface state can be quantitatively and qualitatively decided. The decision data with road surface condition in SMB are temporally stored in Regional Road

Condition Data module and mutually exchanged when the SMB on one vehicle approaches to other SMB. Thus the both SMBs can mutually obtain the most recent road surface state data with just forward road. By the same way, the SMB can also mutually exchange and obtain the forward road surface data from roadside SRS.

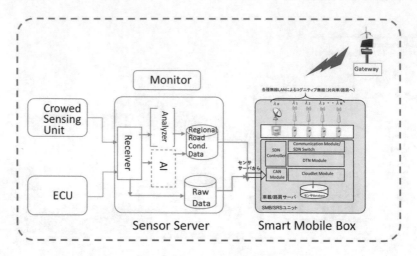

Fig. 2. Sensor server system

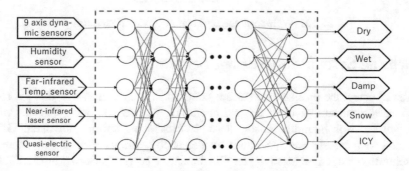

Fig. 3. AI module

5 V2X Communication System

In Fig. 4 shows V2X communication method between the SMB of a vehicle and the SRS of road side server. First, one of the wireless networks with the longest communication distance can first make connection link between SMB and SRS using SDN function. Through this connection link, the communication control data of other wireless networks such as UUID, security key, password, authentication, IP address, TCP port number, socket No. are exchanged. As approaching each other, the second wireless network among the cognitive network can be connected in a short time and actual data transmission can be immediately started. This transmission process can be repeated during crossing each other as long as the longest communication link is connected. This communication process between SMB and SRS is the same as the communication between SMB to other SMB except for using adhoc mode.

On the other hand, the V2X communication between vehicle and the global cloud server on Internet shows in Fig. 8. The sensor data from SMB is transmitted to the SRS in road side server. Then those data are sent to the gateway function unit and the address of those data from local to global address and sent to the global cloud server through Internet. Thus, using the proposed V2X communication protocol, not only Intranet communication among the vehicle network, but also Intranet and Internet communication can be realized.

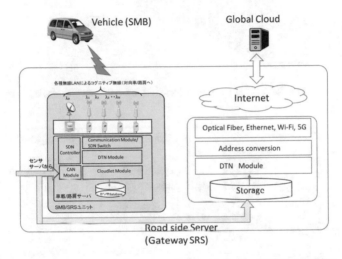

Fig. 4. V2X communication method between the SMBS and SRS

5.1 Prototype System and Evaluation

In order to verify the effects and usefulness of the proposed system, a prototype system is constructed and those functional and performance are evaluated. The prototype system for two-wavelength communication is shown in Fig. 9. We call this prototype system as Smart Mobility Base station (SMB) for mobility and Smart Rely Shelter (SRS) for roadside station. We currently use WI-U2-300D of Buffalo Corporation for Wi-Fi communication of 2.4 GHz as the prototype of two-wavelength communication, and OiNET-923 of Oi Electric Co., Ltd. for 920 MHz band communication respectively. WI-U2-300D is a commercially available device, and the rated bandwidth in this prototype setting is 54 Mbps. On the other hand, the OiNET-923 has a communication distance of 1 km at maximum and a bandwidth of 50 kbps to 100 kbps (Figs. 5 and 6).

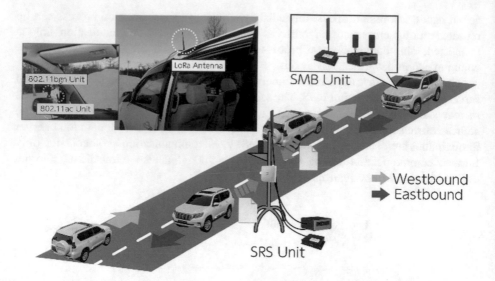

Fig. 5. V2X communication method between the SMB

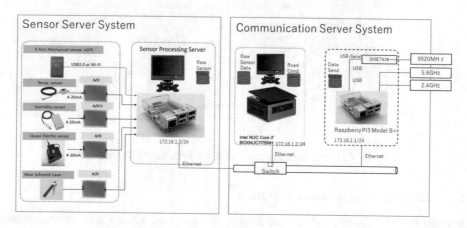

Fig. 6. Realtime road surface data exchange system

On the other hand, in sensor server system, several sensor including BL-02 of Biglobe as 9 axis dynamic sensor and GPS, CS-TAC-40 of Optex as far-infrared temperature sensor, HTY7843 of azbil as humidity and temperature sensor and RoadEye of RIS system and quasi electrical static field sensor for road surface state are used. Those sensor data are synchronously sampled with every 10 ms. and averaged every 1 s to reduce sensor noise by another Raspberry Pi3 Model B + as sensor server. Then those data are sent to Intel NUC Core i7 which is used for sensor data storage and data analysis by AI based road state decision. Both sensor and communication servers are connected to Ethernet switch.

In order to evaluate the sensing function and accuracy of sensing road surface condition, both sensor and communication servers and various sensors are set to the vehicle. We ran this vehicle about 4 h to evaluate decision accuracy in realtime on the winter road with various road conditions such as dry, wet, snowy, damp and icy condition around our campus in winter. In order to evaluate the road surface decision function, the video camera is also used to compere the decision state and the actual road surface state. The Figs. 7 and 8 show a typical sensor data concerned with xyz axis values of acceleration, road surface temperature, humidity and road surface friction. Those values are collected into global cloud as big data through network gateway from SRS and analyzed to predict the future road condition by combining with weather data. Thus, wide area road state information GIS can be realized.

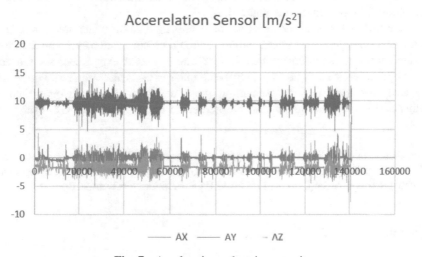

Fig. 7. Accceleration values in xyz-axis

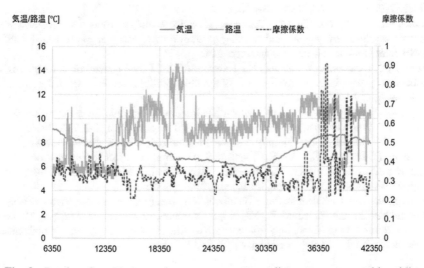

Fig. 8. Road surface friction and temperature, surrounding temperature and humidity

6 Conclusions and Future Works

In this paper, we introduce a wide area road state information platform decision system for not only both urban area but local snow and icy roads area where these states derive very serious traffic accidents. In our system in-vehicle sensor server system can detects and determine various road states such as dry, wet, damp, snowy, icy or even the roughness and friction rate on road surface by AI based decision making method. This road state information can be exchanged and shared with the many vehicles through V2X communication method. Furthermore, this road state information can be collected into the global cloud from smart road stations (SRS) though network gateway as big data to predict the future road state by combining with weather condition. Through the experimentation of our prototyped system, we could confirm that a wide area road state information GIS system can be realized.

As the future works of our research, more sensing data for wider areas, in local and urban areas, on different weather conditions should be corrected as training data for AI decision model to improve the decision accuracy. Social experiment for the actual driving operations such as bus, taxi, renter car enterprises is also required. Furthermore, wide area road state GIS by combining those road state data and open data with weather and geological data should be predicated and indicated as GIS system near future.

Acknowledgement. The research was supported by Strategic Information and Communications R&D Promotion Program Grant Number 181502003 by Ministry of Affairs and Communication, and Strategic Research Project Grant by Iwate Prefectural University in 2019.

References

1. Fujimoto, A., Nakajima, T., Sato, K., Tokunaga, R., Takahashi, N., Ishida, T.: Route-based forecasting of road surface temperature by using Meshi air temperature data. In: JSSI & JSSE Joint Conference, P1-34, September 2018
2. Saida, A., Sato, K., Nakajima, T., Tokunaga, R., Sato: A study of route based forecast of road surface condition by melting and freezing as station using weather mesh data. In: JSSI & JSSE Joint Conference, P2-57, September 2018
3. Fujita, S., Tomiyama, K., Abliz, N., Kawamura, A.: Development of a roughness data collection system for urban roads by use of a mobile profilometer and GIS. J. Jpn. Soc. Civil Eng. **69**(2), I_90–I_97 (2013)
4. Yagi, K.: A measuring method of road surface longitudinal profile from sprung acceleration and verification with road profiler. J. Jpn. Soc. Civil Eng. **69**(3), I_1–I_7 (2013)
5. Nicole, R.: Title of paper with only first word capitalized, J. Name Stand. Abbrev., in press
6. Yorozu, Y., Hirano, M., Oka, K., Tagawa, Y.: Electron spectroscopy studies on magneto-optical media and plastic substrate interface. IEEE Transl. J. Magn. Jpn. **2**, 740–741 (1987). Digests 9th Annual Conf. Magnetics Japan, p. 301 (1982)
7. Young, M.: The Technical Writer's Handbook. University Science, Mill Valley (1989)
8. Takiguchi, K., Suda, Y., Kono, K., Mizuno, S., Yamabe, S., Masaki, N., Hayashi, T.: Trial of quasi-electrical field technology to automobile tire sensing. In: Annual Conference on Automobile Technology Association, 417-20145406, May 2014

Evaluation of End-to-End Performance on N-Wavelength V2X Cognitive Wireless System Designed for Exchanging Road State Information

Akira Sakuraba[1(✉)], Yoshitaka Shibata[1], Goshi Sato[2],
and Noriki Uchida[3]

[1] Regional Cooperative Research Division, Iwate Prefectural University,
Takizawa, Iwate, Japan
{a_saku, shibata}@iwate-pu.ac.jp
[2] Resilient ICT Research Center, National Institute of Information
and Communications Technology, Sendai, Miyagi, Japan
sato_g@nict.go.jp
[3] Faculty of Information Engineering, Fukuoka Institute of Technology,
Fukuoka, Japan
n-uchida@fit.ac.jp

Abstract. Vehicle-to-everything (V2X) communication allows to realize future and safer road traffic environment. By vehicle onboard sensors and their output data analysis, dangerous road areas can be identified such as icy road section. In their application, V2X communication can exchange them among vehicles instantly without depending on any public wireless networks. This paper introduces a cognitive V2X wireless system which exchanges road surface state information among running vehicles and roadside communication units (RSUs). This approach is based on the combination of multiple standard wireless links which have different characteristics. Our system has two different standards of wireless LAN links to deliver road state information by bulk messaging, and one long-range LPWA link in order to exchange network information and node location before the vehicle moves into wireless LAN range. We designed the V2X communication which is independent from public wireless network in order to provide highly availability in case of disaster, not only road surface information providing in normal time. We have conducted and analyzed an end-to-end communication experiment which set up with in Vehicle-to-Road communication using our prototype environment. The result shows which our system design is capable for delivering actual road state information data within several seconds regardless of physical distance of sensed road section.

1 Introduction

Intelligent Transporting Systems (ITS) realize safer road traffic transportation by providing of many types of information which is collected by many types of onboard or roadside sensors such as outside air temperature, vehicle counter, obstacle detection sensor using Light Detection and Ranging (LiDAR) etc. Especially, in the cold district,

L. Barolli et al. (Eds.): EIDWT 2020, LNDECT 47, pp. 277–289, 2020.
https://doi.org/10.1007/978-3-030-39746-3_30

drivers of road vehicle require to understand ahead road surface physically or weather condition to avoid traffic accident as it could be an issue of implementing high-level autonomous vehicle such SAE 5 [1]. Many researchers worked on development of road surface condition estimation method and implemented them using various type of onboard sensors like dynamics sensor, near infrared (NIR) sensor, computer-vision based approach with onboard camera, etc. These estimation methods determine the result on the vehicle in real time however the result does not deliver for other vehicle immediately. Therefore we should consider how a vehicle exchange the information among other vehicles or roadside sensor units.

Vehicle-to-everything (V2X) communication is a combination of vehicle-to-vehicle (V2V) and vehicle-to-road (V2R) in this paper's definition. Wireless communication technology for moving vehicle is widely available today such as 3G or 4G mobile network. However, their coverage area often does not include rural or mountainous area. Thus there is a requirement V2X communication which should be multiplexing with public mobile network. It is considerable challenge that V2X network with service independent design in order to realize the future mobility in terms of resilient network design for enduring natural disaster such earthquake, storm, and flood.

In this paper, we propose a cognitive V2X wireless system which exchanges road surface state information among vehicles and roadside units (RSU) for communication and data storing. This approach is based on combination of multiple wireless standard links which have different wavelengths and characteristics. Our system has two different standard wireless LAN (WLAN) links, one is wireless LAN to deliver road state information (RSI) by bulk messaging, and other is a Low Power Wide Area (LPWA) link which has capability of long distance but low speed wireless communication. This design is intended to deliver network information and exchange of node location before vehicle moves into wireless LAN range. We designed the V2X communication which is independent from public packet wireless network to provide high availability in case of not only road surface information providing in normal time, but for disaster in emergency case. We focus to describe V2R communication on LPWA link in this paper.

We also report and discuss an evaluation experiment of end-to-end communication which is defined it as communication between a pair of network node which has capable to run the RSI application. We have confirmed which the system can deliver RSI resource with $N = 2$ wavelength cognitive configuration to deliver realistic amount of RSI resource data within realistic time.

2   Related Works

A lot of researchers have technological interest in road sensing targeted to understanding of road surface condition using vehicle onboard sensor. Casselgren et al. introduced a road surface condition determination method with near infrared (NIR) sensor [2]. This method uses three different wavelength of NIR laser devices to determine paved road condition into dry, wet, icy, and snowy states.

Modern public mobile network standards such W-CDMA or E-UTRA (LTE) have capability for use of packet exchanging on V2X networking concept among moving vehicles and RSUs on public road. Floating car data is a kind of telemetry technique which remotely collects data or information from moving vehicles via 3G/4G public mobile network. 5G mobile communication between RSU and vehicle which moves at 290 km/h has been confirmed to deliver 1.1 Gbps throughput at effective value [3]. However base station of 5G standard just provides very small coverage area by single base station which is compared with conventional public mobile network.

We can find a large number of proposal implementations for vehicular ad-hoc networking (VANET) which does not depended on any public mobile network service. For example, Su et al. proposed IEEE 802.11 wireless LAN (WLAN) based VANET system [4] which is implemented on Android platform smart devices as an application of traffic accident detection system [5]. As their report, the communication range was within 50 to 60 m.

There is a practical wireless standard named IEEE 802.11p Wireless Access Vehicular Environments (WAVE) which is assigned to use on 5.9 GHz band and designed to deal both V2R and V2V communication, it is announced to equip on commercial vehicles. It is considered to communicate longer communication range which is compared with generic WLAN. Qin et al. implemented on embedded PC based 802.11p testbed system [6]. They measured the maximum range of the testbed within 300 m and availability communication on the move at 18 km/h speed. We have found some works as applications on 802.11p, for instance, traffic light controlling for emergency vehicle [7], delivering traffic accident information and weather information via combination with VANET and V2R communication [8], and providing of video-on-demand service for vehicle from RSU [9].

Low Power Wide Area (LPWA) is spreading wireless technology which allows to deliver packet several kilometers range even single-hop node. Yuze et al. had conducted an experiment for data transmission at very long distance on LPWA based wireless system [10]. The result indicated that sent data can be received at the receiver which is located 37 km away with line-of-sight condition even single-hop node configuration.

3 Proposed System and Configuration

3.1 System Overview

Figure 1 illustrates an overview of our system. Our proposed system can share with highly real-time road states for drivers in manual driving.

Onboard unit is loaded on each vehicle which consists of Sensor Server and SMB (Smart Mobile Box). First, the Sensor Server collects sensor data from multiple onboard sensors. Then the Sensor Server analyzes those data to decide the road states up to the current travelling point corresponding to dried/wet/snowy/icy etc. with granularity. Next, the Sensor Server associates with the road states and their locations.

After that, the SMB delivers those data with raw sensor data to RSU named Smart Relay Shelter (SRS) through N-wavelength V2R cognitive wireless network. On other side, the RSU also collects the road data and states from many other vehicles via the same network. This cognitive network takes V2V communication directly with a vehicle which is heading to the opposite direction from own vehicle. At the same time, SRS also performs to deliver regional road states and acquires other area road surface states with cloudlet on cloud computing and storage resources. Therefore our approach allows to provide road state information service independent on the public cellular network.

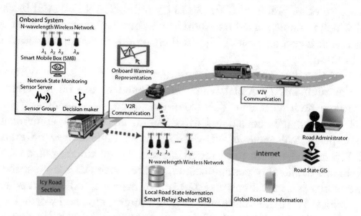

Fig. 1. Overview of proposed system and its functions which contains onboard sensor and communication device, roadside unit, and GIS.

Finally, the system provides visualization capability of road states based on Geographic Information System (GIS) for various users for drivers or operation managers of public transportation system such bus or taxi. For operation managers or driver who is not in driving, our proposed system represents road states as generic GIS designed for using on smart device and PC. Each vehicle equips road surface warning system to represent and warn for driver who is in driving. If ahead road condition would get worse such as icy, whiteout, etc. our system will warn the state visually on display device and playout of warning message by human voice.

Thus, these functions allow acquiring, sharing, and representing road state, to be able to realize safer road traffic even winter season.

3.2   Configuration of System

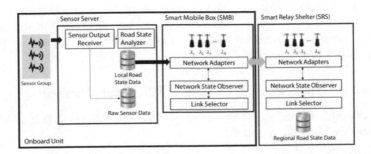

Fig. 2. System configuration and modules on Sensor Server, SMB, and SRS.

Figure 2 illustrates the system configuration of sensor server, SMB, and SRS. The Sensor Server collects sensor data from sensor group which is composed of various and positioning of vehicle to determine road surface state where vehicle is running. Sensor server is based on edge computing architecture and processes acquired sensor values to identify road state by machine learning algorithm. After road state decision has been completed, the result is sent to SMB.

SMB has several wireless networks with different wavelength wireless network device and decision maker which wireless device should be delivered to other SMB or SRS.

4   N-Wavelength V2X Cognitive Wireless System

The system consists of multiple network devices which have multiple different wavelength wireless links. In implementation at this time, we install three different wireless standard based devices which is composed of 5.6 GHz band WLAN and 920 MHz band LPWA in this section.

4.1   Wireless Standards

We have selected following two wireless standards which has different characteristics as cognitive wireless links for our system.

WLAN standard is based on 802.11ac Wave 2, as known as Wi-Fi 5, as the primary data transmission link for exchanging road state information and raw sensor data. This standard has capability for faster data transmission which is compared with conventional 2.4 GHz WLAN standard, by channel bonding technology. Another characteristics is beamforming which can extend longer distance wireless communication by obtaining better signal level. Both characteristics realize higher throughput even in longer range, it helps reliable and massive data transmission while vehicle is moving faster.

LPWA is one of license-free long distance capable wireless communication technology. Network performance of this type wireless technology is designed to enable from hundreds bps to several kbps transmission rate up to several ten kilometers range distance. There is an opinion in which LPWA is not suitable for V2X application due to low transmission rate [11]. However, in our system, LPWA can be used as transmitting controlling message which requires only low data transmission but long distance communication. Our system delivers network configuration information of WLAN connection which delivers road state information before vehicle enters into the range of WLAN access point of SRS.

4.2   Message Sequence on LPWA Link in V2R Communication

Figure 3 describes the message sequence on LPWA link from detecting of other node to terminate the connection gracefully.

Initialization. In this system, beacon delivers own location and node type which identifies RSU or vehicle, as well as notifying own existence for other node. Any nodes broadcast their beacon at initial, when receiving other node's beacon, receiver node will response message to sender immediately. At this time, both node can identify opposite node to communicate.

Network Information Delivering. After identifying opposite node, system is going to exchange network information between both nodes. First, RSU node delivers non-IP network information message which includes SSID of the access point (AP),

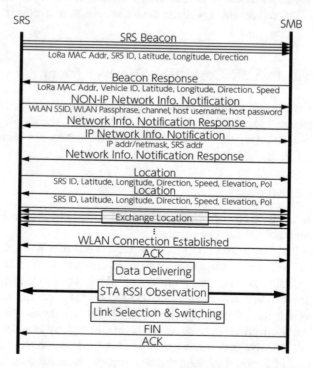

Fig. 3. Message exchange sequence between SRS and SMB in V2R communication.

passphrase of Wi-Fi Protected Access Pre-Shared Key (WPA-PSK), and channel of WLAN. The vehicular node replies ACK message to RSU node when received non-IP network message. Then RSU node sends IP network information message. IP network information message consists of assigned IP address, netmask, and default gateway address. Vehicular node sends ACK message after received IP network information message by the same way as non-IP message receiving.

Exchanging Location. Vehicular node is attempting to connect with AP on WLAN link where is placed on RSU as a station (STA) at this time. While attempting to connect WLAN, both nodes exchange their location information each other until WLAN link is established. A message of location information contains geolocation coordinates, speed, moving direction of vehicle, altitude, and point-of-interest (PoI) which means the requested length of ahead road section and be requested by driver. Location messages are mutually delivered every several seconds.

Road State Information Delivering. The system immediately starts to exchange the road state information on WLAN link when TCP connection on WLAN is established. WLAN link is able to deliver large amount of data which is compared with LPWA link and delivers road state information with bulk messaging procedure. However, time for packet reachable is often quite short, it is a considerable issue to design exchange packets. In our method, system exchanges a part of road state information stored database which depends on PoI, it could reduce data amount to deliver even if packet reachable time is very short. We had already reported that WLAN packet delivery while V2V and V2R communication from running vehicle was feasible by field experiment [12, 13].

Termination. Disruption on WLAN connection invokes session terminating process for LPWA link. The connection will be broken out when vehicle moves out of WLAN range. In order to terminate the WLAN connection correctly, the vehicle node sends FIN message to RSU to terminate the WLAN connection. Then, RSU node receives and replies ACK message. Eventually the whole session by both WLAN and LPWA can be terminated gracefully.

5   Prototype Environment

We built N = 2 wavelength wireless communication unit which uses PLANEX Communication GW-900D USB dongle adaptor for 802.11ac WLAN station for SMB, and Ruckus Wireless T300 access point (AP) for 802.11ac device for SRS.

For 920 MHz band LWPA device, we chose Oi Electric OiNET-928 or OiNET-923 which is confirmed to the requirements of ARIB STD-T108 LoRaWAN standard.

Wireless link switching function on SMB and SRS are processed on ARM based single-board computer. We implemented them for Raspberry Pi3 model B+ with Raspbian Linux system. Application on SRS and SMB is written by C, Ruby 2.5, Python 3.5, and Bash Script.

6 End-to-End Evaluation Over V2R Communication

This section describes the evaluation of our method which is conducted application level data delivering between onboard SMB and roadside SRS.

6.1 Objective and Measuring Target

We must understand how long this proposed cognitive system needs the time from the initiation of communication sequence to the completion of road state information delivering. V2X based communication systems postulate short and very strict initiation time for connection according to moving of vehicular nodes. Therefore, we had an experiment to measure required time of it using actual SMB and SRS unit. We discuss our approach and implementation could be adequate to using V2R communication, which is focused on data delivery time between nodes on each links.

We configured timing module into our $N = 2$ wavelength prototype, the module times a particular message on each wireless links.

In this experiment, on LPWA link, the timer works on SRS and it starts when the beacon response message has been just received from SMB, and measurement will be finished when the second time ACK message has been received message according to IP network information message. As shown in Fig. 4, we define this required time as T_1. Another performance indicator can be measured that timer on WLAN link starts when SMB take a WPA 2 handshake to establish MAC layer connection. The timer will stop to time when SMB received road state information (RSI) completely from SRS. Our definition describes it as T_2, also denoted in Fig. 4.

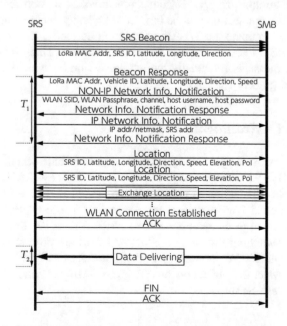

Fig. 4. Measurement time T_1 and T_2 in message exchange sequence.

We measured and analyzed above two different links to estimate the system initiating performance.

6.2 Data to Deliver Between SMB-SRS

Table 1. Fields, data class, and description of delivered RSI resource.

Field	Data class of SQLite3	Description
ID	INTEGER	Unique ID of records
timestamp	TEXT, NOT NULL	Timestamp of this record has been created
latitude	REAL, NOT NULL	Latitude of the position
longitude	REAL, NOT NULL	Longitude of the position
state	TEXT	Determined road state *Dry	damp	wet	slash	snow	ice	unknown*
friction	REAL	coefficient of friction
air_temp	REAL, NOT NULL	Air temperature
road_temp	REAL, NOT NULL	Road surface temperature
humidity	REAL, NOT NULL	Humidity
roughness	REAL	Roughness index calculated from vehicle dynamics analyzing

We also prepared the data to deliver for evaluation on both network links.

On LPWA link, the system delivers Network Information Delivering messages which is described in Sect. 4.2. First of them, NON-IP Network Info message which consists of SSID of access point, WPA passphrase, WLAN channel, SRS host username, and password of SRS. In this experiment configuration, NON-IP Network Info message contains 34 octets payload on our protocol level. Another message is IP Network Info message which is composed of assigned IP address for SMB, netmask, and IP address of SRS host. The massage has 35 octets payload on our protocol level.

The other side, WLAN link delivers road state information which is formatted as a binary file called road state information resource. We implemented RSI resource as SQLite3 database. Table 1 describes field of RSI resource and we estimated that each record has 94 bytes length at the maximum.

Table 2. Pseudo RSI resource file specification assumption the sensor vehicle ran at 50 km/h.

PoI length [km]	Running time of the vehicle [min.]	Numbers of record	File size [KiB]
5	6.0	360	33.0
10	12.0	720	66.1
20	24.0	1,440	132.2
50	60.0	3,600	330.5
100	120.0	7,200	660.9

Our scenario considers several size of RSI resource files to measure variance of delivering time in different distance of PoI assignment. We prepared five different size of RSI resource files which is based on different PoI calculation in advance, it is shown as Table 2. Each file size is determined by number of records which correspond to PoI distance and number of records. We had experiment configuration for PoI which is set to 5. 10, 20, 50, and 100 km and their file contains the number of records as shown Table 2. Each PoI scenario assumed that the data was emulated to be composed by sensor vehicle ran at 50 km/h stable, a record is generated every second, and there is not missing data. Direction of delivering data was from SRS to SMB (R2V) and we implemented a method of file delivering with Secure File Copy.

We had attempted 10 trials per single PoI setting.

6.3   Network Device Setup

We configured the prototype described in Fig. 5.

We assigned a configuration for LPWA device which works transmitting packets with chirp spread spectrum (CSS) modulation. We also chose spreading factor value SF9 which has 1.76 kbps designed throughput. The carrier frequency was f_0 = 920.6 MHz (24ch) with 200 kHz occupied bandwidth.

PHY layer configuration on WLAN link was f_0 = 5.69 GHz (138ch) with 80 MHz occupied bandwidth between access point and station. There was enabled WPA2 authentication and static IP addressing with assigned by network setting information delivered via LPWA link.

We performed this experiment which both SMB and SRS are operated at indoor placement. The distance of both node was about two meters and we removed antenna element from antenna base which is connected to a LPWA end node, this is the reason to decrease the antenna gain as well when both node operated in outdoor.

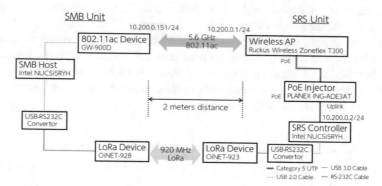

Fig. 5. Network and device setup for this evaluation.

6.4   Result

On LPWA link, we discovered which averaged received signal strength index (RSSI) was -105 dBm (σ = 1.414) and it is equivalent to 300 m distance between both node.

Required time to deliver NON-IP/IP network configuration information T_1 was averagely 1.95 s ($\sigma = 0.46$).

We observed quite stable RSSI on WLAN link in every scenarios and trials and recorded -50 dBm which is converted from percentage of signal level to RSSI value. This is reasonable result due to distance between both node was just few meters.

Figure 6 describes averaged result of file delivering duration T_2 which corresponds to distance of PoI. In every scenarios, there is completely success to obtain RSI resource in every trials.

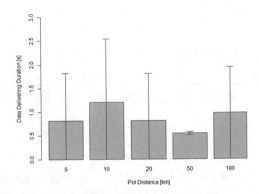

Fig. 6. Averaged duration for RSI resource delivering (T_2) which corresponds to PoI length. Error bar describes standard deviation.

We found interesting result which the averaged delivering time of RSI resource was not much different despite the varying of RSI resource file size, especially it became larger sized one. It spread from 0.55 to 1.2 s averagely. With observing of data delivering duration, it took 4.6 s at the maximum through all trials and scenarios, most of the time accounts for WPA session establishment from the station to the access point before delivering of the file. The handshake of WPA requires a certain time but it is varied for each attempting of connection, thus the result indicates the variance value for duration of data delivering.

7   Discussion

LPWA channel required long time for delivering amount of data. In our protocol, the payload of network information was just 69 octets but it is delivered two seconds. This is a reason of it, we should remind which LoRa standard uses simplex wireless communication. Most use cases of LoRa standard consider uni-directional ad-hoc networking like smart meter implementation, however, our protocol has bi-directional messaging. We considered to implement our protocol with sensing for incoming packet from other node by waiting static sleep time and retrying with timeout, it is not enough setting in current implementation for timeout or sleep time. Therefore, waiting time optimization is required to gain more efficient protocol design.

The result also presented that our approach has realistic delivering performance for current design of RSI resource with short time data delivering. We have analyzed how much the system delivers RSI resource with the effective throughput on application layer which is shown in Table 3. We increased PoI by 20 times however effective throughput did not reach their limit. This result suggested that our system could be capable for delivering more varied types of sensor data which requires massive data, for example, digital image, precisely 3D road shape map as well as wider area RSI resource. Hence, we could implement the future sensor unit with other type onboard sensor to obtain more accurate result, which depends on scalability for data delivering of our system.

Table 3. Length of PoI vs. effective throughput on WLAN file delivering

PoI length	5 km	10 km	20 km	50 km	100 km
Throughput [kbps]	323.3	439.1	1296.9	4,849.1	5,449.4

8 Conclusion

This paper proposed a design for N-wavelength wireless communication system for V2X environment to exchange road surface conditions. This system has three of fundamental functions; observing of physically state of road surface by multiple sensors on the vehicle, analyze these sensor output for decision of road surface conditions, and transmit calculated road condition to roadside device or other vehicle to share road condition which the vehicle travelled until now. Design of our communication subsystem utilizes cognitive wireless network which consists of WLAN for data delivering and LPWA as the control message link.

We have conducted application level V2R performance evaluation in case of both node was static location. We observed which RSI resource was delivered within about two seconds averagely, it is confirmed to realize V2R communication to deliver the information. The data delivering capability of WLAN could be very large, it is enough exchange sensor output which contains massive amount of data, not only wider area RSI resource.

The result also indicated the room for improvement which is more adequate timing of networking protocol LPWA.

In our future work, we are now planning to have a functional and social evaluation with system integration of onboard sensor unit and communication unit. The field of evaluation will be performed on actual winter season on national highway which contains urban and mountainous road section.

Acknowledgments. This research was supported by Strategic Information and Communications R&D Promotion Program (SCOPE) No. 181502003, Ministry of Internal Affairs and Communications, Japan.

References

1. Taxonomy and Definitions for Terms Related to Driving Automation Systems for On-Road Motor Vehicles J3016_201806, SAE International (2018)
2. Casselgren, J., Rosendahl, S., Eliasson, J.: Road surface information system. In: Proceedings of the 16th SIRWEC Conference (2013)
3. Nakamura, T.: Views of the future pioneered by 5G — A World Converging the Strengths of Partners —, NTT Docomo Technical Journal 25th Anniversary, pp. 53–59 (2018)
4. Su, K., Wu, H., Chang, W., Chou, Y.: Vehicle-to-vehicle communication system through Wi-Fi network using android smartphone. In: Proceedings of 2012 International Conference on Connected Vehicles and Expo, pp. 191–196 (2012)
5. White, J., et al.: WreckWatch: automatic traffic accident detection and notification with smartphones. J. Mob. Netw. Appl. **16**(3), 285–303 (2011)
6. Qin, Z., Meng, Z., Zhang, X., Xiang, B., Zhang, L.: Performance evaluation of 802.11p WAVE system on embedded board. In: Proceedings of 2014 International Conference on Information Networking, pp. 356–360 (2014)
7. Sawade, O., Schäufele, B., Radusch, I.: Collaboration over IEEE 802.11p to enable an intelligent traffic light function for emergency vehicles. In: Proceedings of 2016 International Conference on Computing, Networking and Communications, Mobile Computing and Vehicle Communications, pp. 1–5 (2016)
8. Sukuvaara, T., et al.: Wireless traffic safety network for incident and weather information. In: Proceedings of the First ACM International Symposium on Design and Analysis of Intelligent Vehicular Networks and Applications, pp. 9–14 (2011)
9. Begin, T., Busson, A., Lassous, I.G., Boukerche, A.: Video on demand in IEEE 802.11p-based vehicular networks: analysis and dimensioning. In: Proceedings of the 21st ACM International Conference on Modeling, Analysis and Simulation of Wireless and Mobile Systems, pp. 303–310 (2018)
10. Yuze, H., Nebata, S.: A basic study on emergency communication system for disaster using LPWA. In: Proceedings of the Workshops of the 33rd International Conference on Advanced Information Networking and Application, pp. 501–509 (2019)
11. Adelantado, F., et al.: Understanding the limits of LoRaWAN. IEEE Commun. Mag. **55**(9), 34–40 (2017)
12. Sakuraba, A., Ito, T., Shibata, Y.: Network performance evaluation to realize N-wavelength V2X cognitive wireless communication system. In: Proceedings of the Workshops of the 33rd International Conference on Advanced Information Networking and Application, pp. 524–536 (2019)
13. Sakuraba, A., et al.: Evaluation of performance on N-wavelength V2X wireless network in actual road environment. In: Proceedings of 13th International Conference on Complex, Intelligent, and Software Intensive Systems, pp. 555–565 (2019)

Secure Public Cloud Storage Auditing
with Deduplication: More Efficient and Secure

Jiasen Liu, Xu An Wang$^{(\boxtimes)}$, Kaiyang Zhao, and Han Wang

Key Laboratory for Network and Information Security of the PAP,
Engineering University of the PAP, Xi'an 710086, Shanxi, China
`869550196@qq.com, wangxazjd98@vip.163.com`

Abstract. With the advent of the information age and the explosive growth of
data volume, more and more users outsource data storage to cloud storage system
in order to reduce local storage costs. However, the following problem is whether
the cloud storage system is safe and effective, which is mainly reflected in the storage efficiency of cloud servers and the integrity of data storage. Although many
protocols have been proposed, these protocols use complex encryption operations, meet the requirements only in form, and are difficult to apply in practice.
The Proof of Retrievability (POR) and Proof of Data Ownership (PDP) achieve
data integrity auditing, while the Proof of Ownership (POW) maximizes storage
efficiency by removing duplicate data from cloud storage systems. Combining
these two technologies, a new redundancy-based secure public cloud storage audit
is proposed based on distributed string equality check and homomorphic linear
authenticator, which realizes integrity audit and storage redundancy reduction efficiently and securely, and reduces redundancy of the same file from different users
to improve storage efficiency of cloud servers. By using pseudorandom function
and Diffie-Hellman problem, the security of the protocol is realized. Finally, the
effectiveness and scalability of the proposed scheme are verified by theoretical
analysis.

1 Introduction

Cloud storage systems have grown rapidly in recent years as people have spent less on
local storage and computing [1–3]. A variety of cloud storage systems, such as Google
Drive, Microsoft OneDrive and Baidu Cloud, have become an essential part of modern
information life. Users and operators are most concerned about the cloud storage system
of the two performance for data integrity audit efficiency and storage efficiency. First of
all, data loss or damage is inevitable for cloud storage systems. In order to ensure the
security of their own outsourced data, users need to audit the integrity of the outsourced
data regularly. Secondly, in order to improve storage efficiency, cloud storage systems
need to reduce unnecessary redundant data, including the same data uploaded by the
same user and the same data uploaded by different users.

For the security of cloud storage, there are currently POR and PDP technology to
ensure the security of cloud storage system. Proof of Data Possession (PDP) enables
clients to outsource data to a cloud server and perform integrity audits with the cloud

© Springer Nature Switzerland AG 2020
L. Barolli et al. (Eds.): EIDWT 2020, LNDECT 47, pp. 290–300, 2020.
https://doi.org/10.1007/978-3-030-39746-3_31

server [4, 5]. Proof of Retrievability (POR) supports clients to "recover" data that is outsourced to cloud servers.

Proof of Ownership (POW) technology supports data de-duplication for cloud storage efficiency [6, 7]. So far, many companies have begun to develop de-heavy technology. According to EMC [8], 75% of the data in cloud storage systems is duplicated. A file may be uploaded repeatedly by multiple users or by one user. For such files, the cloud storage system holds only one piece of data to reduce storage costs.

Because the separate POW scheme does not support data integrity auditing, it is necessary to link the POR/PDP scheme with the POW scheme in order to achieve high efficiency, high security integrity auditing and storage redundancy reduction. For security reasons, the cloud server needs to verify that the user really owns the data in the deduplication; Users also need to verify that the data in the cloud server is complete after uploading the data due to possible operational errors or malicious users and the cloud. An efficient, effective and secure protocol is needed to implement these processes.

Related Work. For the integrity audit of outsourced data, many protocols have been proposed based on POR scheme and PDP scheme respectively. In terms of security, part is based on standard model implementation and part is based on random oracle model (ROM) implementation. Technically, one is based on basic number theory [1], the other is based on bilinear product on elliptic curve [2, 3]. In reference [9], integrity auditing is implemented simply, efficiently and securely through distributed string equality checking.

For storage redundancy reduction, Halevi et al. [10] proposed the first POW scheme based on the Merkle hash tree, and Pietro reference [14] proposed an improved POW scheme [11] based on the pseudo-random function, and the computational cost was reduced relatively.

For the integration of data integrity audit and de-duplication technology, Zheng et al. [12] A scheme was proposed to provide public data integrity audits and secure storage duplication (POSD). Then Yuan et al. [13] proposed a public and constant cost storage integrity audit scheme (PCAD) based on secure redundancy technology, which is based on polynomial-based authentication tags and homomorphic linear authenticators. POSD has been proved to be unsafe by [16], while the safety and efficiency of PCAD need to be optimized.

Our Contribution. Referring to the existing POR/PDP and POW schemes, this paper proposes a more effective and secure redundancy-based secure public cloud storage audit scheme (SPAD). Efficient implementation of data integrity audit based on distributed string equality check protocol security. Based on Diffie-Hellman problem and the design of pseudorandom functions, the security of this protocol is improved. The SCAD scheme proposed in this paper has the following characteristics:

(1) SPAD uses distributed string equality checking protocol [9] to model outsourced data as vectors in a vector space, which greatly improves computational efficiency while ensuring security.

(2) SPAD can implement public audit and data integrity audit can be implemented by third-party auditor TPA so as to reduce the computational burden of users.

(3) SPAD greatly improves the security of the protocol by calculating the CDH problem [15, 16] and the pseudo-random function.

The rest of this article is organized as follows. In the second section, we model our scheme, and put forward the security model and the design goal. In the third section, the design of public cloud storage audit protocol based on redundancy security is introduced in detail, and the effectiveness, security and performance of the protocol are analyzed in the fourth section. Finally, the fifth section summarizes the present document.

2 Models and Goals

2.1 System Model

In this work, we model the Secure Public Cloud Storage Auditing with Deduplication scheme and design three entities in the scheme, namely, the user (CU), the cloud server (CS), and the third-party auditor (TPA), as shown in Fig. 1.

The CU has a large amount of data that needs to be sent to the cloud server, requires that the integrity of the data stored in the cloud be checked.CS provides remote storage service for CU, which requires strong computing power and storage space to meet the requirements, supports integrity audit requests of users or third-party designers, and deduplication. TPA is commissioned by the CU to complete an integrity audit of the data in the cloud server. TPA is always unable to obtain the user's private data (e.g. private keys).

Secure Public Cloud Storage Auditing with Deduplication system runs as follows: First, CU generates private and public keys to generate digital signatures and identity tags, then uploads the data to CS and deletes local data. Then, in order to verify the integrity of the outsourced data in the cloud server, the CU or TPA sends a challenge message to the CS for integrity audit. For each audit request, the CS should respond a prove information to prove that the data is stored in the cloud server. Finally, the CU or TPA verifies that the outsourced data is complete according to the prove information of

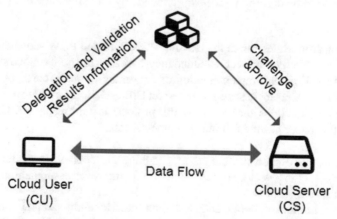

Fig. 1. Model of secure public cloud storage auditing with deduplication.

the CS response. When the user uploads the same file that has been stored in the cloud server, the cloud server verifies whether the user really owns the file, and if not, accepts the new data uploaded by the CU, otherwise notifies the CU that there is no need to upload and lists the CU as one of the owners of the data file.

2.2 Security Model

In the SPAD system, we consider the following factors that may threaten the security:

(1) Data in the cloud storage system is corrupted by an attacker.
(2) An attacker deceives the cloud server into owning a file itself and attempts to become one of its owners.
(3) The cloud server deceives the user that the outsourced data is complete in an attempt to pass the data integrity audit of the TPA, even if the data has been corrupted or lost.
(4) System failure of cloud server or operation error of administrator.

2.3 Design Goals

In order to achieve data integrity audit and storage redundancy reduction safely and effectively, SPAD needs to meet the following objectives:

(1) Effectiveness: The proposed scheme can use the generated public key and private key to verify data integrity and deduplication, according to valid digital signatures and identity tags.
(2) Security: The cloud server cannot spoof the auditor through the data integrity audit by forging the certification information. The attacker cannot spoof the cloud server to obtain the permission to use a certain file by forging the data information.
(3) Efficiency: The cost of calculation, communication and storage of SPAD should be as low as possible.
(4) Functions: Data integrity audit and deduplication are required.

3 Construction of PCAD

3.1 Preliminaries and Notation

Based on the work of Chen [9] and others, this paper firstly signs the data block according to the distributed string equality check. Then the Diffie-Hellman problem and homomorphic linear authenticator are referenced to ensure the security and effectiveness of the protocol.

(1) Classical String Equality Checking Protocol [9]: Assuming that Alice owns the string $x \in \{0, 1\}^n$ and Bob owns the string $y \in \{0, 1\}^n$, the purpose of this protocol

is to check whether the strings owned by Alice and Bob are equal, and the communication cost is $O(\log n)$. There is a common random string pool $S \subseteq \{0, 1\}^n$, and the protocol works as follows: Alice first selects a random string $s \in S$, then sends s and $< x, s >$ mod 2 to Bob. Bob calculate $< y, s >$ mod 2 and determines whet it is equal to $< x, s >$ mod 2. If it is equal, it continues, otherwise Alice is notified to abort that agreement. After 100 iterations of the above validation process, Bob notifies Alice that the two strings are equal if $< x, s >$ mod 2 $=< y, s >$ mod 2 persists. The probability of Bob notifies Alice that the strings are equal when $x \neq y$ is $1/2^{100}$, which is negligible, so the protocol is secure.

(2) Diffie-Hellman problem: We enhance the security of the scheme by exploiting the CDH problem.

Computational Differential-Hellman Problem [14]: G is a cyclic group of order p, g is the generator of G. Given $x, y \xleftarrow{R} Z_p^*$ and (g, g^x, g^y), then g^{xy} cannot be computed.

Static Differential-Hellman Problem [15]: G is a cyclic group of order p, g is the generator of $G, a \xleftarrow{R} Z_p^*$, given $h \in G$ and (g, g^a), then h^a cannot be calculated.

(3) Bilinear mapping: G, G_T is a multiplicative cyclic group of a given prime order p, g is the generator of G, there is a function of $e : G \times G \rightarrow G_T$. For $\forall m, n \in G$ and $x, y \xleftarrow{R} Z_p^*$, there exists an effective algorithm e to make $e(m^x, n^y) = e(m, n)^{xy}$ hold and $e(g, g) \neq 1$.

(4) Pseudorandom function: Suppose $PRF = \{F_K(\cdot)\}$ is a set of deterministic functions generated by K. Because no polynomial-time algorithm can distinguish deterministic function $PRF = \{F_K(\cdot)\}$ from a real random function, it is called pseudorandom function. In practice, pseudorandom function can be a secure encryption algorithm or a keyed hash function.

(5) Mark: $H(\cdot)$ is a one-way hash function, G is a multiplicative cyclic group of prime order p,g is a generator of G, $u \xleftarrow{R} Z_p$. D is that data to be outsource by the user, $F_K(\cdot)$ is a pseudorandom function generated by K and mapped to Z_p.

3.2 Our Construction

In this module, we introduce the structure and content of SPAD in detail.

KeyGen. In this step, the user selects a prime number p randomly according to the security level parameter λ, and the bits of p are longer than the bits of the security level parameter λ. The user then randomly selects two parameters $r \xleftarrow{R} Z_p^*, \varepsilon \xleftarrow{R} Z_p^*$, and generates a key pairs $(spk, ssk) \xleftarrow{R} sign()$. An λ-bit key k_1 is then generated according to the AES encryption scheme for generating a pseudorandom function $F_{K1}(\cdot)$, which is input as an integer and output on Z_p. The end user computes $k = g^\varepsilon$. In this case, the public key and private key of the system are:

$$PK = \{p, g, k, u, spk\}$$
$$SK = \{r, k_1, \varepsilon, ssk\}$$

Setup. When the data D is uploaded, the CU divides the data D into n data blocks $\{d_i\}_{1 \le i \le n}$, where $d_i \in Z_p$. Then a file name $name \xleftarrow{R} Z_p^*$ is randomly selected and the file identifier $\tau \leftarrow name \|n\| Sign_{ssk}(name\|n)$ is calculated with ε. The end user digitally signs, $S_i = \left(g^{rd_i+F_1(\cdot)} \cdot u^{H(name\|i)}\right)^{\varepsilon}$ for each d_i. The user sends all (d_i, S_i) to the CS.

Challenge. In order to verify the integrity of D, the TPA first randomly selects a subset K on $\{1, \ldots, n$ containing k elements. A key K_2 is generated, similar to the KeyGen, a pseudorandom function $F_{K_2}(\cdot)$ is generated, and $(e_1, \ldots, e_k) = F_{k_2}(\cdot) \in Z_p^k$. Define challenge message $\chi = [(1, \ldots, k), (e_1, \ldots, e_k)]$. TPA sends χ to CS for data integrity audit.

Prove. Upon receiving the challenge message, CS calculates $\alpha = \sum_{i=1}^{k} d_i \cdot e_i \mod (p-1)$ and $\beta = \prod_{i=1}^{k} S_i^{e_i} \mod p$ using the public key. In response, CS sends α and β to CU and TPA, respectively.

Verify. Upon receiving the response message, the CU calculates $y = g^{r \cdot \alpha + \sum_{i=1}^{k} e_i \cdot F_{k_1}(\cdot)} \mod p$ and sends y to the TPA. TPA calculates $\eta = u^{\sum_{i=1}^{k} e_i \cdot H(name\|i)} \mod p$ and Verifies:

$$e(\beta, g) \overset{?}{=} e(\eta, k) \cdot e(y, k) \tag{1}$$

If Eq. (1) holds, the TPA output **Accept** indicates a successful validation, otherwise the output **Reject** indicates a failed validation.

Deduplication. When CU uploads a file that is already stored in the CS, CS needs to verify that the user actually owns the file. Similar to **Challenge**, CS randomly selects a subset M on $\{1, \ldots, n\}$ with m elements. A key k_3 is generated, a pseudorandom function $F_{k_3}(\cdot)$ is generated, and $(e_1, \ldots, e_m) = F_{k_3}(\cdot) \in Z_p^m$, and a duplicate checking request $\chi = [(1, \ldots, m), (e_1, \ldots, e_m)]$ is sent to the CU.

After receiving the duplicate checking request, the CU calculates $\mu_i = \left(g^{r \cdot e_i \cdot d_i + e_i \cdot F_{k_1}(\cdot)}\right)^{\varepsilon} \mod p$ and $y' = \sum_{i=1}^{m} \mu_i$, and sends y' to the CS. After the CS receives the response message, it calculates $|\beta' = \prod_{i=1}^{m} S_i^{e_i} \mod p$ and $\eta = u^{\sum_{i=1}^{m} e_i H(name\|i)} \mod p$ and verifies:

$$e(\eta, k) \cdot e(y', k) \overset{?}{=} e(\beta', g) \tag{2}$$

If Eq. (2) holds, then D already exists in the cloud server, CS notifies the CU that there is no need to upload and lists the CU as one of the owners of D. Otherwise, there is no storage redundancy for the file, and CS accepts and stores the file.

3.3 Auditing After Deduplication

In this module, we describe data integrity audits with reduced storage redundancy in detail. Let D' have a total of T owners, where each owner $CU_t (1 \leq t \leq T)$ has his own private public key:

$$PK_t = \{p, g, k_t, u, spk_t\}$$
$$SK_t = \{r_t, k_{1t}, \varepsilon_t, ssk_t\}$$

Where $|r_t, \varepsilon_t \xleftarrow{R} Z_p^*$, $k_t = g^{\varepsilon_t}$ and $(spk_t, ssk_t) \xleftarrow{R} Sign_t()$, there are $\tau_t \leftarrow name\|n\|Sign_{ssk}(name\|n)$ generated by Setup calculation. And $S_{ti} = (g^{r_i \cdot d_i + F_{k1_t}(\cdot)} \mod p \cdot u^{H(name\|i)})^{\varepsilon_i}$ for each d_i in D'.

CU uploads τ_t and S_{ti} to the CS. After receiving information from different CU, CS calculates matrix:

$$S_i = \{S_{ti}\}_{1 \leq i \leq n, 1 \leq t \leq T}$$

When CU_t requires a data integrity audit, TPA sends $\chi = [(1, \ldots, k), (e_1, \ldots, e_k)]$ to CS through **Challenge**, CS computes and sends $\beta_t = \prod_{i=1}^{m} S_{ti}^{e_i} \mod p$, $\alpha_t = \sum_{i=1}^{k} d_i \cdot e_i \mod (p-1)$ to CU_t and TPA, they compute $y_t = g^{r \cdot \alpha_t + \sum_{i=1}^{k} e_t \cdot F_{k_1 t}(\cdot)} \mod p$ $\eta = u^{\sum_{i=1}^{k} e_i \cdot H(name\|i)} \mod p$ and verifies:

$$e(\beta, g) \overset{?}{=} e(\eta, k) \cdot e(y, k) \tag{3}$$

If Eq. (3) holds, the TPA output **Accept** indicates a successful validation, otherwise the output **Reject** indicates a failed validation. This step of integrity auditing is similar to Eq. (1) and can be easily verified, so we will not do so again below.

4 Correctness, Security and Performance Analysis

4.1 Correctness

In this part, we analyze the correctness of the protocol for judging data integrity and redundancy detection by Eqs. (4) and (5):

$$e(\beta, g) = e((u^{\sum_{i=1}^{k} e_i \cdot H(name\|i)} \cdot g^{\sum_{i=1}^{k} e_i \cdot (r \cdot d_i + F_{k1}(\cdot))})^{\varepsilon}, g$$

$$= e(u^{\sum_{i=1}^{k} e_i \cdot H(name\|i)}, g)^{\varepsilon} \cdot e(g^{\sum_{i=1}^{k} e_i \cdot (r \cdot d_i + F_{k1}(\cdot))}, g)^{\varepsilon}$$

$$= e(\eta, k) \cdot e(y, k) \tag{4}$$

$$e(\eta, k) \cdot e(y', k) = e\left(u^{\sum_{i=1}^{m} e_i \cdot H(name\|i)}, g^{\varepsilon}\right) \cdot e\left(g^{\sum_{i=1}^{m} e_i \cdot (r \cdot d_i + F_{k1}(\cdot))}, g^{\varepsilon}\right)$$

$$= e\left(\left(\left(u^{\sum\limits_{i=1}^{m} e_i \cdot H(name\|i)} \cdot g^{\sum\limits_{i=1}^{m} e_i \cdot (r \cdot d_i + F_{k1}(\cdot))}\right)^{\varepsilon}, g\right)\right)$$

$$= e(\beta', g) \tag{5}$$

We verify the correctness of our proposed protocol by Eqs. (4) and (5).

4.2 Security

We can judge that the private key r, k_1, ε can not be forged by the Diffie-Hellman problem and the static Diffie-Hellman problem. If CS falsifies (α^*, β^*), equation $e(\beta^*, g) \overset{?}{=} e(\eta, k) \cdot e(y^*, k)$ will not hold. And a sufficiently secure pseudorandom function can prevent CS from spoofing the data integrity of the user by sending forged certificate information. Next, we analyze the security of data integrity auditing by using pseudo-random functions.

Step 1: When the data in the cloud storage system is corrupted, a probability $Pr[cheat]$ of CS successfully deceiving the user may occur. Assuming that the scheme presented in this paper uses a true random function and $Pr[Unbound]$ represents the probability of CS successfully deceiving CU, then $Pr[Unbound] = negl(\lambda)$ can be obtained by linear algebra theory. In this paper, the scheme uses pseudo-random function, CS can distinguish between true random and pseudo-random, then there has $Adv[PRF] \geq Pr[Cheat] - Pr[Unbound]$. So that $Pr[Cheat] \leq Adv[PRF] + Pr[Unbound]$.

Step 2: If the above-mentioned $Pr[cheat]$ is not negligible, a probability $Pr[Recover]$ of the user recovering the data using the reconstruction algorithm can occur, and $Pr[Recover] = 1 - negl(\lambda)$. The principle is that users can obtain outsourcing data by solving linear equations of challenge vectors and outsourcing data.

The security proof of the deduplication is similar to the security proof of data integrity audit. Because of the unforgeability of private key and the security of pseudo-random function, the attacker can not obtain the right to use the corresponding data file by forging the data file to cheat the establishment of Eq. (2) of CS.

4.3 Performance

In this section we introduce the theoretical performance of SPAD. Let n be the number of data blocks uploaded to the file, λ be the security level parameter, and l be the length of the audit query.

Theoretical Performance. For CU, the processes involved are **SPAD.KeyGen**, **SPAD.Setup**, **SPAD.Verify**, and **SPAD.Deployment**. Since the **KeyGen** and **Setup** are executed only once when data is outsourced, all costs of these two steps can be shared equally among subsequent data integrity audits. When integrity audit queries and deduplication are performed, the user notifies the TPA to perform operations and calculate y for transmission to the TPA, and calculate y' for transmission to the CS. Therefore, the computational cost of the user CU is $O(1)$, and the communication cost is $O(1)$.

Because the user only needs to store his own private and public keys, the storage cost of the user CU is $O(1)$.

For TPA, the processes involved are **SPAD.Prove**, **SPAD.Verify**. When performing a data integrity audit, TPA sends χ and computes η and $e(\eta, k) \cdot e(y, k) \stackrel{?}{=} e(\beta, g)$ at a computational cost of $O(l)$ and a communication cost of $O(l)$. Because TPA only needs to store the public key, its storage cost is $O(1)$.

For CS, α, β needs to be computed and passed during integrity audit. The deduplication is similar to audit query process, so its computational cost is $O(l)$, communication cost is $O(1)$, and storage cost is $O(n)$. The theoretical performance of the SPAD protocol is shown in Table 1.

Table 1. The theoretical performance of the SPAD

	Computation	Communication	Storage
CU	$O(1)$	$O(1)$	$O(1)$
CS	$O(l)$	$O(l)$	$O(n)$
TPA	$O(l)$	$O(l)$	$O(1)$

Performance Comparison. In this part, we compare the computational overhead of our protocols with that of similar protocols [9, 12, 13], and then reflect their performance and advantages respectively. Because of the different protocols, we define s as the elements contained in each data block, *EXP* and *MUL* are the exponentiation and multiplication operations on group *G*, respectively. The following analysis compares the performance of each protocol and is shown in Table 2.

Table 2. Performance comparison of different protocols

	SCS	POSD	PCAD	SPAD
Public	No	No	Yes	Yes
Dedup	No	Yes	Yes	Yes
Secure	Yes	No	No	Yes
Auditing.CS	$O(l)$	$O(ls)MUL + O(l)EXP$	$O(l + s)MUL + O(l + s)EXP$	$O(l)$
Auditing.CU	$O(l)$	$O(ls)MUL + O(l)EXP$	$O(1)MUL + O(1)EXP + O(1)Pairing$	$O(1)$
Auditing.TPA	N/A	N/A	$O(1)$	$O(l)$
Dedup.CS	N/A	$O(ls)MUL + O(l)EXP$	$O(m + s)MUL + O(s)EXP + O(1)Pairing$	$O(m)$
Dedup.CU	N/A	$O(ls)MUL$	0	$O(m)$
Dedup.TPA	N/A	N/A	$O(1)$	0

We find that the efficiency of different protocols depends on the size of the data blocks they partition. SCS protocol has high efficiency and security, but it does not support storage redundancy reduction and public audit. Although POSD supports storage redundancy reduction, it does not support public auditing, which is inefficient and insecure. SPAD supports both public auditing and redundancy reduction, but its efficiency and security need to be optimized.

Because our protocol does not use complex encryption, so the computational complexity is relatively low, and the protocol is effective. Our protocol can use 128-bit modulo p, which is much smaller than RSA and other signatures, so it is more efficient. The main advantages of our protocol are as follows: (1) It does not use complicated encryption method, and its principle and function are simple; (2) our encryption method is simple and effective, and the efficiency is higher; (3) Support simple and effective deduplication. (4) Using CDH, DL and pseudo-random function, the security of the protocol is higher.

5 Conclusion

In order to meet the security requirements of cloud storage system and achieve data integrity audit and de-duplication, we propose an efficient and secure public cloud storage audit protocol based on redundancy, which is based on POR, PDP, POW and SCS, POSD, PCAD schemes. Firstly, we implement data integrity audit based on distributed string equality check protocol, which reduces the cost of the system. Then the security of the system is ensured by calculating Diffie-Hellman problem and pseudo-random function. Finally, we implement the de-duplication technology to reduce the storage cost of the system. Through numerical calculation and analysis of experimental results, we prove the efficiency and safety of SPAD scheme. Compared with other PDP/POR/POW solutions, our solution achieves more complete, efficient, and secure functionality.

Acknowledgments. This work was supported by the National Cryptography Development Fund of China (grant no. MMJJ20170112), Natural Science Basic Research Plan in Shaanxi Province of China (grant no. 2018JM6028), National Natural Science Foundation of China (grant no. 61772550, U1636114, and 61572521), and National Key Research and Development Program of China (grant no. 2017YFB0802000). This work is also supported by Engineering University of PAP's Funding for Scientific Research Innovation Team (grant no. KYTD201805).

References

1. Shi, E., Stefanov, E., Papamanthou, C.: Practical dynamic proofs of retrievability. In: Proceedings of the 2013 ACM SIGSAC Conference on Computer & Communications Security, pp. 325–366 (2013)
2. Xu, J., Chang, E.-C.: Towards efficient proofs of retrievability. In: Proceedings of the 7th ACM Symposium on Information, Computer and Communications Security, pp. 1–23 (2012)
3. Wang, C., Chow, S.S., Wang, Q., Ren, K., Lou, W.: Privacy preserving public auditing for secure cloud storage. IEEE Trans. Comput. **62**(2), 362–375 (2013)
4. Kaaniche, N., Moustaine, E.E., Laurent, M.: A novel zero-knowledge scheme for proof of data possession in cloud storage applications. In: 2014 14th IEEE/ACM International Symposium on Cluster, Cloud and Grid Computing, Chicago, IL, pp. 522–531 (2014)

5. Yuan, J., Yu, S.: Proofs of retrievability with public verifiability and constant communication cost in cloud. In: Proceedings of the ACM ASIACCS-SCC 2013 (2013)
6. Youn, T., Chang, K.: Bi-directional and concurrent proof of ownership for stronger storage services with de-duplication. Sci. China Inf. Sci. **61**(03), 84–94 (2018)
7. Pietro, R., Sorniotti, A.: Proof of ownership for deduplication systems: a secure, scalable, and efficient solution. Comput. Commun. **82**, 71–82 (2016)
8. Gantz, J., Reinsel, D.: The digital universe decade - are you ready?, May 2010. http://www.emc.com/collateral/analyst-reports/idc-digital
9. Chen, F., Xiang, T., Yang, Y.: Secure cloud storage hits distributed string equality checking: more efficient, conceptually simpler, and provably secure. In: Proceedings of the 2014 IEEE Conference on Computer Communications, pp. 2389–2397. IEEE, Piscataway (2014)
10. Halevi, S., Harnik, D., Pinkas, B., Shulman-Peleg, A.: Proofs of ownership in remote storage systems. In: Proceedings of the 18th ACM Conference on Computer and Communications Security, Series CCS 2011, pp. 491–500. ACM, New York (2011)
11. Di Pietro, R., Sorniotti, A.: Boosting efficiency and security in proof of ownership for deduplication. In: Proceedings of the 7th ACM Symposium on Information, Computer and Communications Security, Series ASIACCS 2012, pp. 81–82. ACM, New York (2012)
12. Zheng, Q., Xu, S.: Secure and efficient proof of storage with deduplication. In: Proceedings of the Second ACM Conference on Data and Application Security and Privacy, Series CODASPY 2012, pp. 1–12. ACM, New York (2012)
13. Yuan, J., Yu, S.: Secure and constant cost public cloud storage auditing with deduplication. In: 2013 IEEE Conference on Communications and Network Security (CNS), National Harbor, MD, pp. 145–153 (2013)
14. Diffie, W., Hellman, M.: New directions in cryptography. IEEE Trans. Inf. Theory **22**(6), 644–654 (1976)
15. Brown, D.R.L., Gallant, R.P.: The static Diffie-Hellman problem, Cryptology ePrint Archive, Report 2004/306 (2004)
16. Youngjoo Shin, K.K., Hur, J.: Security weakness in the proof of storage with deduplication, Cryptology ePrint Archive, Report 2012/554 (2012)

A Knapsack Problem Based Algorithm for Local Level Management in Smart Grid

Usman Ali, Usman Qamar$^{(\boxtimes)}$, Kanwal Wahab, and Khawaja Sarmad Arif

National University of Sciences and Technology (NUST), H-12, Islamabad, Pakistan
uthmanmughal@gmail.com, usmanzaman@yahoo.com

Abstract. The world is adapting the renewable energy sources to produce clean energy. Because of this modernization in production, storage and consumers of energy, the conventional grid systems are facing a lot problems e.g. effective control of consumers or recompense of supply instability. The smart grid has the ability to overpower these shortcomings. In this study, we are considering a model of smart grid which has three levels: transmission, micro-grid and local level. We have propose an algorithm for energy management at local level, based on renowned algorithms of scheduling. We model the problem into knapsack problem and then find an optimize solution set using our algorithm which partially based on least cost branch and bound algorithm. This algorithm controls adaptable and shiftable load of smart homes. This algorithm is able to normalize the peak demand and control the preference of home appliances, through distributing energy among appliances depending on their consumption and priority, without exceeding the already decided total energy.

1 Introduction

The electrical grid is a system of different things. It consist of power plant, transmission lines, sub stations and customer. It has no system of communication from power plant to customer and vice versa. Whereas, the Smart Grid (SG) is an intelligent grid system which helps to improve a number of things i.e. production, transmission, distribution, fault recovery and the consumption of electric energy [1]. By introducing the information and communication technology (ICT) to existing conventional electrical grid system, we shall be able to balance the demand and distribution of grid. It has proper communication system between power generators and customers. So, it is a combination of conventional electric grid system and the ICT [2].

The SG technology, specifically, gathers and combine data of producers, customers or in distribution of energy. Also, SG processes the information insightfully. This intelligence is spread over the different levels of network i.e. the production, control and consumption.

Along with objectives of a smart network, we mentioned:

1. An improved integration and controlling of energy from distributed sources in profitable conditions.
2. An improvement in the quality of electricity and management.

3. An enhancement in customer participation (i.e. active consumers and consumption optimization).
4. A better energy efficiency (less losses).
5. A better communication among transport network and network distribution.

The SG is a complex system, which can be divided into multiple layers, and its optimization is a challenging task which requires specific methodologies. The intended purposes of SG are to balance the demand and supply, to decrease the overall consumption, to regularize the consumption curve and to incorporate the new technologies in conventional system [3]. Some of the restrictions of SG are: sending the energy requirements, improve the flow of energy, control, preventive maintenance and reduce the cost variations of electricity to enhance the investments and to get by with renewable energy generation and storage.

Amor, Bui [4] and Ahat et al. [3] have divided the SG into three distinct levels: Transmission and energy Distribution (T&D) network, the micro-grid and the local level. Each of these sub components has its own distinctive behavior and dynamic structure. In this study, we shall be concentrating on local level only. We shall be discussing local level of SG in-depth and propose an algorithm to optimize local level.

The main concern at local level of smart grid is to normalize the consumption peaks and management of home appliances by distributing energy between appliances according to their priorities, denoted by Φ, such that total consumption of energy should not exceed the overall energy received. We have modeled this problem into the knapsack problem. For its solution, we have propose an algorithm.

There are some algorithms and middleware, in the literature, for energy management in SG. They are based on stochastic [4], mixed integer or linear [5] programming models, peer to peer systems [6], auctioning systems based on multi-agents system [7] or game theory [8]. Almost all algorithms uses a distributed algorithm based on multi-agents for energy management. These methodologies require a direct access to appliances and don't consider the privacy of users.

The structure of this papers is as follows: introduction in Sect. 1, model of smart grid discussed in Sect. 2. The Sect. 3 thoroughly explains what is taking place at local level. It describes exactly how the local level problem is modeled into knapsack and it depicts our first algorithm to solve the knapsack problem, which then combined with our second algorithm which finds an optimized solution to the problem.

2 Architecture of Smart Grid

The smart grid integrates ICT with conventional grid so as to produce, transport, distribute and consume electricity in an efficient manner. The SG has all the norms of a complex system [9, 10]. In order to model it later, we must know its actors, goals, structure and its dynamics. An electric grid has three structural levels: producer mesh network, a middle linear network and customer. As soon as electricity is produced, it uses a network of underground and overhead power lines which is comparable to road network. Electricity transits on a network, which depends on following: the total production, the comprehensive consumption and architecture of network.

Following are some of the core objectives of smart grid:

- Self-diagnosis and self-healing from failure.
- Robustness against any attach.
- Active participation of customers.
- Ensure the power quality according to customers need.
- Balance the production according to demand.
- Avoid congestion using efficient transmission and distribution strategies to meet the customer demand.
- Normalize the local and global consumption.
- The T&D deliver the energy to consumer utility poles. It's mainly related to the maximum flow or routing problem.

The Fig. 1 represents the different levels of smart grid. The second ring represents the intermediate level of SG which is known as micro-grid, it links the energy production and energy consumption. It's a tree structure which represents area under a sub-station and it's a wider view of resident customers. It controls the energy distribution and orders a definite quantity of electricity in T&D for local customers. Similarly, the outer dotted ring represents the third level, representing the local entities. These are the customers linked to a sub-station through transmission and distribution network. The isolated buildings represents the local groups linked to one aggregation, i.e. homes and factories which support the energy consumption. In other words, local level in a grid is determined by area controlled by smart meter or by some sort of Home Automation Management System (HAMS).

Fig. 1. Sub components of smart grid (from Siemens)

As, we are more interested in the study of local level. Thus, we shall discuss it in-depth.

3   Local Level Management

The management at this level comprises automation of a house containing different types of devices, for example, lighting, air-conditioner, water heater, washing machine, dishwasher, etc. [11]. In home automation, there is a variable for every unit to reflect its significance for immediate consumption [12]. As an example, during a heatwave air conditioning more important, while heating has no importance. Therefore, a variable is added for the priority of each device on the local level to control the consumption. Thus, a laptop will be briefly deactivate the consumption if the battery is full, an electric vehicle will be charged at night, startup of a washing machine will be delayed in case of over consumption. A lot of appliances can be stopped for a short time or their start-up can be suspended. In housing areas, about the half of the consumption can be controlled without effecting the user's comfort. Lights, desktop computer and refrigerator are some of the devices which should essentially work. Hence, such appliance have default priority value of 0, which guarantee the supply of energy. Other appliances have values according to statistics of use and user behavior.

Home automation controller manages the operating priority of appliances. For instance, smart meters can keep the track of electricity produced, manages flow of power and also provides the customer an understanding of his/her consumption patterns.

A priority value is associated with every appliance. The control of priority variation is crucial for the best operations of SG. There are several management techniques for this:

- Based on usage patterns of electricity, the medium and long term forecasts can be made by several statistics and analysis of possible repetitions, this based on feedback, as it constantly takes data.
- Alternative functions, instead of, simulating the environment. Stochastic patterns then simulate those functions to evade the identical patterns. All appliances can run independently, this is internal managing.

A. Problem Formulation

We modeled our problem into the bounded knapsack problem which tries to find an optimal set of objects from a set of objects, this is called as combinatorial optimization. The bounded knapsack problem allows c number of copies of appliance xi in solution set, c must be a positive integer. It's a classic example of NP-hard problem which has high complexity. It is exemplified by, a biker who has a list of objects and a bag. Every object has certain utility value and takes certain space in the bag. The bag has limited space, all of the objects can't be fitted into the bag. So the problem is: how do we maximize the utility by remaining within the bag limit?

Let us say, we have n objects $\{l \ldots n\}$, every object o, in the set, has a utility U_o and a weight W_o associated. The bag has a capacity C. Now the problem is, how can we pack the bag without exceeding its capacity C such that the utility is maximize? [13].

In order to answer this question, we have to resolve this linear program:

$$x \in arg \ (maximum \sum U_i \cdot x_i \cdot \sum U_i \cdot x_i \ \text{stress:} \ \sum w_i \cdot x_i \leq C.$$ Then x, where $x = (x_i; \ 0 \leq i \leq n)$ where $i \in \{0 \ldots n\}$, is the answer. Vector x has binary value i.e.

0 or 1 only, where x_i will be 0 if object is selected and $x = 1$ if object is not selected. Now, we have to select a sub-set so that the following function is maximize:

$$\sum_{i=1}^{n}(u_i x_i) \tag{1}$$

$$\sum_{i=0}^{n}(w_i x_i) \; subject \; to \tag{2}$$

$$0 \leq x_i \leq c$$

There are two types of solutions for any knapsack problem: approached and exact. The approached method has low time and space complexity but it selects a local optimal and hopes that it will lead to some global optimal. Exact method has high time and space complexity. However, exact method provides no less than one global optimal solution with prove [15]. In this situation, we shall be using the exact method to resolve our problem of local level of smart grid. To get the optimized solution, we proposed an algorithm bases on least cost branch and bound algorithm.

B. Proposed Algorithm

As stated early, our goal is to manage the priority of home appliances and normalize the consumption peaks, by distributing energy among appliances at best according to their priorities Φ and their total consumption should not exceed the overall incoming electricity (Fig. 2).

Our algorithm is consist of two segments, which execute successively. First part of algorithm do the priority management, which do the analysis of home appliances and determines the priorities of appliances. The second half of algorithm, resolves the issue of sub charges. As soon as consumer needs to use many appliances having same priority, this algorithm manages it and tries to meet the needs of consumer as closely as possible without exceeding the total received energy (Fig. 3).

In the follow-up, we have explained this algorithm. As described, it has two sub parts. Let us explain the different variables used in the algorithm.

The input variable are following:

- c: array of maximum no. of copies for an item.
- n: total number of manageable appliances.
- Φ: array having the priorities of all appliances.
- U: array having utility values of appliances.
- C: total energy capacity all owed to use.
- α: the set of selected appliances to run.
- ω: array having the consumption of appliances.
- L: list of appliances.
- F: list of frequency of items.

The algorithm gives a set of selected appliances as output. The utility of some home appliance e, denoted by U_e, is given below:

$$U_e = (\omega_{max} \times \Phi_{max}) - (\omega_e \times \Phi_e) - \omega_e \tag{3}$$

Algorithm A
Variable: $C, c, U, n, \omega, x, \Phi, L$
Output: $\alpha = \emptyset$
loop $i = 1$ to n
  $count = 0$
  loop $j = i$ to n
    if $L[i] == L[j]$
      if $count < c[i]$ (same items have same Frequency)
        if $priority\ of\ item[i] = 0$
          $\alpha = $ add $L[i]$ to α (selected Items)
          $\omega_r = \omega_r + \omega[i]$ (consumption)
          $count = count + 1|$
  if $\sum_{i=1, i \notin \alpha}^{n} \omega[i] \leq (C - \omega_r)$
    loop $i = 1$ to n
      if $i \notin \alpha$
        $\alpha = \alpha \cup \{i\}$
  else
    loop $k = 1$ to n
      loop $i = 1$ to n
        if $\sum_{i=1, i \notin \alpha}^{k-1} \omega[i] \leq C - \omega_r < \sum_{i=1}^{k} \omega[i]$
          $l = k$
        call Algorithm B

Fig. 2. Algorithm A based on knapsack problem and it manages the appliances based on their utility values.

i. *Priority Management:* The first part of algorithm manages the priority variable. If the value of priority variable is *0*, then corresponding appliance starts working immediately. After that, the algorithm deducts the consumption of appliance from total energy capacity.

ii. *Optimization:* The second part of our algorithm makes its choice by carrying out a binary search based upon least cost branch & bound optimization. This algorithm has an acceptable overall complexity of $O(n^2)$.

4 Simulation and Results

To test our algorithm and to prove its significance and feasibility, we tested it on a real world data. We used a publicly available REDD dataset from Massachusetts Institute of Technology. We simulated our algorithm in Python. We used data of a house having seven appliances as following: (d1: refrigerator, d2: cooker, d3: washing machine, d4: television, d5: iron, d6: dishwasher, d7: computer). We assigned id to each appliance for

Algorithm B

solution set = {}

UB = upper bound = $-\infty$

$UB_a, C = 0$

sort (n) (on the basis of consumption)

do

$\quad C = C + C_i$

while$(C \leq C_{rem})$

$UB = C$

if $(C \leq C_{rem})$

$\quad C = C + \frac{U_i}{C_i} * (C_{rem} - C)$

for $(i = 1\ to\ n)$

$\quad j = 0, C = 0$

$\quad$*while* $(C \leq C_{rem})$

$\quad$*if* $(j\ != i)$

$\quad\quad C = C + C_i$

$\quad j{+}{+}$

$\quad$*if* $(C \leq C_{rem})$

$\quad\quad C = C + \frac{U_i}{C_i} * (C_{rem} - C)$

if $(C < UB)$

$\quad$add appliance to solution set

Fig. 3. Algorithm B based on least cost branch and bound and it finds the optimal solution set.

easy identification. The Table 1 shows list of appliances of the house with essential data required for our simulation.

The variables in the Table 1 are following:

- C_i: consumption of appliance i in watts/hour.
- U_i: utility of appliance i.
- Φ_i: priority of appliance i.

As, the priority of appliance is user defined so, we manually determined the working priority Φ_i of appliances. If Φ_i is zero then appliance must be active instantly, or else it

Table 1. All appliances and their priorities, utilizations and consumptions in watts.

	d_1	d_2	d_3	d_4	d_5	d_6	d_7
Φ_i	0	1	1	1	1	0	0
U_i	45	45	30	50	45	35	10
C_i	300	200	600	500	400	800	700

may be left for later. The utility U_i of each appliance shows its utility importance, thus, any appliance with the greater utility value should be served first.

If an appliance d_i is active, then its x_i value will be 0 or else it will be 1. The objective is to maximize the co-efficient $\sum_{i=1}^{n} U_i \times X_i$ with respect to the restriction $\sum_{i=1}^{n} C_i \times X_i \leq C_{max}$, here C_{max} represents the maximum allowed consumption of electricity from grid (Table 2).

Table 2. Schedulable appliances and their priorities, utilizations and consumptions in watts.

	d_2	d_3	d_4	d_5
Φ_i	1	1	1	1
U_i	45	30	50	45
C_i	200	600	500	400

In the first half of algorithm serves appliances who have priority 0 and their total consumption is 1800 W/h, which is less than the C_{max}.

There are two distinct stages of this algorithm:

a. Priority management: in this phase, algorithm serves the appliance having the priority 1. In this case, it will serve computer, cooker and iron. Then, it will minus the total consumption of these appliance from C_{max}, $C_{rem} = C_{max} - C_1 - C_6 - C_7$.

b. Optimization: The sum of consumptions of all remaining appliances is greater than C_{rem}, i.e. $C_2 + C_3 + C_4 + C_5 > C_{rem}$. The list of unserved appliances will be passed to algorithm B and it will try to find an optimal solution which maximizes the utility such that the consumption of remaining appliances shall not go beyond C_{max}, maximum capacity.

This algorithm generates a binary tree of potential solutions. We used the greedy approach to find the solution so it will explore only those potential solutions which has least cost. It saves the best found solution so far in the memory. During assessment of sub sets, if the value of a sub set U_b is greater than the existing best solution, then it will be discard and would not be solve any further.

The solution set is divided on the basis of value x_i. There are 16 possibilities, when we have 4 appliances. Once the evaluation is done, then two further cases are possible:

a. If $\sum_{i=1}^{n} C_i \times X_i > C_{max}$, then sub set S_i will not has the optimal solution. So, it is useless to further explore it.

b. If $\sum_{i=1}^{n} C_i \times X_i \leq C_{max}$, then solution is evaluated and verified for its aptness. We update the set, if its evaluation is acceptable or else we keep exploring till one of above mentioned conditions is satisfied.

In Fig. 4, increasing possible solutions are represented in the form of tree. According to our algorithm the dishwasher, television and washing machine will operate.

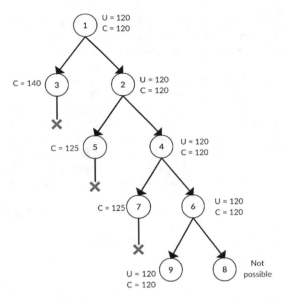

Fig. 4. Exploration tree of finding optimal solution set using greedy approach.

5 Conclusion

There are three distinctive levels of smart grid: T&D, micro-grid and local level. As it is a complex system and its optimization is a challenging task. In this we paper, we suggested an algorithm for the autonomous management of home appliances, based on their priorities. This algorithm finds an optimal solution set which try to meet the predefined maximum allowed energy consumption as close as possible. This optimization helps to avoid the peak load and smooths the total consumption curve. The future work of this study can be algorithm for management at the street level or at micro-grid level of smart grid.

References

1. Shum, C., et al.: Co-simulation of distributed smart grid software using direct-execution simulation. IEEE Access **6**, 20531–20544 (2018)
2. Technology in micro-grid control. Electron. Sci. Technol. Appl. **5**(1) (2018)
3. Ahat, M., Amor, S., Bui, M., Bui, A., Guérard, G., Petermann, C.: Smart grid and optimization. Am. J. Oper. Res. **03**(01), 196–206 (2013)
4. Liang, Y., He, L., Cao, X., Shen, Z.: Stochastic control for smart grid users with flexible demand. IEEE Trans. Smart Grid **4**(4), 2296–2308 (2013)
5. Parisio, A., Glielmo, L.: A mixed integer linear formulation for microgrid economic scheduling. In: 2011 IEEE International Conference on Smart Grid Communications (SmartGridComm), Brussels, pp. 505–510 (2011)
6. Abdella, J., Shuaib, K.: Peer to peer distributed energy trading in smart grids: a survey. Energies **11**(6), 1560 (2018)
7. Panda, R., Tiwari, P.: Economic risk-based bidding strategy for profit maximisation of wind-integrated day-ahead and real-time double-auctioned competitive power markets. IET Gener. Transm. & Distrib. **13**(2), 209–218 (2019)

8. Noor, S., Guo, M., van Dam, K., Shah, N., Wang, X.: Energy demand side management with supply constraints: game theoretic approach. Energy Procedia **145**, 368–373 (2018)
9. Guérard, G., Amor, S., Bui, A.: Survey on smart grid modelling. Int. J. Syst. Control Commun. **4**(4), 262 (2012)
10. Yoldaş, Y., Önen, A., Muyeen, S., Vasilakos, A., Alan, I.: Enhancing smart grid with microgrids: challenges and opportunities. Renew. Sustain. Energy Rev. **72**, 205–214 (2017)
11. Colak, I., Kabalci, E., Fulli, G., Lazarou, S.: A survey on the contributions of power electronics to smart grid systems. Renew. Sustain. Energy Rev. **47**, 562–579 (2015)
12. Wright, A., Firth, S.: The nature of domestic electricity-loads and effects of time averaging on statistics and on-site generation calculations. Appl. Energy **84**(4), 389–403 (2007)
13. Ali, U., Arif, K.S., Qamar, U.: A hybrid scheme for feature selection of high dimensional educational data. In: 2019 International Conference on Communication Technologies (ComTech), Rawalpindi, Pakistan, pp. 71–75 (2019)
14. Dantzig, G.B.: Discrete-variable extremum problems. Oper. Res. **5**, 266–288 (1956). Rand Corporation, Santa Monica, California
15. Horowitz, E., Sahni, S.: Computing partitions with applications to the knapsack problem. J. ACM **21**(2), 277–292 (1974)

Automatic Generation of E-Learning Contents Based on Deep Learning and Natural Language Processing Techniques

Yiyi Wang[1(✉)] and Koji Okamura[2]

[1] Graduate School of Information Science and Electrical Engineer,
Kyushu University, Fukuoka, Japan
`wang.yiyi.849@s.kyushu-u.ac.jp`
[2] Research Institute for Information Technology, Kyushu University,
Fukuoka, Japan
`oka@ec.kyushu-u.ac.jp`

Abstract. With the development of automation in all industries, E-Learning automation: automatically generate E-Learning contents gives E-Learning teachers new opportunities to implement effective management of E-Learning systems. And many eLearning professionals already use E-Learning templates and online asset libraries to create cost effective eLearning courses. Moreover, all E-Learning Automation methods have one thing in common—algorithms: used to control various aspects of e-learning course creation. Additionally, another important thing is online exams which are used to get an immediate feedback from learners. This research focuses on automation generation of E-Learning contents, which can be mainly divided into three parts. Firstly, automatically summarize relevant documents using natural language processing and deep learning techniques. Secondly, detecting keywords in generated summaries and then delete them from input summaries. Finally, rearrange output results and fill them into E-Learning system.

Keywords: Natural Language processing · E-Learning contents generation · E-Learning Automation

1 Introduction

E-Learning Automation has been a hot research topic recently and it is a broad term with several interpretations based on different domains. However, there is a common thing in all automatic E-Learning systems that is the algorithm used to generate and manage E-Learning contents, which is a set of computer codes and instructions. Such computer codes and instructions are always used to allow learning teachers to personalize their E-Learning contents as well as effectively manage their E-Learning systems. Christoforos Pappas of E-Learning Industry, gives a good example of automation in E-Learning, where "a particular piece of E-Learning content displays when an online learner passes the E-Learning assessment. If they don't meet the minimum requirements, the code may generate a resource list they can use to improve." [1] For instance, what contents can be used as learning materials to teach learners and fill the system as teaching books. Moreover, what contents can be used to test learners' in order to show up learners' assessments before and after using the learning systems. Besides, the online quiz is also important for determining the subject mattering to learners' level in order to find some appreciate learning courses.

What's more, above discussed algorithms can also be used to generate quiz and learning materials so that the teachers can keep track of learners throughout the course and effectively monitor their progress. Moreover, advances in deep learning and natural language processing techniques have led huge interesting in automatic e-learning systems to satisfy the needs of E-Learning systems. In this article, we mainly focus on automatically generate E-Learning contents based on deep learning and natural language processing techniques. Specially, the techniques of automatic summarization can be used to automatically generate summary from long and boring passages to get a short and pithy abstract. Then fill e-learning system with generated abstracts as learning materials. Additionally, test analysis techniques can be used to extract keywords from generated summaries. Finally, all of the generated outputs should be rearranged to form questions.

2 Related Work

Automatically generate contents with algorithms in E-Learning fields is not a totally new concept because there are already some existed methods proposed by the history researches. And those proposed methods can be used to relive the stress of effectively manage E-Learning systems to some extent. One efficacious class of those approaches focuses on the motivation by addressing the needs of teachers and learners, as well as technically constrained to respect the special requirements of educational data and algorithms. For instance, using natural language processing (NLP) methods to automatically develop text-based dialogue tutoring systems and process text from website in order to personalize instructional materials to the interests of individual students, automate the creation of questions for teachers and automate the authoring of educational technology systems [2].

Recently, the aspect on automatic questions generation of this approach has been extended to more complete implementations by combining the benefits of semantics-based method with the surface-form flexibility of a template-based way. Although semantic and template-based strategies can be used to reduce the grammar mistakes in generated questions, the quality is still significantly low because of hand-crafted templates. For simple templates written in simple semantic patterns are naturally simple and fast to write, while more complete templates using more difficult semantic rules and more ambiguous. However, for generating high quality questions, the complete and diverse question templates are significant. In other words, generation of high-quality questions by semantic methods is still a time-consuming task.

For example, in our previous research, we used ontology-based method combined with java code to generate questions [3]. And the main defect for that method is that is not appreciated for every domain knowledge. Moreover, most of time is used to create ontology and question templates. Thus, a deep learning combined with NLP method-based algorithm is much more suitable for high coverage and less attractive for applications of questions and contents generation. In this research, our approach uses a method that combines the benefits of automatic text summarization approach with the text analysis, and text mining approach.

3 Design and Implementation of the Prototype

3.1 Design

The whole research is designed mainly into three parts: automatic summarization, keywords extraction and results rearrangement which can be shown more specifically in Fig. 1.

Automatic Summarization

Text summarization refers to the technique of shortening long pieces of text to create a coherent and fluent summary having the main points outlined in the source document [4]. At present, the main methods used for automatically text summarize are extractive based method and abstractive based method. Extractive summarization is essentially understanding the importance of all sentences and their relations with each other and then pick out the best sentences to represent as its summary. On the other hand, abstractive summarization assembles summaries by understanding the content of the text and then summarize by creating its own sentences.

In this research, for the various difficult questions level we use both those methods, where extractive summarize method can be used to generate easy questions because of the features of without making any changes to the source texts. This means that this kind of questions can be easily understood, and learners can find answer from the

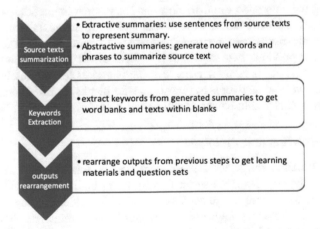

Fig. 1. Design of the whole research which are mainly divided into three parts like automatic text summarization, keywords extraction and outputs rearrangement.

source texts. While because of the characteristics of novel words and phrases generation, the abstractive based method is used for generating difficult questions. To answer
this kind of questions, learners should totally understand the course contents. Meanwhile, those summaries can be used as learning materials to fill the E-Learning systems,
and then take summaries as inputs of the next steps.

Keywords Extraction

Keyword extraction can be considered as the core technology of all automatic processing for documents because it was used to extract a specific group of words or
keywords which can highlight the main content of the documents [5]. Also, there are
some existed useful methods which can be used to extract keywords from texts, such as
word frequency, word collocations and co-occurrences, TF-IDF (short for term
frequency-inverse document frequency) and RAKE (Rapid automatic keyword
extraction). The task of this research is to extract keywords from generated summaries
which are some short texts. Thus, the RAKE method a well-known method which uses
a list of stop words and phrase delimiters to detect the most relevant words or phrases
in a piece of text, is chosen for keywords extraction purpose.

We aim to use this technique to extract keywords from our generated summaries
from summarization part in order to get word bank questions, as well as the fill in
blanks questions. First of all, take summaries generated from summarization part as
input source text. Secondly, apply the RAKE algorithm to extract keywords from these
texts. Thirdly, putting those extracted keywords into word lists and then delete them
from source texts to get texts within several blanks. The output of this part will be a
word bank with all keywords appeared in source texts and some texts within some
blanks. And then take them as inputs of the next steps.

Results Rearrangement and Question Generation

After finishing the automatic summarization and keywords extraction parts we get the
outputs: relevant text summaries, word banks and summary texts without keywords.
Subsequently, this results rearrangement part means to rearrange all of the outputs that
we got from the previous two steps and use them to fill the e-learning system. For
example, take some of the summaries as learning materials for easy understanding and
take those texts without keywords as questions with blanks. Furthermore, combine the
word banks with texts within blanks to let learners choose correct words fill the blanks.

3.2   Implementation and Prototype

This section describes the specific implementation of this research. As showing in
Fig. 2, it basically includes three steps: Firstly, using automatic summarization techniques to summarize source text to get shorter and easier understanding version text for
learners. Second, extracting keywords from generated summaries, this is a middle
bridge for getting word banks and texts within requirements. Finally, rearranging the
outputs generated from the previous steps to generate questions.

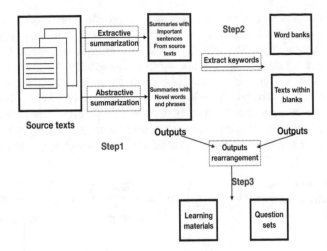

Fig. 2. Basic workflow of implementation part of this research where two kinds of summarization methods used for different difficult level of the questions and text analysis techniques used to extract keywords from summaries.

Step1. Automatic Summarization

Text summarization means using some techniques to produce a concise and precise summary of voluminous texts and convey useful information without losing the overall meaning. Broadly, there are two approaches to summarize in NLP techniques: extractive and abstractive based methods. The extractive method identifies the important sentences or phrases from the original text and extracts only those from the text to represent summary while abstractive method generates new words or phrases to summarize source texts. In this research, to distinguish different level of questions, the extractive method is used for easy level while abstractive method is used for a little difficult level, for it generates novel words or phrases different from source texts.

The baseline of extractive summarization method can be shown in the following Fig. 3: input source texts→find the most important words from source texts→calculate the sentence scores based on the important words found in last step→choose the most important sentences based on the sentence scores→form the summary using the important sentences.

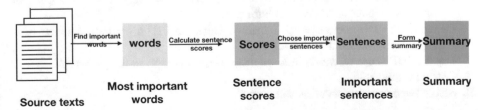

Fig. 3. Baseline of extractive based summarization methods: most important words finding, sentence scores calculation, important sentences chosen and summaries formalization.

Essentially, extractive based methods can be broadly classified as Unsupervised Learning and Supervised Learning methods [6]. Recent works are well implemented by Unsupervised Learning methods which do not need human summaries in deciding the important features of the document and easy to be implemented rather than Supervised Learning methods. So in this research, we used TF-IDF (short for term frequency-inverse document frequency) algorithm which is a kind of Unsupervised Learning methods and is a numeric measure that is used to score the importance of a word in a document based on how often did it appear in that document and a given collection of documents. And more specifically process of TF-IDF algorithm is shown in Fig. 4.

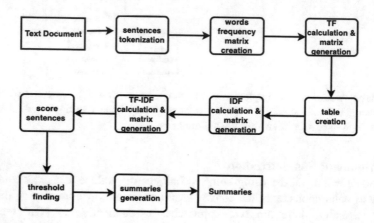

Fig. 4. Basic workflow of TF-IDF extractive summarization method used in this research including data preprocessing, tokenization, frequencies and threshold calculation and finally summaries generation.

More specifically, as showing in Fig. 4, the TF-IDF including following basic workflow:

1. Tokenize the sentences in order to give those sentences weight.
2. Create the frequency matrix of the words in each sentence.
3. Calculate Term Frequency for each word in a paragraph. Here TF(t) = (Number of times term t appears in a document)/(Total number of terms in the document)
4. Creating a table for every word in the documents.
5. Calculate IDF and generate a matrix. IDF(t) = log_e(Total number of documents/Number of documents with term t in it)
6. Generate TF-IDF and generate a matrix. Here TF-IDF algorithm is made of 2 previous algorithms multiplied together.
7. Use TF-IDF score of words in a sentence to give weight to the paragraph.
8. Calculate threshold.
9. Select important sentences as summaries.

The example summary result generated from this extractive summarization method is shown in the following Fig. 5.

```
● ● ●                          input1.txt
In a SYN Flood, a victim server, firewall or other perimeter defense receives (often spoofed
and most often from a botnet) SYN packets at very high packet rates that can overwhelm the
victim by consuming its resources to process these incoming packets.  In most cases if a
server is protected by a firewall, the firewall will become a victim of the SYN flood itself
and begin to flush its state-table, knocking all good connections offline or even worse –
reboot.  Some firewalls in order to remain up and running, will begin to indiscriminately
drop all good and bad traffic to the destination server being flooded. Some firewalls
perform an Early Random Drop process blocking both good and bad traffic. SYN floods are
often used to potentially consume all network bandwidth and negatively impact routers,
firewalls, IPS/IDS, SLB, WAF as well as the victim servers.Syn Flood Attack

A SYN-flood DDoS attack (see the accompanying figure) takes advantage of the TCP
(Transmission Control Protocol) three-way handshake process by flooding multiple TCP ports
on the target system with SYN (synchronize) messages to initiate a connection between the
source system and the target system.

The target system responds with a SYN-ACK (synchronize-acknowledgement) message for each SYN
message it receives and temporarily opens a communications port for each attempted
connection while it waits for a final ACK (acknowledgement) message from the source in
response to each of the SYN-ACK messages. The attacking source never sends the final ACK
messages and therefore the connection is never completed. The temporary connection will
eventually time out and be closed, but not before the target system is overwhelmed with
incomplete connections.
```

```
● ● ●                          summary.txt
In a SYN Flood, a victim server, firewall or other perimeter defense receives (often
spoofed and most often from a botnet) SYN packets at very high packet rates that can
overwhelm the victim by consuming its resources to process these incoming packets. Some
firewalls in order to remain up and running, will begin to indiscriminately drop all good
and bad traffic to the destination server being flooded.
```

Fig. 5. Summary result generated from extractive summarization method, where the above one shows up the source text form website and the rest one is our summarization result generated based on the extractive summarization method described in this paper.

As described above, for getting more difficult questions we need questions with the form of some novel words and phrases which don't show up in the source texts. Because this kind of questions requires learners to be totally understand the source texts. Because of the characters of abstractive summarization methods which was discussed above, it can be used to satisfy the needs described before.

For the perfect abstractive summary, the model has to truly understand the document firstly and then try to express its understanding in short using new words and phrases. This means naturally abstractive summarization model should have complex capabilities like generalization, paraphrasing and in-corporation real-world knowledge [7]. So, compared with the extractive summarization methods, abstractive approaches are harder to be implemented. Based on the survey from internet, two methods proposed by the previous research made a great achievement in abstractive summarization field: the approach using Pointer-Generator Network together with the baseline sequence-to-sequence attention model [8] and the approach proposed by Romain Paulus in [9].

In this research, to effectively abstractive summarization, we combined those two approached together:

LSTM based Sequence-to-Sequence model for abstractive summarization.
Pointer-generator network for handling OOV (out of vocabulary) and copy words from source texts via pointer.
Intra-temporal and Intra-decoder attention mechanism to handle the repeated words. Because repetition is a common problem in the traditional sequence-to-sequence model.
Self-critical policy gradient training along with MLE training to train the model.

Although till now the techniques used to implement abstractive summarization are more mature than before. However, since the difficulty to naturally automatic summarize text like what human do, getting human liked abstract still need a lot to do. Till now, using abstractive text summarization approach to automatically summarize in this research is not very ripe, and this will be the future work which we will focus on.

Step 2. Keywords Extraction to Get Word Banks and Required Form Texts.
Keyword extraction is an automatic process using machines to analyze huge sets of data and extract the most relevant words and expressions from source texts. And to some extents, the keywords extraction is a part of extractive summarization. As what we discussed above, there are some existed algorithms that can be used to implement keywords extraction.

In this research, the Rapid Automatic Keywords Extraction (RAKE) which uses a list of stop words and phrase delimiters to detect the most relevant words or phrases in the source texts was used to implement the request of keywords extraction [10]. The specific process can be described as following.

1. Splitting source texts into a list of words and remove stop words from that list. This step will return a list of content words.
2. Creating candidate expressions based on the phrase delimiters and stop words split by this algorithm.
3. Creating a matrix of word co-occurrences after splitting the text Here in the generated matrix each row shows the number of times that a given content word's co-occurrences.
4. Each word is given a score which can be calculated as the degree of a word in the matrix.

```
In a SYN Flood, a victim server, firewall or other perimeter defense receives (often
spoofed and most often from a botnet) SYN packets at very high packet rates that can
overwhelm the victim by consuming its resources to process these incoming packets. Some
firewalls in order to remain up and running, will begin to indiscriminately drop all good
and bad traffic to the destination server being flooded.
```

↓ keywords extraction

```
[('perimeter defense receives', 9.0),
('high packet rates', 9.0),
('syn flood', 4.0),
('syn packets', 4.0),
('incoming packets', 4.0),
('indiscriminately drop', 4.0),
('bad traffic', 4.0),
('destination server', 4.0)]
```

Fig. 6. Results of extracted keywords result, generated summaries from previous step used as input in this step and used the RAKE algorithm to extract keywords form those summaries. The above picture shows the input text and the following is the extraction results. This result shows with two columns: left column shows the keywords and right column shows the keyword score.

5. Using Beam Search method to choose the keywords and phrases whose scores belong to the top N scores, where N is the number of keywords that will be extracted from the source texts.

For the above example, the extraction keyword result is shown as, the following Fig. 6 where the above part is the source text used as input into this abstract system and the lower one is the keywords extracted result from the methods used in this research.

Step 3. Outputs Rearrangement to Get E-Learning Materials and Online Quiz.
The outputs generated from the previous steps are source text relevant summaries, word lists with keywords extracted from the summaries and the rest texts without keywords. In other words, the previous two steps used in this research can be seen as middle bridge used to do some must preparation for the last question and online learning contents generation processing step. Here those generated summaries can be directly used as online learning martials, because they are short version of the source texts which contain the core information of source texts but are easy to be understand for learners compared to the source long and boring texts.

However, to generate online learning use questions, we still need to rearrange the outputs which we got from the previous steps. Since till now this research is still undergoing, we s didn't get a satisfied result of generated questions. Till now what we have already done is writing some codes to combine the outputs together. Such as write a python code to write both the keywords bank and texts within blanks into another new file. But it still under debugging phrase and the prototype which we want to get can be showing as below (Fig. 7).

```
Please choose correct words to fill the following blanks.

high packet rates     incoming packets     perimeter defense receives     bad traffic
SYN packets

In a SYN Flood, a victim server, firewall or other _____ (often spoofed and most often
from a botnet) _____ at very _____ that can overwhelm the victim by consuming its
resources to process these_____. Some firewalls in order to remain up and running, will
begin to indiscriminately drop all good and _____ to the destination server being
flooded.
```

Fig. 7. Prototype of the final generate questions with the form of fill in the blanks including three parts: stem, word bank and question body.

4 Discussion and Evaluation

The development of this research is still undergoing, and till now the parts what we have finished are extractive based summarization using TF-IDF algorithm and keywords extraction using RAKE algorithm. While the abstractive summarization part is still undergoing and will be the main task in the following time. We also evaluated this research in two parts: evaluate precision of summaries and keywords extraction results.

According to evaluation method proposed in [11], the evaluation process of text summarization is performed by using three parameters which are precision, recall, and f-measure. Table 1 represents the evaluation of extractive based summarization method using the proposed parameters. The TF column shows the number of sentences that are extracted by the system and human; the FP column shows the number of sentences that are extracted by the system, and the FN column shows the number of sentences extracted by the human. The precision describes a ratio between the total of the relevant information and information which can be relevant or irrelevant to the system. And recall is a ratio between the total of the relevant information given by the system and the total of the relevant information which occurs inside the collection of information. Finally, f-measure is a relationship between recall and precision which represents the accuracy of the system. According to the following Table 1, the average of those parameters is around 0.61, so the result accuracy still needs to be improved in the future.

Table 1. Extractive based method summarization evaluation.

Input text	TF	FP	FN	Precision	Recall	F-Measure
1	6	2	3	0.75	0.75	0.75
2	8	8	6	0.5	0.571	0.533
3	5	2	2	0.714	0.714	0.714
4	10	9	9	0.526	0.526	0.364
5	9	6	7	0.6	0.563	0.581
Average				0.618	0.628	0.588

Based on the summarization results, the summaries generated from the extractive method can keep the important meanings of the source texts but still has the problem of grammar mistakes. While the abstractive based method can be used to relieve this kind of problems. However, because of the complicate implementation, till now the abstractive summarization method is still undergoing.

Besides, another thing should be considered is to evaluate the precision of the keywords extraction results. We evaluate the extraction results based on the approach described by Mihalcea and Tarau in [12]. Based on the extraction results, we can extract keywords with 80% accuracy by the RAKE extraction method. However, the shortcoming of RAKE algorithm is it will face some problem when deal with other languages such as Chinese. For example, when we tried to use the RAKE algorithm to deal with Chinese, we faced the problem that several words are glued together. That means the RAKE algorithm used in this research cannot be used for all languages, we still need to improve the RAKE algorithm used in our research and also fix the problem of improving the accuracy of extract results.

5   Conclusion and Future Work

This research mainly aims at solving the problem that E-Learning teachers spend a lot of time on course creation, so we proposed the system that can be easily and effectively used for course creation management. The proposed system will exploit the advantages of the deep learning and natural language processing techniques, those techniques will be used for automating the process of the E-Learning contents generation. As a result of the proposed system: firstly, we generated summaries by automatic summarization techniques, then, based on the generated summaries we extracted keywords, and finally, we deleted those keywords from generated summaries to form questions within blanks.

The advantage of this method is it can be applied to generate question forms different from the multiple chooses questions which we generated in our previous works [3]. Moreover, this method can be used in more generic topics. Because as described in previous research [3], the questions are generated based on the domain ontology that means for different domains we need to build a different ontology. Obviously, it is not very generic methods for more generic topics.

What's more, from the result of Table 1, we can see that the defect is that till now the accuracy of summarization part is only around 60%, and the generated summaries can't have the form of human summarization. Consequently, in the future, we will focus on the implementation of abstractive summarization approach in order to generate more ideal summary results.

Furthermore, after finishing completing making online courses. We will find some participants to conduct some experiments. For example, in order to evaluate our system, we will use our system to study and do online tests to get feedback for the efficient of our system.

Acknowledgments. This research was supported by Strategic International Research Cooperative Program, Japan Science and Technology Agency (JST) SICORP Grant Number JPMJSC16H3 and JSPS KAKENHI Grant Number JP16K00480.

References

1. Pappas, C.: The definitive guide to e-learning automation (2016)
2. Litman, D.: Natural language processing for enhancing teaching and learning. In: Proceeding of the Thirtieth AAAI Conference on Artificial Intelligence (AAAI-2016) (2016)
3. Wang, Y., Allakany, A.: Automatically generate e-learning quizzes from IoT security ontology (2019)
4. Garland, M.J.: A quick introduction to text summarization in machine learning, 19 September 2018
5. Hasan, H.M.M., Sanyal, F.: An empirical study of important keyword extraction techniques from documents (2017)
6. Moratanch, N., Chitrakala, S.: A survey on extractive text summarization. In: IEEE International Conference on Computer, Communication, and Signal Processing (2017)
7. Singhal, S., Bhattacharya, A.: Abstractive text summarization (2017)
8. See, A., Liu, P.J., Manning, C.D.: Get to the point: summarization with pointer-generator networks (2017)
9. Paulus, R., Xiong, C., Socher, R.: A deep reinforced model for abstractive summarization (2017)
10. Rose, S., Engel, D.: Automatic keyword extraction from individual documents (2010)

11. Nedunchelian, R., Muthucumarasamy, R., Saranathan, E.: Comparison of multi document summarization techniques. Int. J. Comput. Appl. **11**(3), 155–160 (2011)
12. Mihalcea, R., Tarau, P.: Textrank: bringing order into texts. In: Lin, D., Wu, D. (eds.) Proceedings of EMNLP 2004, Barcelona, Spain, pp. 404–411. Association for Computational Linguistics (2004)

Identification of Manual Alphabets Based Gestures Using s-EMG for Realizing User Authentication

Hisaaki Yamaba[1]([✉]), Shotaro Usuzaki[1], Kayoko Takatsuka[1],
Kentaro Aburada[1], Tetsuro Katayama[1], Mirang Park[2], and Naonobu Okazaki[1]

[1] University of Miyazaki, Miyazaki, Japan
yamaba@cs.miyazaki-u.ac.jp
[2] Kanagawa Institute of Technology, Atsugi, Japan

Abstract. At the present time, since mobile devices such as tablet-type PCs and smart phones have penetrated deeply into our daily lives, an authentication method that prevents shoulder surfing attacks comes to be important. We are investigating a new user authentication method for mobile devices that uses surface electromyogram (s-EMG) signals, not screen touching. The s-EMG signals, which are generated by the electrical activity of muscle fibers during contraction, can be detected over the skin surface, and muscle movement can be differentiated by analyzing the s-EMG signals. Taking advantage of the characteristics, we proposed a method that uses a list of gestures as a password in the previous study. In order to realize this method, we have to prepare a sufficient number of gestures that are used to compose passwords. In this paper, we adopted fingerspelling as candidates of such gestures. We introduced manual kana of the Japanese Sign Language syllabary and compared the identification performance of some candidate sets of feature values with adopting support vector machines.

1 Introduction

This paper presents an evaluation of manual alphabets as candidates of gestures used in the user authentication method for mobile devices by using surface electromyogram (s-EMG) signals, not screen touching.

An authentication method that prevents shoulder surfing, which is the direct observation of a users personal information such as passwords, comes to be important. At the present time, mobile devices such as tablet type PCs and smartphones have widely penetrated into our daily lives. So, authentication operations on mobile devices are performed in many public places and we have to ensure that no one can view our passwords. However, it is easy for people who stand near such a mobile device user to see login operations and obtain the users authentication information. And also, it is not easy to hide mobile devices from attackers during login operations because users have to see the touch screens of

L. Barolli et al. (Eds.): EIDWT 2020, LNDECT 47, pp. 323–333, 2020.
https://doi.org/10.1007/978-3-030-39746-3_34

their mobile devices, which do not have keyboards, to input authentication information. On using a touchscreen, users of a mobile device input their authentication information through simple or multi-touch gestures. These gestures include, for example, designating his/her passcode from displayed numbers, selecting registered pictures or icons from a set of pictures, or tracing a registered one-stroke sketch on the screen. The user has to see the touch screen during his/her login operation; strangers around them also can see the screen.

To prevent this kind of attack, biometrics authentication methods, which use metrics related to human characteristics, are expected. In this study, we investigated application of surface electromyogram (s-EMG) signals for user authentication. S-EMG signals, which are detected over the skin surface, are generated by the electrical activity of muscle fibers during contraction. These s-EMGs have been used to control various devices, including artificial limbs and electrical wheelchairs. Muscle movement can be differentiated by analyzing the s-EMG [1]. Feature extraction is carried out through the analysis of the s-EMGs. The extracted features are used to differentiate the muscle movement, including hand gestures.

In the previous researches [2–9], we investigate the prospect of realizing an authentication method using s-EMGs through a series of experiments. First, several gestures of the wrist were introduced, and the s-EMG signals generated for each of the motion patterns were measured [2]. We compared the s-EMG signal patterns generated by each subject with the patterns generated by other subjects. As a result, it was found that the patterns of each individual subject are similar but they differ from those of other subjects. Thus, s-EMGs can confirm ones identification for authenticating passwords on touchscreen devices. Next, a method that uses a list of gestures as a password was proposed [3,4]. And also, a series of experiments was carried out to investigate the performance of the method extracting feature values from s-EMG signals adopted in [5] and the methods to identify gestures using dynamic data warping (DTW) [6] and support vector machines (SVM) [7]. As for the selection of gestures used in the method, we introduced manual alphabets used in Japanese Sign Language [8,9].

In this paper, we investigated the performance of some candidates of feature values to identify manual alphabes based gestures with adopting SVMs.

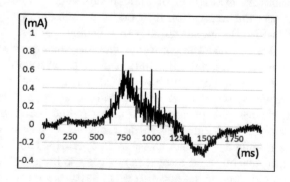

Fig. 1. A sample of an s-EMG signal

2   Characteristics of Authentication Method for Mobile Devices

It is considered that user authentication of mobile devices has two characteristics [2].

One is that an authentication operation often takes place around strangers. An authentication operation has to be performed when a user wants to start using their mobile devices. Therefore, strangers around the user can possibly see the user's unlock actions. Some of these strangers may scheme to steal information for authentication such as passwords.

The other characteristic is that user authentication of mobile devices is almost always performed on a touchscreen. Since many of current mobile devices do not have hardware keyboards, it is not easy to input long character based passwords into such mobile devices. When users want to unlock mobile touchscreen devices, they input passwords or personal identification numbers (PINs) by tapping numbers or characters displayed on the touchscreen. Naturally, users have to look at their touchscreens while unlocking their devices, strangers around them also can easily see the unlock actions. Besides, the user moves only one finger in many cases. So, it becomes very easy for thieves to steal passwords or PINs.

To prevent shoulder-surfing attacks, many studies have been conducted. The secret tap method [10] introduces a shift value to avoid revealing pass-icons. The user may tap other icons in the shift position on the touchscreen, as indicated by a shift value, to unlock the device. By keeping the shift value secret, people around the user cannot know the ture pass-icons, although they can still watch the tapping operation. The rhythm authentication method [11] relieves the user from looking at the touchscreen when unlocking the device. In this method, the user taps the rhythm of his or her favorite music on the touchscreen. The pattern of tapping is used as the password. In this situation, the users can unlock their devices while keeping them in their pockets or bags, and the people around them cannot see the tap operations that contain the authentication information.

3   User Authentication Using s-EMG

3.1   Authentication Process of the Proposed Method

The s-EMG signals (Fig. 1) are generated by the electrical activity of muscle fibers during contraction and are detected over the skin surface [2]. Muscle movement can be differentiated by analyzing the s-EMG. [12] developed a wearable sensor to passively recognize people using bioimpedance. Some other studies have explored the use of s-EMG signals for user authentication. For example, [13] used four electrodes attached on a forearm and investigated the recognition accuracy of several hand gestures using root mean squares and fast Fourier transforms, while [14] used eight electrodes attached around a user's wrist to investigate false acceptance rates (FARs) when extending the hand using a convolutional neural network (CNN). But these studies mentioned a few devices to put s-EMG-based authentication method in practical use. This study proposes the use of "pass-gesture," which is a set series of hand gestures, as a reliable authentication method.

In the previous research, the method of user authentication by using s-EMGs that do not require looking at a touchscreen was proposed [3,4]. The s-EMG signals are measured, and the feature values of the measured raw signals are extracted. We estimate gestures made by a user from the extracted features. In this method, combinations of the gestures are converted into a code for authentication. These combinations are inputted into the mobile device and used as a password for user authentication.

1. At first, pass-gesture registration is carried out. A user selects a list of gestures that is used as a pass-gesture (Fig. 2(a)).
2. The user measures s-EMG of each gesture, extracts their feature values, and register the values into his mobile device (Fig. 2(b)).
3. When the user tries to unlock the mobile device, the user reproduces his pass-gesture and measures the s-EMG.
4. The measured signals are sent to his mobile device.
5. The device analyzes the signals and extracts the feature values.
6. The values are compared with the registered values.
7. If they match, the user authentication will succeed (Fig. 2(c)).

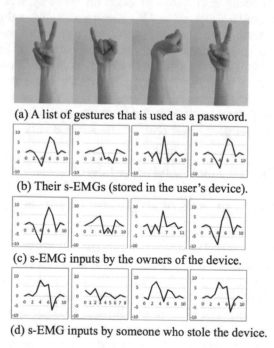

(a) A list of gestures that is used as a password.

(b) Their s-EMGs (stored in the user's device).

(c) s-EMG inputs by the owners of the device.

(d) s-EMG inputs by someone who stole the device.

Fig. 2. A list of gestures used as a password

8. On the other hand, an illegal user authentication will fail because a list of signals given by someone who stole the device (Fig. 2(d)) will not be similar with the registered one.

Adopting s-EMG signals for authentication of mobile devices has three advantages. First, the user does not have to look at his/her device. Since the user can make a gesture that is used as a password on a device inside a pocket or in a bag, it is expected that the authentication information can be concealed. No one can see what gesture is made. Next, it is expected that if another person reproduces a sequence of gestures that a user has made, the authentication will not be successful, because the extracted features from the s-EMG signals are usually not the same between two people. And then, a user can change the list of gestures in our method. This is the advantages of our method against other biometrics based methods such as fingerprints, an iris, and so on. When authentication information, a fingerprint or an iris, is stolen, the user can't use them because he/she can't change his/her fingerprint or iris. But the user can arrange his/her gesture list again and use the new gesture list.

3.2   Collecting Gestures for the Authentication Method Using s-EMG

In order to realize the proposed authentication method using s-EMG, we have to prepare many gestures that represent characters of passwords. Such gestures have to be easy to tell each of them from others.

We are planning to adopt such gestures referring to fingerspelling. Fingerspelling is the representation of the letters using hands. The set of manual signs is used to help the communication using sign languages. For example, proper nouns such as names of persons are represented by fingerspelling. Sign languages such as American Sign Language, French Sign Language, British Sign Language, and so on, have there own manual alphabet.

We adopted the Japanese Sign syllabary (see Fig. 3) as the candidates of gestures used in our authentication method. They are called *yumimoji*, which means "finger letters." Comparing with manual alphabets representing Latin alphabet, there are larger numbers of manual kana in *yumimoji*, 46 letters.

By adopting gestures referring fingerspelling, a password that is made up of gestures corresponds to a string that is made up of letters. It is expected that this helps users to remember their passwords.

To select *yumimoji* that are suitable for the use of this study, we examined their performances as pass-gesture by measuring their s-EMG signals. [8]. From the results of experiments that were carried out in the study, some manual alphabets were expected to be used as pass-gestures. And, we classified some gestures into five groups shown below.

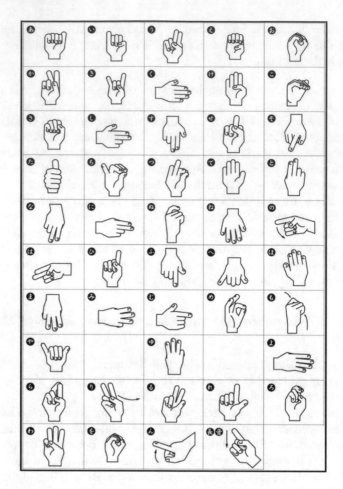

Fig. 3. The Japanese Sign syllabary. (https://upload.wikimedia.org/wikipedia/commons/d/dc/JSL-AIUEO.jpg)

Type 1: Yubimoji that bends one's wrist forth
   This kind of yubimoji bend one's wrist forth. Some of their fingers are extended pointing to the side. Amplitude of s-EMG signals are quite large.
Type 2: Yubimoji that points downward
   This kind of yubimoji bend one's wrist forth but their extended fingers point downward. Amplitude of s-EMG signals are large.
Type 3: Yubimoji that extends only one finger
   This kind of yubimoji extends only one finger. Also they don't bent one's wrist. Amplitude of s-EMG signals measured are very slight. It seems hard to use these letters as elements of pass-gestures.

Type 4: Yubimoji that extends more than one finger
This kind of yubimoji extends more than one finger including index finger and middle finger. Also they don't bent one's wrist. Amplitude of s-EMG signals measured are slight.

Type 5: Yubimoji that turns the forearm
This kind of yubimoji turns the forearm. They don't bent one's wrist. Amplitude of s-EMG signals are clear.

Type 3 gestures that are simple one are not suitable because amplitude of vibration of signals generated by such gestures are quite small. Type 1 gestures that bends forth are promising because they generate larger amplitude. However, it has not been examined that we can distinguish gestures in this type. Type 1 gestures use muscles where electrode sensors are attached. So, it is expected that distinct signals can be obtained by attaching several sensors appropriate positions for each gesture.

3.3   User Authentication System Using s-EMG Signals

In order to realize our user authentication system using s-EMG, gesture distinction has to be carried out by a computer program. So, we extract some feature values that represent characteristics and the user authentication system identify a gesture and a user using such feature values.

To realize a computer-based system that can identity gestures, we adopted SVMs. SVMs are one of the pattern recognition models of supervised learning. Linear SVM was proposed in 1963, and extended to non-linear classification in 1992. A support vector machine builds a classifier for sample data that belong to one of two classes. An SVM trains the separation plane that has the largest margin, and samples on the margin arc called support vectors. An SVM is one of the recognition method that has the highest performance.

We used the pair of the maximum value and the minimum value of raw s-EMG signals as the feature value in our previous study [4]; however its performance were not enough. We attempted to apply more complex sets of features used in [6,7] in this work.

Also, in the previous studies [6], we found that each person had some good gesture for his/her pass-gesture in the experimental set of gestures. So, we should explore suitable gestures for each person.

4   Experiments

4.1   Purpose

A series of experiments was carried out to investigate the prospect of the authentication method using *yubimoji*. Concretely, we investigated the performance of some candidates of feature values to identify manual alphabes based gestures.

And also, we attempted to select prospect yubimoji and explore characteristics of such yubimoji.

First, we compared three candidates of a set of feature values:

FV1. We adopted a pair of maximum value and minimum value in the series of signal values.
FV2. We adopted these eleven values used in the previous study. sum total, standard deviation, mean, sum of squares, skewness, kurtosis, 5 number summary in the series of signal values.
FV3. In the previous study, we divided one series of signal data into 10 pieces and we extracted these 11 feature values from every pieces. We obtain 110 values from one series of signal data.

Next, we also examined whether some gestures are good one as a pass-gesture of some person. In the previous studies, we found that each person had some good gesture for his/her pass-gesture in the experimental set of gestures [6].

4.2 Conditions

In these experiments, the set of DL-3100 and DL-141 (S&M Inc.) that was an electromyograph used in the previous researches also used to measure the s-EMG of each movement pattern in this study (Fig. 4.) Experimental subjects made gestures by their left hand. Subjects sit on an arm chair and put his forearm on the armrest. Electrode sensors were put on the palm side of the forearm. The measured data were stored and analyzed on a PC.

Six students of University of Miyazaki participated as experimental subjects. Gestures used in this study were based on *yubimoji* shown in Fig. 3.

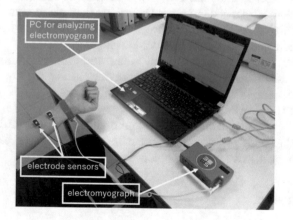

Fig. 4. Measuring an s-EMG signal

Table 1. Comparison of average correction rates

Set of feature values	FV1	FV2	FV3
Average correction rate	0.6375	0.6517	0.5458

However, since a subject put his forearm on an armrest, we arranged some *yubimoji*. When we make a *yubimoji* like *TE*, the palm faces forward. But in these experiment, the palm faces upward.

Before making each *yubimoji*, subjects clenched their left fists. S-EMG signals making *yubimoji* from a clenched fist were measured. The subjects repeated each *yubimoji* ten times and their s-EMG signals were recorded. This measurement was carried out 3 times and 30 signals were obtained for each subject and for each *yubimoji*.

And four gestures ("HU", "MU" "NE", "ta") were used for the experiments. They were selected from Type 1, 2 and 5. Gesture in these groups generate signals of large amplitude. And we selected this yubimoji "hu" to compare with "mu" and "ne". "hu" and "mu" extend same finger but in different type. "hu" and "ne" are in the same group but extend different fingers.

In this research, The programming language "R" was used. SVM function of the programming language R can classify data into several categories. We prepared a SVM for each experimental subject and gave s-EMG signals of gestures as training data. This SVM selects one gesture for given s-EMG signals. Ten test signal data of the four *yubimoji* based gestures are given to each SVM, and correction rates are obtained.

Table 2. Cmparison of average correction rates

#1	HU	MU	NE	TA
HU	8	1	2	3
MU	1	**9**	3	2
NE	0	0	5	0
TA	1	0	0	5

#2	HU	MU	NE	TA
HU	**10**	0	2	0
MU	0	8	2	3
NE	0	0	4	2
TA	0	2	2	5

4.3 Results

Table 1 shows the results of the comparison of the candidates of feature values sets. The results of FV2 was the best one but FV1 (max and min) was as good as FV2(11 values). And FV3(11 values of 10 segments) that was usend in the previous study was worse than other two.

And also, we examined the results of each of the experimental subjects and found that each subject has his good gesture. Table 2 shows the results of subject #1 and #2. "MU" and "HU" were suitable for subject #1 and #2 respectively. Other subjects had also a suitable gesture. From these results, it is expected that selection of appropriate gestures increases the performance of the proposed method.

SVM can be used to identify gesture from s-EMG signals, and it is expected that each person can select suitable pass-gesture from *yubimoji* based gestures. But the performance was not as good as our previous studies. The gestures used in the previous studies were simple and easy to learn compared with gestures based on *yubimoji*. This results shows that we should adopt a set of *yubimoji* that are easy to learn and use.

5   Conclusion

We investigated a new user authentication method that can prevent shoulder-surfing attacks in mobile devices. To realize the authentication method using s-EMG, we examined the characteristics of the *yubimoji*, the Japanese Sign Language syllabary, as the candidate of the element of pass-gestures. A series of experiments was carried out to compare the performance of candidates of feature values used to identify gestures from s-EMG signals. Results of the experiments showed that the performance of the feature values used in our previous studies were not same with the previous studies using simple gestures when using the *yubimoji* based gestures. But It is expected that *yubimoji* based gestures can be used to select suitable pass-gesture for each person.

We are planning to examine characteristics of *yubimoji* using a lot of s-EMG data from many people including younger and elder people and attempt to explore gestures that are good in the view point of reproducibility. Also, We would like to explore an appropriate set of feature values. Finally, We are planning to adopt other sets of features for training SVMs and to apply other machine learning methods such as Deep Learning to identify gestures from s-EMG signals.

Acknowledgements. This work was supported by JSPS KAKENHI Grant Numbers JP17H01736, JP17K00139, JP17K00186, JP18K11268.

References

1. Tamura, H., Okumura, D., Tanno, K.: A study on motion recognition without FFT from surface-EMG. IEICE Part D **J90–D**(9), 2652–2655 (2007). (in Japanese)
2. Yamaba, H., Nagatomo, S., Aburada, K., et al.: An authentication method for mobile devices that is independent of tap-operation on a touchscreen. J. Robot. Netw. Artif. Life **1**, 60–63 (2015)
3. Yamaba, H., Kurogi, T., Kubota, S., et al.: An attempt to use a gesture control armband for a user authentication system using surface electromyograms. In: Proceedings of 19th International Symposium on Artificial Life and Robotics, pp. 342–245 (2016)
4. Yamaba, H., Kurogi, T., Kubota, S., et al.: Evaluation of feature values of surface electromyograms for user authentication on mobile devices. Artif. Life Robot. **22**, 108–112 (2017)
5. Yamaba, H., Kurogi, T., Aburada, A., et al.: On applying support vector machines to a user authentication method using surface electromyogram signals. Artif. Life Robot. (2017). https://doi.org/10.1007/s10015-017-0404-z
6. Kurogi, T., Yamaba, H., Aburada, A., et al.: A study on a user identification method using dynamic time warping to realize an authentication system by s-EMG. In: Advances in Internet, Data & Web Technologies (2018). https://doi.org/10.1007/978-3-319-75928-9_82
7. Yamaba, H., Aburada, A., Katayama, T., et al.: Evaluation of user identification methods for an authentication system using s-EMG. In: Advances in Network-Based Information Systems (2018). https://doi.org/10.1007/978-3-319-98530-5_64

8. Yamaba, H., Inotani, S., Usuzaki, S., et al.: Introduction of fingerspelling for realizing a user authentication method using s-EMG. In: Advances in Intelligent Systems and Computing (2019). https://doi.org/10.1007/978-3-030-15035-8_67
9. Yamaba, H., Usuzaki, S., Takatsuka, K., et al.: Evaluation of manual alphabets based gestures for a user authentication method using s-EMG. In: Advances in Intelligent Systems and Computing (2019). https://doi.org/10.1007/978-3-030-29029-0_56
10. Kita, Y., Okazaki, N., Nishimura, H., et al.: Implementation and evaluation of shoulder-surfing attack resistant users. IEICE Part D J97–D(12), 1770–1784 (2014). (in Japanese)
11. Kita, Y., Kamizato, K., Park, M., et al.: A study of rhythm authentication and its accuracy using the self-organizing maps. In: Proceedings of the DICOMO, pp. 1011–1018 (2014). (in Japanese)
12. Cornelius, C., Sorber, J., Peterson, R., et al.: Who wears me? Bioimpedance as a passive biometric. In: Proceedings of the 3rd USENIX Workshop on Health Security and Privacy, HealthSec 2012, pp. 1–10 (2012)
13. Cannan, J., Hu, H.: Automatic user identification by using forearm biometrics. In: 2013 IEEE/ASME International Conference on Advanced Intelligent Mechatronics, pp. 710–715 (2013)
14. Shoji, R., Ito, S., Ito, M., Fukumi, M.: Personal authentication based on wrist EMG analysis by a convolutional neural network. In: 2017 Proceedings of the 5th IIAE International Conference on Intelligent Systems and Image Processing, pp. 12–18 (2017)

Classification of Malicious Domains by Their LIFETIME

Daiji Hara[1]([⊠]), Kouichi Sakurai[1], and Yasuo Musashi[2]

[1] Graduate School, Kyushu University, Fukuoka, Japan
{daiji.hara,sakurai}@inf.kyushu-u.ac.jp
[2] CMIT, Kumamoto University, Kumamoto, Japan
musashi@cc.kumamoto-u.ac.jp

Abstract. In this study, we look for malicious domains in the logs of the primary DNS server of Kumamoto University using a malicious domain check tool (Virus Total), We then classify them according to their LIFETIME (LT) and investigate their main attack applications. The following results were obtained from the experiment: (1) Ransomware, phishing, and DDoS attacks were the 3 most frequent attacks. (2) We obtained two sets of LIFETIME by plotting the number of malicious domains according to their frequency (3) The frequency distribution obtained on ransomware, phishing, and DDoS attacks show that the LT distribution of ransomware and phishing is similar, however, the frequency of DDoS attacks is shorter. (4) From these results, we learn that the attack method can be determined by measuring the LT. The LT shows to be a good parameter to be used with machine learning to detect malicious domain names.

1 Introduction

It is undeniable that we are living in an informatized society. This originates from the development and popularization of the Internet which is expected to continue to expand all over the world. The informatized society is intricately intertwined with computer networks as well as human relationships (The more people need to connect to each other, the more the networks will expand due to high demand), these networks can change or have new possibilities added to them. New problems will arise as the information society expands. The biggest application of Networking, Internet, is used by many people in many situations, and various information are transmitted as digitized data. Such a big flow in digitalized data will create the need to think about the importance, sensitivity and security of these information as important data such as personal information may spread unintentionally or unknowingly through attacks from others through the Internet. Due to the above, cybersecurity technologies have been regarded as an important step to take before exchanging sensitive information.

Recently, Fully Qualified Domain Name (FQDN) obtained from analyzing malware such as viruses and worms is called a malicious domain name and has been actively studied [1–5, 9–12]. In this study, we used a DNS packet capturing data of the primary DNS server of Kumamoto University and Virus total [8] for 7 days from August 1, 2018. The malicious domain names were classified according to a LIFETIME (LT) parameter

© Springer Nature Switzerland AG 2020
L. Barolli et al. (Eds.): EIDWT 2020, LNDECT 47, pp. 334–341, 2020.
https://doi.org/10.1007/978-3-030-39746-3_35

and the main uses of the malicious domain were investigated, and as a result, 16,658 malicious domains were obtained.

When this malicious domain was investigated, the following results were obtained:

(1) Ransomware, phishing, and DDoS attacks were the 3 most present attacks according to the frequency of attack methods.
(2) Two sets of LT were obtained by plotting the frequency of attacks through time.
(3) The frequency distribution on DDoS, Ransomware and phishing attacks show that they have mostly similar LT but the LT for DDoS attacks is smaller.
(4) From these results, we learn that the attack method can be determined by measuring the LT. Also, the LT prove itself as an effective candidate for machine learning to detect malicious domains.

2   Related Works

Related research include Detecting Malicious Domains Using Virus Total (an Integrated Malware Analysis Service) [1], Real-time Detection of Short-term Malicious Domains [2], and Analysis of Malicious Domain Name Usage focused on DNS Name Resolution [3]. However, here are some of the things that are deeply related to our study.

2.1   Analysis of Malicious Domain Name Usage Focused on DNS Name Resolution [4]

In this research, they analyzed from the viewpoint of how IP addresses obtained by DNS name resolution could be used. As a result, more than half of the domains were operated using a predetermined IP address. We have observed the domain from the IP address obtained from the DNS log, and this study confirmed the effectiveness of the method. Also, a LIFETIME is defined using the characteristics of a malicious domain: domain name is discarded in a short span and IP address is constant.

2.2   Malicious Domain Name Detection Based on Extreme Machine Learning [5]

In recent years, machine learning has been increasingly used to determine whether a domain is benign or malicious, and the features used for that learning are various. This research used a domain lifespan. This is registered with WHOIS, defined by the domain registration expiry date and the domain creation date interval. However, WHOIS is valid for all domains. Besides, although a lifespan of this domain is calculated in days, it can be seen that some malicious domains have ultra-short-term domains whose domain LIFETIME is shorter than one day [2]. The domain LIFETIME we propose does not require WHOIS information. It can be calculated even if the LIFETIME of the domain is shorter than one day. We are check if the LIFETIME can be used as one of the features in machine learning to determine if a domain name is benign or malicious.

2.3 We Know It Before You Do: Predicting Malicious Domains [6]

Since malicious domains are only used for a very short period, blacklist-based counter-measures are not effective enough at present, so this study proposes a system that predicts the mass malignancy of domains. The approach focuses on the life cycle of the domain, which consists of (1) the preparation phase, (2) the activation phase, (3) the deactivation phase, and the preparation and activation phases. To make a malicious domain available, an attacker needs to perform three actions: domain name selection, domain name registration, and DNS record creation. The time interval of the action predicts which domain is benign or malicious. This life cycle is defined by the flow from once being blacklisted to being stopped and then being registered again. However, LIFETIME we propose is defined on the assumption that the malicious domain is sequentially renamed to escape detection before being blacklisted.

2.4 EXPOSURE: A Passive DNS Analysis Service to Detect and Report Malicious Domains [7]

Many malicious activities rely on a DNS to manage a large distributed network of infected machines. Also, they constructed EXPOSURE to detect malicious domains using machine learning. EXPOSURE has selected a time-based feature as one of its features, where a time-based feature is a change in the number of requests in the domain. Its lifespan is shortened, and the number of requests for the domain involved in phishing increases rapidly as many victims access the site. Therefore, to determine whether it is malicious or not, observe whether the lifespan of the domain is short or if the number of requests for the domain has changed abruptly. Our definition of LIFETIME is different, it is determined by the time elapsed between the domain's first observed time and the last observed time. Therefore, if the target domain has been deleted, LIFETIME can still be calculated.

3 LIFETIME Definition

A domain always has a lifespan. This determines how long the domain is valid. The lifespan of this domain is set to approximately 1–2 years for benign domains. However, it is known that the lifespan of a malicious domain is shorter than that of a benign domain. However, a lifespan of a domain is an information that only the registrant of the domain or the domain administrator knows, and an accurate value is difficult to calculate. In order to get an approximation, we defined a LIFETIME which is the difference between the time when a domain requested a name resolution from the DNS server once and the time where the domain was later deleted and a name resolution request was made again with another domain name but with the same IP Address. Figure 1 shows how the LIFETIME is calculated.

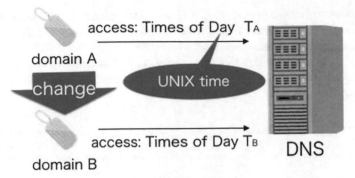

Fig. 1. A LIFETIME definition

4   Experiment

4.1   Experiment 1

We diagnosed benign and malicious domains using Virus Total for the IP addresses existing in the log for 7 days from August 1st to 7th, 2018. Therefore, a LIFETIME of 16,658 domains judged to be malicious was measured. Based on this, a logarithmic frequency distribution was created. The frequency distribution created is shown in Fig. 2 below. From Fig. 2, it was found that two feature points can be selected (LT 1, LT 2).

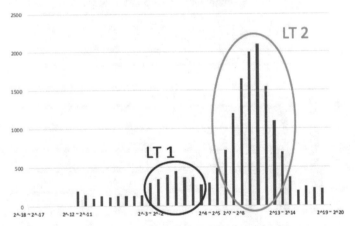

Fig. 2. Horizontal axis: LIFETIME. Vertical axis: frequency.

4.2   Experiment 2

The two groups (LT 1 and LT 2) obtained in Experiment 1 were analyzed in order to know the type of attacks they performed. By referring to the information available on the internet about known malicious domains, we investigated our set of malicious domains whose use was unknown after deleting the virtual machine used for detection and access to the addresses registered in that domain. The investigation consisted of classifying the types of attacks and is shown in Figs. 3 and 4 below. The frequency distribution obtained by the results is also shown.

From Fig. 2, it can be seen that the number of DDoS attacks is the highest in the range of the LT1, and from Fig. 3, the LT2 group has the highest number of malicious domains used for phishing.

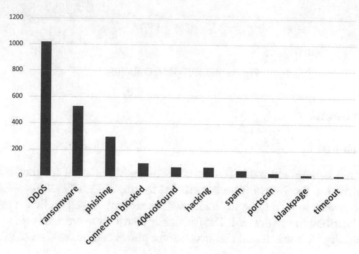

Fig. 3. Frequency of LT 1 attack types

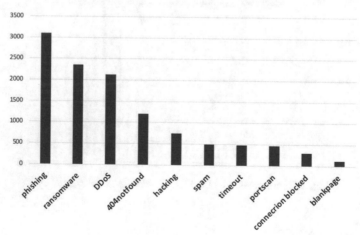

Fig. 4. Frequency of LT 2 attack types

4.3   Experiment 3

In Experiment 3, we confirmed the frequency of each malicious domain attack method obtained by this DNS log analysis. The created frequency distribution is shown in Fig. 5 below. As a result, we found that there were many ransomware, phishing, and DDoS domains. To know the characteristics, a frequency distribution was created for each. The created frequency distribution is shown in Figs. 6, 7 and 8 below. Based on these results, the ransomware and phishing domains were similar by having a peak in similar positions and were widely distributed elsewhere. Also, DDoS has a peak at a position similar to the other two types, though the other two types are also characterized in that feature points can be seen in other places where the LIFETIME is short. It turns out that different attacks have different properties.

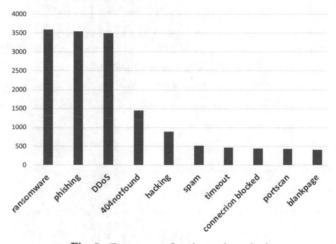

Fig. 5. Frequency of each attack method

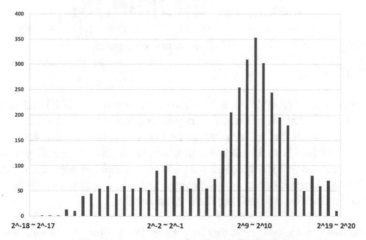

Fig. 6. Ransomware frequency distribution

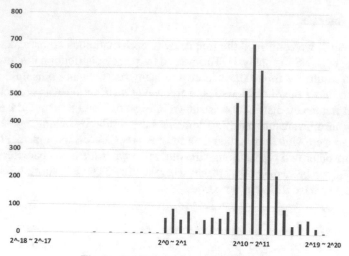

Fig. 7. Phishing frequency distribution

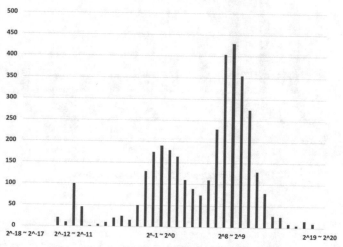

Fig. 8. DDoS frequency distribution

5   Conclusion

The experiment confirmed that there were two peaks from the LIFETIME-based loga-rithmic frequency distribution. From the frequency distribution for each attack method, the LIFETIME distribution of ransomware and phishing was similar, however, DDoS was different from these two. It was found that the LIFETIME proposed by us was useful and a possible candidate for automated analysis using machine learning. By combining LIFETIME and Machine Learning, we believe that it is possible to tell if a domain is malicious or not but also determining the type of attacks.

Acknowledgments. We would like to express our gratitude to Professor Kenichi Sugitani, Pro-fessor Yuji Nakano, Professor Masashi Toda, Associate Professor Shinichiro Kubota, and assistant professor Masahiro Ueda from the Kumamoto University CMIT Laboratory for their advice.

References

1. Tanabe, R., Mori, H., Harada, K., Yoshioka, K., Matsumoto, T.: Detecting malicious domains using virus total an integrated malware analysis service. Inf. Process. Soc. Jpn. **59**(9), 1610–1623 (2018)
2. Kumamoto University: CMIT, Graduation thesis Ryo Okahara. Real-time Detection of Short-term Malicious Domains, February 2018
3. Chiba, D., Yagi, T., Akiyama, M., Shibahara, T., Yada, T., Mori, T., Goto, S.: DomainProfiler: discovering domain names abused in future. In: 2016 46th Annual IEEE/IFIP International Conference on Dependable Systems and Networks, pp. 491–502 (2016)
4. Mizutani, M.: Analysis of malicious domain name usage focused on DNS name resolution, Computer Security. IBM Japan, Tokyo Research
5. Shi, Y., Chen, G., Li, J.: Malicious domain name detection based on extreme machine learning. Neural Process. Lett. **48**(3), 1347–1357 (2018)
6. Xu, W., Sanders, K., Zhang, Y.: We Know It Before You Do: Predicting Malicious Domains. In: Virus Bulletin Conference, pp. 73–77. Palo Alto Networks, Inc., September 2014
7. Bilge, L., Sen, S., Balzarotti, D., Kirda, E., Krugel, C.: Exposure: a passive DNS analysis service to detect and report malicious domain. ACM Trans. Inf. Syst. Secur. (TISSEC) **16**(4) (2014). Article 14
8. https://www.virustotal.com/gui/
9. Antonakakis, M., Perdisci, R., Lee, W., Vasiloglou II, N., Dagon, D.: Detecting malware domains at the upper DNS hierarchy. In: Proceedings of the 20th USENIX Conference on Security, San Francisco, CA, 08–12 August 2011, p. 27 (2011)
10. Zhao, G., Xu, K., Xu, L., Wu, B.: Detecting APT malware infections based on malicious DNS and Traffic analysis. IEEE Access **3**, 1132–1142 (2015). https://doi.org/10.1109/ACCESS.2015.2458581
11. Grosse, K., Papernot, N., Manoharan, P., Backes, M., McDaniel, P.: Adversarial examples for malware detection. In: Computer Security – ESORICS 2017, pp. 62–79 (2017)
12. Saxe, J., Berlin, K.: Deep neural network-based malware detection using two-dimensional binary program features. In: 2015 10th International Conference on Malicious and Unwanted Software (MALWARE), pp. 11–20, October 2015

Design of a DSL for Converting Rust Programming Language into RTL

Keisuke Takano[1]([✉]), Tetsuya Oda[2], and Masaki Kohata[2]

[1] Graduate School of Engineering, Okayama University of Science, 1-1 Ridai-cho,
Kita-ku, Okayama-shi, Okayama, Japan
takano@kpc1.ice.ous.ac.jp
[2] Faculty of Engineering, Okayama University of Science, 1-1 Ridai-cho, Kita-ku,
Okayama-shi, Okayama, Japan
{oda,kohata}@ice.ous.ac.jp

Abstract. Recent research has focused on a large amount of processing such as streaming processing, big data, deep learning and so on. Since the processing time of these processes increases in proportion to the amount of calculation, an arithmetic unit that can increase the speed is required. In this situation, Field Programmable Gate Array (FPGA) has been attracting attention because it can speed up processing and reduce power consumption. However, Hardware Description Language (HDL) such as Verilog used when developing FPGA increases the development time, but also makes it difficult to guarantee memory safety. In this paper, we propose a Register Transfer Level (RTL) designing Domain Specific Language (DSL) for Rust programming language convert to RTL.

1 Introduction

Field Programmable Gate Arrays (FPGAs) have begun to be used in various fields because of researchers, developers, engineers and so on attention to performance per electric power [1,2]. Different research has shown that FPGAs can be used to implement logic circuit based systems with better power efficiency and processing performance compared to Central Processing Units (CPUs) and Graphics Processing Units (GPUs) [3,4]. It is particularly popular in fields such as image processing and deep learning because applications with high parallelism can be implemented.

When developing software on a general computer, it is possible to select and develop a programming language and framework according to the usage, such as the policy of developer and operating status. On the other hand, languages can be used for development in logic circuit are limited. Generally, Hardware Description Language (HDL) such as Verilog [5] and VHDL [6] is used for hardware development. However, HDL is a low level programming language, and it tends to take a long time to develop an hardware. A research and development on HDL purposed at improving the efficiency of logic circuit development using existing programming languages has been conducted in research field on FPGA [7–9].

L. Barolli et al. (Eds.): EIDWT 2020, LNDECT 47, pp. 342–350, 2020.
https://doi.org/10.1007/978-3-030-39746-3_36

When implementing applications that use FPGAs, if the development programming language and method can be selected according to the purpose and preferences as with software, the range of logic circuit development will be further expanded. Therefore, we propose a hardware design domain specific language using Rust, which is a programming language with few syntax failures. We describe the design and experimental results of Domain Specific Language (DSL) for converting Rust programming language into Register Transfer Level (RTL).

The rest of the paper is organized as follows. Section 2 presents the overview of DSL, RTL and Rust programming languages. In Sect. 3, we show the description and design of the proposed system. Experimental results are shown in Sect. 4. Finally, conclusions and future work are given in Sect. 5.

2   DSL, RTL and Rust Programming Languages Overview

2.1   DSL

DSL refers to a language specialized for a specific task. An examples of DSL are Unified Modeling Language (UML) [10] and SQL [11]. So far, efforts purpose at simplification and higher efficiency in hardware design have been actively pursued [12]. In order to achieve the purpose are many developments targeting existing programming languages. For example, A hardware design DSL using programming languages such as Java and Python has been proposed. Veriloggen [13] is a hardware design DSL in Python. Veriloggen can be built directly as an Abstract Syntax Tree (AST) of layered Verilog HDL without using the AST module of Python. Chisel [14] is a hardware design DSL in Scala, it is a programming language implemented as an embedded DSL like Veriloggen. JHDL [15] is a hardware design DSL by Java, SFL [16], which is a visually testable tool, is HDL used in the PARTHENON of logic synthesis system. In addition, a DSL and it is compiler have been developed for the purpose of efficient processing of streaming data in FPGA [17,18].

2.2   RTL

RTL is an abstraction for defining of a design of logical circuits. A design method of logic circuits in RTL can be simplified logic circuit design by combinatorial a different stateful circuit (the smallest part dealing with sequential circuit) [19]. Application Specific Integrated Circuits (ASICs) and FPGAs are usually designed based on RTL models using HDL. HDL includes Verilog [5] and VHDL [6], which are used generally.

FPGA is composed of logic elements, input/output elements, wiring elements, and the like, and an arbitrary logic circuits are formed by connecting these elements. The logic elements is composed Look-Up Table (LUT), Flip-Flop (FF),

Block Random Access Memory (BRAM), and so on. The logic circuit design that reduces the logic elements used is required in FPGA [20]. When designing a large-scale circuit, it is necessary to design at the RTL level where data can be assigned to registers and signals can be defined for each clock.

2.3 Rust Programming Language

Rust programming language [21,22] is an open source system programming language supported by Mozilla. Development of Rust programming language is underway with the goal of achieving speed, safety, and concurrency. Development purpose of Rust programming language as follows:

1. The processing speed is optimized from the source code to the target platform and the binary is generated, so it can be calculated at a speed equivalent to or faster than C/C++.
2. Safety is realized by the compiler generating safe binaries by performing static verification of resources in advance and preventing access to illegal memory areas.
3. Concurrent processing is achieved by including parallel functions in the language standard library.

3   Proposed System

Figure 1 shows the proposed system is a DSL for converting Rust programming language into RTL [23]. The Verilog Module (`VModule`) can build an Verilog AST (`VAST`), so `VAST` construction procedure and code generation. AST can use as intermediate processing in compilers and interpreters, and the proposed system are building `VAST` on Rust programming language. `VAST` implementation is executed on Rust programming language by structure equivalent Verilog syntax. AST is tree-structured data in which only the information required for the target programming language is extracted. Code of Listing 1 can be output by calling

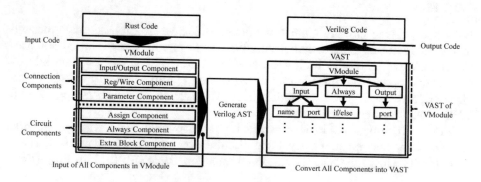

Fig. 1. AST construction procedure by proposed system.

the implemented `endmodule` method from the `VModule` that has completed construction of the AST. Verilog code can be generated from standard output is output after compiled. The generation syntax is based on Verilog2001 [5], and implemented as a library consists of synthesizable logic.

`VAST` is implemented definition of a structure equivalent Verilog syntax using Rust programming language. The generation syntax of `VAST` is based on Verilog2001, it is implemented as a library consisting of syntax that can be synthesized. Listing 1 show a code of LED lighting circuit using Rust programming language. In this code, `VModule` that has completed construction of the AST can be output by calling the implemented `endmodule` method.

Listing 1 Example code of LED circuit using proposed system.

```
1     let mut m = VModule::new("LED");
2
3     let clk = m.Input("CLK", 1);
4     let rst = m.Input("RST", 1);
5     let btn1 = m.Input("BTN1", 1);
6     let btn2 = m.Input("BTN2", 1);
7     let mut led = m.Output("LED", 8);
8
9     let mut fsm = Clock_Reset(clk.clone(),rst.clone())
10        .State("State")
11        .AddState("IDLE").goto("RUN", F!(btn1 == 1))
12        .AddState("RUN").goto("END", F!(btn2 == 1))
13        .AddState("END").goto("IDLE", Brank!());
14    let run = fsm.Param("RUN");
15    let fstate = m.FSM(fsm);
16
17    m.Assign(led._e(_Branch(F!(fstate == run), _Num(8),
          _Num(0))));
18    m.endmodule();
```

3.1 Module Creation, Signal Line and Parameter Generation

In proposed system, it generate a top module (`VModule`) for storing syntax, at first. `VModule::new()` generates an AST structure according to the code shown in lines 1–2 of Listing 1. A module object is created by passing a module name as a string as an argument. The connection components of `VModule` methods can add parameters, Input/Output (I/O) ports, `reg` and `wire` of Verilog. I/O ports can be defined as shown in lines 2–6 in Listing 1. The connection components are using type arguments that strings of signal wire names and `i32` type of wire widths.

3.2   Combination Circuit and Sequential Circuit

The circuit component can generates combination circuit and sequential circuit. The assign syntax or function syntax can implement combination circuit. **Assign** method at line 15 in Listing 1 show created combination circuit code equivalent to the Verilog assign statement. The **func** method can be writing the equivalent code as function syntax in Verilog. Method of **VAST** can add a port and branch syntax such as **if/else** and **case** methods.

The **always** [5] in Verilog HDL syntax can create a sequential circuit that operates based on a physical clock. In proposed system, the equivalent description is possible by using the method of ".**Always** ()". Listing 2 shows a example code of a sequential circuit using proposed system.

Listing 2 An example code of sequential circuit.

```
1  m.Always(Posedge(clk.clone()))
2    .If(F!(rst != 1), Form(F!(done =0)))
3    .Else_If(F!(btn0 == 1), Form(F!(data = 10)))
4    .Else(Form(F!(data = 20))
5      .Form(F!(done = 1)))
6    .Case(select.clone())
7      .S(_Num(1),Form(F!(o_1 = 1)))
8      .S(_Num(2),Form(F!(o_1 = 2)))));
```

3.3   Finite State Machine

The proposed system can implement a FSM using state and conditional expressions. An code example of FSM implementation is shown in Listing 1, lines 7–13. It is implemented in three states: IDLE, RUN, and END.

3.4   Verilog Code Output

The proposed system can be output the code by calling the **endmodule** method from the **VModule** that has completed the generation of **VAST**. Listing 3 shows the result of proposed system outputting a Verilog code of the LED lighting circuit.

Listing 3 An example code of Verilog output result for LED circuit using proposed system.

```verilog
1  module LED (
2      input CLK,
3      input RST,
4      input BTN1,
5      input BTN2,
6      output [7:0] LED
7  );
8      // ----Generate local parts----
9
10     localparam IDLE = 0;
11     localparam RUN = 1;
12     localparam END = 2;
13     reg [31:0] State;
14     reg [31:0] State_Next;
15
16     // ----Generate assign component----
17
18     assign LED = (State==RUN)? 8: 0;
19
20     // ----Extra component set----
21
22     always@(posedge CLK or posedge RST) begin
23         if (RST == 1) begin
24             State <= IDLE;
25         end
26         else begin
27             State <= State_Next
28         end
29     end
30     always@(posedge CLK) begin
31         case(State)
32             IDLE : begin
33                 if(BTN1==1&&RST!=1)
34                     State_Next <= RUN;
35             end
36             RUN : begin
37                 if(BTN2==1)
38                     State_Next <= END;
39             end
40             END : begin
41                 State_Next <= IDLE;
42             end
43         endcase
44     end
45  endmodule
```

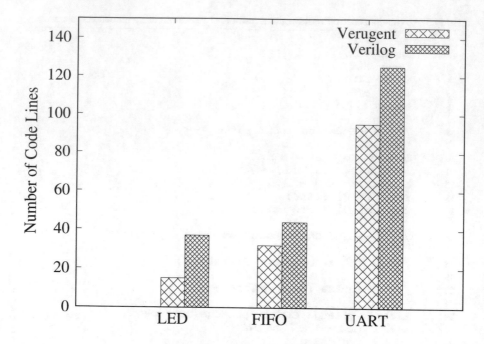

Fig. 2. Number of code lines.

Table 1. Number of code lines and reduction rate.

Circuit	$\frac{ProposedSystem}{Verilog}$ [%]
LED	40.5
FIFO	72.7
UART	76.0

4   Experimental Results

4.1   Experimental Scenario 1

For evaluation, we compare that the total number of source code lines and the number of output Verilog HDL code lines generated using proposed system. We measure without blank and comments lines in Verilog HDL generated by proposed system. In addition, LED blinking circuit is shown in Listing 1, the First-In First-Out (FIFO) and Universal Asynchronous Receiver/Transmitter (UART) transmission/reception circuits is implemented by proposed system.

Figure 2 and Table 1 shows experimental results that the number of proposed system and Verilog source code lines, and the reduction rates. Figure 2 shows that the LED circuit uses the FSM in proposed system so the code lines is shortened. Because, The code lines amount of FIFO and UART circuits are about 70 [%]. Also, LED circuit using the FSM function be able to reduce the code amount to

Table 2. Usage of the slice and resources in FPGA

Resource/Process	FF	LUT	Memory LUT	BRAM
Synthesis	230 (0.49 [%])	262 (0.22 [%])	0 (0.0 [%])	0 (0.0 [%])
Implementation	230 (0.49 [%])	262 (0.22 [%])	0 (0.0 [%])	0 (0.0 [%])

about 40 [%] like the code in the Listing 3. For Table 1, it is shown that the circuit can be designed with a description shortened to about $\frac{2}{3}$ of the HDL code. Since the FSM structure automatically generates state definition parameters when code is output, it is considered that the circuit can be implemented efficiently with a smaller amount of code lines in proposed system than Verilog.

4.2 Experimental Scenario 2

For this scenario, we evaluate if proposed system can implement circuit for FPGA. Table 2 shows the performance of the Verilog code of the hill climbing method using proposed system. As an experiment, the code generated by the proposed system is implemented on FPGA. For implementation, Xilinx FPGA Zybo Z7-20 and development software Vivado 2018.2 are used. In this circuit, the FPGA resources used in Synthesis (logic synthesis) and Implementation (place and route) both use 230 blocks of FF and 262 blocks of LUT. Memory LUT and BRAM are not used because they are not circuits that refer to memory.

5   Conclusion

In this paper, we proposed DSL for convert Rust programming language into RTL. The proposed system show that can be reduce the amount of code lines compared to Verilog. The experimental results showed the amount of code lines using the proposed system is about $\frac{2}{3}$ of code lines using Verilog. We also confirmed that a circuit of hill climbing that can be mounted on an FPGA generated possible.

Future work will focus on performing experiment with different applications to effectively utilize proposed system.

References

1. Wang, T., et al.: A survey of FPGA based deep learning accelerators: challenges and opportunities. arXiv:1901.04988, pp. 1–10 (2018)
2. Vipin, K., et al.: FPGA dynamic and partial reconfiguration: a survey of architectures, methods, and applications. ACM Comput. Surv. **51**(4), 1–39 (2018)
3. Asano, S.: Performance comparison of FPGA, GPU and CPU in image processing. In: Proceedings of FPL-2007, pp. 126–131 (2009)
4. Gomez-Pulido, J., et al.: Accelerating floating-point fitness functions in evolutionary algorithms: a FPGA-CPU-GPU performance comparison. Genet. Program. Evolvable Mach. **12**, 403–427 (2011)

5. IEEE Std. 1364-2001: IEEE Standard for Verilog Hardware Description Language. IEEE SA, pp. 1–590 (2001)
6. ModelSim: Foreign Language Interface. User Manual, Version 5.6d, 13 November 2019
7. Windh, S., et al.: High-level language tools for reconfigurable computing. Proc. IEEE **103**(3), 390–408 (2015)
8. Kapre, N., Bayliss, S.: Survey of domain-specific languages for FPGA computing. In: Proceedings of FPL-2016, pp. 1–12 (2016)
9. Saravanakumaran, B., Joseph, M.: Survey on optimization techniques in high level synthesis. In: Proceedings of CCSIT-2017, pp. 11–21 (2017)
10. The Unified Modeling Language, 13 November 2019. https://www.uml-diagrams.org/
11. Chuan, C., et al.: A survey of SQL language. J. Database Manag. (JDM) **4**(4), 4–16 (2019)
12. Seo, Y.: High-level hardware design of digital comparator with multiple inputs. Integration **68**, 157–165 (2019)
13. Veriloggen, 13 November 2019. https://github.com/PyHDI/veriloggen
14. Bachrach, J., et al.: Chisel: constructing hardware in a scala embedded language. In: Proceedings of DAC-2012, pp. 1212–1221 (2012)
15. Bellows, P., Hutchings, B.: JHDL - an HDL for reconfigurable systems. In: Proceedings of IEEE FCCM-1998, pp. 175–184 (1998)
16. Parthenon home page, 13 November 2019. http://www.parthenon-society.com/archive/NTT/
17. Jocelyn, S., François, B., Sameer, A.: Implementing stream-processing applications on FPGAs: a DSL-based approach. In: Proceedings of FPL-2011, pp. 130–137 (2011)
18. Sano, K.: DSL-based design space exploration for temporal and spatial parallelism of custom stream computing. In: Proceedings of FSP-2015, pp. 29–34 (2015)
19. Kuon, I., et al.: FPGA architecture: survey and challenges. Found. Trends Econometrics **2**(2), 135–253 (2007)
20. Chen, Z., Su, A., Sun, M.: Resource-efficient FPGA architecture and implementation of Hough transform. IEEE Trans. Very Large Scale Integr. (VLSI) Syst. **20**(8), 1419–1428 (2012)
21. Matsakis, N., Klock II, F.: The rust language. In: ACM SIGAda HILT-2014, vol. 34, no. 3, pp. 103–104 (2014)
22. The rust programming language, 13 November 2019. https://www.rust-lang.org/
23. Verugent, 13 November 2019. https://github.com/RuSys/Verugent/
24. Oda, T., et al.: Analysis of node placement in wireless mesh networks using Friedman test: a comparison study for Tabu Search and Hill Climbing. In: Proceedings of IMIS-2015, pp. 133–140 (2015)

Usage-Oriented Resource Allocation Strategy in Edge Computing Environments

Tsu-Hao Hsieh, Kuan-Yu Ho, Meng-Yo Tsai, and Kuan-Chou Lai[✉]

Department of Computer Science, National Taichung University of Education,
No. 140, Minsheng Rd., West Dist., Taichung City 40306, Taiwan
kclai@mail.ntcu.edu.tw

Abstract. Edge-computing using distributed computing architecture has solved the problem of massive data transmitting. By letting multiple nodes being computed locally and upload afterward to clouds, data processing gained great progress efficiently. Edge-computing involved virtualizing techniques, but due to the heaviness of virtualization, people prefer another lighter technique, the Container. Furthermore, while any node is installed with dockers, the node can normally execute containerized applications. This article simulates the edge-computing environment through Kubernetes. By adding a new predicate strategy to ensure the efficiency of job allocating, two built-in algorithms in the Kubernetes are compared with the proposed approach. Experimental results show the performance improvement after adopting the proposed approach.

Keywords: Edge computing · Kubernetes · Container · Resource allocation

1 Introduction

The distributed computing structure of edge computing has solved the cost problem associated with the massive data transmission in cloud computing. In edge computing, a group of nodes performs computations to accomplish a common goal by first dividing the data, which are to be subjected to considerable computing, into multiple segments. Next, computations are performed separately at each node, and the results are uploaded to the cloud. This computing structure not only increases the speed of data processing but also enables analysis and clear planning when designing an algorithm. Images in a container are smaller than those in a virtual machine, which renders image management and the activation and closing of applications faster and more convenient. Through the Docker container engine, container management and operation processes are greatly simplified. In the case of a mass container operation, orchestration tools such as Kubernetes, Docker Swarm, and Marathon can be used. After examining Kubernetes' structure and its scheduler execution process, this study developed a distributive computing structure for containers according to the container management and dispatches used by Kubernetes systems.

Kubernetes is an open source engine developed by Google for automating deployment, scaling, and management of containerized applications. The components of Kubernetes are divided into "masters" and "nodes." A master (master node) mainly provides resource scheduling for clusters, manages application deployment, and serves

© Springer Nature Switzerland AG 2020
L. Barolli et al. (Eds.): EIDWT 2020, LNDECT 47, pp. 351–360, 2020.
https://doi.org/10.1007/978-3-030-39746-3_37

as an API server. A node (worker node) is responsible for executing a container. A "pod" is the smallest deployable unit in Kubernetes and comprises a group of one or more containers, and containers within a pod share an IP address and port space. "Scheduler," a module in Kubernetes, is used to deploy pods, and the mission of a scheduler is to filter suitable nodes for pods that need to be deployed. If a suitable pod is identified for a node, the information of the pod is bound to the node. Unmatched pods stay in the pending phase until a suitable node appears in the cluster, and the scheduler schedules the pod to that node.

Figure 1 depicts the process from the creation of a pod to the deployment of a pod to a node. First, in Step 1, the API server REST API creates a pod, and then the API server writes the received data into the ETCD. In Step 2, the kube-scheduler continuously monitors for changes in resources. After a new pod that has not been bound to any node appears, the kube-scheduler uses a series of complicated scheduling strategies to choose a suitable node to be deployed. In Step 3, the kube-scheduler binds the pod with the chosen node and updates the bound data to the ETCD. In Step 4, once the kubelet in the chosen node detects that a new pod has been scheduled to be deployed to the node, it transmits the pod's related data to the Docker to execute that pod. In Step 5, the kubelet obtains information regarding the state of the pod through container runtime, updates related information in the API server, and writes the data into the ETCD. The kube-proxy is responsible for managing visits to services, including visits from pods to services in the cluster. The aforesaid steps describe the user's process, spanning from the creation of a new pod through the API server REST API to the deployment of the pod to a selected node through a scheduler.

In this paper, a heterogeneous container orchestration tool for processing resources is introduced. We use Kubernetes as the foundation and improve performance using the proposed algorithm. The algorithm performs rating and judgement according to the computing resource required for the task and the total computing resources the machine currently has. The purpose of the algorithm is to increase the utilization rate of nodes and reduce the average waiting time. The "complement" approach entails putting together the parts of the task that require more resources and the parts that requires fewer resources. Accordingly, resources are used more effectively, reducing waste. For example, a task that uses more CPU but less memory and the task that uses less CPU but more memory are put together to reduce waste in resource usage.

Section 2 presents a literature review. Section 3 introduces the Kubernetes algorithm process and introduces the proposed algorithm. Section 4 presents the experimental results and analysis. Section 5 is the conclusion.

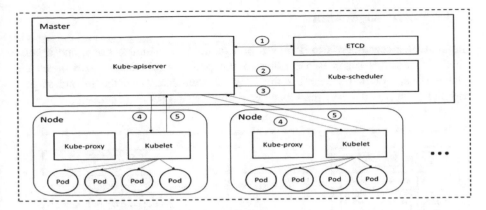

Fig. 1. Generation flow chart of Pods

2 Related Work

Edge computing is a concept developed from cloud computing. Data computation is performed near to the location of resources, thereby reducing the cost of data transmission to the cloud. Although the computing performance of machines in edge computing is poorer than that of cloud computing, edge computing is advantageous for the large quantity of data involved, dense data distribution, and high expansibility. Another advantage of edge computing is that when one machine malfunctions or is not connected to the Internet, related tasks can be redeployed using other machines near the same cluster to increase the stability of the whole system. In [1], the concept of edge computing was introduced and used to design an edge computing framework to process multitenant user applications. In [2], the applications of mobile edge computing in the Internet of Things were mentioned, and an edge computing structure applicable to large-scale tasks was designed by analyzing the structure of mobile crowdsensing. [3] discussed the prevalence of the Internet of Things, which has resulted in an increase in the numbers of devices and the amount of data. The programming framework that enables users to decide how to process data streams according to the data content and locations was proposed. In addition, the transmission and transfer between edge computing machines were described. The algorithm in this paper was designed on the basis of these references; moreover, the method used for designing an algorithm to deploy containers is called FreeContainer, which is further discussed in [4]. This algorithm increases the overall throughput by up to 90% and lowers the cost of communication. To understand orchestration tools, articles [5–7] on Docker and Docker Swarm, apart from Kubernetes, were studied. [5] presented basic details of Docker and Docker Swarm, and [6] introduced the three current basic scheduling strategies: spread, binpack, and random. [7] discovered, through conducting tests on I/O, CPU, and GPU, that the cost of running data-intensive computing on CPU or GPU is quite low. This result indicated that Docker containers can be applied to deep learning applications. After acquiring a thorough understanding of orchestration tools and Docker, the new algorithm was designed.

3 Proposed Algorithm

The goal of a scheduler is to deploy a pod to the most suitable node, and a kube-scheduler performs searches according to the pod's requirements. Kubernetes dispatches consist of two stages: the first stage is Predicates, and the second stage is Priorities. Figure 2 displays the scheduler's filtering and scoring process.

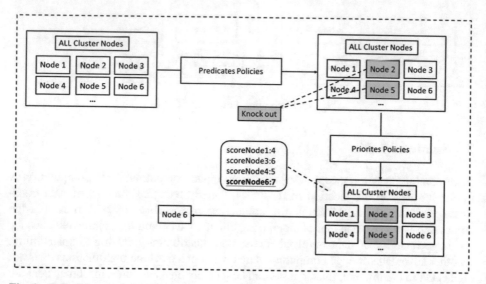

Fig. 2. Scheduler module filters and scores the node to show which node is optimal to be deployed with a Pod.

In Predicates, a kube-scheduler first executes a series of predicate functions and checks whether the conditions set by the user are met. For example, whether a pod matches the Kubernetes kit on a node is determined. If the node does not qualify, it is filtered and the scheduler enters the next stage. In this process, there is no association between data, so screening for qualified nodes using multiple filter conditions set by the user are considered independent tasks. Therefore, new filter modes can be expanded with flexibility.

In the Priorities stage, a kube-scheduler rates all the nodes that have passed the Predicates stage according to the conditions set by the user and then chooses the highest scoring node. After that, the scheduler binds it to the pod and then runs the pod.

Compared with the built-in algorithm of Kubernetes, the proposed algorithm focuses more on balancing CPU load. Under the heterogeneous conditions of each machine, when a new task pod is created, the algorithm assesses it to determine to which machine it should be deployed to be run. Table 1 lists the symbols used in this paper and their definitions. Formula (1) is the Predicates algorithm created in this study. This algorithm can judge task placement, and resources can be used more effectively by comparing with other nodes.

The computing process is shown in Fig. 3. Under the circumstances where a certain amount of known tasks are to be completed, we first run all the tasks once in every node to determine the required CPU amount, memory, and run time of all the tasks at different nodes. Next, the required amount of CPU and memory of all the tasks are added and then divided by the number of total tasks, yielding an average number of CPU and memory to be used by the tasks. After obtaining the two required parameters, the tasks can be placed in two-dimensional quadrants and matched according to the number of required CPU and memory. As shown in Fig. 4, the tasks in the first and third quadrant are matched, and the tasks in the second and fourth quadrant are matched. After matching, the combination that uses the most resources is placed in the matching sequence to determine whether a node with enough resources exists for the whole set of tasks to be placed in it. After confirming the nodes suitable for placement, one of the tasks is assessed by Formula (1) to determine which node is suitable for it to be placed in. After placing said task, the remaining tasks are also placed in that node. The total run time of the whole set of tasks in that node is ranked and then compared with the run times in other nodes. If the ranking is below average, the node is eliminated and the search for the next node continues.

Table 1. Key symbols and their definition.

Symbol	Definition
$CPU_{capacity}$	The origin CPU resources in the node
$CPU_{avaliable}$	The available CPU resources in the node
$CPU_{requested}$	The CPU resources required by the new pod
$Memory_{capacity}$	The origin memory resources in the node
$Memory_{avaliable}$	The available memory resources in the node
$Memory_{requested}$	The memory resources required by the new pod

ResourceComplement

$$= \frac{\dfrac{(CPU_{avaliable} - CPU_{requested}) * 10}{CPU_{capacity}} + \dfrac{(Memory_{avaliabe} - Memory_{requested}) * 10}{Memory_{capacity}}}{2} \quad (1)$$

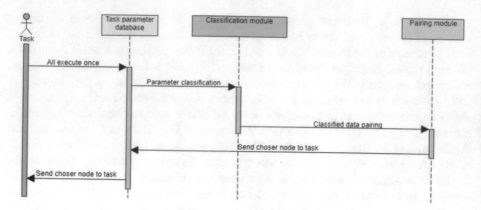

Fig. 3. Algorithm flow chart

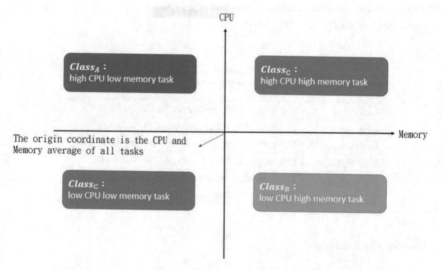

Fig. 4. Take the average and then classify via their property

4   Evaluation

In this paper, an edge computing environment was set up with four HP servers. The specifications of the machine in the cluster are presented in Table 2. One HP ProLiant served as the master in Kubernetes, and three virtual machines opened by other HP ProLiant served as the slave nodes in Kubernetes using different IPs. All machines used Ubuntu 16.04 as the operating system. Kubernetes 1.13.3 and Docker were used as the environment. The control group used the LeastRequestedPriority algorithm in Kubernetes. LeastRequestedPriority uses the formulas in (2) to grade the ratio of usable resources of a node and the resources required for the tasks.

$$LeastRequestedPriority = \frac{\frac{\left(CPU_{avaliable} - CPU_{requested}\right)*10}{CPU_{avaliable}} + \frac{\left(Memory_{avaliable} - Memory_{requested}\right)*10}{Memory_{avaliable}}}{2} \quad (2)$$

Table 2. Hardware specification

Devices	CPUs	Memory (GB)
HP ProLiant DL360 G6	Intel Xeon E5520*16@2.27 GHz	36 GB

Table 3. Virtual machine configuration in the experiment

Virtual machines	CPUs	Memory (GB)
VM slave A	14	15
VM slave B	16	16
VM slave C	17	15

In this paper, ten tasks were applied to run the simulation process, and the amounts of CPU and memory required for the tasks are shown in Fig. 5. No preruns were conducted for the run time for each machine.

Figures 6 and 7 depict the processes of task assignment in scheduling for LeastRequestedPriority and the proposed algorithm. The assignment conditions of Machine A, Machine B, and machine C after the first assignment at 0 s are shown in Fig. 6. For Machine A, the remaining CPU resource was 0, and the remaining memory resource was 10. For Machine B, the remaining CPU resource was 5, and the remaining memory resource was 6. For Machine C, the remaining CPU resource was 6, and the remaining memory resource was 5. The assigned tasks are shown in the upper part of Fig. 6, and numbers in the parentheses indicate the required CPU and memory resources; the assigned machines and total required time (in minutes) are shown in the back.

LeastRequestedPriority took 19 min to finish all the tasks, and the new algorithm required only 12 min. LeastRequestedPriority executed the tasks by the order of entrance when scheduling; however, the new algorithm sorted the tasks according to the execution time before scheduling, which reduced the total time needed to finish all the tasks.

In addition, the differences in resource usage at the first assignment are shown in Figs. 8 and 9. The figures show the resource usage of each machine after the first assignment at 0 s. The same colors indicate the ratios of the occupied resources of the same tasks in the machine, and the black parts indicate the unused and wasted resources. For LeastRequestedPriority, after the first assignment when no space was available for processing another task, the CPU usage was 100% and the memory usage was 33% for Machine A. The CPU usage was 69% and memory usage was 63% for Machine B. The CPU usage was 65% and the memory usage was 67% for Machine C. The average resource usage was 68%. As for the proposed algorithm, the CPU usage was 86% and the memory usage was 80% for Machine A. The CPU usage was 75% and the memory usage was 75% for Machine B. The CPU usage was 65% and the memory usage was 73% for Machine C. The average resource usage was 76%. The resource usage in the first assignment suggests that the proposed algorithm can increase the efficiency of machine resource usage and reduce resource waste.

task order	Required CPU	Required Memory	Execution time on A	Execution time on B	Execution time on C
1	4	7	8	6	9
2	2	8	5	7	6
3	8	3	5	9	8
4	9	2	7	6	4
5	7	3	3	5	5
6	6	2	4	2	6
7	9	3	6	7	8
8	2	9	4	5	7
9	3	9	6	7	9
10	10	4	14	6	10

Fig. 5. Add in the required CPU and memory for the tasks

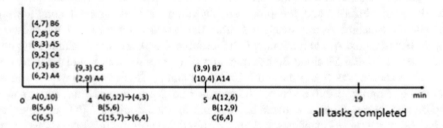

Fig. 6. Scheduling process of LeastRequested Priority

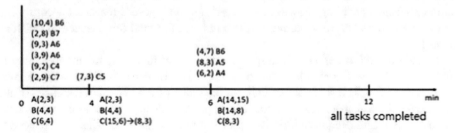

Fig. 7. Scheduling process of ResourceComplement Priority

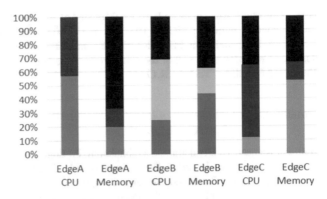

Fig. 8. First allocation resource usage of LeastRequestedPriority

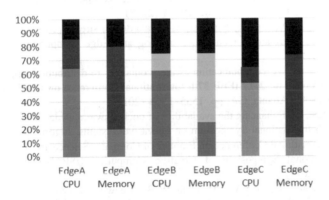

Fig. 9. First allocation resource usage of ResourceComplement

5   Conclusions

When the data increase in volume and become more complex, the centralized structure of conventional cloud computing gradually fails to respond. This article mainly discusses the simulation of a heterogeneous environment in edge computing using the Kubernetes structure and proposed a new algorithm with CPU and memory as parameters. The experimental results indicated that the CPU and memory required for each task were determined using pre-run and were utilized for scheduling in the algorithm introduced in this paper. Figures 6 and 7 display the scheduling processes and indicate that the new algorithm required less time to finish all tasks than did LeastRequestedPriority. Figures 8 and 9 show that when executing the assignment of the first task, the utilization rate of the new algorithm was higher than that of LeastRequestedPriority. The resource waste was thus reduced, and resources were utilized more efficiently. Accordingly, the proposed algorithm can improve resource utilization by using a complement approach to reduce the total time required for task completion.

that he infects a vulnerable IoT device, for instance, by exploiting the default password or an unnecessary open port of the device, with his malware and alters it as an agent machine called "bot" that performs DDoS attack. After that, he sends a command to the bots in order to attack a targeted system. Because he spreads the malware and infects many IoT devices with it, a large number of bots attack the targeted system simultaneously according to his commands. As a result, the targeted system cannot maintain their service due to the overload caused by the simultaneous attack from the bots. IoT DDoS attack is frequently used to obstruct an online service such as website, e-commerce service, SNS service, etc. As reported by Krebs, a famous IoT malware "Mirai" [7] attacked his website causing network traffic exceeding 600 Gbps [8]. Since IoT DDoS attack is causing such large-scale security incidents, it is one of critical cyberattacks that we need to take countermeasures.

Fig. 1. An example scene of the proposed method of network packets visualization.

One of the effective solutions of IoT DDoS attack is to reduce vulnerable IoT devices which hold potentiality to be altered to bots. However, it is difficult to make an environment where IoT devices are not infected with any malware. Although security "prevention" from the infection is of course important, security "response" in the case that IoT devices are infected is also important. If we notice our IoT device is infected with an IoT DDoS malware, we can take several countermeasures, e.g. to disconnect the device from Internet, to turn off the device, to restore the device as factory default state, to take the device to a network specialist, etc. However, one of sticky problems of IoT DDoS attack is that it is difficult to notice that our IoT device is infected and altered to a bot. Unlike general personal computers, most IoT devices do not have their own monitoring device, thus, it becomes hard to inspect the status of the device. As the result, a lot of IoT devices are left running as bots. To solve the situation, we propose a method for us to easily notice that our IoT devices are altered to bots that perform a DDoS attack. To achieve the purpose, we selected a method to visualize network packets. Figure 1 is an example scene of our proposed visualization method. The proposed visualization method provides flow of network packets between a device to device by using mixed reality technology. The reason why we adopted the two key technologies of visualization and mixed reality is as following. The visualization technology supports for us to easily notice the unusual situation by seeing the visualized scene that our IoT devices are spreading a large number of packets and the mixed reality technology supports for us to easily find which device is spreading the network packets by merging real-world and virtual-world.

To evaluate whether our visualization method achieves the research purpose, we set the following three metrices of M1, M2 and M3. The M1, M2 and M3 are metrics to evaluate for "distinction", "discoverability" and "immediacy", respectively. This paper explains the details of the visualization method and also shows the evaluation results.

(M1) To clearly realize unusual situation that DDoS attack is happening
(M2) To easily find which device is performing as a bot of DDoS attack
(M3) To immediately notice the beginning of DDoS attack in real-time

The remaining of this paper is as follows. The next section introduces several related works about network visualization. Section 3 explains the method and implementation of the proposed visualization system. Section 4 shows several evaluation results of the proposed visualization system. Lastly, we conclude this research and also explain the remaining tasks of this research.

2   Related Works

Since network communication is invisible, there are strong needs to visualize it in order to comprehend the communication situation. Shiravi et al. well-surveyed network visualization systems [12]. Also, they classified network visualization methods into several categories in consideration with its use-case. According to their classification, our proposed visualization method would be in the category of Internal/External monitoring. Although they introduced several visualization systems of this category, our proposed method is different from them and provides another new viewpoint to the network visualization field.

Most common motivation of network visualization would be for security protection, as is our motivation too. As visualization systems for real-time security thread in global scale, there are famous ones; CYBERTHREAT REAL-TIME MAP [6], Digital Attack Map [4] and Fortinet Threat Map [3]. The systems take a common visualization method to describe trajectories of cyberattacks from a source point to a destination point. Although our proposed visualization method follows this manner and takes a similar expression for the visualization, our research target is different to their works. Our visualization method targets security tread in local scale, e.g. smart home. As a visualization system in the local scale, NIRVANA [14] is well-designed visualization system to observe security threat. Although our proposed visualization method has similar concept of monitoring security thread in local scale, the mechanism to find the suspicious computer is totally different of the visualization system.

To find the suspicious computer, mixed reality technology supports our proposed visualization method. As network visualization systems using mixed reality technology, there are Magic lens [9] and the visualization system developed by Beitzel et al. [1]. Although our proposed visualization method uses the same technology, the purpose and expression for visitation are different of them.

3   Method and Implementation

In this section, we explain our proposed visualization method and also introduce a visualization system implementing the method. Figure 1 shows an example scene seen through the visualization system. We named the visualization system as PACK-UARIUM. The name comes from its landscape that the visualization system provides in its mixed reality environment. In the environment, we can see a beautiful scenery that visualized network packets are drifting in the air like fishes in an aquarium. The scenery is like an aquarium for network packets. For that reason, we made the name of PACKUARIUM by combining "packet" and "aquarium". The following subsections provide the method and implementation details.

3.1   System Architecture

To visualize the network packets, we need to prepare a packet capturing system and a data visualization system. System architecture of the proposed visualization method is described in Fig. 2. Because we supposed that this visualization system is used in smart home with some IoT devices settled, the basic components of the system are designed to run on local area network (LAN). The basic components are the followings; "Router", "IoT Device" and "Smart Glasses" denoted in Fig. 2. The Router has a role to transfer network packets sent from all IoT devices connected in the LAN as the gateway between the LAN and the Internet. The Router runs two applications as its own processes apart from its main role of the network packets routing. One is the capture application (CA in Fig. 2) which captures network packets and stores the packet information into the database (DB in Fig. 2). Since all network packets to access a computer outside of the LAN require to pass through the Router (because of the gateway), the CA is able to capture all network packets sent from the IoT Device. The other one of the applications is the web application (WA in Fig. 2) to serve the packet information to a client computer who requests to get them. Another component of IoT Device is connected to the Router and running its own roles. The IoT device is supposed to be communicating a cloud server in the Internet as a kind of sensor device. Therefore, the IoT Device is supposed to regularly send network packets to the server. Although Fig. 2 describes only one IoT Device as a component of the system, there is no problem even if some IoT devices are in the LAN within the range of connectable units of the Router's configuration. The other component is Smart Glasses in Fig. 2. The Smart Glasses is also connected to the Router and runs the viewer application (VA in Fig. 2) for visualizing network packets. The process for visualizing network packets is denoted in Fig. 2 as step (1x) and (2x). The step (1a) means that the IoT Device is communicating to a cloud server through the gateway (Router). The dot line expresses a route of network packets for the communication. As aforementioned, all the packet information are stored into the DB by the CA (1b). The step (2a) means that the VA accesses to the WA to get the packet information. Then, the WA pulls the records of the packet information from the DB (2b) and sends them to the VA (2c). Eventually, the VA visualizes the network packets according to the packet information.

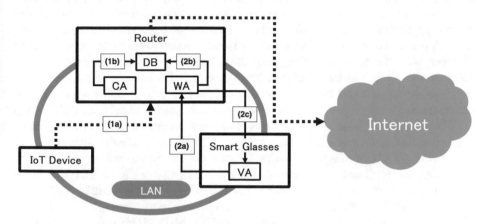

Fig. 2. System architecture of the proposed visualization method.

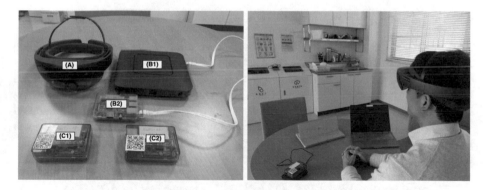

Fig. 3. Devices for the proposed visualization system and usage scene of the system.

We implemented PACKUARIUM, the proposed visualization system, according to the system architecture. The left image of the Fig. 3 is the actual devices to execute PACKUARIUM. The (A) in the image is a visualization display device which corresponds to the Smart Glasses in Fig. 2. We adopted Microsoft HoloLens, a kind of smart glasses with holographic functions, as the Smart Glasses. We implemented the VA running on the HoloLens by using Unity, a software development environment called game engine. The VA regularly requests to get packet information stored in the DB by polling process of every second. When the VA sends the request to the WA, the WA responses to send packet information that not visualized yet. The (B1) and (B2) correspond to the Router in Fig. 2. The (B1) is a network router of a commercial product of BUFFALO INC., and is the gateway device to connect the LAN with the Internet. The (B2) is also a network router that makes the LAN for communication among devices in the LAN. The reason why we prepared the two router devices is that the proposed system needs to run the CA and WA in a network router. However, we could not run them in the commercial product because the product was not designed for users to be able to customize. For that reason, we added one more network router for running the two applications. We adopted Raspberry Pi 3 B+, a single-board computer, as the network router. To alter the single-board computer to a network router, we adopted the two softwares; "hostapd" and "Dnsmasq". hostapd can append functions as an access point to a network interface. Because Raspberry Pi 3 B+ has one wireless network interface and one ethernet interface, we assigned the functions to the wireless network interface and practiced port forwarding with network address translation (NAT) from the wireless network interface to the ethernet interface for transferring network packets outside of the LAN. Dnsmasq is charge of the DHCP server to provide IP addresses to devices in the LAN. We implemented the CA and WA by Python and adopted MySQL as the DB. The (C1) and (C2) correspond to the IoT Device in Fig. 2. The devices are communicating with each other in the LAN and also communicating with other computers outside of the LAN according to several programs running on the devices. The right image of Fig. 3 is a usage scene of PACKUARIUM. He wears HoloLens on his head and is seeing visualized images of network packets transferred from the device on the trash box in front of him to the device on the table he is on.

3.2 Network Packets Capture

As aforementioned, the CA has the roll to capture network packets and to store the packet information into the DB. To detect DDoS attack from the information, we require, at least, packet information of source IP address, destination IP address, source port number, destination port number and time when the network packet was generated. Although there are several methods to capture network packets, we adopted "tcpdump" as the capturing tool. tcpdump is a famous packet analyzer that can be run with a command line interface. Although tcpdump has many functions, PACKUARIUM used only the following command (1).

$$tcpdump \; -i \, NETWORK\_INTERFACE \; -nn \, src \, host \, IP\_ADDRESS \qquad (1)$$

The i option is for selecting a network interface to capture network packets. The nn option means not to resolve hostnames or ports. The $src \, host$ is a kind of filter to dump only information of network packets which matches the condition. The command (1) filters condition of a source IP address. Since the purpose of PACKUARIUM is to detect a bot performing DDoS attack, we need to identify the source point that is generating a large number of network packets. By executing the command (1), tcpdump outputs the following packet information.

$$TIME\ PROTOCOL\ SOURCE\_IP\_ADDRESS.SOURCE\_PORT >$$
$$DESTINATION\_IP\_ADDRESS.DESTINATION\_PORT: \qquad (2)$$
$$FLAGS,\ COMMUNICATION\_INFORMATION$$

Before to store the packet information, the CA filters them according to the following two conditions. One of the conditions is whether the *PROTOCOL* is IP or not. In other words, the condition means to remove ARP (Address Resolution Protocol) packets. ARP packets are for resolving the link layer address, i.e. MAC address, and are frequently transferred in LAN. Also, the packets are not related to DDoS attack. If the system does not remove ARP packet information, we would be confused with a large number of network packets generated from ARP and cannot realize situation of DDoS attack. The other condition is whether the *FLAGS* is [.] or not. In other words, the condition mans to remove packet information of SYN, PUSH, FIN, RTS. These packets are related to handshake communication. On TCP/IP communication, the handshake communication frequently happens. Therefore, if the system visualizes the packet information, we would be difficult to realize DDoS attack because a large number of network packets are always visualized. After the filtering, the CA stores the information of *TIME, SOURCE_IP_ADDRESS, SOURCE_PORT, DESTINATION_ IP_ADDRESS and DESTINATION_PORT* into the DB.

3.3 Visualization Method

Since the role of PACKUARIUM is for a user to realize whether DDoS attack is happening or not, we required to consider effective design to clearly express the difference between usual communication and DDoS attack communication by using appropriate visualization method. One of remarkable difference of the two situation is the number of network packets generated from communication. Therefore, we designed to directly express the number of network packets as objects of computer graphics (CG). Namely, one record of packet information corresponds to one CG object. We expressed the object as arrow shape for clearly understanding the direction of packets movement and designed the arrow shape to move on a trajectory connecting between a source device and a destination device. The design of packet movement was inspired by the visualization method of NICTER, a security threat analyzing system [11]. The trajectory of packets movement is drawn by using Bézier curve. Each trajectory of packet movement has its own control points of Bézier curve. PACKUARIUM had 50 control points on each trajectory. Also, to avoid the situation to overlap the arrow shapes, the control point is randomly shifted around the basic trajectory. PACK-UARIUM randomly shifted trajectories along with X, Y and Z axis in range from −1.0 m–1.0 m, 1.0 m–2.0 m and −1.0 m–1.0 m, respectively. In addition to that, we colored the arrow shapes with transparency in consideration with mixed reality environment. By seeing background in real-world through the arrow shapes, we can feel as if we are in a fantastic world where the real-world and virtual-world are mixed. Since PACKUARIUM is supposed to be used in smart home, the beautiful scenery of packets movement may be also suitable as an installation art at home.

3.4 Device Location Identification

To provide mix reality environment, we need to accurately overlap CG objects of network packets on an IoT device that generates actual network packets. Therefore, we required to identify IoT devices and to tie the locations of each device in real-world with locations in virtual-world. There are several methods to identify an object in real-world. As the communication visualization method of Mayer et al., they identified devices by recognizing feature points of them [9]. This method is convenient from the viewpoint of preparation that we need not to set a kind of marker to identify the object. However, in a situation of smart home, we may have the same shape of IoT device. For example, we have several air conditioners of the same type in different rooms. This situation may happen other IoT devices such as TV monitor, light or something wearable. For that reason, we adopted QR code to identify IoT devices. Although the method is inconvenient and may spoil interior design, it would be one of realistic solutions at the moment. We used ZXing.Net as QR code scan library. The procedure to tie a device location in real-world with a location in virtual-world is that we watch an IoT device and the VA scans a QR code attached with the device. The VA tied the location of the scanned device in real-world with the location where certain distance away from forward direction of the virtual-world camera (viewpoint in virtual-world). PACKUARIUM locates the position 30 cm away from the forward distance of the camera. The distance value is just heuristic one. By embedded each IP address of IoT devices into each QR code, the VA can tie the IP address and the device location.

4 Experiment and Evaluation

4.1 Experiment Environment

To evaluate the proposed visualization system, we conducted several experiments. In the experiments, we executed pseudo DDoS attack. The source and target device were the (C1) and (C2) in Fig. 3, respectively. If the (C1) actually altered to a bot and executed DDoS attack to a cloud server located outside the LAN, the system would visualize network packets moving on the trajectory from the source device (C1) to the network router (B2). For the reason why we clearly denote understandable images, in this experiment, we executed the DDoS attack from the (C1) to (C2). This substitution of the destination target of the (B2) to the (C2) does not affect to visualization results itself. Also, in this experiment, we used HTTP communication for the pseudo DDoS attack.

4.2 Experiment Results and Evaluation

Since the purpose of PACKARIUM is to detect a bot IoT Device that is performing DDoS attack, we evaluated the visualization system according to the metrics of M1, M2 and M3 in Sect. 1.

4.2.1 Evaluation for Distinction

The content of M1 is whether we can clearly realize unusual situation that DDoS attack is happening. In other words, this metric evaluates for "distinction" whether a user can distinguish the difference between usual communication and unusual communication of DDoS attack. The evaluation result is related to whether the expression of the proposed visualization method is appropriate or not. Figure 4 shows an example scene that visualizes network packets of usual HTTP communication between the left-side device to the right-side device. We used cURL, a command line tool that we can communicate by using various network protocol, for the HTTL communication. The visualized result was generated from a situation that the left-side device accessed only one time to the right-side device by using HTTP communication. There are only a few visualized network packets in the air. On the other hand, as we already introduced before, Fig. 1 shows an example scene that visualizes network packets that emulated a situation of DDoS attack by repeatedly accessing between them with HTTP communication for every 0.1 s. Since the visualization results of the two situations are clearly difference, almost users can clearly realize the unusual situation that DDoS attack is happening.

Fig. 4. Example scene of visualizing network packets of usual HTTP communication.

4.2.2 Evaluation for Discoverability

The content of M2 is whether we can easily find which device is performing as a bot of DDoS attack. In other words, this metric evaluates for "discoverability" whether a user can find the infected device in real world or not. The evaluation result is related to whether the proposed method provides appropriate expression of CG overlay by mixing a location in real-world and a location in virtual-world. Figure 5 shows example scenes of the proposed visualization in different cases. The (1-a) and (1-b) are example scenes of network packets visualization in the same situation of Fig. 1, which were captured from different viewpoints. Differ from a naive augmented reality technology to overlay CG on a camera image, the proposed visualization method overlays the CG objects on the actual location in real-world. Therefore, even when we change viewpoint, the source point which generates visualized packet objects is not shifted. The (2-a) and (2-b) are example scenes of network packets visualization in an experiment we supposed the case of a wearable IoT device. The (2-a) is supposed a situation that the wearable IoT device (the right-side device in the hand) is the source of DDoS attack. Because the VA always keeps running the program to scan QR cedes, the system can identify location of the device in real-world even if we move the device to another place. On the other hand, the (2-b) is another situation that the wearable IoT device (the left-side device in the hand) is targeted as the destination of DDoS attack. Although the purpose of the proposed visualization method is to detect the source of DDoS attack, the method is also able to detect the situation that our IoT device is targeted of DDoS attack. Since the proposed visualization method accurately identifies a device location and provides well-mixed GG overlay, almost users can easily find which device performing as a bot of DDoS attack.

Fig. 5. Example scenes of the proposed visualization captured from different viewpoints.

4.2.3   Evaluation for Immediacy

The content of M3 is whether we can immediately notice the beginning of DDoS attack in real-time. In other words, this metric evaluates for "immediacy" of visualized results. Whenever we use network communication, time-delay is definitely occurred. To immediately detect the beginning of DDoS attack, we need to reduce the time-delay between the actual beginning time of DDoS attack and the beginning time of the visualization. Figure 6 denotes experimental results about the time-delay. The (A-1) and (A-2) show time-delay in the case that network packets were sent every 0.1 s. The (A-1) shows the difference between the beginning time of network packets transferred (the blue line) and the beginning time of the packet information is visualized (the orange line). The horizontal axis in the figure is count of step when we generated a large number of network packets. In the experiment, as we already mentioned above, we regularly sent network packets (every 0.1 s). The step means the timing when we started communication. The vertical axis in the figure means timestamp of Unix time (only showing the last two digits). The space between the blue line and orange line represents the time-delay. The (A-2) shows the details of the time-delay. The delayed time in each step is plotted in the figure. The horizontal axis in the figure is the same value of the step in (A-1). The vertical axis in the figure means time (sec.). The delayed time is in range from 2.0 to 8.0 s. The average and variance of the delayed time are 5.07 and 0.87, respectively. However, according to the number of steps increases, the delayed time also gradually increases. This phenomenon seemed to be caused from the performance of the network router (Raspberry Pi). Because a large number of network packets were transferred to the network router when we executed the DDoS attack, the network router device seemed gradually become overloaded. To proof the evidence, we had another experiment of DDoS attack in which we sent network packets every 1.0 s (1/10 frequency of the first experiment). The results are (B-1) and (B-2). The (B-1) seems almost same situation of (A-1). On the other hand, the (B-2) seems different result of the (A-2). The delayed time did not increase even when the number of steps increased. The delayed time is keeping controlled the range from 1.0 to 8.0 s. The average and variance of the delayed time are 4.32 and 1.77, respectively. The result shows evidence of the hypothesis that the network router had become overload seems correct. Although the delayed time is not precisely real-time, the delayed time seems enough to notice the beginning of DDoS attack.

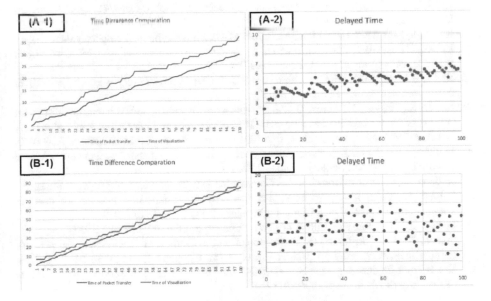

Fig. 6. Results of time-delay until visualizing network packets.

5 Conclusion

This paper explained a method of network packets visualization using mixed reality technology to detect an IoT device that has been altered to a bot performing DDoS attack. As a concrete system implemented the method, this paper introduced PACK-UARIUM. The visualization system was evaluated by several metrics. The evaluation results showed better performances in each metrics.

As a remaining task, we need to improve the method to identify an IoT device. In this research, we used QR code to identify it. However, this method is inconvenient because a user requires to prepare QR codes for each IoT device. We will try to discover a better method to do it.

Acknowledgement. This research was supported by Strategic International Research Cooperative Program, Japan Science and Technology Agency (JST) regarding "Security in the Internet of Things Space".

References

1. Beitzel, S., Dykstra, J., Toliver, P., Youzwak, J.: Exploring 3D cybersecurity visualization with the microsoft HoloLens. In: Nicholson, D. (ed.) Advances in Human Factors in Cybersecurity. AHFE 2017. Advances in Intelligent Systems and Computing, vol. 593, pp. 197–207. Springer, Cham (2018)
2. European Network and Information Security Agency: Major DDoS Attacks Involving IoT Devices (2016). https://www.enisa.europa.eu/publications/info-notes/major-ddos-attacks-involving-iot-devices
3. FORTINET: Fortinet Threat Map. https://threatmap.fortiguard.com/
4. Google and Arbor Networks: Digital Attack Map. http://www.digitalattackmap.com/
5. Gu, Q., Liu, P.: Denial of Service Attacks. Technical Report. http://s2.ist.psu.edu/paper/DDoS-Chap-Gu-June-07.pdf
6. Kaspersky: CYBERTHREAT REAL-TIME MAP. https://cybermap.kaspersky.com/
7. Kolias, C., Kambourakis, G., Stavrou, A., Voas, J.: DDoS in the IoT: Mirai and other Botnets. IEEE Comput. **50**(7), 80–84 (2017)
8. Krebs, B.: Krebsonsecurity hit with record DDoS (2016). https://krebsonsecurity.com/2016/09/krebsonsecurity-hit-with-record-ddos/
9. Mayer, S., Hassan, Y., Sörös, G.: Magic lenses for revealing device interactions in smart environments. In: SIGGRAPH Asia 2014 Mobile Graphics and Interactive Applications (SA 2014). ACM, New York Article 9, 6 p. (2014). https://doi.org/10.1145/2669062.2669077
10. McAfee Labs Threats Report, August 2019. https://www.mcafee.com/enterprise/en-us/assets/reports/rp-quarterly-threats-aug-2019.pdf
11. National Institute of Information and Communications Technology. https://www.nicter.jp
12. Shiravi, H., Shiravi, A., Ghorbani, A.: A survey of visualization systems for network security. IEEE Trans. Visual Comput. Graphics **18**(8), 1313–1329 (2012)
13. Sparks, P.: The route to a trillion devices, white paper of ARM Limited (2017)
14. Suzuki, K., Eto, M., Inoue, D.: Development and evaluation of NIRVANA: real network traffic visualization system. J. National Institute Inf. Commun. Technol. **58**, 61–77 (2011)
15. The Internet Engineering Task Force (RFC4732): Internet Denial-of-Service Considerations (2006). http://tools.ietf.org/html/rfc4732

A Handover Challenge of Data Analytics: Multi-user Issues in Sustainable Data Analytics

Toshihiko Yamakami[✉]

ACCESS, Tokyo, Japan
Toshihiko.Yamakami@access-company.com

Abstract. Data analytics is capturing attention from research and industries with the advances of algorithms, tools, platforms, GPUs, cloud computing and increased data availability. Increased penetration widens coverage, complexities, time frames of data analytics. These advances social and organizational challenges as well as technological ones. Such challenges include multi-user issues of data analytics. The author performs interview with data analysts to identify issues in a handover. The author categorizes those issues in order to build a framework to deal with a handover in data analytics. The author describes a dimensional model and its applications to deal with handover issues.

1 Introduction

Data Analytics is a concept related to pattern and relevant knowledge discovery from large amounts of data. Data analytics is increasing industrial applications. Data availability, computational resources, advances in tools and platforms, and research advances increase applicability of data analytics.

Increased penetration of data analytics extends opportunities of multi-user data analytics. Multi-user data analytics provides social and organizational challenges as well as technological ones.

Collaborative process of data analytics is studied according to the growth of size and complexity of data analytics. A time-dimensional aspect of data analytics is relatively unexplored. For example, a handover take place when a data analytics work is transferred from one person to another. It is a superficially simple act, but it involves multiple deep factors embedded in data analytics.

Data analytics incorporates deep learning these days. Data analytics incorporates data handling to deal with volume, velocity, variety and veracity, so called 4V. These emerging trends bring special expertises and analyst-specificity. This broken-into-silos status is challenged by an increasing demand of flexible work load assignment.

Emerging new platforms like SageMaker enables easier work piece transition. It also provides new challenges because it facilitates wider participation in the data analytics.

The author considers the challenges of handovers in the data analytics work. Then, the author discusses data analytics-specific factors of multi-user handovers.

© Springer Nature Switzerland AG 2020
L. Barolli et al. (Eds.): EIDWT 2020, LNDECT 47, pp. 373–383, 2020.
https://doi.org/10.1007/978-3-030-39746-3_39

2   Background

2.1   Purpose of Research

The aim of this research is to identify the framework of multi-user handover in data analytics.

2.2   Related Work

Research on multi-user data analytics include (a) collaborative data analytics, (b) data analytics in organizational contexts, and, (c) multi-user tools of data analytics.

First, in regards to collaborative data analytics, Tucker et al. discussed Tucker et al. discussed a framework for collaborative data analytics [8]. Banerjee et al. discussed a framework and system for collaborative data analytics for IoT [1]. Bhardwaj et al. discussed a central repository for data analytics by many individuals and teams [2].

These collaborative data analytics research focuses on synchronic collaboration with multiple roles.

Second, in regard to data analytics in organizational contexts, Khalifa et al. discussed a taxonomy of data analytics ecosystem to deal with organizational needs [5]. Toreini et al. discussed an attentive dashboard issuing visual feedback to deal with work resumption after interruption [7]. Schwee et al. discussed privacy protection in cross-organizational data analytics [6]. Callinan et al. discussed open data co-creation in a government open data [3].

Third, in regard to multi-user tools of data analytics, Horak et al. discussed a framework of visualization with heterogeneous cross-devices [4].

Even though the size and complexity of data analytics continue to increase, multi-user issues of data analytics are largely unexplored to the author's knowledge. The originality of this paper lies in its identification of handover challenge in data analytics in an organizational context in a time dimension.

2.3   Definition

The definitions are provided in Table 1.

Table 1. Definitions

Term	Definition
Handover in data analytics	A work transfer from one person to another person in data analytics
Multi-user issue	Issues arised when an activity is performed by more than one users

Usually, handover is used when a mobile phone in mobility switches base stations. It is not related in this paper.

3   Method

This study started with the following conversation between data analysts.

> Data Analyst A: I built a stand alone server. Does anyone want to use it?
> Data Analyst B: No, thank you. The environment which others built is difficult to use.

This conversation inspired the author to perform this study. There was no preceding examples of framework. Therefore, the author started from the unstructured interview.

The author performs the following steps:

- perform unstructured interview with data analysts on handover issues,
- identify categorize to analyze handover issues,
- identify a dimensional model to position handover issues,
- discuss applications from the proposed model.

4   Observation

4.1   Multi-user Issues

General multi-user issues are depicted in Table 2.

Table 2. General multi-user issues

Issue	Description
Intention	Different intentions of multiple users need harmonization or negotiation
Goal and rule sharing	When there is any collaboratory work, it need goal and rule sharing
Trust	Trust and distrust affect multi-user contexts
Role taking	When there are multiple roles, role taking impacts the multi-user interaction
Understanding of context	When multiple users are involved, understanding of context is not consistent and causes problems if there is any discrepancy
Token passing	When there is any exclusivity, token passing is needed as conversation

These deal with synchronic collaboration in most cases. Time dimensional collaboration, like a handover, needs to be addressed in the emerging demands of data analytics because it requires more flexible work load sharing.

4.2   Interview

Multi-user issues in data analytics were largely ignored in the past. There is an increasing incoming flow of data analysts into industries. Despite the increase of the number of data analytics in industrial landscapes, the size and complexities of big data drives more multi-user data analytics than before. This promotes the needs to deal with multi-user issues.

One of the multi-user issues is handover. There are 15 data scientists in our office. I performed interview with 3 senior data scientists on the handover issue. The question is "What are the issues when a person cannot complete a data analytics work for whatever reasons, holidays, sick leave, quitting or workload?" They were unstructured interviews and free format responses were recorded.

4.3   3 Dimensions of Handover Issues

From those, I categorized the issues in 3 dimensions depicted in Fig. 1.

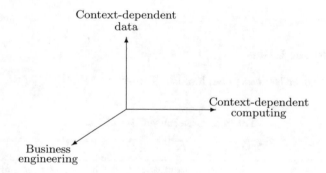

Fig. 1. 3 dimensional model

Issues in the business engineering dimension are depicted in Table 3.

Table 3. Issues in the Business engineering dimension

Item	Description
Goal	The goal of the specific data analytics work needs to be shared. It is preferable for a face-to-face meeting is held with the stakeholders
Intention	Intention of previous sequences of data analytics work needs to be shared
Unaccomplished work	A handover take place because there are some unaccomplished parts of work. These unaccomplished work should be shared with requirements and failed processes

All three data analysis emphasize the importance of intention sharing more than sharing issues in technical infrastructure. It is partly because it is common to use the same technical infrastructure in the same team. However, it also addresses the important aspect of data analytics. It is a common that the end result of data analytics is unpredictable. Data analytics is useful because it brings some unexpected results. This exploratory characteristic brings focus on the intention rather than procedures.

Issues in the context-dependent data dimension are depicted in Table 4.

Table 4. Issues in the Context-dependant data dimension

Item	Description
Code sequence	Code sequence should be recorded with the coder's intention. Reproduce-able code should be shared
Data design	Procedures and rationales for data structure and preprocessing methods should be shared
Data cleansing	Preceding data cleansing operation should be shared
Data characteristics	Data characteristics found in the previous work should be shared as knowhow to deal the datasets
Data dynamics	In some cases, data is constantly coming in. In those cases, data is not frozen. Data dynamics should be properly handled at a handover

During phases of gathering, preprocessing, analysis, and modeling, data are manipulated in a context dependent way. For example, data preprocessing is not a one time process. In processes following the data preprocessing, new insights on data gathering and data characteristics are found and feedbacked to the preprocessing to increase fidelity of data. Without understanding this total work sequences in details, it is difficult to continue.

Data does not stand as neutral or universal. No data is immune to processing from gathering to storing. Data is implicitly bound to the preceding procedures, explicit or implicit. Especially implicit context behind data is sometimes critical to be covered for successful handovers.

Issues in the context-dependent computing dimension are depicted in Table 5.

Deep learning, an important part of data analytics these days, is rapidly progressing. Open source software libraries are available and useful, but the complicated version dependency is a challenge, especially when a handover requires reproduction over a certain time span. This presents issues similar in large system integration. Environment management increases its importance and is a major issue in a handover process.

In some sense, there is new coming data and a data analyst constantly performs the whole process of data analytics. In that sense, a data analyst to succeed the work can totally renew the whole processing including the underlying technical infrastructure as far as the goals can be satisfied. This is a unique characteristic

Table 5. Issues in the Context-dependent computing dimension

Item	Description
Programming language	The programming language and its version should be shared
Libraries	The libraries, versions, version dependency, configuration for version dependency should be shared
Toolchains	Complicated combination of tools and services requires toolchain management. Automated process hides the important details and those details should be shared
Platform	Underlying platform for the previous sequences should be shared. Sharing of configuration, toolchains and executables for reproduction needs consideration
Data store	Choice and rationales of data store should be shared
Configuration	Each analyst has their specific environment configuration. For example, when environment variables capture this configuration, they should be shared. Configuration on the underlying technical infrastructure needs sharing
Taste	There are many ways to deal with data. Each person has one's own taste and its impact on data and processing. It is better to share understanding of taste to deal with data

of data analytics. When there is an upgrade of learning models or when there is a requirement to reduce costs, a data analyst may replace a part of data analysis process entirely, which usually does not happen in system integration.

4.4   Comparison

The author compares handover issues in data analytics with the following engineering

- system integration
- requirement engineering
- workflow tasks

Comparison between system integration and data analytics is depicted in Table 6.

Table 6. Comparison between system integration and data analytics

Domain	Characteristics
System integration	System integration explicitly defines requirement capturing. During requirement definition, a requirement document is produced to eliminate any face-to-face meeting with the After the requirement definition, it does not require direct interaction with customers. Software development proceeds step by step, approaching the end goal little by little
Data analytics	Data analytics is similar to the system integration without requirement engineering best practices. In addition, a part of data analytics, machine learning does not proceed step by step. In many cases, data analyst does not know in advance how to achieve the end learned model

These differences highlight the characteristics and challenges of data analytics in an IT corporative perspective. Data analytics need continuous data handling directly related to the real world. And it needs patience and uncertainty in the project management. These characteristics need to be considered in a handover.

Comparison between requirement engineering and data analytics is depicted in Table 7.

Table 7. Comparison between requirement engineering and data analytics

Domain	Characteristics
Requirement engineering	Requirement engineering aims at well defined requirement documents. It is independent to the process work to satisfy the requirements and technology choice to deploy. Handover is done within the documentation process
Data analytics	Data analytics aims at underlying technological infrastructure and detailed execution of analytics process. Handover is done based on the performed code sequence

This comparison indicates more context and history-bound characteristics of data analytics. Best practices need to be developed to deal with these aspects.

Comparison between workflow tasks and data analytics is depicted in Table 8. It indicates the dynamic characteristics of data analytics, which makes handover challenging.

Table 8. Comparison between workflow tasks and data analytics

Domain	Characteristics
Data analytics	Data analytics still remains person-specific process. It lacks solid process of handover. It partially depends on rapid progress of tools and platforms which obsoletes previous procedures
Workflow tasks	Business tasks have its workflow. Handover is done on the predefined workflow with predefined interfaces. Use of tools defines interfaces between accomplished tasks and tasks to be addressed

From these comparisons, they show that the data analytics has similarities with other process, and that they highlight the data analytics-specific characteristics.

5   Discussion

5.1   Advantages of the Proposed Method

There are three advantages of the proposed method.

- (a) raising awareness of handover issues
- (b) providing a framework with three major components to diagnose handover issues
- (c) highlighting differences to related IT work activities

For (a) raising awareness of handover issues, the author presents multiple handover issues to bring research and industrial attention to handover issues in data analytics. In early days, data analytics was an isolated special field where each expert exercise their own discretion. When two different specialized persons perform handovers, every aspect of handover is bilateral agreement.

One reason that we have to bring attention to handover issues is that we have more open data infrastructures which combine data analytics and everyday business such as DataOps. In these cases, businesses need to put more attention to the handover process with a more systematic methodology to maintain sustainability of business operation.

Another reason is that we have more and more data analytics requirements in today's business. It leads to more demands of flexible workload adjustment, which includes handovers. Collaborative data analytics done by a team is common. However, a time dimensional analysis such as a handover has been unexplored in the multi-user management domain.

For (b) providing a framework to diagnose handovers, the author provides three major components as dimensions. Some of handover issues may have overlapped over multiple dimensions. However, the dimensions provide the base to make analyze the handover problems. It is useful to build best practices by subdividing issues into different dimensions. It is also helpful to identify the root causes when a problem takes place after a handover.

For (c) highlighting differences with other IT activities, using this breakdown by dimension makes it easier, to compare and highlight issues with other domains such as system integration or requirement engineering. The comparison clarifies the data analytics-unique handover issues, which can be helpful to prepare for handover issues.

The age of DataOps, when data analytics is tightly coupled with business operation, requires us to prepare handovers in data analytics in a systematic manner. Many handovers in the past were done in bilateral and implicit manners. It was same in the early days of software development. As the data analytics gets mature and complicated, it requires systematic process management not only in technological aspects but also in organizational aspects such as handovers.

The proposed model provides a stepping stone toward constructing a systematic methodology.

Also, a handover is one of the many issues in multi-user data analytics, which is a multidisciplinary domain between data analytics and organizational management. This research area will need more studies in the near future.

5.2   Applications

The multi-user issues in data analytics are expected to increase from the reasons depicted in Table 9.

Table 9. Reasons why multi-user issues will increase in data analytics

Reason	Explanation
Size and complexity of data analytics	When data analytics work increases its size and complexity, it needs multi-user involvement
Need of second opinion	Data analytics work is perceived as a black box work. End users who do not have expertizes to understand the data analytics work needs second opinion from an external expert similar to the difficult medical cases
DataOps	DataOps is integration of data analytics and everyday business operations. Direct integration to the daily business increases malfunction and misleading risks. Therefore, it needs multi-user check more thoroughly

The proposed framework has the promising applications as depicted in Table 10.

Table 10. Applications

Application	Description
Best practices	Based on the proposed model, each dimension can be used to create dimension-specific best practices of handover. Specific measures can be deployed to suppress the issues
Checklists	Based on the proposed model, each dimension can be used to create checklists to deal with handover issues
Training materials	Based on the proposed model, each dimension can be utilized to develop training materials for data analytics handovers
Guidelines	Based on the proposed model, each dimension can be used for creating a framework to accommodate guidelines to deal with a handover

5.3   Limitations

This research is qualitative and exploratory.

The number of interview is small. The interview is unstructured and does not follow any systematic method.

This research lacks any quantitative measures to verify. The proposed model has no in-depth discussion in real use cases. The model lacks any comparative or quantitative analysis.

Detailed analysis of handover process is not studied in this paper. Social and organizational relationship in the stakeholders is not discussed.

The applications of the proposed model are not implemented in the real world. Real world deployment requires future research. Detailed analysis of differences of skills, experiences, and motivations is not performed.

This research does not cover organizational aspects of data analytics handover. Cultures of data analytics are not considered. The cultural impacts of handover is not covered in this paper.

6   Conclusion

Outburst of data and advances in methods, tools, technological infrastructure facilitate growth of size, coverage, and complexity of data analytics. This leads to team work in data analytics. Even though collaborative data analytics with different roles and skills have been researched, the time dimensional collaboration is relatively unexplored field.

When data analytics work is transferred to one person to the other, there are multiple issues to cover. This handover is one of the multi-user issues of data analytics.

The author performed interview with data analysts to gather data for potential issues in handovers. Then, the author categorizes the issues in three dimensions.

The proposed model provides a base of framework to deal with multi-user issues in data analytics. It provides a skeleton for best practices, checklists, training, and guidelines to facilitate handovers.

Rapid progress of methods, tools, and platforms continue to obsolete data analytics process, especially the parts related to deep learning. This context provides a reason why handover process is rigidly defined. There is a tradeoff between well defined process and flexibility to adopt new methodologies.

The proposed model provides a measure to keep balance between the two conflicting factors.

Collaborative data analytics is catching attention in research and industries. A shift from customized and proprietary data analyzing platform to common and shared one facilitates analysis of collaborative data analytics and development of best practices. The proposed model is a building block for research to empower the multi-user data analytics.

References

1. Banerjee, S., Chandra, M.G.: A software framework for procedural knowledge based collaborative data analytics for IoT. In: Proceedings of the 1st International Workshop on Software Engineering Research & Practices for the Internet of Things, SERP4IoT 2019, pp. 41–48, IEEE Press, Piscataway (2019)
2. Bhardwaj, A., Deshpande, A., Elmore, A.J., Karger, D., Madden, S., Parameswaran, A., Subramanyam, H., Wu, E., Zhang, R.: Collaborative data analytics with DataHub. Proc. VLDB Endow. 8(12), 1916–1919 (2015)
3. Callinan, C., Scott, M., Ojo, A., Whelan, E.: How to create public value through open data driven co-creation: a survey of the literature. In: Proceedings of the 11th International Conference on Theory and Practice of Electronic Governance, ICEGOV 2018, pp. 363–370. ACM, New York (2018)
4. Horak, T., Mathisen, A., Klokmose, C.N., Dachselt, R., Elmqvist, N.: Vistribute: distributing interactive visualizations in dynamic multi-device setups. In: Proceedings of the 2019 CHI Conference on Human Factors in Computing Systems, CHI 2019, pp. 616:1–616:13. ACM, New York (2019)
5. Khalifa, S., Elshater, Y., Sundaravarathan, K., Bhat, A., Martin, P., Imam, F., Rope, D., Mcroberts, M., Statchuk, C.: The six pillars for building big data analytics ecosystems. ACM Comput. Surv. 49(2), 33:1–33:36 (2016)
6. Schwee, J.H., Sangogboye, F.C., Kjærgaard, M.B.: Anonymizing building data for data analytics in cross-organizational settings. In: Proceedings of the International Conference on Internet of Things Design and Implementation, IoTDI 2019, pp. 1–12. ACM, New York (2019)
7. Toreini, P., Langner, M., Maedche, A.: Use of attentive information dashboards to support task resumption in working environments. In: Proceedings of the 2018 ACM Symposium on Eye Tracking Research & Applications, ETRA 2018, pp. 92:1–92:3. ACM, New York (2018)
8. Tucker, I., Gil-Garcia, J.R., Sayogo, D.S.: Collaborative data analytics for emergency response: identifying key factors and proposing a preliminary framework. In: Proceedings of the 10th International Conference on Theory and Practice of Electronic Governance, ICEGOV 2017, pp. 508–515. ACM, New York (2017)

4 Observation

Transitions of user experience are depicted in Fig. 1. The final shift toward real-world interface provides immersive, real-time, entertaining, and useful. It also presents the real world privacy risk to XR applications.

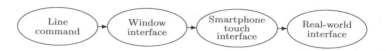

Fig. 1. Transitions of user experience

XR interfaces is one of the next generation user experience. It is cutting-edge, so it also means to be uncharted yet. Privacy of XR is still to be explored in theory and practices.

One of XR applications is AR applications. AR applications are always-on, always-sensing real-time applications.

The privacy protection targets to be considered in XR are depicted in Table 1.

Table 1. Privacy protection targets

Target	Description	Risks
User	Primary users. People who use the AR applications	A user may be tracked, or identified. Privacy-sensitive information of the user (e.g. medical records) may be exposed
Related people	People who communicate with the people who use the AR applications	Families or friends may be tracked, or identified. Their privacy-sensitive information may be exposed
Passer-by	Secondary, incidental, passive users. People who are captured in the AR applications even though they don't use the AR applications in a direct manner	Passers-by may be tracked or identified. Their privacy-sensitive information may be exposed

Users should be clearly informed about security risks. At the same time, privacy of related people and passers-by should be considered.

Aspects of privacy threats are depicted in Table 2. These dimensions require instant or long-term privacy protection mechanisms.

Behavior data such as poses enables potential identification or tracking. Tracking presents potential threats of identification in an implicit manner. Intensive sensor use for regeneration of the real world provides more opportunities

Table 2. Aspects of privacy threats

Aspect	Description
Content	Expression, surrounding data, and so on. They are captured by cameras
Static attributes	Device identification (Capabilities). They are captured by setting, configuration or user input during the use
Dynamic attributes	Location, moving routes, vital information, weather information. They are captured by cameras and sensors
Behavior	Contacts, interactions and so on. They are captured by cameras and sensors

Table 3. Perception of privacy protection status

Item	Description
Status perception	Awareness of privacy status
Status change notification	Awareness of privacy status change

Table 4. Privacy threats in sensor APIs

Threat	Description
Location tracking	Using the location tracking and behavior information from other sources, it can identify the user without his/her explicit consent
Eavesdropping	Using the interception of audio data from human and environmental sources, it can identify the user without his/her explicit consent
Keystroke monitoring	Using the keystroke monitoring, it can detect the user with sensitive information
Device fingerprinting	Using information of capabilities of augmented reality devices, it can identify the owner with filtering out other possibilities
User identifying	Using sensor information, it can identify users with various behavior patterns

for privacy-related information. It also provides opportunities to use multiple sources to track a user.

There are two methods to increase privacy protection. One is to use automatic processing. The other is to notify users about privacy protection status. Awareness of privacy protection status will ensure that the user will behave to protect one's own privacy. It can also notify surrounding people about the status of application use to some extent. Perception of privacy protection status is depicted in Table 3.

Table 5. LINDDUN threat categories

Target	Description
Linkability	Linkability of entity
Identifiability	Identifiability of entity
Non-repudiation	User cannot repudiate their behavior
Detectability	Device can be fingerprinted by its capabilities and function declaration
Disclosure of information	Unintentional leakage of privacy information
Unawareness	Privacy information is shared by a third-party without awareness of a user
Non-compliance	Declared or selected privacy policy is violated. Consent is not complied

Table 6. Attackers

Attacker	Description
AR attacker	Attacks on augmented reality content
Ad attacker	Attacks on advertisement
Web attacker	Attacks on web services to host an augmented reality content
Service attacker	Attacks on an augmented reality service
Network attacker	Attacks on connectivity

Privacy threats in Sensor APIs are depicted in Table 4 [11]. Privacy threats in sensor APIs are applicable to XR privacy.

In an XR application context, there are varieties of risks in device fingerprinting. In order to improve user experience, an XR application needs to identify the capabilities of devices that a user carries. It also requires to obtain permissions of a user to render XR applications using multiple sensors. These permissions lead to potential exposure of device fingerprinting.

LINDDUN threats categories are depicted in Table 5. These threats should be incorporated in API design including standardization and application design.

Attackers in XR are listed in Table 6. XR applications may incorporate external web services in their context in order to enhance user experience. Non XR-specific attacks are possible in XR applications in such contexts.

Implementation-related privacy issues are depicted in Table 7. These threats need to be considered when performance boost is implemented in networks or edges. Also, new device APIs for XR should be designed to deal with privacy issues.

Consent issues are depicted in Table 8. XR applications are a compound application with multiple sensors. This leads to a challenge of easy-to-use consent forms. When granularity of content and implementation does not match, it leads to either too coarse content or too detailed difficult-to-use consent.

Table 7. Implementation-related privacy issues

Item	Description
Edge offloading	Data for augmented reality is processed at edge nodes, but sometimes the data is transferred to the more powerful cloud for performance improvement
Network spoofing	Data can be stolen when a malicious party impersonates another device or user on a network
Application level XR APIs	In addition to sensor APIs, application level XR APIs can be targets for privacy attack

Table 8. Consent issues

Item	Description
Complexity	When there is a complexity in a consent form, it is difficult to use in a consistent and comprehensive manner
Intrusiveness	When a privacy consent form is too intrusive, it breaks user satisfaction and usability of a service
Persistence	When a user content is tentative and volatile, it is difficult to use in a usable manner

Table 9. Privacy dimensions in social contexts

Dimension	Description
Reserve	One would like to avoid exposure to others
Isolation	Desire of no interaction
Solitude	One behaves as a standalone without relationship
Intimacy with family	One behaves as a persistent social relationship which each person cannot modify
Intimacy with friends	One behaves as a persistent social relationship
Anonymity	One behaves as a non-identifiable temporary person

Privacy dimensions in social contexts are depicted in Table 9. This is largely unexplored field and how to deal with it remains future research.

5   Differential Approach

5.1   Differential Analysis

The differential analysis shows the XR-specific aspects of privacy threat in comparison with the similar applications. The differences between XR applications and camera-equipped applications are shown in Table 10.

Table 10. Privacy threat differences between XR applications and camera-equipped applications

Item	Description
Disclosure at the application level	XR applications (especially AR and MR applications) use the captured real world image to render application-overlapped views. It increases exposure risk of sensitive information. Its realtime-ness increases exposure risk due to the processing time constraint
Disclosure at the network and edge	When XR applications are rendered by edge computing, it increases risk of disclosure at network and edge
Disclosure at the API level	XR applications use XR-level APIs which discloses underlying hardware capabilities

When a new level of development and deployment is added, it should be added in this perspective. Generally, a compound development environment in XR systems to pursue the real-world-ness adds a new threat component.

The differences between XR applications and sensor applications are shown in Table 11. From this analysis, additional devices and application-level user interaction in XR systems provides another layer of privacy threats. It should be noted that the additional layer of privacy protection can indicate further tracking means for privacy threats.

Table 11. Privacy threat differences between XR applications and sensor applications

Item	Description
Device fingerprinting	Use of coordinated multiple sensors increases device fingerprinting risk
User identification	Identification of user characteristics such as poses and interpupillary distance increases risk of user identification
Consent information	XR applications require multiple user contents to organize a device-specific XR content, which exposes user-trackability

5.2   Differential Dimensional Model

From the above differential analysis, the author proposes a 3-dimensional privacy threat model as depicted in Fig. 2.

In addition to Individual treat to each underlying enabling devices and networks, XR applications add two dimensions of privacy threat: Realtime operation threat and compound UX threat. These two threats are additional threats which is added in the XR application context.

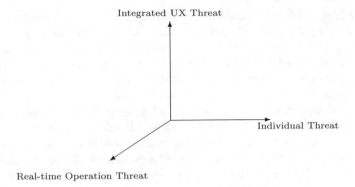

Fig. 2. 3-dimensional privacy threat model

The dimensions are depicted in Table 12. When APIs and applications are built, the privacy threat in the specific environment can be tracked by this proposed model. These two new dimensions require appropriate assumptions and mitigations with XR applications. New emergence of technology is captured in the individual threat dimension. The other two dimensions capture the independent threats separated by technology adoption.

Table 12. Privacy threat dimensions

Item	Description
Individual threat	Threats from each component (camera, sensor, client application, underlying OS, et al.)
Integrated UX threat	Threats from coordinated and integrated nature of XR application UX
Real-time operation threat	Threats from realtime processing nature such as real time processing of captured images and edge computing

6   Discussion

6.1   Advantages of the Proposed Method

XR application technology is a combination of hardware and software engineering. The complexity brings new combined privacy risk.

The multi-perspective threat analysis provides a cue for new threats for XR applications, e.g. the XR API level threat.

Risk analysis of XR applications are rather an unexplored research field. The differential comparison shows unique privacy threats for XR applications. The author presents a 3-dimensional privacy risk threat model. It shows the new two dimensions: integrated UX threat and real-time operation threat.

Emergence of new technologies provide new privacy threats in emerging applications. The model deals with the high level encapsulation of individual threat to allow expansion of threat coverage for adoption of new technologies.

Integrated UX threat is unique for XR applications which attempts to customize rendering to customize underlying technical components. Real-time operation threat has its origin for immersive applications. Immersiveness requires real-time capturing the real world, which leads to privacy leakage according to the real-time location tracking of an user.

6.2   Limitations

This research is qualitative and descriptive. It is also exploratory with only a theoretical framework. The paper does not present exhaustive coverage of XR applications.

No quantitative analysis is performed. No quantitative measures for threats or mitigation are discussed. Detailed studies of empirical evaluation is not presented in this paper.

This paper provides a general framework, not the concrete methodology to deal with XR application privacy risks.

No detailed use cases with existing XR applications are presented.

Concrete privacy protection guidelines or best practices from the proposed framework are not covered and remain future research. Mitigation techniques and strategies are not covered in this paper.

7   Conclusion

There are many hardware that enables virtual reality and augmented reality user experience. They require high-precision and low-latency communication to enable great user experience with immersive-ness. They provide unique and rich user experience with new opportunities in entertainment, education, and social interactions.

The opportunities are tightly coupled with challenges. The challenges include privacy threats.

The past privacy risk analysis depends on functions of each component. The author presents a 3-dimensional risk threat model which focuses the dynamic and integrated nature of XR applications. The proposed model separates (a) individual component threat, (b) combined threat, and (c) real-time processing threat. These three dimensions capture the different aspects of XR application privacy risks.

The proposed model copes with emerging new hardware, software, and platform with the descriptive and abstract dimensions. It provides a building block for guidelines and best practices for improved privacy protection in emerging XR applications.

It provides a stepping stone to analyze emerging XR applications with a high level view.

References

1. Desai, B.C.: IoT: imminent ownership threat. In: Proceedings of the 21st International Database Engineering & Applications Symposium, IDEAS 2017, pp. 82–89. ACM, New York (2017)
2. Dong, R., Ratliff, L.J., Cárdenas, A.A., Ohlsson, H., Sastry, S.S.: Quantifying the utility-privacy tradeoff in the internet of things. ACM Trans. Cyber Phys. Syst. **2**(2), 8:1–8:28 (2018)
3. Ferdous, M.S., Chowdhury, S., Jose, J.M.: Privacy threat model in lifelogging. In: Proceedings of the 2016 ACM International Joint Conference on Pervasive and Ubiquitous Computing: Adjunct, UbiComp 2016, pp. 576–581. ACM, New York (2016)
4. Jones, B., Waliczek, N.: WebXR device API. In: W3C Working Draft, 10 October 2019. https://www.w3.org/TR/2019/WD-webxr-20191010/
5. Kettani, H., Cannistra, R.M.: On cyber threats to smart digital environments. In: Proceedings of the 2nd International Conference on Smart Digital Environment, ICSDE 2018, pp. 183–188. ACM, New York (2018)
6. Lebeck, K., Kohno, T., Roesner, F.: How to safely augment reality: challenges and directions. In: Proceedings of the 17th International Workshop on Mobile Computing Systems and Applications, HotMobile 2016, pp. 45–50. ACM, New York (2016)
7. Moustaka, V., Theodosiou, Z., Vakali, A., Kounoudes, A.: Smart cities at risk!: privacy and security borderlines from social networking in cities. In: Companion Proceedings of the Web Conference 2018, WWW 2018, pp. 905–910. International World Wide Web Conferences Steering Committee, Republic and Canton of Geneva (2018)
8. Olejnik, L., Novak, J.: Self-review questionnaire: security and privacy, December 2018. https://w3ctag.github.io/security-questionnaire/
9. Siby, S., Maiti, R.R., Tippenhauer, N.O.: IoTScanner: detecting privacy threats in IoT neighborhoods. In: Proceedings of the 3rd ACM International Workshop on IoT Privacy, Trust, and Security, IoTPTS 2017, pp. 23–30. ACM, New York (2017)
10. Soh, L., Burke, J., Zhang, L.: Supporting augmented reality: looking beyond performance. In: Proceedings of the 2018 Morning Workshop on Virtual Reality and Augmented Reality Network, VR/AR Network 2018, pp. 7–12. ACM, New York (2018)
11. Waldron, R.: Generic sensor API, November 2018. https://w3c.github.io/sensors/#security-and-privacy
12. Wuyts, K., Van Landuyt, D., Hovsepyan, A., Joosen, W.: Effective and efficient privacy threat modeling through domain refinements. In: Proceedings of the 33rd Annual ACM Symposium on Applied Computing, SAC 2018, pp. 1175–1178. ACM, New York (2018)

13. Zhang, R., Zhang, N., Du, C., Lou, W., Hou, Y.T., Kawamoto, Y.: From elec-
    tromyogram to password: exploring the privacy impact of wearables in augmented
    reality. ACM Trans. Intell. Syst. Technol. **9**(1), 13:1–13:20 (2017)
14. Zhang-Kennedy, L., Mekhail, C., Abdelaziz, Y., Chiasson, S.: From nosy little
    brothers to stranger-danger: children and parents' perception of mobile threats.
    In: Proceedings of the 15th International Conference on Interaction Design and
    Children, IDC 2016, pp. 388–399. ACM, New York (2016)

Human Mobility and Message Caching in Opportunistic Networks

Tomoyuki Sueda[1] and Naohiro Hayashibara[2(✉)]

[1] Graduate School of Frontier Informatics, Kyoto Sangyo University, Kyoto, Japan
i1888088@cc.kyoto-su.ac.jp
[2] Faculty of Computer Science and Engineering, Kyoto Sangyo University,
Kyoto, Japan
naohaya@cc.kyoto-su.ac.jp

Abstract. Opportunistic communication is one of the key technologies for advertisement, information sharing, disaster evacuation guidance in delay-tolerant networks (DTNs), vehicular ad hoc networks (VANETs) and so on. The efficiency of message delivery in opportunistic communication is correlated to the routing protocols and the movement pattern of mobile entities. In this paper, we focus on message delivery based on opportunistic communication by pedestrians who move based on Lévy walk and Random walk movement patterns on the road network of a city. Moreover, we introduce roadside units located in the city, which play a role in the distributed message cache. So, we analyze the influence of roadside units, pedestrian walk patterns and the routing protocols of opportunistic communication on message delivery. We assume Random walk and Lévy walk as pedestrian mobility models and Epidemic, Spray and Wait and PRoPHET as the routing protocols. Our simulation results clarify the difference of routing protocols and show that the roadside units have a significant impact on message delivery with a small number of pedestrians.

1 Introduction

It is well-known that random walks become an important building block for designing and analyzing protocols in mobile ad hoc networks [8,9,26], searching and collecting information [3,5] in P2P networks, and solving global optimization problems [28,29]. In particular, Lévy walk (also called Lévy flight) has recently attracted attention due to the optimal search in animal foraging [10,27] and the statistical similarity of human mobility [19]. It is a mathematical fractal which is characterized by long segments followed by shorter hops in random directions. The pattern has been proposed by Paul Lévy in 1937 [16], but the similar pattern has also been evolved and sophisticated as a naturally selected strategy that gives animals an edge in the search for sparse targets to survive [10].

Although most of the works on Lévy walk have been done on a continuous plane, we assume unit disk graphs as the underlying environment. Unit Disk Graphs are widely used for routing, topology control, analysis of virus spreading in ad hoc networks (e.g., [2,15,26]). Roughly speaking, the difference between a graph and a continuous plane is the freedom of movement. It means that the link structure of the

© Springer Nature Switzerland AG 2020
L. Barolli et al. (Eds.): EIDWT 2020, LNDECT 47, pp. 395–405, 2020.
https://doi.org/10.1007/978-3-030-39746-3_41

graph restricts the movement. We suppose an agent as a mobile entity. Agents can only move to a neighbor node of the current node on the graph, although agents move anywhere on the continuous plane. It means that agents are restricted their movement by the link structure of a graph. There are several research results on random walks on graphs [14, 18] but there are a few result on Lévy walk on graphs [21].

We assume roadside units (RSUs) as a distributed message cache, which are located in a city. We also assume a delay-tolerant network (DTN) formed by human pedestrians who have mobile devices such as smartphones. Thus, messages are sent, carried, and forwarded by devices of pedestrians who move based on Random walk and Lévy walk. The structure of the road network that is often modeled as a graph restricts the movement of pedestrians in a city.

In this paper, we analyze the influence of pedestrian walk patterns and the routing protocols of opportunistic communication on the efficiency of message delivery by opportunistic communication in DTNs. Moreover, we measure the average delivery ratio and the latency in our simulations in the presence of RSUs.

2 Related Work

We introduce several research works on Lévy walk and random walks.

Birand et al. proposed the Truncated Levy Walk (TLW) model based on real human traces [4]. The model gives heavy-tailed characteristics of human motion. Authors analyzed the properties of the graph evolution under the TLW mobility model.

Valler et al. analyzed the impact of mobility models including Lévy walk on epidemic spreading in MANET [26]. They adopted the scaling parameter $\lambda = 2.0$ in the Lévy walk mobility model. From the simulation result, they found that the impact of velocity of mobile nodes does not affect the spread of virus infection.

Thejaswini et al. proposed the sampling algorithm for mobile phone sensing based on Lévy walk mobility model [24]. Authors showed that proposed algorithm gives significantly better performance compared to the existing method in terms of energy consumption and spatial coverage.

Fujihara et al. proposed a variant of Lévy walk which is called Homesick Lévy Walk (HLW) [11]. In this mobility model, agents return to the starting point with a homesick probability after arriving at the destination determined by the power-law step length. As their result, the frequency of agent encounter obeys the power-law distribution.

Mizumoto et al. measured the encounter rate of Lévy walk and Brownian walk regarding mating of sexual dimorphism (e.g., male and female) on the one-dimensional and the two-dimensional infinite and borderless space with limited lifespans [17]. They also analyzed the relationship of the scaling parameter of Lévy walk, the encounter rate and the lifespan based on the simulation result. From the result, Lévy walk is efficient regarding the encounter rate in a short lifespan, whereas Brownian walk becomes efficient by the increase of the lifespan.

There are several papers on random walks on finite graphs. Ikeda et al. showed the impact of local degree information of nodes in a graph [14]. They proposed a random walk, called the β random walk, which uses the degree of neighboring nodes and moves to a node with a small degree in high probability compared to the ones with a high degree. The hitting time becomes $O(n^2)$ and the cover time becomes $O(n^2 log\ n)$ in the β random walk despite the hitting and cover times are both $O(n^3)$ in the simple random walk without local degree information.

Nonaka et al. presented the hitting time and the cover time in the Metropolis walk which obeys the transition probability produced by the Metropolis-Hastings algorithm [18]. It is a typical random walk used in Markov chain Monte Carlo methods. They showed that the hitting time is $O(n^2)$ and the cover time $O(n^2 log\ n)$.

Despite of many articles on Lévy walk, they evaluated it on continuous fields, and hardly any results on graphs have been available. Shinki et al. newly defined the algorithm for Lévy walk on a unit disk graph and evaluated the efficiency of message dissemination compared with the random walk, β random walk and the metropolis walk [21]. According to the evaluation, the Lévy walk movement pattern is efficient for message dissemination compared to other random walks.

Sueda and Hayashibara clarified the impact of RSUs and movement patterns with Epidemic routing protocol in opportunistic communication [23]. According to the simulation results, the message delivery ratio improved more than 50% with the small number of pedestrians (up to 20 pedestrians) by RSUs both Lévy walk and Random walk. The latency of message delivery of Lévy walk reduced by 7% to 10% and that of Random walk reduced by 4% by RSUs.

3   System Model

In this paper, we assume delay-tolerant networks (DTNs) consists of human pedestrians who have mobile devices that have a short-range communication capability such as Bluetooth, ad hoc mode of IEEE 802.11, Near Field Communication (NFC), infrared transmission, and so on. Moreover, we assume a city road network and roadside units (RSUs) located randomly in the road network. They also exchange messages with pedestrians and help to distribute messages by storing some messages temporally. They follow FIFO policy on message replacement. Whenever new messages arrive at a RSU, older messages are deleted from the storage of it.

There exist several routing protocols for message delivery in DTN. We use Epidemic [25], Spray and Wait [22] and PRoPHET [12] as routing protocols in our simulation. It is a flooding-style messaging protocol where every pedestrian sends copies of all messages that are stored in the mobile device of the pedestrian to others in the communication range. To prevent exhausting storage resources in the network, we assume TTL (Time-to-Live) for every message.

4   Routing Protocols of Opportunistic Communication

We introduce several routing protocols of opportunistic communication, which we use in the simulations.

4.1 Epidemic

Epidemic routing [25] is a flooding-based protocol. Every node copies and forward messages to newly encountered nodes which do not already hold a copy of the message. This protocol aims to realize a higher message delivery ratio and lower latency. On the other hand, it induces a higher overhead ratio and requires a large storage space of each node as a consequence of the number of copies of messages flooded in the network. Whenever two nodes n_i and n_j exchange messages, n_i sends a Summary Vector, that contains the digest of the messages stored in n_i, to n_j and vice versa. Then, n_i and n_j send a request for messages, which are not in own storage, to each other node. Finally, they send messages according to the request. This procedure prevents to receive duplicated messages.

4.2 Spray and Wait

Spray and Wait protocol [22] developed by Spyropoulos et al. to optimize the resource utilization. Due to multiple copies of message, maximum resources are used in Epidemic protocol while Spray and Wait protocol limits the number of replicated messages to N. This protocol consists of two stages; the spray stage and the wait stage. When a new message arrives at a node, it generates N copies of the message and transmits them to N different nodes in the spray stage. Whenever a node receives the copy, the protocol enters in the wait stage. It holds the message copy until the destination node is come in direct communication range.

4.3 PRoPHET

PRoPHET [12] is a probabilistic routing protocol. First of all, it calculates delivery predictability which includes two properties; a history of encounters and transitivity. The former one indicates how many times any two nodes come across and the latter one is the probability calculated from the history in terms of message forwarding. Each node initially has an initialization constant of delivery predictability and it is updated based on encounters. The predictability is reduced if no encounters occur in a certain period of time.

5 Pedestrian Movement Patterns

The efficiency of message broadcasting by opportunistic communication depends on the movement pattern of agents [13]. We focus on Lévy walk and Random walk as the movement pattern of pedestrians. Both are often used as human mobility models. Lévy walk in particular statistically resembles human mobility traces [19]. It has a parameter λ that determines the behavior of agents. In general, the ballistic trajectory frequently appears on agents' movement with a decrease of λ. Each pedestrian is assumed to move at the same velocity.

5.1 Random Walk

Random walk originally came from the description of the random motion of particles. There are lots of variations; here we describe a simple implementation of random walks. In the simple random walk, agents move to a neighbor of the current node at uniformly random. Thus, the probability $Pr(u, v)_{RW}$ that each agent, which is located at the node u, moves to its neighbor v is defined as follows.

$$Pr_{RW}(u, v) = \begin{cases} \frac{1}{deg(u)} & (v \in N(u)) \\ 0 & (otherwise) \end{cases} \qquad (1)$$

The hitting time and the cover time of the simple random walk is $O(n^3)$ [1]. This random walk is also called as Brownian walk.

5.2 Lévy Walk

As we mentioned in Sect. 3, we assume a city road network modeled as a unit disk graph. So, we use the algorithm for Lévy walk on unit disk graphs proposed by Shinki et al. [21].

Lévy walk is a variation of random walks where each node selects a direction uniformly from within $[0, 2\pi)$ and a step length of a walk is determined the Lévy probability distribution described as follows.

$$p(d) \propto d^{-\lambda} \qquad (2)$$

d is a step length and λ is the scaling parameter to draw the different shape of the probability distribution.

According to the figures, the maximum step length is getting longer as λ decreases. In general, Lévy walk becomes similar to the random walk in terms of its behavior when λ is greater than 3.0 [6].

We describe the algorithm for Lévy walk on unit disk graphs.

The main difference between continuous planes and graphs is freedom of mobility. Agents can only move to the neighbor node of the current node in graphs. It means that the movement of agents in a graph is more restricted than the one in a continuous plane.

We explain the algorithm for Lévy walk on unit disk graphs proposed in [21].

In general, an agent selects a destination node v from a set of neighbors $N(u)$ of the current node u randomly and moves to v in random walks on graphs. In contrast, an agent determines the orientation o from $[0, 2\pi)$ at random, and the step length d by the power law distribution (see Eq. 2) at the beginning of Lévy walk. Then, it selects v according to the information and moves to it d times. Note that d is the distance (hops) in a graph.

We now define *a walk* as the sequence of the movement in a direction and *a step* as the movement with the step length $d = 1$. Thus, a walk consists of a sequence of steps in Lévy walk though a walk equals to a step in random walks.

In every walk, each agent determines the step length d by the power-law distribution described in Eq. 2, and selects the orientation o of a walk randomly from $[0, 2\pi)$.

Each agent can obtain a set of neighbors $N(c)$ and a set of possible neighbors $PN(c) \subseteq N(c)$, to which agents can move, from the current node c. In other words, a node $x \in PN(c)$ has the link with c that the angle θ_{ox} between o and the link is smaller than δ which is a given parameter, called a *permissible error*.

Each agent selects the next node $v \in PN(c)$ which has the minimum θ_{ov} and move to v. Thus, each agent moves towards the determined o with δ, d times in a walk.

6   Performance Evaluation

The performance evaluation aims to clarify the difference between the routing protocols in terms of the reachability and the latency of message delivery in the presence of RSUs as message caches. In our previous work, we clarify the impact of RSUs with different walk pattern (i.e., Lévy walk and Random walk) using Epidemic routing protocol in opportunistic communication. In this paper, we compare Spray and Wait and PRoPHET routing protocols with Epidemic protocol regarding the message delivery ratio and the latency in the presence of RSUs. We have conducted simulations by using The One simulator, which is a simulator for opportunistic network environments implemented in Java [7].

We assume the Helsinki city scenario as a simulation field included in The ONE simulator.

6.1   Parameters

We configure the following parameters for our simulations.

- The communication range of each pedestrian is 10 m. Two pedestrians can communicate each other when the distance between them is less than 10 m.
- n is the number of pedestrians located in a field. Each pedestrian is located at a unique position and the position of it is determined at random. We set $n = [10; 320]$.
- The number of RSUs $r \in \{0, 50, 100, 200\}$. $r = 0$ means that there is no RSUs that temporally hold messages.
- Scaling parameter λ is a parameter for the Lévy walk to determine the trajectory of agents. The ballistic trajectory is emphasized if λ increases. We set $\lambda = 1.2$ because it is efficient for a search on unit disk graphs [20] and it can cover a wider area on a graph.

Then, we explain parameters depending on the routing protocols. We set TTL = 300 (min) in Epidemic protocol to prevent exhausting storage resources. We set the number of copies of a message $N = 6$ in Spray and Wait protocol. We configure the initialization constant = 0.75, the transitivity scaling constant = 0.25 and aging constant = 0.98 on the delivery predictability of PRoPHET protocol.

6.2   Results

We measured the average delivery ratio and the average latency of message delivery by pedestrians based on Lévy walk and Random walk movement pattern. We analyze the impact of RSUs as message caches on the delivery ratio and the latency. We also compare Lévy walk and Random walk from these points of view. Each result is plotted as an average value of 100 simulation runs with 95% confidence interval.

6.2.1 Average Delivery Ratio

Figures 1 and 2 show the average delivery ratio on messages with the number of pedestrians based on Random walk and Lévy walk, respectively. Each line corresponds to the number of RSUs from 0 to 200. We use them as a baseline for comparison.

Figures 3 and 4 are the average delivery ratio using Spray and Wait protocol and Figs. 5 and 6 are that using PRoPHET protocol based on Random walk and Lévy walk.

Obviously, Epidemic protocol is the best regarding the average delivery ratio among the protocols that we use because it is flooding-based protocol. However, it imposes a heavy load in the network and storage.

Spray and Wait does not depend on the number of pedestrians regarding the average delivery ratio. It is 20 to 30% with Random walk and 20 to 40% with Lévy walk. RSUs do not help improve the delivery ratio in this protocol since the protocol limits the number of copies of each message and each node (pedestrian) may recognize RSUs as a relay nodes and stores several copies into RSUs. As a result, RSUs as message caches are effective in terms of the message delivery ratio in flooding-based protocols.

The average delivery ratio of PRoPHET protocol is 40 to 50% less than the one of Epidemic protocol with $n \leq 40$. On the other hand, It increases according to the increment of the number of pedestrians and it comes close to that of Epidemic with $160 \geq n$. It means that PRoPHET could be an alternative if the number of pedestrians is sufficiently large. In this protocol, RSUs help improve the delivery ratio by 40 to 80% with $n \leq 40$. The impact is gradually reduced according to the increment of the number of pedestrians.

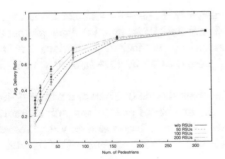

Fig. 1. Average delivery ratio in epidemic based on Random walk.

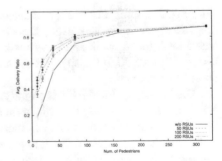

Fig. 2. Average delivery ratio in epidemic based on Lévy walk.

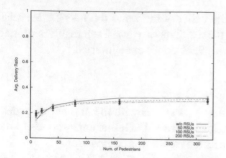

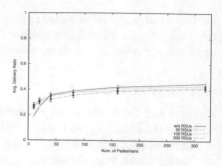

Fig. 3. Average delivery ratio in Spray and Wait based on Random walk.

Fig. 4. Average delivery ratio in Spray and Wait based on Lévy walk.

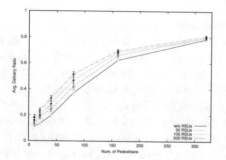

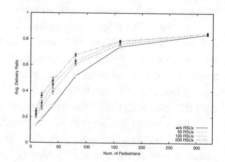

Fig. 5. Average delivery ratio in PRoPHET based on Random walk.

Fig. 6. Average delivery ratio in PRoPHET based on Lévy walk.

6.2.2   Latency of Message Delivery

We have measured the latency of message delivery. Figures 8 to 12 show the average latency on messages using Epidemic, Spray and Wait, and PRoPhet protocols based on Random walk and Lévy walk, respectively. Number of pedestrians n is set $n = \{10, 20, 40, 80, 160, 320\}$. Each result is plotted as an average latency with 95% confidence interval.

The latency of PRoPHET is slightly better than the one of Epidemic with $n \leq 40$. It is increasing according to the increment of the number of pedestrians and it is more than 40% greater than the one of Epidemic with $160 \geq n$. The reason why it increases is the calculation of transitivity. It might not be efficient in the presence of RSUs with a significant large number of pedestrians.

On the walk pattens, the latency of PRoPHET based on Lévy walk is slightly better than the one based on Random walk while the latency of Spray and Wait does not have a significant difference in both movement patterns (Figs. 7, 9, 10 and 11).

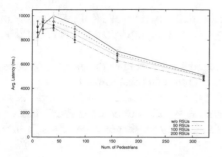

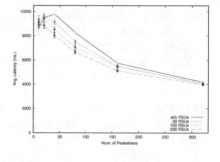

Fig. 7. Latency in epidemic based on Random walk with various number of RSUs

Fig. 8. Latency in epidemic based on Lévy walk with various number of RSUs

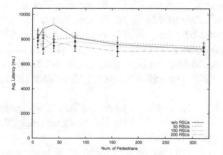

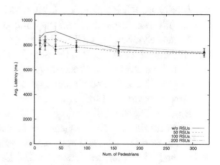

Fig. 9. Latency in Spray and Wait based on Random walk with various number of RSUs

Fig. 10. Latency in Spray and Wait based on Lévy walk with various number of RSUs

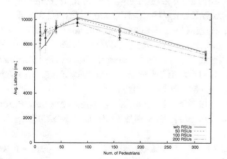

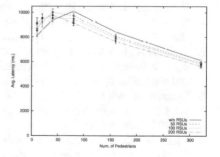

Fig. 11. Latency in PRoPHET based on Random walk with various number of RSUs

Fig. 12. Latency in PRoPHET based on Lévy walk with various number of RSUs

7   Conclusion

In this paper, we assumed RSUs as message caches located in a city and clarify the difference of routing protocols on the message delivery ratio and the latency of messages delivery by opportunistic communication in DTNs that consist of human pedestrians.

The impact of RSUs is significant on the average delivery ratio of PRoPHET, especially with a small number of pedestrians (e.g., $n \leq 40$). On the other hand, It is not efficient for Spray and Wait protocol.

The latency of PRoPHET protocol is increasing according to the number of pedestrians. RSUs are not efficient for improving the latency in this protocol.

We believe that this result would be useful for the advertisement for shops, events and public services in a city with mobile devices. We plan to develop PRoPHET protocol to improve both the delivery ratio and the latency in the presence of RSUs in the future.

References

1. Aleliunas, R., Karp, R.M., Lipton, R.J., Lovasz, L., Rackoff, C.: Random walks, universal traversal sequences, and the complexity of maze problems. In: Proceedings of the 20th Annual Symposium on Foundations of Computer Science (SFCS 1979), pp. 218–223 (1976)
2. Alzoubi, K.M., Wan, P.J., Frieder, O.: Message-optimal connected dominating sets in mobile ad hoc networks. In: Proceedings of the 3rd ACM International Symposium on Mobile Ad Hoc Networking and Computing, MobiHoc 2002, pp. 157–164. ACM, New York (2002). https://doi.org/10.1145/513800.513820
3. Baldoni, R., Beraldi, R., Quema, V., Querzoni, L., Tucci-Piergiovanni, S.: Tera: Topic-based event routing for peer-to-peer architectures. In: Proceedings of the 2007 International Conference on Distributed Event-based Systems, pp. 2–13 (2007)
4. Birand, B., Zafer, M., Zussman, G., Lee, K.W.: Dynamic graph properties of mobile networks under levy walk mobility. In: Proceedings of the 2011 IEEE Eighth International Conference on Mobile Ad-Hoc and Sensor Systems, MASS 2011, pp. 292–301. IEEE Computer Society, Washington, DC (2011). https://doi.org/10.1109/MASS.2011.36
5. Bisnik, N., Abouzeid, A.A.: Optimizing random walk search algorithms in P2P networks. Comput. Netw. 51(6), 1499–1514 (2007). https://doi.org/10.1016/j.comnet.2006.08.004
6. Buldyrev, S.V., Goldberger, A.L., Havlin, S., Peng, C.K., Simons, M., Stanley, H.E.: Generalized lévy-walk model for dna nucleotide sequences. Phys. Rev. E 47(6), 4514–4523 (1993)
7. Desta, M.S., Hyytiä, E., Keränen, A., Kärkkäinen, T., Ott, J.: Evaluating (Geo) content sharing with the one simulator. In: Proceedings of the 14th ACM Symposium Modeling, Analysis and Simulation of Wireless and Mobile Systems (MSWiM) (2013)
8. Dolev, S., Schiller, E., Welch, J.L.: Random walk for self-stabilizing group communication in ad hoc networks. IEEE Trans. Mobi. Comput. 5(7), 893–905 (2006). https://doi.org/10.1109/TMC.2006.104
9. Draief, M., Ganesh, A.: A random walk model for infection on graphs: Spread of epidemics & rumours with mobile agents. Discrete Event Dyn. Syst. 21(1), 41–61 (2011). https://doi.org/10.1007/s10626-010-0092-5
10. Edwards, A.M., Phillips, R.A., Watkins, N.W., Freeman, M.P., Murphy, E.J., Afanasyev, V., Buldyrev, S.V., da Luz, M.G.E., Raposo, E.P., Stanley, H.E., Viswanathan, G.M.: Revisiting lévy flight search patterns of wandering albatrosses, bumblebees and deer. Nature 449, 1044–1048 (2007)
11. Fujihara, A., Miwa, H.: Homesick lévy walk and optimal forwarding criterion of utility-based routing under sequential encounters. In: 2013 Proceedings of the Internet of Things and Inter-cooperative Computational Technologies for Collective Intelligence 2013, pp. 207–231 (2013)

12. Grasic, S., Davies, E., Lindgren, A., Doria, A.: The evolution of a DTN routing proto-col - PRoPHETv2. In: Proceedings of the 6th ACM Workshop on Challenged Networks, CHANTS 2011, pp. 27–30. ACM, New York (2011). https://doi.org/10.1145/2030652.2030661
13. Helgason, Ó., Kouyoumdjieva, S.T., Karlsson, G.: Opportunistic communication and human mobility. IEEE Trans. Mob. Comput. **13**(7), 1597–1610 (2014)
14. Ikeda, S., Kubo, I., Yamashita, M.: The hitting and cover times of random walks on finite graphs using local degree information. Theor. Comput. Sci. **410**(1), 94–100 (2009)
15. Kuhn, F., Wattenhofer, R.: Constant-time distributed dominating set approximation. In: Proceedings of the Twenty-second Annual Symposium on Principles of Distributed Computing, PODC 2003, pp. 25–32. ACM, New York (2003). https://doi.org/10.1145/872035.872040
16. Lévy, P.: Théorie de L'addition des Variables Aléatoires. Gauthier-Villars (1937)
17. Mizumoto, N., Abe, M.S., Dobata, S.: Optimizing mating encounters by sexually dimorphic movements. J. R. Soc. Interface **14**(130), 20170086 (2017)
18. Nonaka, Y., Ono, H., Sadakane, K., Yamashita, M.: The hitting and cover times of metropolis walks. Theor. Comput. Sci. **411**(16–18), 1889–1894 (2010)
19. Rhee, I., Shin, M., Hong, S., Lee, K., Kim, S.J., Chong, S.: On the levy-walk nature of human mobility. IEEE/ACM Trans. Netw. **19**(3), 630–643 (2011). https://doi.org/10.1109/TNET.2011.2120618
20. Shinki, K., Hayashibara, N.: Resource exploration using lévy walk on unit disk graphs. In: The 32nd IEEE International Conference on Advanced Information Networking and Applications, AINA-2018. Krakow, Poland (2018)
21. Shinki, K., Nishida, M., Hayashibara, N.: Message dissemination using lévy flight on unit disk graphs. In: The 31st IEEE International Conference on Advanced Information Networking and Applications, AINA 2017, Taipei, Taiwan ROC (2017)
22. Spyropoulos, T., Psounis, K., Raghavendra, C.S.: Efficient routing in intermittently connected mobile networks: the multiple-copy case. IEEE/ACM Trans. Netw. **14**, 77–90 (2008)
23. Sueda, T., Hayashibara, N.: Opportunistic communication by pedestrians with roadside units as message caches. In: Proceedings of the 22nd International Conference on Network-Based Information Systems, NBiS 2019, pp. 167–177 (2013)
24. Thejaswini, M., Rajalakshmi, P., Desai, U.B.: Novel sampling algorithm for human mobility-based mobile phone sensing. IEEE Internet Things J. **2**(3), 210–220 (2015)
25. Vahdat, A., Becker, D.: Epidemic routing for partially-connected ad hoc networks. Technical report. CS-2000-06, Duke University (2000)
26. Valler, N.C., Prakash, B.A., Tong, H., Faloutsos, M., Faloutsos, C.: Epidemic spread in mobile ad hoc networks: determining the tipping point. In: Proceedings of the 10th International IFIP TC 6 Conference on Networking - volume Part I, NETWORKING 2011, pp. 266–280. Springer-Verlag, Heidelberg (2011). http://dl.acm.org/citation.cfm?id=2008780.20088
27. Viswanathan, G.M., Afanasyev, V., Buldyrev, S.V., Murphy, E.J., Prince, P.A., Stanley, H.E.: Lévy flight search patterns of wandering albatrosses. Nature **381**, 413–415 (1996)
28. Yang, X.S.: Cuckoo search via Lévy flights. In: Proceedings of World Congress on Nature & Biologically Inspired Computing, NaBIC 2009, pp. 210–214 (2009)
29. Yang, X.S.: Firefly algorithm, lévy flights and global optimization. Research and Development in Intelligent Systems XXVI, pp. 209–218 (2010)

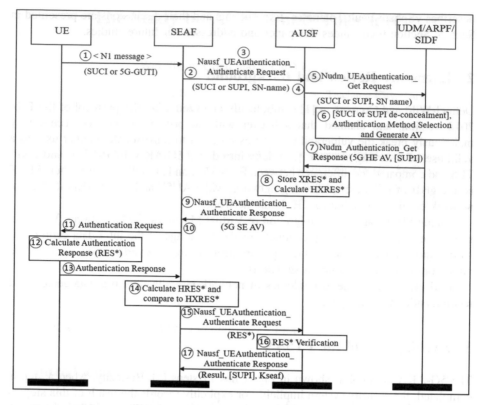

Fig. 1. The sequence chart of 5G-AKA [5].

EPC has been virtualized as a virtual EPC (vEPC) by using SDN and NFV. In the FT5AS, the AUSF pool, managed by SDN Controller, provides programmable APIs with which users can monitor the pool's operations. The interface between two arbitrary AUSFs is N_{ausf}. It can be a bus connecting all AUSFs, like that of an Ethernet, or all AUSFs are tightly connected to a complete graph. In other words, if n AUSFs are there, the amount of N_{ausf} links utilized is $n(n-1)/2$.

Also, 5G users request high-QoS services. Service interruption is a serious problem. As mentioned above, AUSF records the results of a step when the step is completed. Our opinion is that the granularity of original steps can be refined. In this study, we rearrange the steps and divide a step concerning AUSF into three sub-steps, aiming to lower the chance of executing unnecessary sub-steps again.

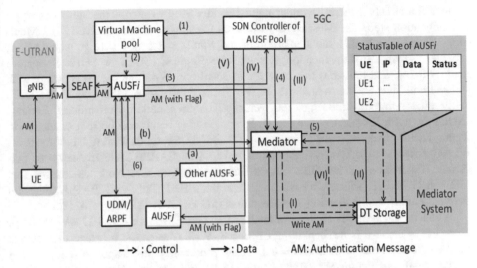

Fig. 2. Architecture of the FT5AS and its data transmission and control procedures (Steps (1)–(6) for waking up a VM; Steps (a)–(b) show the heartbeat delivery process; Steps (I)–(VI) present the AUSF fault procedure when AUSF$_j$ fails).

3.1 Mediator

Mediator consists of Receiver and Writer. Receiver takes charge of receiving messages from other entities, e.g., the commands sent by SDN Controller and authentication messages transmitted by AUSF. Writer is responsible for writing the massages/data that Receiver receives in Distributed-Topology storage. Also, when Mediator discovers that an AUSF fails, e.g., AUSF$_i$, $1 \leqq i \leqq m$, where m is amount of AUSFs now in AUSF Pool, it delivers a message to SDN Controller telling it about this event. SDN Controller will allocate the UEs authenticated by the failed AUSF to a part of or all of the other $n - 1$ AUSFs.

3.1.1 AUSF to Distributed Topology Storage Through Mediator

When AUSFi cannot work properly, as mentioned above, an authentication step is divided into three sub-steps, including receiving a message, processing the message and sending another message. The symbol A.Y means the sub-step B of Step A, A = 0, 4, 8 or 16, B = 1, 2, 3, because in 5G-AKA, AUSF only the processes Steps 4, 8 and 16 (see Fig. 1).

A. Data table and Action-Storage Procedure. In the Distributed Topology Storage, we establish a data table for each AUSF, denoted by StatusTablei. When AUSFi sends a message to Mediator, a flag, called FG, is added. The value of FG can only be 0 and 1 with initial value which is 0. When Receiver receives a message delivered by AUSFi, if FG = 1, representing that AUSF$_i$ is authenticating an UE, Writer saves the message in the field, called Message, in StatusTable$_i$.

In a StatusTable, there is a field named Status used to record the sub-steps of an authentication step. The Message field records the message/data received from Mediator. When FG = 1 and a sub-step is completed, Status = Status + 0.1e.g., when $AUSF_i$ receives Nausf_UEAuthentication_Authenticate Request (SUCI or SUPI, SN-name) sent by SEAF and transmits this message to Mediator, Writer stores the message with FG = 1 and Status = 4.1 in $StatusTable_i$, i = 1,2,...m, where m is the amount of AUSF. After processing this message, i.e., Nausf_UEAuthentication_Authenticate Request (SUCI or SUPI, SN-name) - see Fig. 1, the processing result will be sent to Mediator. Then, Writer changes Status from 4.1 to 4.2. As AUSF sends Nudm_UEAuthentication_Get Request (SUCI or SUPI, SN name) to UDM/ARPF/SIDF in Step 5 of 5G-AKA, Writer changes the UE's authentication Status to 4.3. Now we assume that UDM/ARPF/SIDF selects 5G-AKA, rather than EAP-TLS and EAP-AKA', and delivers Nudm_Authentication_Get Response (5G HE AV, [SUPI]) to AUSF.

When AUSF receives Nudm_Authentication_Get Response (5G HE AV, [SUPI]), since FG = 1 and it is the beginning of Step 8, Writer keeps the message in Message field of $StatusTable_i$ with Status = 8.1. After processing this response (e.g., storing XRES* and calculating HXRES*, Status = 8.2) and sending Nausf_UEAuthentication_ Authenticate Response (5G SE AV) to SEAF in Step 9, Status = 8.3. When AUSF receives Nausf_UEAuthentication_ Authenticate Request (RES*) from SEAF (Step 15), Status = 16.1. AUSF sends RES* to Mediator and verifies whether or not RES* = XRES*, Status = 16.2 and FG = 1. Following that, AUSF notifies SEAF with Nausf_UEAuthentication_Authenticate Response (Result, [SUPI], Kseaf) to serve underlying UE and the authentication is completed, i.e., Status = 0 and FG = 0.

B. AUSF Failure. After receiving Nausf_UEAuthentication_Authenticate Request (SUCI or SUPI, SN-name), if $AUSF_i$ fails before requesting Mediator to save the message in its Data field, since $StatusTable_i$ does not keep any data, i.e., Status = 0, the takeover AUSF, e.g., $AUSF_j$, needs to authenticate this UE from the very beginning. That is, UE has to resend <N1 message> (SUCI or 5G-GUTI) (Step1 of Fig. 1) to initiate the authentication process. On the other hand, if $AUSF_i$ fails after Mediator successfully records the receiving message, i.e., Nausf_UEAuthentication_Authenticate Request (SUCI or SUPI, SN-name), in $StatusTable_i$, since the recorded status is 4.1, rather than 0, $AUSF_j$ will retrieve the Nausf_UEAuthentication_Authenticate Request (SUCI or SUPI, SN-name) from $StatusTable_i$, rather than from $StatusTable_j$, and process this message. If Mediator successfully records the processing result in $StatusTable_i$, i.e., in Data field, then Status will be 4.2. After that, if $AUSF_j$ fails, the takeover AUSF, e.g., $AUSF_k$, then retrieves this processing result from $StatusTable_i$, sends Naudm_UEAuthentication_Get Request (SUCI or SUPI, SN-name) to UDM/ARPF/SIDE and Mediator to record the transmitted message. Now Status = 4.3. On receiving this message, UDM/ARPF (note that SIDF is used to decrypt SUCI to SUPI) starts authenticating this UE (Step6 of 5G-AKA) and then sends one Authentication vector (AV), rather than n Authentication vectors, which is carried in Nudm_UEAuthentication_Get Response (5G HE AV, [SUPI]) to $AUSF_k$ (Step 7 of Fig. 1).

After that, if $AUSF_k$ fails, the takeover AUSF, e.g., AUSFl, will check the Status of this UE. If Status = 8.1, meaning that before $AUSF_k$ fails, it has received Nudm_UEAuthentication_Get Response (5G HE AV, [SUIP]), AUSF1 continues Step

8 of 5G-AKA, i.e., storing XRES* and calculating HXRES*. However, if Status = 4.3, and $AUSF_k$ has not received Nudm_UEAuthentication_Get Response (5G SE AV. [SUPI]) from UDM/ARPF (Step 7 of Fig. 1), AUSF1 resends Nudm_UEAuthentication_Get Request (SUCI or SUPI, SN-name) retrieved from $StatusTable_i$ to UDM/ARPF and waits for UDM/ARPF to reply an AV carried in Nudm_Authentication_Get Response (5G HE AV, [SUPI]), i.e., Step 7 of 5G-AKA protocol.

However, after taking over for $AUSF_k$ and retrieving authentication data from $StatusTable_i$, if Status = 4.3 and AUSF1 has received Nudm_Authentication_Get Response (5G HE AV, [SUPI]) from UDM/ARPF, AUSF1 will send this message to SEAF and Mediator to continue the authentication.

4 Fault Tolerance Mechanism for VM

For the use of HTML5, VMware starts with the vSphere6.5 version, abandoning the vSphere Client window management tool developed by using C#. Besides, the vSphere Web Client adopts the Flash plug-in components, while the rest of vSphere, including ESXi and vCenter, employs HTML5 technology.

The network topology of the FT5AS is built by using mininet installed in ESXi6.5. Users can access the IP address of the ESXi Web Client via a browser. Figure 3 shows an example of ESXi Web Client. As shown, you can see the ESXi web client Home page which illustrates the details of ESXi and performs tasks on the ESXi host. Navigator will help users to directly navigate to the Host, Virtual machines, Storage and Networks.

Fig. 3. VMware ESXi web client

4.1   VSphere Fault Tolerance

vSphere Fault Tolerant mechanism is employed to ensure normal operation of the VM. A protected VM is referred to as a primary VM, and a redundant VM (i.e., a secondary VM), executed in the same manner as that of the primary VM, can be established and executed on another host. When the primary VM fails, the secondary VM as shown in Fig. 4 takes over for the failed at any interruption point to ensure an uninterrupted operations and user services. Generally, the primary and secondary VM continuously monitor the status of one another to ensure that Fault Tolerance is maintained. If the host running the primary VM fails a transparent failover occurs, since the secondary VM is immediately activated to substitute for the primary. Meanwhile, a new secondary VM is started to automatically reestablish the Fault Tolerance redundancy. If the host running the secondary VM fails, the same substitution will also immediately perform. In both cases, users experience no service interruption and no loss of delivered and processed data.

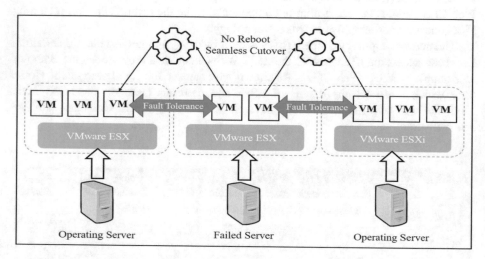

Fig. 4. Fault Tolerant mechanism of our system

5   Simulation and Discussions

This study proposes a fault tolerant mechanism for VM. All VMs are running on the VMware ESXi platform. With a vSphere client, users can conveniently install, manage, and access VMs. Here, P2V (Physical to Virtual) refers to the migration of a physical machine to virtual machines (VMs), and V2V (Virtual to Virtual) is the migration of an operating system (OS), an application program or data, from a virtual machine to another virtual machine. In VMware ESXi, P2V and V2V are both performed by vCenter Converter.

Table 1 shows our tested VM environment, in which ESXi is the VM platform. Compared to previous version, ESXi 6.5.0 provides administrators with an embedded host-client web interface, making VM management, including VM status checking, performance monitoring, VM's start and stop operations, and VM migration, more efficiently.

Table 1. VM environment

Item	Description
Operating system	Ubuntu linux 16.04
Physical deploy	RAM 8G
vCenter server	6.5
ESXi version	6.5.0
vSphere client	6.5

5.1 Experiments

This study simulates the situation when the primary VM is shut down, aiming to test whether the VM can execute vMotion to the secondary VM immediately to continue this interrupted services. The size of a packet is fixed to 1.5 KB. The number of packets sent per second is between 80 and 42 K, and the bandwidth is set to 200 Mbps (=25 MB/sec). We would like to know the feasibility when installing a vSphere to our proposed system and observe the service difference between the two cases when transferring the control from the primary VM to the secondary VM and when the vSphere Fault Tolerant mechanism is unused.

5.1.1 Performance Analysis

Figure 5a illustrates the throughputs. When number of packets sent per second exceeds 16 K, the occupied bandwidth is 24 MB/sec (not shown). Throughputs of the only UE began to decline because high data rates seriously congest network links. Figure 5b shows end-to-end delays. Although between 80 and 12 K, the delays slightly increase with the number of packets sent. After reaching 16 K, the delays and packet loss rates rise sharply (see Fig. 5c), because the data rates exceed the bandwidth, resulting in higher delays and packet loss rates.

Figure 5 compares network QoS between vSphere FT and FT5AS. Throughputs (Fig. 5a), end-to-end delays (Fig. 5b), and packet loss rates (Fig. 5c) of the two tested schemes are themselves similar to each other, demonstrating that vSphere FT which provides a fault-tolerant system for VMs makes our proposed system more reliable.

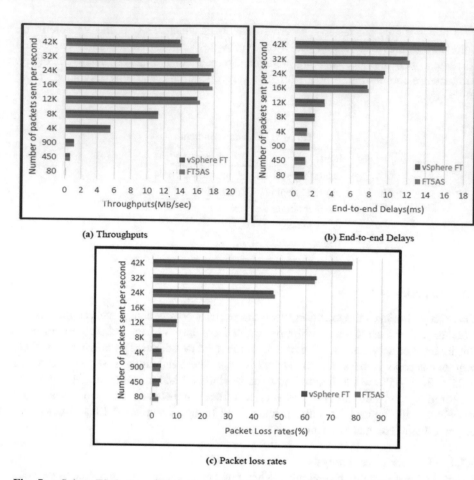

(a) Throughputs   (b) End-to-end Delays

(c) Packet loss rates

Fig. 5. vSphere FS QoS on different numbers of packet sent per second which range from 80 to 42 K

6   Conclusion and Future Studies

Generally, this study proposes a fault-tolerant mechanism to solve the fault problem of AUSF. The experimental results also show that our approach is feasible. However, there are still some issues that have to be explored in the near future. For example, if AUSF is down in the process of handover, how other AUSFs take over for it to continue the handover process. This may conduct a trust issue or the authorization problem of accessing data in Common Storage. In the future, we will also propose a more complete handover mechanism, and continue improving this system. We would also like to derive its behavior model and reliability model so that users can comprehend its behaviors and reliabilities before using it. These constitute our future studies.

References

1. Dehnel-Wild, M., Cremers, C.: Security vulnerability in 5G-AKA draft (3GPP TS 33.501 draft v0.7.0). Department of Computer Science, University of Oxford (2018)
2. Informed Inside. A Comparative Introduction to 4G and 5G Authentication. WINTER 2019. https://www.cablelabs.com/insights/a-comparative-introduction-to-4g-and-5g-authentication. Accessed 23 Aug 2019
3. Adami, D., Giordano, S., Pagano, M., et al.: Virtual machines migration in a cloud data center scenario: an experimental analysis. In: 2013 IEEE International Conference on Communications (ICC), pp. 2578–2582. IEEE (2013)
4. Bari, M.F., Boutaba, R., Esteves, R., et al.: Data center network virtualization: a survey. Commun. Surv. Tutor. IEEE **15**(2), 909–928 (2013)
5. GPP TS 33.501 version 15.5.0 Release 15 (2019-07) Security architecture and procedures for 5G System. https://www.etsi.org/deliver/etsi_ts/133500_133599/133501/15.05.00_60/ts_133501v150500p.pdf. Accessed 2 Sept 2019

Classification and Regression Based Methods for Short Term Load and Price Forecasting: A Survey

Hira Gul, Arooj Arif, Sahiba Fareed, Mubbashra Anwar, Afrah Naeem,
and Nadeem Javaid[✉]

COMSATS University Islamabad, Islamabad, Pakistan
hiragul001@gmail.com, nadeemjavaidqau@gmail.com

Abstract. Due to increase in electronic appliances, electricity is becoming basic necessity of life. Consumption of electricity depends on various factors like temperature, wind, humidity, weekend, working days and season. In electricity load forecasting, many researchers perform data analysis on electricity data provided by utilities to extract meaningful information. Smart Grid (SG) is power supply network which allows consumers to monitor their energy usage. It integrates different components of electricity like variety of operations, smart appliances, data collected from smart meters and efficient energy sources. To reduce the consumption of electricity, accurate prediction is compulsory. A good forecasting model makes an acceptable use of all characteristics of electric loads based data and also reduces dimensionality of that data. Various machine learning techniques are proposed for load forecasting in literature. In this research, we present a survey based on different short term electricity forecasting techniques for price and load. We broadly categorized different types of techniques into traditional machine learning and deep learning techniques.

Keywords: Load forecasting · Support Vector Machine · Smart Grid · Price forecasting · Deep learning · Neural network

1 Introduction

In this era of technology, energy consumption is expanding exponentially. Demand of electricity is increasing day by day as number of electronic appliances are increasing. Consumption of electricity is divided into six areas: residential, industrial, commercial, agriculture, transport and other government. Residential area is consuming 65% of total electricity [1]. Traditional grid is used as electromechanical technology, which faces several difficulties like one-way communication, central distribution, manual monitoring, few sensors and manual restoration. Smart Grid (SG) is established to resolve the above-mentioned issues. It plays a leading role in balancing the consumption, generation and transmission of electricity [2]. SG is power grid which monitors generation, transmission and consumption of electricity [3]. SG helps producers to produce required

© Springer Nature Switzerland AG 2020
L. Barolli et al. (Eds.): EIDWT 2020, LNDECT 47, pp. 416–426, 2020.
https://doi.org/10.1007/978-3-030-39746-3_43

electricity according to consumers need. Electricity forecasting is necessary for utilities to balance between demand and supply of electric load. Accurate electricity forecasting helps to improve the working of SG [4].

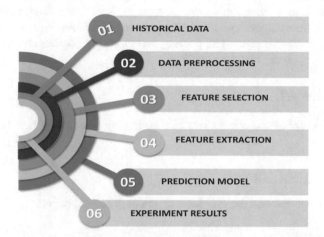

Fig. 1. General forecasting model

Demand of electricity is dependent on numerous factors like humidity, wind, temperature, season, weekend and weekdays. It also depends on number of householders and their daily routine. Data analytic is procedure of assess the data by using statistical software to obtain useful information. In electricity load forecasting, many researchers do data analysis on electricity data to extract meaningful information. Many machine learning techniques are proposed for load forecasting. Large amount of big data is required for prediction. This data contains sensitive information with thousands of entries. Accurate load forecasting reduces the electricity price as it minimizes the consumption in peak hours [5].

Based on time interval, electricity load forecasting is split into three parts. First is Short Term Load Forecasting (STLF), it forecasts the electricity from a time period of 24 h to one week. Second is Medium Term Load Forecasting (MTLF), it forecasts load from a period of one week to one year. Third is Long Term Load Forecasting (LTLF), it forecasts the load from a time period of one year to more then two years [6]. Many load forecasting approaches developed in literature like Support Vector Machine (SVM), regression and neural network. In deep learning, Convolutional Neural Network (CNN) is most commonly used method for load forecasting model. CNN consists of input layer, multiple hidden layers and output layer. Hidden layer contains multiple layers like convolutional layers and pooling layers. Inside pooling layer, there are max pooling, average pooling and sum pooling [7].

Forecasting models consist of six steps including historical data, preprocessing, feature selection, feature extraction, prediction model and experimental results as shown in Fig. 1. Load forecasting is classified into three types based on time intervals. First is Short Term Load Forecasting (STLF), it forecasts the electricity form a time period of 24 h to one week. Second is medium term load forecasting, it forecasts the load from a period of one week to one year. Third is long term forecasting, it forecasts the load from a time period of one year to more then two years [6]. Many load forecasting approaches have been developed like SVM, regression and neural network. In deep learning, CNN is most commonly used method for load forecasting. CNN consists of input layer, multiple hidden layers and output layer [8]. Hidden layer contains multiple layers like convolutional layers and pooling layers. Inside pooling layer, there are max pooling, average pooling and sum pooling.

2   Related Work

This section describes the related work of electricity forecasting for short term load forecasting.

2.1   Problem Addressed

In this section, we describe the problem statement of related papers.

2.1.1   Machine Learning Verses Deep Learning

Machine learning techniques are used for electric load forecasting. These techniques are Random Forest (RF) and Gradient Boosting (GB). Computational complexity of RF is very high, as it takes very long time to build decision tree. Error rates are very high in RF [2]. In [3], authors discussed the problems of data redundancy in feature selection and extraction. Existing models used only univariate price data for price forecasting. Price data is not sufficient for accurate forecasting. There is need of big data for electricity price forecasting yet computational complexity of big data is very high. Over fitting problem occurs in decision tree. Over fitting means that a model performs well on training model and its performance degrades on testing data.

In existing studies, features directly move into model without any preprocessing. According to authors, parameter tunning is considered to be important for accurate forecasting. So, there is need of feature selection and extraction technique for load forecasting [4]. In [5], Previous Recurrent Neural Network (RNN) models do not used future hidden state vector and available past information. If a state vector is generated incorrectly at any specific time then cannot be corrected, as it is very important for enhanced forecasting in specific time. In [6], many authors used optimization techniques still there is need to optimize the hybrid models. Authors consider the problem of back propagation in hybrid approach composed of DWT, EMD and RVFL. Back propagation is an algorithm which consists of two main problems: slow convergence and trapped in local minimum. In [9], authors discussed drawbacks of ANN in DCANN. ANN has some disadvantages like slow convergence, less generalizing performance, trapped at local minimum and poor initialization of parameters.

2.1.2   Traditional Forecasting Methods

In [10], WD only used for pre-processing and reconstruction results based on independent prediction. Such combination are not sufficient for Support Vector Regression

(SVR). Moreover, the purpose of Multi-resolution Wavelet Decomposition (MWD) is needed to consider with SVR. In existing models, traditional approaches were used for load forecasting. These techniques are: time series and linear regression. They mainly focus on aggregated electric load demand pattern. Now authors are combining machine learning techniques with deep learning approaches. So, there is need to consider the integrated deep learning techniques with combination of machine learning [11]. Short term day ahead load forecasting is very challenging job, because it depends on external environment factors like temperature, wind and humidity. Existing day ahead load forecasting models improve the accuracy by compensating the computational cost. Computational complexity of hybrid model is a big challenge [12]. In [13], authors addressed the issue of data redundancy among features in minimal Redundancy Maximal Relevance (mRMR). Authors also discussed that influential factors like weather changes, day type and economical condition effects the load forecasting. Authors ensemble different models to addressed above mention problems. In [14], Linear regression and ARIMA gives best results with linear problems while these models perform unsatisfactory with non-linear time series data. Therefore, authors proposed a hybrid model which works better for non-linear time series data. Time-Varying parameter Regression (TVR) has a problems of price instability, fluctuating input variables, managing availability of data and complex data inputs [15]. In [16], authors discussed the limitations of ANN like over-fitting problem, trapped in local minimum and poor initialization of parameters. To solve these issue, authors used combination of different models for short term load forecasting. In [17], authors discussed that load forecasting is affected by unstable factors like temperature, price, policy management and holidays data. Due to fluctuation in these factors, noisy data is generated which gives inaccurate results. Authors used EMD to decompose original data into IMFs and forecast the electric load. In [18], authors discuss the issue of cyber security in load forecasting. Hackers injected false information in original dataset. Data integrity problems are relatively new in this domain.

2.2 Solution Proposed

This section describes the proposed solution of related papers.

2.2.1 Machine Learning Verses Deep Learning

In [2], authors proposed Grey Correlation Analysis (GCA) for feature selection to remove the repetition of features. They combined Kernel Function with Principle Component Analysis (KPCA). They used Differential Evolution (DE) which is based on Support Vector Machine (SVM) classifier for the classification of price forecasting. In [3], authors proposed probability density forecasting model for short term power load forecasting. Authors proposed Multi-Layer Perceptron (MLP) function with deep neural network. Kernel density estimation method consist of three sub modules deep learning, loss function and quantile regression.

In [4], authors proposed a deep learning model called Stack Denoising Auto encoders (SDAs) for feature extraction. This model train SVR to forecast the electric load of day ahead. By using heterogeneous deep model, accuracy of forecast is better with less errors. In [5], authors proposed RICNN which is combination of RNN and

1- dimensional Convolutional Neural Network (1-D CNN). 1-D CNN inception module is used to adjust the prediction time and hidden state vector values. They calculate the hidden state vector from adjacent time steps. As a consequence, RNN generates optimized and robust network through prediction time and hidden state vector. In [6], authors proposed hybrid incremental learning method. In this method three techniques are combined: Discrete Wavelet Transform (DWT), Empirical Mode Decomposition (EMD) and Random Vector Functional Link network (RVFL). RVFL is very useful as it generate weights randomly among input layer and hidden layer. It also provide nearest possible solution for parameters calculation. In incremental learning, when they add DWT and EMD it impressively increases accuracy and effectiveness of short term load forecasting. In [9], authors proposed a model named Dynamic choice artificial neural network (DCANN) which selects input dynamically for ANN. DCANN is hybrid model which consists of supervised and unsupervised learning problems. This approach selects unsupervised learning technique to choose input variable for individual output. Authors train correspondence inputs and outputs through supervised learning.

2.2.2   Traditional Forecasting Methods

In [10], authors proposed hybrid incremental learning method. In this method, three techniques are combined: DWT, EMD and Random Vector Functional Link network (RVFL). RVFL is very useful as it generate weights randomly among input layer and hidden layer. It also provide nearest possible solution for parameters calculation. In incremental learning, when authors add DWT and EMD it impressively increases accuracy and effectiveness of short term load forecasting. In [11], authors proposed a algorithm which is based on deep neural network for short-term load forecasting (STLF). In this algorithm there are input layers and output layers. Input layers denote the past data while output layer denote the future energy load. There is a deep energy which has two main processes: feature extraction and forecasting. In future extraction, there are three more layers of convolutional layers and three pooling layers. In [12], authors proposed a hybrid ANN day ahead based load forecasting model for smart grid. Proposed model consists of three components: pre-processing module, forecast module and optimization module. In preprocessing module, irrelevant variable and data redundancy are removed from input sample. In forecast module, Ann is used with sigmoid function and multivariate auto regression algorithm.

In optimization module, heuristic problem solving approach is used to reduce error. For this purpose they used enhanced differential evolution algorithm. In [13], authors proposed hybrid model EMD-mRMR-FOA-GRNN which is combination of Empirical Mode Decomposition (EMD), minimal Redundancy Maximal Relevance (mRMR), fruit Fly Optimization Algorithm (FOA) and General Regression Neural Network (GRNN). Firstly, authors divide load series data into Intrinsic Mode Functions (IMFs), secondly, mRMR is applied to select the best features. Finally, FOA is used to enhance the factors in GRNN. In [14], authors proposed AS-GCLSSVM hybrid model for short term load forecasting. AS-GCLSSVM is stands for Autocorrelation feature selection and Least Squares Support Vector Machine wolf algorithm and cross validation. In [15], authors proposed Hybrid Iterative Reactive Adaptive (HIRA) model. This model consist of two steps. In first step, they identified only those parameters which effect the electricity prices.

In second step, selected variables are used in price forecasting by applying hybrid app-roach. This HIRA model is combination of statistical models and neural network tools. In [16], authors proposed Wavenet ensemble model for STLF. In this model, authors combined different components like input parameters, mean, median, mode, cross-validation, selection and algorithms. Authors predicted hourly load forecasting through one step ahead strategy. In [17], authors proposed a short term forecasting model named as EMD Mixed ELM. In this model, authors combined empirical mode decomposition (EMD) and extreme learning machine (ELM). Authors used EMD for denoising and normaliza-tion of complicated features from a large dataset. The performance of ELM is depend upon type of kernel they used. Authors used mixed kernel for ELM. Mixed kernel is combination of UKF kernel and RBF kernel. In [18], authors proposed framework for load forecasting named as systematic data integrity simulation.

2.3   Future Challenges/Open Research Issues

This section discusses the limitation and future challenges of literature.

2.3.1   Machine Learning Verses Deep Learning

In [2], authors used RF for decision tree. Computational complexity of RF is very high, as it takes very long time to build decision tree. DE is used for tunning parameters, however it is prove to converge to local optima. In future, authors will apply real time requirements in the proposed model. In [3], authors used RNN model which do not used future hidden state vector and available past information. If a state vector is generated incorrectly at any specific time then cannot be corrected, as it is very important for enhanced forecast-ing in prediction time. In [4], authors did not specify the condition on which basis they extract the features. authors did not consider many important parameter like temperature, wind and humidity. In [5], authors used EMD and it has mode mixing problem. ELM is feedforward neural network and drawback of ELM is, it does not update weights and parameters. In future, authors will combined proposed model with ANN and SVR and applied on real time applications to evaluate the performance [6]. In [9], main limitation of the paper is computational power which is very high. It consume more resources than existing model. Authors did not validate that generated inputs are corrected or not.

2.3.2   Traditional Forecasting Methods

In [10], there is need to tunning the parameters for better results. There should be differ-ent selection and extraction techniques used for different buildings. For seasonal data prediction, there should be more sample data required to increase the consistency of training data. In future, this model will applied on region to attain the accurate electric forecasting of a region. In [11], authors concluded that due to complex neuron structure in neural network the computational power is very high as compared to existing models. Three layers of pool makes the model more complex. Over fitting problem arises which affect the training data. In [12], authors will use enhance signal processing techniques for features selection and extraction. They will also apply optimization techniques on scheduling based application. In [13], prediction performance reduced due to poor gen-eralization capability of GRNN. Computational complexity of proposed model is very

Table 1. Related work.

Problem identified	Proposed solution	Results	Limitations
In [2], computational complexity of RF is very high, as it takes very long time to build decision tree	GCA for feature selection is proposed to remove the repetition of features and they combined KPCA	Proposed model gives 98% accuracy and shows more robustness as compared to NB and DT	DE is used for tuning parameters, however, it shows convergence at local optima
In [3], over fitting problem occurs in decision tree, which means decision tree performs good in training but not in prediction	Probability density forecasting model is proposed	The error rate of proposed model is less than RF and GB	RNN models do not used future hidden state vector and available past information
In [4], features directly move into model without any preprocessing, for accurate results tunning parameter is necessary	SDAs model is proposed by authors	Error rate of proposed model is less than plain SVR and ANN	Authors did not explain the rules on which basis they perform features selection and extraction
In [5], previous Recurrent Neural Network (RNN) models do not used future hidden state vector and available past information	Authors proposed RICNN	Value of MAPE is 4.779 while the values of MAPE in benchmark techniques are 8.084, 7.371 and 5.634. These values show that proposed model is more accurate than existing models	RNN consumes more time for training the dataset and deduction of the results as compared to MLP and CNN
In [6], the problem of back propagation in hybrid approach composed of DWT, EMD and RVFL	Authors proposed hybrid incremental learning method. In this method three techniques are combined: DWT, EMD and RVFL	Value of RMSE is 218.329 while the values of MAPE in benchmark techniques are 355.503 for GLMLF, 307.892 for SHLFN, 278.511 for RF and 244.820 for EMD-RVFL. These values show that proposed model is more accurate and efficient than existing models	This model will decomposed with various other model like deep learning, support vector regression and kernel ridge regression

Table 1. (*continued*)

Problem identified	Proposed solution	Results	Limitations
In [9], drawbacks of ANN in DCANN. ANN has some disadvantages like slow convergence, Less generalizing performance, trapped at local minimum and poor initialization of parameters	DCANN is proposed which selects input dynamically for ANN	Results of proposed model with dynamic selection are 10.71% and 8.39%. These values show that proposed model is more accurate and than existing models	Main limitation of the paper is computational power which is so high
In [10], WD only used for pre-processing and reconstruction results based on independent prediction. Such combination are not sufficient for SVR	Authors proposed hybrid incremental learning method. In this method, three techniques are combined: DWT, EMD and RVFL	Authors compared pure SVR and hybrid SVR and experiment shows that addition of MWD increases the accuracy of forecasting	For seasonal data prediction, there should be more sample data required to increase the consistency of training data
In [11], need to consider the integrated deep learning techniques with combination of machine learning	A algorithm which is based on deep neural network is proposed for STLF	The results of previous models are 9% and 11% while, Results of proposed model are 9.77% for MAPE and 11.66% for RMSE. Authors claimed that they achieved high accuracy	Due to too much neural network the computational power is very high as compared to existing models
In [12], existing day ahead load forecasting model improves the accuracy by compensating the computational cost. Computational complexity of hybrid model is a big problem	Authors proposed a hybrid ANN day ahead based load forecasting model for smart grid	Value of MAPE is 1.23 while values of existing models are 3.18 and 2.31	FS technique is not performs satisfactory, it needs more refine ment
In [13], authors addressed the issue of data redundancy among features in mRMR	Authors proposed hybrid model EMD-mRMRFOA-GRNN	Performance of EMD-mRMR-FOA-GRNN model is better than existing models	Prediction performance reduced due to poor generalization capability of GRNN

Table 1. (*continued*)

Problem identified	Proposed solution	Results	Limitations
In [14], Linear Regression and ARIMA gives best results with linear problems while these models perform unsatisfactory with non-linear time series data	Authors proposed AS-GCLSSVM hybrid model	Values of MAPE, MAE and B^2 are 0.5596, 32.2088 and 0.9952. Based on results authors concluded that proposed model is better than existing model	Computational complexity of proposed model is high because of GWO algorithms
In [15], over-fitting problem, trapped in local minimum and poor initialization of parameters	Wavenet ensemble model for STLF	Performance of WNN ensemble model is better than existing models	Authors will compare ensemble learning algorithms with base learner class like deep neural network or ELM
In [16], the limitation of ANN like overfitting problem, trapped in local minimum and poor initialization of parameters	Proposed Wavenet ensemble model for STLF	Performance of WNN ensemble model is better than existing models	Authors will compare ensemble learning algorithms with base learner class like deep neural network or ELM
In [17], EMD is used to denoise original signals. However, it has mode mixing problem and ELM cannot update the weights and biases	Authors combined EMD and ELM	Values of MAE, RMSE, MAPE and TIC are 7.3550, 9.5823, 08093 and 0.0052 perspectively, Performance of EMD-mRMR-FOA-GRNN model is better than existing models	Prediction performance reduced due to poor generalization capability of GRNN
In [18], Linear Regression and ARIMA gives best results with linear problems while these models perform unsatisfactory with non-linear time series data	Authors proposed AS-GCLSSVM hybrid model for short term load forecasting	Values of MAPE, MAE and B^2 are 0.5596, 32.2088 and 0.9952, Based on results authors concluded that proposed model is better than existing model	Computational complexity of proposed model is high because of GWO algorithms

high as compared to existing model. In [14], computational complexity of proposed model is high because of GWO algorithms. Authors consider AutoCorrelation Function (ACF) relationship between two parameter. However, authors should add external similar parameters like temperature, weather, holidays and festival related parameters.

In [15], computational complexity is very hight of proposed model. It is because four different models are combine together. This model is for big data and it will not perform well in small datasets. In [16], authors will compare ensemble learning algorithms with base learner classes like deep neural network or ELM. Authors also validate the effect of feature selection algorithms. Authors will apply model to other applications like fault tolerance. Authors used EMD and EMD had mode mixing problem. ELM is feedforward neural network and drawback of ELM is it do not update weights and parameters [17]. In [18], When attacks are serve all four forecasting models are fail to generate robust forecast (Table 1).

3   Conclusion

Demand of electricity is increasing exponentially as number of electronic appliances increasing day by day. Different forecasting models were proposed in last decade. In this paper, we conduct a survey based on load and price forecasting. We discussed traditional machine learning approaches and deep learning approaches. In this research, we compare the performance of different load and price models to observe the best results. The research has been moving towards new and more efficient techniques and replacing old approaches. There is a clear move towards hybrid techniques. Hybrid techniques are efficient, better computational complexity and more flexible. This research brings open challenges for future.

References

1. Masip-Bruin, X., Marin-Tordera, E., Jukan, A., Ren, G.J.: Managing resources continuity from the edge to the cloud: architecture and performance. Future Gener. Comput. Syst. **79**, 777–785 (2018)
2. Guo, Z., Zhou, K., Zhang, X., Yang, S.: A deep learning model for short-term power load and probability density forecasting. Energy **160**, 1186–1200 (2018)
3. Wang, K., Xu, C., Zhang, Y., Guo, S., Zomaya, A.Y.: Robust big data analytics for electricity price forecasting in the smart grid. IEEE Trans. Big Data **5**(1), 34–45 (2017)
4. Tong, C., Li, J., Lang, C., Kong, F., Niu, J., Rodrigues, J.J.: An efficient deep model for day-ahead electricity load forecasting with stacked denoising auto-encoders. J. Parallel Distrib. Comput. **117**, 267–273 (2018)
5. Kim, J., Moon, J., Hwang, E., Kang, P.: Recurrent inception convolution neural network for multi short-term load forecasting. Energy Build. **194**, 328–341 (2019)
6. Qiu, X., Suganthan, P.N., Amaratunga, G.A.: Ensemble incremental learning random vector functional link network for short-term electric load forecasting. Knowl.-Based Syst. **145**, 182–196 (2018)
7. Wang, H.Z., Li, G.Q., Wang, G.B., Peng, J.C., Jiang, H., Liu, Y.T.: Deep learning based ensemble approach for probabilistic wind power forecasting. Appl. Energy **188**, 56–70 (2017)
8. Razavi-Far, R., Saif, M., Palade, V., Zio, E.: Adaptive incremental ensemble of extreme learning machines for fault diagnosis in induction motors. In: International Joint Conference on Neural Networks, IJCNN 2017, pp. 1615–1622. IEEE, May 2017

9. Wang, J., Liu, F., Song, Y., Zhao, J.: A novel model: dynamic choice artificial neural network (DCANN) for an electricity price forecasting system. Appl. Soft Comput. **48**, 281–297 (2016)
10. Chen, Y., Tan, H.: Short-term prediction of electric demand in building sector via hybrid support vector regression. Appl. Energy **204**, 1363–1374 (2017)
11. Kuo, P.H., Huang, C.J.: A high precision artificial neural networks model for short-term energy load forecasting. Energies **11**(1), 213 (2018)
12. Ahmad, A., Javaid, N., Mateen, A., Awais, M., Khan, Z.: Short-term load forecasting in smart grids: an intelligent modular approach. Energies **12**(1), 164 (2019)
13. Liang, Y., Niu, D., Hong, W.C.: Short term load forecasting based on feature extraction and improved general regression neural network model. Energy **166**, 653–663 (2019)
14. Yang, A., Li, W., Yang, X.: Short-term electricity load forecasting based on feature selection and Least Squares Support Vector Machines. Knowl.-Based Syst. **163**, 159–173 (2019)
15. Cerjan, M., Petricic, A., Delimar, M.: HIRA Model for Short-Term Electricity Price Forecasting. Energies **12**(3), 568 (2019)
16. Ribeiro, G.T., Mariani, V.C., dos Santos Coelho, L.: Enhanced ensemble structures using wavelet neural networks applied to short-term load forecasting. Eng. Appl. Artif. Intell. **82**, 272–281 (2019)
17. Chen, Y., Kloft, M., Yang, Y., Li, C., Li, L.: Mixed kernel based extreme learning machine for electric load forecasting. Neurocomputing **312**, 90–106 (2018)
18. Moon, J., Kim, K.H., Kim, Y., Hwang, E.: A short-term electric load forecasting scheme using 2-stage predictive analytics. In: 2018 IEEE International Conference on Big Data and Smart Computing, BigComp, pp. 219–226. IEEE, January 2018

Electricity Price and Load Forecasting Using Data Analytics in Smart Grid: A Survey

Mubbashra Anwar, Afrah Naeem, Hira Gul, Arooj Arif, Sahiba Fareed, and Nadeem Javaid[✉]

COMSATS University Islamabad, Islamabad, Pakistan
Mubbashraanwar@gmail.com, nadeemjavaidqau@gmail.com

Abstract. Smart grid (SG) is bringing revolutionary changes in the electric power system. SG is supposed to provide economic, social, and environmental benefits for many stakeholders. A smart meter is an essential part of the SG. Data acquisition, transmission, processing, and interpretation are factors to determine the success of smart meters due to the excess amount of data in the grid. Electricity price and load are considered the most influential factors in the energy management system. Moreover, electricity price and load forecasting performed through data analytics give future trends and patterns of consumption. The energy market trade is based on price forecasting. Accurate forecasting of electricity price and load improves the reliability and management of electricity market operations. The aim of this paper is to explore the state of the art proposed for price and load forecasting in terms of their performance for reliable and efficient smart energy management systems.

1 Introduction

Smart grid (SG) is an electricity distribution network. Benefits provided by SG include: maintaining the balance between electricity supply and demand, managing energy peak load, reliability, cost efficiency, and two-way communication [1]. SG includes smart meters, renewable energy resources, and energy-efficient resources. Smart meters are being installed in many countries.

Advanced metering infrastructure (AMI) is an essential component of SG. AMI is the system that measures, collects, analyzes, transfers and stores the power related information and communicates with network. AMI collects energy data from smart meter and transfers it to utility. Hence, it creates a communication link between consumers and utility.

Electricity price and load forecasting plays an important role in decision making and operations of SG and energy industry [2,3]. It helps utilities to plan their capacity and different operations in order to reliable supply to all consumers. Electricity load and price forecasting categorization is based on time scale: short-term (ST), mid-term (MT) , and long-term (LT) forecasting [3,4]. ST covers days, MT includes months, and LT covers years. LT forecasting is used for

© Springer Nature Switzerland AG 2020
L. Barolli et al. (Eds.): EIDWT 2020, LNDECT 47, pp. 427–439, 2020.
https://doi.org/10.1007/978-3-030-39746-3_44

power system planning, while MT and ST are used in operations. Moreover, LT forecasting plays a significant role in the strategy-making of energy trading [3]. In particular, accurate forecasting maximizes the profit and minimizes the cost of electricity [5]. On the other hand, inaccuracy leads to electricity shortage, blackouts and high prices [6]. Therefore, there is a need for an effective and efficient approach to achieve accuracy.

Forecasting includes an entire set of steps, which are not often linear from top to bottom. It starts with defining the goal, getting the data, exploring the data, pre-processing the data, applying forecasting methods, comparing the results, evaluating the performance, and finally implementing some forecasting system. The whole forecasting process is depicted in Fig. 1. Integration of renewable energy sources, high volatility, uncertainty, huge data and different types of external and internal factors are the key challenges to electricity load and price forecasting. Season and weather parameters affect electricity load [1]. Various factors have an impact on electricity price, such as weather, population growth, fuel price, and different economic attributes [4,7]. These factors are considered as input features for forecasting. Furthermore, non stationary, nonlinear, and high volatility nature of electricity data make load and price forecasting complicated and challenging.

As accurate electricity price and load forecasting is essential for the stability of power industry. To improve the accuracy, researchers have proposed many approaches and models. Comprehensive literature exists in the context of electricity price and load forecasting. Though, existing models are classified into different categories based on the method used for forecasting, such as data mining, time series, artificial intelligence, machine learning and hybrid methods.

Various factors have an impact on electricity price and load such as, weather, population growth, fuel price, and different economic attributes [1,4,7]. These factors are considered as input features for training of the model. Furthermore, non stationary, nonlinear, and high volatility nature of electricity price and load dataset makes forecasting complicated and challenging.

Accurate electricity forecasting is essential for the stability of grid. Machine learning is considered well-known technique that is used for electricity forecasting. Deep learning is considered as the most promising technique as it has received success in different areas: medical, computer vision and image processing. This paper aims to provide a review on electricity price and load forecasting using different techniques. It also compares the performance of various techniques to see which technique gives best results. The limitations and future challenges are also discussed for further research.

The order of remainder paper is arranged as follows: in Sect. 2, the problems of each paper are addressed. In Sect. 3, the main contributions of each paper are summarized. In Sect. 4, performance evaluation metrics are reviewed. Limitations and future research challenges are discussed in Sect. 5.

2   Problems Addressed

This section describes the problems that have been addressed by different authors.

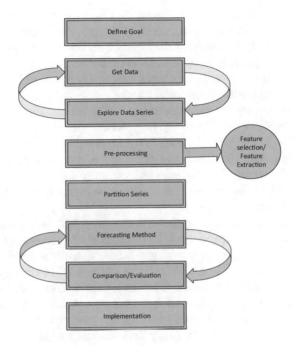

Fig. 1. The forecasting process

2.1   Electricity Price Forecasting

There exists multiple relations between input and output. According to [3], unified models used for the day-ahead electricity price forecasting (DAEPF), identify single relation between historical data and future data. It leads to errors due to fluctuations in data. Existing approaches have achieved satisfactory results for different conditions. Now it is hard to improve the accuracy considering only optimization techniques. For the further improvement of forecasting accuracy, main focus should be the formulation of the prediction problem. Overall accuracy can be improved by predicting a specific pattern.

According to [7], big data is not considered for electricity price forecasting. Furthermore, existing systems having difficulties to manage the excess amount of cost data in the grid. There is a need of an integrated system to regulate energy system. Data preprocessing is not performed well that leads to inefficient and inaccurate prediction. Accuracy could be raised by addressing these problems.

Different types of techniques have been used by researchers for forecasting power load and price [8]. Feature selection plays a significant role in forecasting models. Some researchers proposed feature selection techniques to get optimal results. The major drawback of the existing techniques is the same and fixed number of features. These features are fed to model as input. While each output have different suitable input and the other problem is the reliability of selected features that are used to train the model.

In [9], different problems are addressed. The time-series models are linear and perform well for low-frequency data whereas, electricity price signals are nonlinear and are of high-frequency that causes errors in price prediction. Moreover, the stationary process is unable to capture non-stationary features from price signals. Tuning of irrelevant features and less amount of input-output data decreased the performance of proposed model.

The main challenges for the electricity price forecasting are high voltage and non-linear behavior of electricity [10]. Moreover, external factors are also difficult to handle. These uncertainties are addressed in this work. Non-linearity in electricity price data questions the accuracy of prediction. The accuracy of forecasting models depends on the selected variables for input. Many techniques have been proposed for feature selection. How to select robust feature is the main issue that is addressed in paper [11].

All the approaches proposed to forecast the electricity price, have some drawbacks [12]. In data-driven approaches, initialization of parameters is difficult due to their sensitivity. In machine learning models, there is a need for new methods to improve accuracy. Season affects the pattern of electricity demand. ST load forecasting is more helpful for the electricity market to make a decision rather than long term electricity cost forecasting. Considering the non-linearity of the cost signal, it is very difficult to select the best feature to train the data. Robust feature can enhance the accuracy and efficiency of the forecasting models (Table 1).

2.1.1   Electricity Load Forecasting

In [2], different models have been proposed for electricity load forecasting. Explanatory variables: ambient temperature, solar radiation, relative humidity, wind speed, and weekday index are used to train the regression models. These variables varies building to building. Some of them are restricted to only one building that causes overfitting problem. It deceases the performance of the model. To overcome this problems deep learning is used as it has gained fame in many other fields due to flexible nature.

According to [13], for the last decades, researchers and scholars are proposing different models and approaches for electricity demand forecasting. Accuracy of the forecasting is affected by some factors: holiday, season, price, day and hour, economy, policy, electricity power data, and management policy. Existing techniques predict electric load without considering different factors and relationship between them. That leads to inaccuracy and difficulty in forecasting electricity load. Taking into account, pre-processing of original data enhances the accuracy of prediction that is used in the proposed technique.

Table 1. Electricity price forecasting

Task	Dataset	Data Pre-processing	Proposed Model	Evaluation	Benchmarks
Day-ahead electricity price forecasting [3]	PJM	–	k-mean clustering WVM	RMSE MAPE SDE	ARIMA RBFNN SVR ENN ELM
Short-term electricity price forecasting [4]	Iberian	–	ARIMA with other forecasting models	MAPE	ARIMA ARIMA-SVM ARIMA-RF ARIMA-LOWESS.
Electricity price forecasting [5]	New South Wales France Germany	–	DE-LSTM	MAE RMSE MAPE	BPNN DE-BPNN SVM RNN
Short-term electricity price forecasting [7]	ISO NE-CA	GCA RF Relief-F KPCA	DE-SVM	–	NB DT PCA Radial KPCA
Electricity price forecasting [8]	AEM PJM	IBSM	Updated DCANN	MAPE MAE	LSSVM BPANN GARCH DCANN
Day-ahead electricity price forecasting [9]	NYISO CAPITL GENESE	EGT-cluster HANT Time lagged analysis	BRNN	MAPE RMSE Forecast skill	BRNN + K-means BRNN + SOM BRNN + NG BRNN + EGT-Cluster
Electricity price forecasting [10]	ERCOT	–	Rerouted method (Polynomial regression SVR DNN)	MSE	Simulations over four months
Electricity price forecasting [12]	Ontario	MOBBSA	ANFIS-BSA	RMSE RACF MAPE Absolute error	ANFIS-PSO ANFIS-GA MLP-PSO MLP-GA MLP

Limited predictability of weather factors raises operational challenges for power systems in supply and demand of electricity [14]. ST forecasting (up to 12 h-ahead) improve operational efficiency of power systems. In literature there are many existing models but have low efficiency than the proposed VAR model.

Feature selection is very important for the models used for prediction [15]. If data is fed without preprocessing it increases computational time and overfitting problem. Although many researchers proposed different techniques for feature selection, on the contrary, a small number of them merged and explored feature section tools and prediction models. In different studies, it is stated that application of response concept in decentralized energy system can reduced electricity cost. Whereas, techniques used for prediction of electricity demand are not well explored. Most of the methods forecast electricity demand only on large-scale and are unable to predict specific electric power at lower-scale, that is equally important for efficient and reliable energy system.

Nonlinear and complicated patterns are the hurdles for the electricity load forecasting [16]. A semi-parametric additive model was proposed to forecast short term electricity load that is not efficient for the nonlinear and serial correlation. Artificial neural network is considered as a good approach for regression and classification, which is trained mostly using back-propagation (BP). BP can easily be trapped and also have slow convergence rate that are considered as drawbacks. The network fails if the entire data set is fed to model, but incremental learning succeeds. Empirical mode decomposition (EMD) has been used in different research fields due to good performance. To improve accuracy, efficiency, and to overcome existing problems a hybrid approach is proposed. For the short-term load forecasting many researchers proposed different models [17]. As climate, economy, holidays and other random variables are the challenges for electricity load forecasting. This paper combines a set of models that perform better than others (Table 2).

Table 2. Electricity load forecasting

Task	Dataset	Data Pre-processing	Proposed Model	Evaluation	Benchmarks
Day-ahead electricity load forecasting [1]	California Los Angeles New York Florida	SDA	SVR	MAPE	Plain SVR ANN
Short-term electricity load forecasting [2]	Nanjing Lianyungang Suzhou	–	DNN Probability density forecasting (deep learning model, quantile regression, and kernel density)	MAE MRPE MAPE	GBM RF
Short-term electricity load forecasting [6]	Ireland	–	TCMS-CNN (MS-CNN and time-cognition models)	MAE RMSE MAPE	RM-LSTM DM-GCNN DM-MS-CNN
Short-term electricity load forecasting [13]	New South Wales Queensland Victoria	EMD	EMD-Mixed-ELM (RBF kernel and UKF kernel)	AE MAE RMSE MAPE	RBF-ELM UKF-ELM Mixed-ELM
Short-term electricity load forecasting [16]	Australian Energy Market Operator	DWT-EMD	DWT-EMD based RVFL	RMSE MAPE	GLMLF-B SLFN RF RVFL EMD-SLFN EMD-RF EMD-RVFL
Short-term electricity load forecasting [17]	Italy-North	Feature-fusion module	CNN-LSTM	MAE MAPE RMSE	DT RF DE CNN LSTM

3   Contributions

In the following section, contributions made by different authors in terms of electricity price and load forecasting are described.

3.1   Electricity Price Forecasting

The basic objective targeted in [7], is to forecast the accurate electricity cost using big data. The proposed model is the combination of three modules. Initially, features are selected from random data using a combination of random forest (RF) and Relief-F algorithm. Redundancy is removed from features using grey correlation analysis (GCA). The kernel function and principal component analysis (KPCA) is used to reduce the dimensions. Finally, classification is performed, using differential evolution (DE) based on support vector machine (SVM) classifier to forecast the electricity cost.

A novel model, dynamic choice of artificial neural network (DCANN) is proposed in [8]. To overcome the problem of feature selection, index of bad sample matrix (IBSM) is proposed that identifies bad features and selects appropriate input for the corresponding output. DCANN gives input to the artificial neural network (ANN). The unsupervised learning approach is used to select input for corresponding output and a supervised learning approach is used for the training of these inputs.

To enhance the performance of clustering, enhanced game theoretic (EGT) cluster is proposed [9]. Load and price both time series are considered through 2-dimensional input selection. For the selection of robust cluster a persistence approach is introduced. Bayesian recurrent neural network (BRNN), data clustering and time-lagged signal analysis are used in proposed hybrid method.

Soft computing techniques are used to forecast the real-time market [10]. Initially, mapping is performed between historical wind power generation and historical price. The second step is forecasting rule for wing power generation. For the selection of the best model, different machine learning approaches are discussed. A hybrid forecasting framework is proposed. Systematic analysis is performed to extract the hidden features of electricity price data. To choose the input feature, the hybrid feature selection (HFS) method is developed by integrating singular spectrum analysis (SSA), cuckoo search algorithm, and SVM [11].

It revealed that classification modeling is better than unified modeling for mapping relations [3]. As the unified model considers a single relation between historical data and future data while in classification modeling, a model is established to identify the multiple relations for each pattern. A DPP is proposed that can be used with different time series forecasting algorithms. The main idea of DPP is to predict the price pattern through the conventional forecasting method and then classification is performed to improve the accuracy of the forecasting. To improve the accuracy of the pattern prediction a weighted voting mechanism (WVM) method is proposed. In WVM multiple pattern predictions are merged.

In [12], a hybrid electricity cost method is proposed in this paper. A combination of backtracking search algorithm (BSA) and adaptive neuro-fuzzy inference system (ANFIS) approach is presented to improve the accuracy of forecasting. Multi-objective binary-valued backtracking search algorithm (MOBBSA) and ANFIS are merged for optimal feature selection. The main contribution of this paper is, solution of the feature selection problem, a multi-objective feature selection approach is proposed. Feature selection approach is made of MOBBSA and ANFIS. MOBBSA selects feature subsets from different combinations of inputs and ANFIS evaluates feature subset based on the performance.

3.2 Electricity Load Forecasting

In this paper [13], EMD and ELM are used to overcome the problems of existing systems. The EMD-Mixed-ELM method, initially decomposes the electricity load data to cut the noise. EMD decomposes data into IMFs and residues, known as a trend. This de-noised data is used by ELM to forecast the electricity load. A combination of kernels, RBF and UKF kernel are integrated with ELM because only one kernel can not extract all the features of electricity demand data.

Main focus of paper [17] is to forecast the one day ahead load using hybrid incremental learning model. DWT, EMD, and RVFL are used in this model. The problem of EMD is frequency mixing that is solved by using DWT. Defined incremental DWT-EMD based on RVFL network improved both accuracy and efficiency.

To forecast short-term electricity, deep learning-based model is proposed that discovers all key factors and also probability density prediction [2]. Initially, a deep neural network is used to forecasting of electric load. DNN consist on MLP used as a deep learning model.Feature selection is performed using feature engineering and then to get electricity load pattern data visualization is employed. for load foresting purposes. Probability density forecasting is applied by merging deep learning model, quantile regression, and kernel density estimation method.

In [1], a deep learning-based multi-modal is proposed that uses four types of data sets: i.e., season parameter, previous day vector, previous three days'vector, and weather parameter. SDAs is used in deep feature extraction module. Then extracted features are used to train the SVR to predict the day-ahead electricity load.

VAR model is employed to capture correlation of three different variables [14]. Regression technique is used to capture the autocorrelation of the variables. Furthermore, VAR model also useful to capture time-lagged correlation of different correlation. Time series approach used in the proposed model, capture different characteristics of weather: linear trend, seasonal component, and stochastic component.

A hybrid feature selection based on machine learning is proposed, to get the most applicable feature for the accurate prediction of electric load in a decentralized energy system [15]. BGA is used for the feature selection. GPR is used to measure the importance of the features. Main contributions are as follows:

investigation and suggestion related to FS for accurate prediction of electric load.

To overcome the above mentioned problems [16], this paper introduced a multi-scale convolution neural network with time-cognition. Initially, a CNN network model based on MS-CNN is used for the robust feature selection. A unique periodic code is introduce to improve the accuracy of sequential model regarding time cognition. The combination of multi-scale and time cognitive feature builds an multi-step deep learning model.

It describes the use of wavelet ensembles to build a forecast model [17]. Main focus of the paper is to build a pipeline from raw data and compare its results with existing techniques. Wavelet is used in this pipeline to extract the features and for learning process.

4   Validation

In this section performance evaluation of the existing literature is discussed. Performance metrics: mean absolute percentage error (MAP), absolute percentage error (APE), root mean square error (RMSE), and mean absolute error (MAE) are widely used in the following papers.

4.1   Electricity Price Forecasting

In [7], several simulations are performed to evaluate the performance of the proposed model using real-world data. For this purpose, a simulator is developed that uses hourly electricity cost data and power generation data of ISO new England control area (ISO NE-CA) from 2010 to 2015. Naive bayes (NB) and decision tree are used as a benchmark to compare the efficiency of DE-SVM. KPCA performance is compared with principal component analysis (PCA). Simulation results show that the proposed model performs better than other benchmarks.

The performance of the proposed model [8] is validated, using original data of electricity price from PJM. Moreover, MAE and MAPE are used as performance metrics. Other benchmarks used for the comparison of the proposed model are back propagation artificial neural network (BPANN), radial basis function network (RBFN), least squares support vector machine (LSSVM), and SVM. Numerical results show that the proposed model DCANN has the best performance considering PJM dataset. Furthermore, only LSSVM and DCANN give relevant results, in high volatility electricity price. DCANN is an efficient model compared with other benchmarks.

In [9] evaluation of the proposed EGT-cluster is performed using MSE, MAPE, RMSE, and forecast-skill used to evaluate the performance. EGT-cluster is compared with k-means, original SOM and NG algorithms using different price series data. Data used to evaluate the performance of the forecasting algorithm is New York independent system operator (NYISO) market price data from 2008 to 2014.

A comparison is conducted to the conventional method verifies the higher accuracy and reliable relation between input and output [10]. Simulations results verified that the performance of the proposed model is effective and efficient. The noise was added to check the stability of the proposed model, results prove it stable model.

4.2   Electricity Load Forecasting

In this paper [13], EMD and ELM are used to overcome the problems of existing systems. The EMD-Mixed-ELM method, initially decomposes the electricity load data to cut the noise. EMD decomposes data into intrinsic mode functions (IMFs) and residues, known as a trend. This de-noised data is used by ELM to forecast the electricity load. A combination of kernels, radial basis function kernel (RBF) and UKF kernel are integrated with ELM because only one kernel can not extract all the features of electricity demand data. In [3], Time Series Cross-Validation is used to evaluate the performance. Several Metrics such as MAPE, RMSE, and MAE are also used to represent prediction accuracy. MAPE is used as a performance indicator in [4]. In [5], the accuracy of model examined via RMSE and MAPE. In [6], four metrics such as Mean Error (AE), MAE, MAPE, and RMSE are presented. MAPE is applied to evaluate the performance in [7]. In [8], accuracy metrics such as RMSE, Normalized Mean Square Error (NMSE), MAE, and MAPE are used for evaluation. In [9], NYISO and AEMO data set is used for testing the model and MAPE, RMSE, MAE, and mean percentage error (MPE) are used for performance metrics.

5   Future Work

This section enlightens the limitations of existing approaches that can be addressed in future for further improvement.

5.1   Electricity Price Forecasting

The effectiveness of the proposed model [3] is verified using data from same market, while data from different markets have different characteristics can affect the results. In time series forecasting model there are many factors with different ratio of impact and there exist a relation between them. In feature selection impact of these feature plays an important role to improve the accuracy. So correlation analysis should be used to analyze the impact of the feature. The proposed method performs multiple steps for pattern prediction of next day, accuracy of final pattern depends on the accuracy of each step. So to avoid this limit a state prediction method can be used to predict the pattern.

Computational sources used in the proposed model [7] increases the cost of the model, so it is costly to implement it on real-time data. DE-SVM is used in this paper do not give optimal results because it does not cover whole search space for the tuning of the parameters so the accuracy rate decreases.

Random forest is used for the extraction that takes random values and gives result, as in the proposed method big data is used so random forest can randomly select garbage value that affects output. Computational time increases with the complexity of the model and the proposed model is complex.

Dynamic choice artificial neural network (DCANN) is an extended form of ANN, so it has same limitations as ANN. Complications of the model increases computational time moreover, it do not cover whole search space that lead to inefficient forecasting. DCANN has maximum computational time compared to other benchmarks [8]. The proposed method in [9] is complex in nature, data scalability becomes a problem for the model by increasing the amount of data. In [10], to achieve better results features with different types of parameters should be considered. Selected features have causal relation towards electricity price.

In the proposed model [11], only electricity price data is considered that increases the complexity of forecasting process. Other factors should also be integrated in forecasting model. Proposed framework is designed using seasonal characteristics of the electricity price.

5.2 Electricity Load Forecasting

CNN and recurrent neural networks based on deep learning and can handle large and complex datasets, that can be used for better results. Location is also a significant factor that can be used for the improvement of accuracy of forecasting methods [2]. As in the proposed model [13], we are predicting electric load and claiming the accuracy and performance of the model, some external factors affect the consumption of electricity and are ignored by this model. EMD used in proposed method decomposes high-frequency signals into low frequency. Results lead to overshoot or undershoot and that shows, EMD is incapable of separating components of closely spaced frequencies.

Efficiency of the proposed model decreases in terms of small dataset. Due to complexity of the model it is expensive to train the data [15]. Proposed model in [17], works only on electricity load data, by merging proposed decomposition technique with other learning models electricity cost data can also be analyzed. Use of classification methods with proposed model can provide result in the form of intervals, scenarios, and density functions. The complexity of the model increases computational time.

6 Conclusion

Smart grids are replacing traditional grids. It involves two-way communication, generation, transmission, distribution and consumption of electricity. This paper provides a limited review, that represents the different techniques used for electricity load and price forecasting. Accurate electricity price and load forecasting is essential for the stability of grid. Machine learning is a well-known technique that is used for electricity price and load forecasting. The paper is classified in terms of problems addressed, proposed solutions, limitations, and open research challenges. Results are shown in the form of tables. This paper provides enough information, insights, and references in the area of load and price forecasting. We hope that this paper will help scientific community, researchers, and practitioners to assist in future development and to contribute to this new challenging and important area.

References

1. Tong, C., Li, J., Lang, C., Kong, F., Niu, J., Rodrigues, J.J.: An efficient deep model for day-ahead electricity load forecasting with stacked denoising auto-encoders. J. Parallel Distrib. Comput. **117**, 267–273 (2018)
2. Guo, Z., Zhou, K., Zhang, X., Yang, S.: A deep learning model for short-term power load and probability density forecasting. Energy **160**, 1186–1200 (2018)
3. Wang, F., et al.: Daily pattern prediction based classification modeling approach for day-ahead electricity price forecasting. Int. J. Elect. Power Energy Syst. **105**, 529–540 (2019)
4. Chinnathambi, R.A., et al.: A multi-stage price forecasting model for day-ahead electricity markets. Forecasting **1**(1), 26–46 (2019)
5. Peng, L., Liu, S., Liu, R., Wang, L.: Effective long short-term memory with differential evolution algorithm for electricity price prediction. Energy **162**, 1301–1314 (2018)
6. Deng, Z., Wang, B., Xu, Y., Xu, T., Liu, C., Zhu, Z.: Multi-scale convolutional neural network with time-cognition for multi-step short-term load forecasting. IEEE Access **7**, 88058–88071 (2019)
7. Wang, K., Xu, C., Zhang, Y., Guo, S., Zomaya, A.Y.: Robust big data analytics for electricity price forecasting in the smart grid. IEEE Trans. Big Data **5**(1), 34–45 (2017)
8. Wang, J., Liu, F., Song, Y., Zhao, J.: A novel model: dynamic choice artificial neural network (DCANN) for an electricity price forecasting system. Appl. Soft Comput. **48**, 281–297 (2016)
9. Ghayekhloo, M., Azimi, R., Ghofrani, M., Menhaj, M.B., Shekari, E.: A combination approach based on a novel data clustering method and Bayesian recurrent neural network for day-ahead price forecasting of electricity markets. Elect. Power Syst. Res. **168**, 184–199 (2019)
10. Luo, S., Weng, Y.: A two-stage supervised learning approach for electricity price forecasting by leveraging different data sources. Appl. Energ. **242**, 1497–1512 (2019)
11. Zhang, X., Wang, J., Gao, Y.: A hybrid short-term electricity price forecasting framework: Cuckoo search-based feature selection with singular spectrum analysis and SVM. Energ. Econ. **81**, 899–913 (2019)
12. Pourdaryaei, A., Mokhlis, H., Illias, H.A., Kaboli, S.H.A., Ahmad, S.: Short-term electricity price forecasting via hybrid backtracking search algorithm and anfis approach. IEEE Access **7**, 77674–77691 (2019)
13. Chen, Y., Kloft, M., Yang, Y., Li, C., Li, L.: Mixed kernel based extreme learning machine for electric load forecasting. Neurocomputing **312**, 90–106 (2018)
14. Yixian, L.I.U., Roberts, M.C., Sioshansi, R.: A vector autoregression weather model for electricity supply and demand modeling. J. Mod. Power Syst. Clean Energ. **6**(4), 763–776 (2018)

15. Eseye, A.T., Lehtonen, M., Tukia, T., Uimonen, S., Millar, R.J.: Machine learning based integrated feature selection approach for improved electricity demand forecasting in decentralized energy systems. IEEE Access **7**, 91463–91475 (2019)
16. Tan, T.Y., Zhang, L., Lim, C.P.: Adaptive melanoma diagnosis using evolving clustering, ensemble and deep neural networks. Knowl.-Based Syst. **187**, 104807 (2019)
17. Qiu, X., Suganthan, P.N., Amaratunga, G.A.: Ensemble incremental learning random vector functional link network for short-term electric load forecasting. Knowl.-Based Syst. **145**, 182–196 (2018)

Genetic Algorithm Based Bi-directional Generative Adversary Network for LIBOR Prediction

Xiao Tan[✉]

Department of Computer Science, University of Hong Kong, Hong Kong, China
U3546092@connect.hku.hk

Abstract. LIBOR (London Inter Banking Offered Rate) is one of the most important indicators of global currency liquidity risk. LIBOR market, involving top 18 member banks (including HSBC, Citibank, Bank of Tokyo-Mitsubishi UFJ, Credit Suisse etc.) and thousands non-member banks crossing different continents, is the huge market for banks to keep liquidity and currency flow globally. Because LIBOR is so important, decided by so many huge banks together and impacted by both current demand and supply of monetary currency and the forecast of future market, therefore the prediction is quite challenging. This paper is to introduce genetic algorithm ("GA") based bi-directional generative adversary network ("BiGAN") to predict the LIBOR in USD. Both the pro and cons of the algorithm will be discussed, with fitness values and Mean Squared Error ("MSE"). 50 test cases are executed randomly to verify the performance of the predictions. The target variance between predication and actual value is no more than 0.015.

1 Introduction

LIBOR (London Inter Banking Offered Rate) is an important measure of liquidity risk in global monetary market. LIBOR market offers international active banks to get funds from the wholesale and unsecured funding market with the particular LIBOR in different currencies and tenors. LIBOR is the reference interest rate for transactions in offshore Eurodollar markets, with the product scope covering the derivatives, bond and loans, and consumer lending instruments including mortgages and student loans. The major LIBOR products include the maturity of 1 day, 1 week, 1 month, 2 months, 3 months, 6 months and 12 months with the five currencies of USD, EUR, GBP, JPY, CHF. Because it is not only linked with the supply and demand of the money currency in the market, but also impacted by the estimation of the changes of the external macro environments in different regions and countries, the market expectation of central bank interest rate, liquidity premium in the money markets and the indicator of liquidity health of the banking system. Therefore, the prediction of LIBOR is quite challenging.

This research is to propose the new algorithm of genetic algorithm based bi-directional generative adversary network ("BiGAN") to predict the future LIBOR. With genetic algorithm as the optimizer, the BiGAN is adopted to predict the time

© Springer Nature Switzerland AG 2020
L. Barolli et al. (Eds.): EIDWT 2020, LNDECT 47, pp. 440–447, 2020.
https://doi.org/10.1007/978-3-030-39746-3_45

series dataset, e.g.: LIBOR with higher accuracy than traditional recurrent neural network. Such performance is assessed and proven in experiment phase with fifty random test cases using the dataset of LIBOR with the term of 1 Month in USD.

1.1 Generative Adversarial Network ("GAN")

Generative adversarial network (GAN) was introduced first by Goodfellow in 2014, which is a deep neural net architecture comprised of a pair of "adversarial" models called the generator and the discriminator. The generator is a generation process to map a latent distribution to a data distribution; and the discriminator evaluates the generation quality and, with some policy, detects whether the data distance is true or fake. The formula of GAN is shown as below

$$\max(V(G, D)) = Ex \sim pdata[logDG(X)] + Ex \sim pg[log(1 - DG(X))]$$

Where pdata is the generated value by train dataset; pg is the generated value by latent space. $Ex \sim pdata[logDG(X)]$ is the discriminator to distinguish the valid value generated by the generator while $Ex \sim pg[log(1 - DG(X))]$ is the discriminator to distinguish the fake value generated by the generator. The architecture of a standard GAN is shown as Fig. 1.

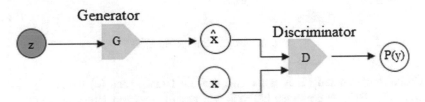

Fig. 1. Architecture of standard GAN

The faked x value generated from latent space by generator, with the train dataset as of x as the valid value, are discriminated by discriminator to get the predicted values.

1.2 Bidirectional Generative Adversarial Network ("BiGAN")

As an extension of GAN, BiGAN, as proposed by Donahue et al. in 2016 and Dumoulin et al. in 2016, is constructed with three models. Except for generator and discriminator, encoder E is added as a learning process to invert the generator G. In the architecture of BiGAN, the discriminator D discriminates not only in the data space (x versus G(z)) which is for discrimination in standard GAN, but also in data and latent spaces (tuples (x, E(x)) versus (G(z), z)), where the latent component is either the encoder output E(x) or generator input z. The objective function of BiGAN is defined as:

$$\min_{G,E} \max_{D} V(D,E,G) = E_{x \sim p_x} \underbrace{E_{z \sim p_{E(\cdot|x)}}[logD(x,z)]}_{logD(x,E(x))}$$

$$+ E_{z \sim p_z} \underbrace{E_{x \sim p_{G(\cdot|z)}}[1 - logD(x,z)]}_{log(1-D(G(z),z))}$$

Where $logD(x, E(x))$ is the non-linear discrimination function of the distribution generated by encoder, and $log(1 - D(G(z), z))$ is the discrimination of the distribution generated by generator.

The architecture of BiGAN is shown as Fig. 2:

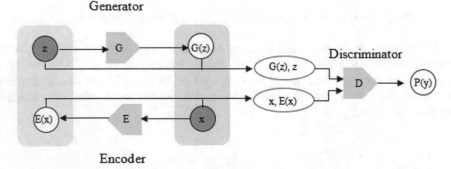

Fig. 2. Architecture of BiGAN

Unlike the standard GAN which maps from latent space (z) to data (x) as one-direction, the BiGAN conducts bidirectional mappings: from latent space (z) to data (x) by generator, and from data (x) to latent space (z) adversarially via encoding process. Both the input of latent space (z) and the output of data (E(x)) are the targets to be discriminated by the discriminator.

When predicting LIBOR, the predicted value is not Libor on t + 1 but the average of the latest 8 values of log 2 of the difference between predicted value on t and actual value on t. It can be described as formula showed as below:

$$P(X)_{t+1} = \text{Average}(\log 2(Y_{t-7} - X_{t-7}), \log 2(Y_{t-6} - X_{t-6}), \dots \log 2(Y_t - X_t))$$

Where $P(X)t + 1$ is the value to be trained in BiGAN model, Yt is the predicted value, Xt is the actual value.

The architecture of BiGAN for this dissertation is shown as Fig. 3.

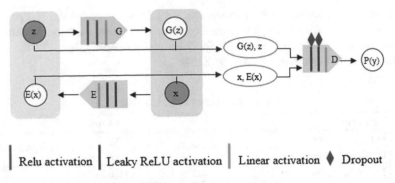

Relu activation | Leaky ReLU activation | Linear activation ◆ Dropout

Fig. 3. Architecture of BiGAN for LIBOR prediction

1.3 Genetic Algorithm as the optimizer

Genetic Algorithm is integrated with BiGAN as the optimizer. The specific deployment in this algorithm is that the predicted value needs to be calculated as $\exp2(P(X)_{t+1}) + X_t$. The actual value, predicted value, difference between predicted and actual value, $\log10$ of the difference and the reward, as five genes, construct the one-dimension chromosome, then construct the populations to enter the optimization process. After the population is built up, the genetic algorithm initiated following the four steps:

1. Fitness: To assign value to the gene of "reward" based on whether the difference between the output and the actual value is less than the targeted value as of 0.015 or not. If the difference is less than 0.015, the "reward" is assigned to "1", otherwise it is "0".
2. Random selection: The chromosome datasets after the "Fitness" step are selected randomly for the next generation;
3. Crossover: Two chromosome datasets crossover with the uniform random probability and generate the child with same gene structure but different values;
4. Mutate: A uniform random value is added to each child with the probability less than mutation power, then generate the new population.

Genetic algorithm can help to generate the optimized predictions.

2 Experiment Setup

2.1 Dataset and Preprocessing

The Libor in USD with maturity of 1month is adopted as the dataset for the experiment. The dataset is downloaded with the file name as of "USD1MTD156N" from Federal Reserve Economic Database (https://fred.stlouisfed.org/). The null value of LIBOR is filtered out during preprocessing.

When building up the experiment for BiGAN, the chosen training dataset size is set as 1000 and test dataset size is set as 200. To avoid overfitting, the training dataset and test dataset are separated. After training, the next 15 values after the test dataset are predicted. The test dataset is picked up from the total dataset randomly and run the experiments to generate the test results for 50 times consecutively.

The parameters are given as below (Table 1):

Table 1. Parameters of GA + BiGAN.

No.	Parameters	Values
1	Population size	215
2	Crossover probability	0.6
3	Mutation power	0.015
4	Evaluation rate (Difference between actual and predicted values/actual value)	0.015
5	Latent dimension	1
6	Batch size	50
7	Iterations	8000
8	Leaky ReLU slope	0.1
9	Dropout	0.2
10	Optimizer	RMSE (0.00002, clipvalue = 1, decay = 1e−8)

2.2   Computer Module and OS Parameters

All computations were executed on 2.3 GHz quad-core processors (Turbo Boost up to achieve 3.8 GHz) PC computer with 8 GB RAM iOS.

3   Results and Analysis

Two types of experiments are executed for performance evaluation. The single sample experiment is to use same dataset for prediction with different algorithms: GA based BRNN and GA based BiGAN. Similar to the experiments in Sects. 1 and 2, fitness value and Mean Squared Error ("MSE") are adopted as the measurements of the performance evaluations. The Libor for prediction is from 13th Apr 2018 to 4th May 2018. And the evaluation results are listed as below:

Table 2. Single example: comparison between two algorithms

Evaluation item	GA + BRNN	GA + BiGAN
Fitness	0.2666666666666666	0.5333333333333333
MSE	0.00046446	0.00035463
Graph of distribution		

(*continued*)

Table 2. (*continued*)

Evaluation item	GA + BRNN	GA + BiGAN
Graph of the comparison between actual and predicted values		
Execution time	20 min	12 min

In Table 2, it can be observed that comparing GA based BRNN and GA based BIGAN, although the fitness value and MSE of BiGAN is better than the ones of BRNN, BRNN shows better performance when simulating the wave shape and trend. This is because of the architecture of bidirectional recurrent neural network, which predicts the future value based on the historic sequence values.

However, GA based BRNN spends longer time for execution than BIGAN. The former needs 20 min to complete the training and prediction while BIGAN needs 12 min for execution.

The second experiment is to execute 50 test cases randomly with the test dataset size of 215. The main measures of the algorithm performance, including Mean Squared Error ("MSE"), the fitness value, the average of differences and the standard deviation values are listed as the Table 3. Fitness value in this experiment is calculated based on the difference between the actual LIBOR and the predicted LIBOR. For a specific timepoint, if the difference is less than 0.015, the fitness value is 1, else it is recorded as 0.

Table 3. Summary of performance evaluation

Type of algorithms	Number of test cases	Average fitness	Average MSE	Average execution time
GA + BiGAN	50	0.592	0.000667739	12 min

Table 3 shows that the average fitness is 0.592, which means 59.2% of the predictions with the value precision higher than 98%. And average MSE is 0.000667739. Comparing the performance indexes of GA based RNN, the fitness value of GA based BiGAN got improved, but the average MSE is not so satisfying.

The output details, including evaluation values and graph of each test scenario are listed as the Table 4.

Table 4 shows the prediction performances of fifty random test cases. Comparing those results with the ones generated by GA based BRNN, the key issue of the GA based pure BiGAN can be observed that without RNN to be deployed as discriminator or generator or encoder, the time trend or wave shapes cannot be simulated quite well even though the difference between the prediction and the actuals is quite close.

Proposal of an Interactive Brainstorming Environment for Various Content Sharing and Meeting Progress

Ryo Nakai[1]([✉]) and Tomoyuki Ishida[2]

[1] Ibaraki University, Hitachi, Ibaraki 316-8511, Japan
18nm728a@vc.ibaraki.ac.jp
[2] Fukuoka Institute of Technology, Fukuoka, Fukuoka 811-0295, Japan
t-ishida@fit.ac.jp

Abstract. In this study, we have implemented the interactive brain brainstorming environment to support brainstorming that generates unique and new ideas. This system provides integrated support for the generation of ideas and the convergence of ideas in the brainstorming field. This system provides users with specialized functions for each phase support by seamlessly switching the functions during the divergent phase (generating ideas) and the convergent phase (organizing ideas) of brainstorming.

1 Introduction

In recent years, brainstorming is often used as an idea generation method in companies. Brainstorming is a method of generating original ideas while giving opinions freely in groups. Usually, brainstorming involves grouping and organizing highly related ideas after giving them ideas. After grouping ideas, draw a final conclusion based on a pre-set theme. In such idea generation methods, sticky notes and whiteboards are often used. However, many of the ideas obtained by the idea generation method are reusable, and a new sticky note must be created again when using a sticky note with a reusable idea. For this reason, electronic brainstorming that manages ideas by digitizing sticky notes has been proposed [1–3].

2 Purpose of This Study

In this study, we propose a "CrossIT" that supports which supports the brainstorming (divergence phase) and the convergence phase (organizing ideas) after brainstorming using a large display and a tablet. CrossIT realizes CrossIT VIS, which interactively visualizes ideas obtained from participants on the large display and organizes the ideas. CrossIT also realizes CrossIT WEB, which shares ideas from the web browser of the tablet to CrossIT VIS. In addition, our proposal system can share various multimedia contents in addition to text-based ideas during the divergent phase. CrossIT provides users with flexible brainstorming support by seamless switching between the divergent and convergent phases.

L. Barolli et al. (Eds.): EIDWT 2020, LNDECT 47, pp. 448–457, 2020.
https://doi.org/10.1007/978-3-030-39746-3_46

3 System Configuration

Cross IT system configuration is shown in Fig. 1. CrossIT consists of CrossIT WEB (web application used by meeting participants), CrossIT VIS (desktop application that displays content transmitted from CrossIT WEB on the large display), and Idea Sharing Server.

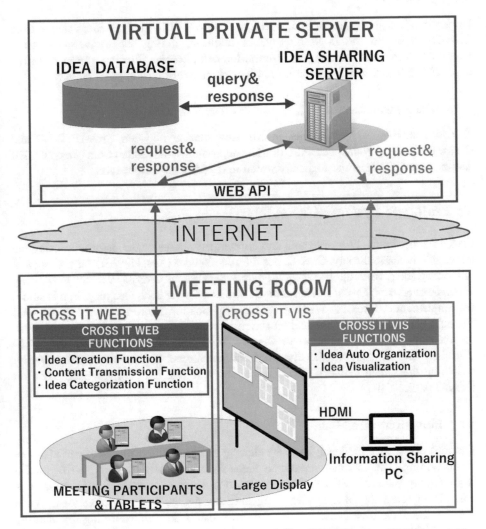

Fig. 1. CrossIT system configuration consisting of CrossIT WEB, CrossIT VIS, and Idea Sharing Server.

3.1   CrossIT WEB

During the divergent phase of brainstorming, CrossIT WEB transmits contents (images, PDFs, URLs) and ideas on the tablet to CrossIT VIS via the idea sharing server. In addition, during the convergent phase, CrossIT WEB provides users with a category creation function and an idea classification function for shared ideas.

3.2   CrossIT VIS

During the divergent phase of brainstorming, CrossIT VIS visualizes ideas shared from CrossIT WEB on the large display. In addition, during the convergent phase, CrossIT VIS receives the category information and idea classification transmitted from CrossIT WEB and visualizes the ideas for each category on the large display.

3.3   Idea Sharing Server

The idea sharing server manages information sharing between CrossIT WEB and CrossIT VIS using Web Socket API. In addition, contents transmitted from CrossIT WEB and proposed ideas are automatically stored in the idea database server.

4   Prototype System "CrossIT"

CrossIT supports the idea divergent and convergent phases of brainstorming using the tablet and the large display. CrossIT has the functions of CrossIT WEB for sharing and organizing ideas from the tablet on the large display and CrossIT VIS for visualizing ideas received from CrossIT WEB on the large display for each category. In this study, we implemented CrossIT WEB as a web application running on the tablet and CrossIT VIS as a desktop application running on the PC. After starting CrossIT VIS, the meeting participants connect to CrossIT WEB from the tablet and begin brainstorming. CrossIT supports the divergent and convergent phases of brainstorming by switching the screen between idea divergence mode and idea convergence mode on CrossIT VIS.

4.1   Idea Divergence Mode

The idea divergence mode supports the sharing of ideas during the divergent phase of brainstorming. The user can share their ideas from the tablet (CrossIT WEB) on the large display (CrossIT VIS) by starting the idea divergence mode. In this way, it is possible to reuse the idea by digitizing and storing the brainstorming idea. When starting the idea divergence mode, the user can start brainstorming by starting CrossIT VIS on the PC and selecting the "Start" button on the login screen shown in Fig. 2.

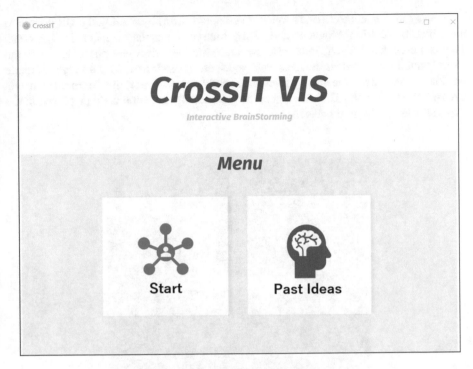

Fig. 2. CrossIT VIS startup screen.

When CrossIT VIS idea divergence mode is started, QR code and authentication code are displayed on the large display as shown in Fig. 3. This authentication code is used for connection authentication from CrossIT WEB.

Fig. 3. QR code and authentication code for connecting to CrossIT WEB.

Users can connect to CrossIT WEB (Web application) by reading the QR code into the tablet. In addition, "Switch" and "Sort" buttons were placed under the QR code. The user uses the "Switch" button for switch to the idea divergent phase, and uses the "Sort" button for sort ideas, images, web pages, etc. shared in the idea divergence mode on the screen. The meeting participants authenticate from the login screen via CrossIT WEB (Fig. 4). The user enters the 4-digit authentication code displayed in the CrossIT VIS on the login screen.

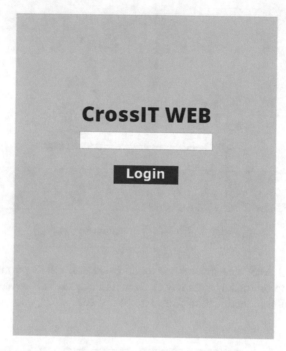

Fig. 4. CrossIT WEB login screen. The user enters the 4-digit authentication code on this screen.

When the user enters the authentication code on the login screen of CrossIT WEB and selects the "Login" button, an authentication request is transmitted to the idea sharing server. The idea sharing server collates the transmitted authentication request with the CrossIT VIS authentication code, and transmits authentication success information to CrossIT WEB. When CrossIT WEB receives authentication success information, the top screen is displayed (Left of Fig. 5). On the top screen of Cross IT, the user can move and enlarge/reduce idea contents, images and web pages shared on the large display. The user moves the content by scrolling the screen and enlarge/reduce the content by pinch-in/pinch-out. In addition, the menu screen is displayed by selecting the menu button on the top right of the top screen (Right of Fig. 5).

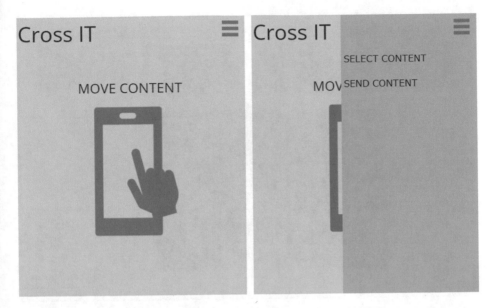

Fig. 5. Left is Cross IT WEB top screen, and right is Cross IT WEB menu screen.

Fig. 6. Switching operation of the content sharing screen. The user selects the button at the bottom of the screen according to the content.

On the menu screen, the user can select a content selection screen for switching content to be remotely controlled and a content sharing screen for sharing content such as ideas and web pages on the large display. The user can share text (devised ideas), images, web pages, and document files on the content sharing screen. The user switches the screen according to the shared content (Fig. 6).

The user can interactively share content by selecting content such as images and flicking content on the tablet. When the user transmits content by flicking from the content sharing screen, the content is transmitted to CrossIT VIS. After receiving the content, CrossIT VIS visualizes the transmitted content on the large display in real time (Fig. 7).

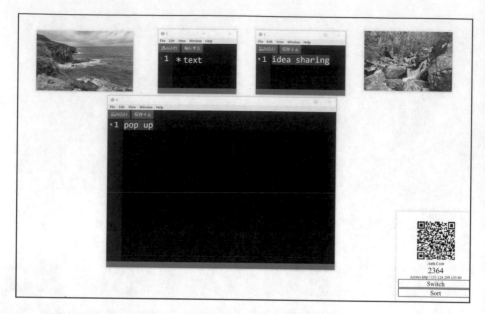

Fig. 7. Screen on the large display when sharing text content to CrossIT VIS.

4.2 Idea Convergence Mode

The idea convergence mode supports the organization of various ideas by using the category classification function of ideas. The user can switch from the idea divergence mode to the idea convergence mode by selecting the "Switch" button in Fig. 3 described above. When the idea convergence mode is started, all the contents displayed on the large display are once deleted from the shared screen. At the same time, the items on the menu screen of CrossIT WEB are switched (Fig. 8).

We implemented a function to categorize ideas shared by users participating in brainstorming in the idea divergence mode. When the user categorizes ideas using CrossIT WEB, the uncategorized content list screen is used (Left of Fig. 9). After the user selects the content to be classified on the uncategorized content list screen, the content categorization screen is displayed (Right of Fig. 9).

When the user selects a category from the content categorization screen, the idea of the selected category is displayed on the information sharing screen of CrossIT VIS (Fig. 10). When the user adds an idea displayed on the uncategorized content list screen to a new category, enter the category name in the input form of the content categorization screen. This creates a new category window on the information sharing screen of CrossIT VIS.

On the CrossIT WEB category list screen (Fig. 11), the created category labels and categorized contents (ideas) are listed. The user can re-register a category name by selecting the correct button located to the right of each category name on the category list screen.

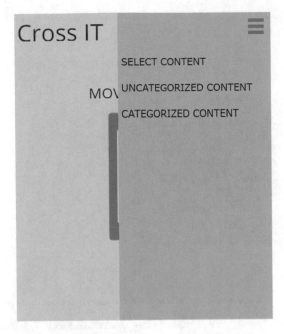

Fig. 8. CrossIT WEB menu screen in idea convergence mode.

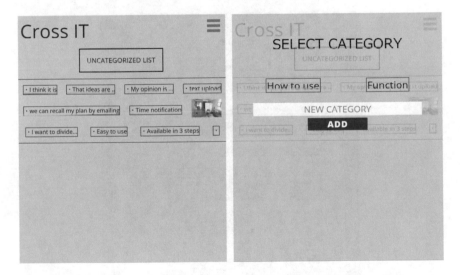

Fig. 9. Left is uncategorized content list screen, and right is content categorization screen.

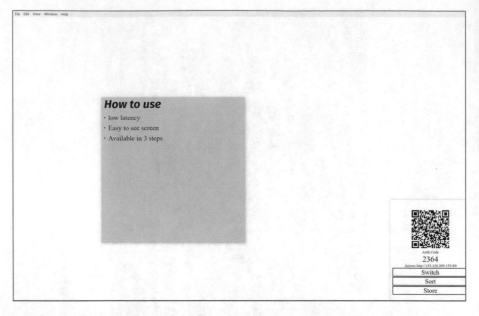

Fig. 10. Information sharing screen when "How to use" category is selected on the content categorization screen.

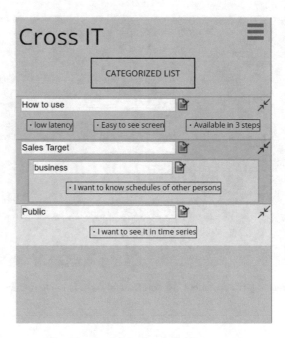

Fig. 11. Categorized content list screen.

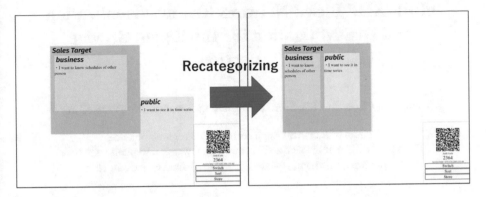

Fig. 12. Associating minor categories with major categories.

When the user selects each category on the category list screen, the content categorization screen is displayed, and the selected category can be associated with another category. Figure 12 shows the information sharing screen when the "public" category is associated with the "Sales Target" category. With this function, multiple minor categories can be treated as major categories.

5 Conclusion

In this study, we have implemented an interactive brainstorming environment "CrossIT" for various content sharing and meeting progress. CrossIT consists of CrossIT VIS, which visualizes and organizes brainstorming ideas on the large display, and CrossIT WEB, which transmits ideas from the tablet terminal to CrossIT VIS. CrossIT provides meeting participants with a new idea generation method by seamlessly switching between the idea divergent phase and the idea convergent phase during brainstorming.

References

1. Chan, J., Dang, S., Dow, S.P.: IdeaGens: enabling expert facilitation of crowd brainstorming. In: Proceedings of the 19th ACM Conference on Computer Supported Cooperative Work and Social Computing Companion, pp. 13–16 (2016)
2. Ito, E., Suzuki, T., Niwa, Y., Ozono, T., Shintani, T.: Developing a recommendation mechanism to reuse brainstorming cases. In: Proceedings of the 31st Annual Conference of the Japanese Society for Artificial Intelligence, pp. 1–4 (2017)
3. Munemori, J., Sakamoto, H., Itou, J.: Development of idea generation consistent support system that includes suggestive functions for preparing concreteness of idea labels and island names. IEICE Trans. Inf. Syst. **E101-D**(4), 838–846 (2018)

eBPF/XDP Based Network Traffic Visualization and DoS Mitigation for Intelligent Service Protection

YoungEun Choe, Jun-Sik Shin, Seunghyung Lee, and JongWon Kim[✉]

School of Electrical Engineering and Computer Science,
Gwangju Institute of Science and Technology, Gwangju, Republic of Korea
{yechoe,jsshin,shlee,jongwon}@nm.gist.ac.kr

Abstract. Kubernetes is expected to be widely used for intelligent services, leveraging multiple machines working as a cluster. When machines are connected through the internet, they exploit security issues regarding the network such as DoS attacks. We intend to apply eBPF/XDP to monitor and filter the packets for an intelligent service running on a Kubernetes cluster of two physical machines. This paper validates the performance of eBPF/XDP packet filtering against a HTTP get, post request flood DoS attack. Furthermore, we visualize the network status of network interfaces based on the number of incoming packets retrieved by our eBPF/XDP program.

1 Introduction

With the emergence of AI (Artificial Intelligence), which requires a massive amount of computing power, it is becoming more common to use HPDC (High Performance Distributed Computing). By using HPDC, a user can leverage multiple servers as a cluster to support computation of an intelligent service and use the servers as a storage as well for the data to be processed. Also, to aid intelligent services' maintenance, it is becoming more popular to use containers to prevent service failures from version updates of intelligent service components. For this purpose, an open-source container-orchestration system for automating application deployment, scaling, and management such as Kubernetes is being used by corporations and developers. Most companies are either running or considering Kubernetes as a platform for their workloads, in which AI takes an important part [1]. As intelligent services gain popularity, there will be more intelligent services that use Kubernetes to run multiple physical machines as a cluster. Despite all of the benefits, having physical machines connected through the internet will propose various security issues.

A DoS (Denial of Service) attack can be considered as one of the significant security issue that an intelligent service using Kubernetes possesses. DoS attacks have been performed countless times throughout the history of the Internet since the first DoS attack by 13-year-old David Dennis in 1974. DoS attacks prevent servers or networks from providing their services [2]. One of the common ways to launch a DoS attack is to send extensive amount of HTTP get or post requests, saturating the computation resources of

© Springer Nature Switzerland AG 2020
L. Barolli et al. (Eds.): EIDWT 2020, LNDECT 47, pp. 458–468, 2020.
https://doi.org/10.1007/978-3-030-39746-3_47

the victim server. We implement eBPF (extended Berkeley Packet Filter)/XDP (eXpress Data Path) for DoS mitigation and measure its performance. In addition, we visualize the network traffic status of the network interfaces of our test environment based on incoming packet numbers using data retrieved from our eBPF/XDP program. We expect that having each server run their own eBPF/XDP programs to monitor and filter their incoming packets instead of having just a single firewall will contribute to evading DoS attacks and protecting data that an intelligent service facilitates.

2   eBPF/XDP Packet Filtering and Monitoring

2.1   eBPF/XDP Actions

eBPF and one of its modes, eBPF/XDP are part of the Linux Foundation's Open-source project, IOVisor. An eBPF program initiates a virtual machine in the Linux kernel. eBPF provides a user with the ability to write programs in restricted C which is then executed when a set of kernel hook points invoke the eBPF program [3]. eBPF/XDP allows the eBPF code to be executed within the device driver context [4] and with an appropriate hardware, on the NIC (Network Interface Card) itself [5]. eBPF/XDP supports five actions, XDP_PASS, XDP_DROP, XDP_TX, XDP_REDIRECT and, XDP_ABORT [6] and their operations are as depicted in Table 1.

2.2   eBPF/XDP Modes

eBPF/XDP can only process incoming packets. The way received packets are processed by eBPF/XDP depends on the mode that the eBPF/XDP operates on, which are generic, native and hardware offload. Generic mode executes an eBPF/XDP program after a socket buffer is allocated to a packet [7]. This mode takes the longest time to process a packet but it does not have any hardware requirements and can be executed on any NIC. Native mode executes eBPF/XDP program in the networking driver's early receive path [8]. Native mode has better performance than generic mode, however, native mode requires a network driver support and not all NIC can execute this mode. With hardware offload mode, eBPF/XDP program is offloaded entirely into a NIC instead of being executed on the host CPU [8]. Hardware offload mode shows the best performance amongst the modes available in eBPF/XDP. Nonetheless, hardware offload mode demands NIC hardware support and is typically implemented in Smart NICs.

Table 1. eBPF/XDP actions

Action	Operation
XDP_PASS	Pass the packet to the normal network stack
XDP_DROP	Instruct the driver/NIC to drop the packet
XDP_TX	Bounce the packet-page back out the same NIC
XDP_REDIRECT	Redirect RAW frames to another device
XDP_ABORT	eBPF program error (drop the packet)

2.3 eBPF Map

An eBPF map is used to keep state between invocations, to share data between eBPF kernel programs, and between kernel and user space applications [9]. According to [10], eBPF map is a generic data structure for storing various data types. Multiple eBPF maps can be created and accessed by multiple eBPF programs in parallel. eBPF map has four attributes: type, maximum number of elements, key size in bytes, and value size in bytes. eBPF provides commands to access the maps, and eBPF maps are accessed via file descriptors.

2.4 BPF Compiler Collection

We adopt BCC (BPF Compiler Collection) to make a eBPF/XDP Packet Monitoring and Filtering program. BCC is a toolkit for kernel tracing and manipulation programs which leverages eBPF. The kernel space program is written in C language while the user space program can be written in Python or Lua language [11]. We used python to make our user space program of the eBPF/XDP program.

3 eBPF/XDP Kernel Space Program for Packet Monitoring and Filtering

3.1 eBPF/XDP Program Load onto a Network Interface

When our eBPF/XDP program is initiated, the kernel space program of our eBPF/XDP program is loaded onto a network interface as hardware offload mode. Our program uses two eBPF maps. The first map named packets is used to record the number of incoming packets. Another map, ip_address, keeps track of the source IP addresses of the incoming packets.

3.2 IPv4 Protocol Check

When a packet reaches the network interface that our eBPF/XDP program is attached to, our program confirms that the incoming packet is long enough to contain a full Ethernet header. If not, the packet is dropped. Otherwise, our program reads the value of h_proto of struct ethhdr defined in the Linux header, if_ether.h. The value of h_proto is compared with ETH_P_IP, a constant value designating IPv4. Before the comparison is made, our program converts the value of ETH_P_IP from host byte order to network byte order. This step is necessary since the network byte order is always defined as big-endian while the host byte order could differ based on the machine being used [12]. Our eBPF/XDP program is designed for IPv4 packets. If h_proto does not match with ETH_P_IP, the packet is dropped.

3.3 Source IP Address Retrieval

After verifying the protocol of the packet, our program facilitates iphdr struct pointer defined in the Linux header, ip.h. Our program sets the pointer to the address where the IPv4 header begins in the packet and checks the length of the IPv4 header of the packet. If the address of the first byte after the IPv4 header is larger than the address of the first byte after the packet, it indicates that the IPv4 header's address that is being considered exceeds the address range of the IP packet. In this case, our kernel space program drops the packet. Our kernel space program reads the source IP address value saved in saddr of struct iph.

3.4 Packet Filtering Based on Source IP Addresses

Once the kernel side program retrieves the source IP address from the incoming packet, the value is compared with the source IP address block list, which a list of source IP addresses that we wish to block. If the source IP address of the packet belongs to the list, the packet is dropped. Otherwise, the source IP address of the packet is updated into the eBPF map, ip_address and the value stored in the eBPF map, packets is incremented by one.

4 eBPF/XDP User Space Program for Packet Monitoring and Filtering

4.1 Kafka

Before the eBPF/XDP program is initiated, we have a Kafka zookeeper, and three Kafka brokers containerized and running with a Kafka topic created and specified in our program. The Kafka consumer part that uses the Kafka brokers and the Kafka topic stated earlier resides in our visualization program. When the user space program is initiated, it starts a Kafka producer connected to the Kafka brokers, topic, and the zookeeper.

4.2 IP Address Conversion from a Machine Friendly Form to a Human Friendly Form

When our user space program receives the source IP address of a packet from the eBPF map, ip_address, the source IP address value is not written in a human friendly form (i.e. 192.168.1.1). Instead, it is written in a string of integers (i.e.14145922). Our user space program translates the IP address into a human friendly form. Upon receiving the source IP address from the kernel space, our user space program converts the received integer string into a binary string and inspects if the converted binary string contains twenty-eight digits. If not, zeros are filled in front of the string until the string is twenty eight digits long. Then, the binary string is divided into four segments, each segment eight digits long. Then the four segments are translated into decimal integer numbers. Afterwards, the order of the segments are reversed with dots inserted in between the segments.

4.3   Data Transmission via Kafka

The user space program retrieves the value of the eBPF map, packets and adds it at the end of the translated source IP address with a space in between. The data from our user space program is transmitted via Kafka every five seconds. When there is no packet received on the network interface that the eBPF/XDP program still is attached to, the user space program transmits '0.0.0.0 0' value to the visualization program.

5   Onion-Ring Visualization Program

5.1   Plotly Sun Burst Charts

We use Plotly Python open source graphing library for our visualization program. According to [13], Plotly is used to make interactive graphs that can be used for publications. Plotly supports various graphs such as scatter plots, line charts, bar charts, pie charts, error bars, heatmaps, and etc. We use one of their supported charts, the sun burst chart. It is useful for visualizing hierarchical data. When the visualization program is executed, the network interface status is visualized on a web browser such as FireFox or Chrome.

5.2   Retrieving the Source IP Address and Incoming Packet Number

Our visualization program is built with Python. When our visualization program receives the data sent by our eBPF/XDP program, our visualization program separates the incoming packet number from the source IP address. Two data are saved into two separate variables.

5.3   Coloring Criteria Based on the Incoming Packet Number

Our visualization program inspects the incoming packet number. Inside the program, we define a criteria as to how the chart should be colored. Our visualization program allocates a corresponding color to the number group following the criteria shown in Table 2.

Table 2. Coloring criteria

Color	The incoming packet number domain
Gray	0
Yellow Green	1–700
Light Green	701–1400
Green	1401–2100
Red	2101–2800

5.4 Source IP Address

Our visualization program shows the source IP address of the incoming network traffic with a mouse over function. When the user places the mouse pointer on the top of the figure that represents each network interface, all source IP addresses corresponding to the traffic is displayed. The data that is received by the visualization program is also printed in the terminal that the program is being executed on as well. The final out come of the visualization program is as shown in Fig. 1.

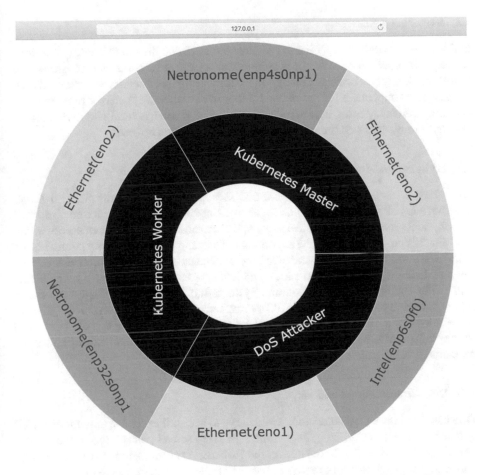

Fig. 1. Visualization program showing three physical servers (*Kubernetes Worker, Kubernetes Master, and DoS Attacker*) and their network interfaces. There is only Ethernet network traffic (*yellow green*) and no network traffic in other NICs (*gray*) in each server.

5.5 Onion-Ring

Our Onion-Ring visualization program shows the entire machine in the test environment and their network interfaces. Physical servers are colored black in the first layer right above the center circle. Network interfaces are depicted in the second layer, each attached to the physical machines they belong to. Network interfaces are colored according to our coloring criteria.

6 eBPF/XDP Based DoS Mitigation Performance

6.1 Kubernetes Service

In this paper, we utilize a specific IoT-Cloud service called the Smart Energy IoT-cloud service [14] based on microservices architecture by using Kubernetes orchestration as a DoS attack target. The Smart Energy IoT-Cloud service is a service that recommends the adequate temperature for the air conditioner in our laboratory's server room in terms of energy efficiency using machine learning. Also, the service visualizes and monitors the server room status on the dashboard accessed via a web browser.

6.2 Kubernetes Master, Kubernetes Master and DoS Attacker

This paper considers a scenario where a DoS attacker with a single physical server launches a HTTP get, post flood attack against a Kubernetes cluster service comprised of two physical servers. Three physical servers, Kubernetes master, Kubernetes worker, and DoS attacker are facilitated for the experiment. Their Ethernet ports, using public IPs are connected to the M4300 switch manufactured by 'Netgear Inc.' which is connected to the internet. In addition, Kubernetes master and Kubernetes worker each have a single 10G NIC card manufactured by 'Netronome Systems Inc.' that supports hardware offload XDP connected to the Netgear M4300 switch using local IP addresses. DoS attacker server also has a 10G NIC manufactured by 'Mellanox Technologies Ltd' connected to the Netgear M4300 switch using a local IP address. The environment set up for the experiment in this paper is shown in Fig. 2.

6.3 DoS Attack Using Goldeneye

DoS attacker uses a DoS attack tool, Goldeneye to transmit excessive amount of HTTP get, and post requests utilizing five workers with ten socket connections to the Kubernetes cluster service. Due to the attack, the Kubernetes cluster service consumes more hardware resources to return replies to the HTTP requests made by the attacker. Goldeneye does not support IP spoofing, which means that the incoming attack at the Kubernetes cluster service has a constant source IP address. To verify the performance of the eBPF/XDP program's packet filtering, the average CPU usage is measured three times. First, without any eBPF/XDP packet filtering and without any DoS attack. Second, without any eBPF/XDP packet filtering and with DoS attack. Third, with eBPF/XDP hardware offload mode and with DoS attack. We compare the results to verify that eBPF/XDP program can be used to mitigate HTTP get, post flood attack with minimum

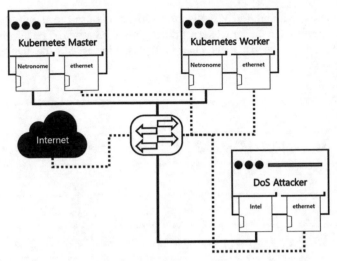

Fig. 2. Kubernetes Master, Kubernetes Worker, and DoS Attackers are connected to the internet through their Ethernet port (*straight line*) and the three servers are connected by Netronome NIC and Intel NIC with local IP addresses (*dotted line*).

hardware resource consumption. Detection of such DoS attack is not a topic to be considered in this paper, so we assume that the detection has already been made and filter all packets from the physical server, DoS attacker.

6.4 Average CPU Usage Comparison

The first experiment measures CPU usage without any eBPF/XDP program and without any DoS attack. The result is depicted on Table 3. Since the second experiment has no means to mitigate DoS attack, it shows the highest average CPU usage among the three experiments. The result of the experiment is shown in Fig. 3. When eBPF/XDP program was leveraged during DoS attacks, every packet sent by the DoS attack were dropped.

Table 3. Average CPU usage

Experiment	Dos attack	eBPF/XDP	Average CPU usage	Maximum CPU usage	Minimum CPU usage
1	X	X	2.24561	8.63	0.68
2	O	X	17.9154	23.62	15.94
3	O	O	2.44902	4.35	1.02

During the experiment, our visualization program is used to monitor the status of the network interfaces in the target server during the DoS attack. The terminal output is depicted in Fig. 4 and the visualized chart is depicted in Fig. 5 respectively.

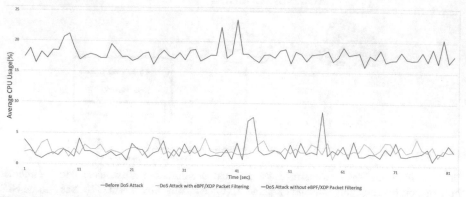

Fig. 3. Average CPU Usage (*Y axis*) in accordance with time elapsed (*X axis*). Experiment 2 (*Red line*) shows the highest average CPU consumption. Experiment 1 (*blue line*) shows higher average CPU usage compared experiment 3 (*gray line*).

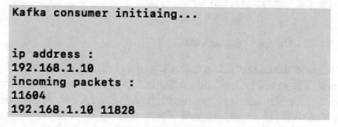

Fig. 4. Terminal output during the DoS attack.

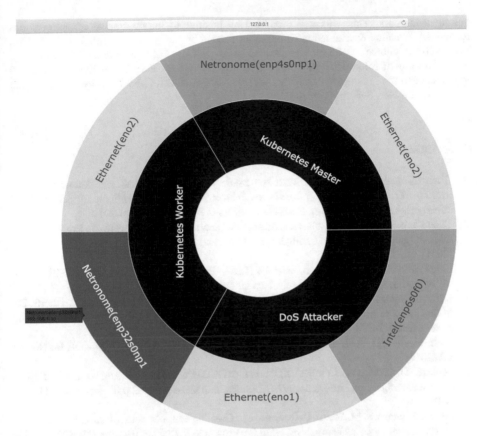

Fig. 5. Visualization program visualizing the Netronome NIC interface (*Red*) of the physical server, Kubernetes Worker under the DoS attack. The source IP address of the physical server, DoS attacker is displayed on top of the network interface that the DoS attacker was sending packets to.

7   Conclusion

In this paper, we leverage eBPF/XDP to visualize network interface status and to mitigate DoS attack. In our experiment, DoS attack packets are filtered using an eBPF/XDP program. In addition, we visualize network traffic status of network interfaces based on the data monitored by our eBPF/XDP program. By using our Onion-Ring visualization program, a user will be able to monitor network traffic of network interfaces and which physical servers the network interfaces belong to. In further works, we intend to expand the visualization program in order to visualize source IP addresses and their individual network traffic.

Acknowledgments. This work was supported by Institute of Information & communications Technology Planning & Evaluation (IITP) grant funded by the Korea government(MSIT) (No. 2015-0-00575, Global SDN/NFV OpenSource Software Core Module/Function Development), and by Institute of Information & Communications Technology Planning & Evaluation (IITP) grant funded by the Korea government (MSIT) (No. 2017-0-00421, Cyber Security Defense Cycle Mechanism for New Security Threats).

References

1. Rimi, C.: Kubernetes and AI: Marriage Made in IT Heaven, datacenterknowledge.com, para. 6, 31 October 2018. https://www.datacenterknowledge.com/industry-perspectives/kubernetes-and-ai-marriage-made-it-heaven#menu. Accessed 30 Sept 2019
2. Chao-yang, Z.: DOS attack analysis and study of new measures to prevent. In: Proceedings of International Conference on Intelligence Science and Information Engineering, pp. 426–429 (2011)
3. Linux Networking programming with P4 (Linux Plumbers 2018 Fabian Ruffy, William Tu, Mihai Budiu, VMware Inc. and University of British Columbia)
4. Høiland-Jørgensen, T., Brouer, J.D., Borkmann, D., Fastabend, J., Herbert, T., Ahern, D., Miller, D.: The eXpress data path: fast programmable packet processing in the operating system kernel. In: ACM CoNEXT (2018)
5. Kicinski, J., Viljoen, N.: eBPF hardware offload to SmartNICs: cls bpf and XDP. In: Netdev 1.2 (2016)
6. Beckett, D., Kicinski, J.: eBPF/XDP. In: SIGCOMM 2018, 26 August 2018. https://conferences.sigcomm.org/sigcomm/2018/files/slides/hda/paper_1.2.pdf. Accessed 11 Oct 2019
7. Miano, S., Betrone, M., Risso, F., Tumolo, M.: Creating complex network service with eBPF: experience and lessons learned. In: High Performance Switching Routing (HPSR), vol. 18, p. 5. IEEE, August 2018
8. Cilium: BPF and XDP Reference Guide, cilium.readthedocs.io, n.d. https://cilium.readthedocs.io/en/v1.0/bpf/#bpf-guide. Accessed 27 Sept 2019
9. eBPF Maps, n.d. https://prototype-kernel.readthedocs.io/en/latest/bpf/ebpf_maps.html. Accessed 11 Oct 2019
10. Kerrisk, M.: Linux Programmer's Manual BPF(2), n.d. http://man7.org/linux/man-pages/man2/bpf.2.html. Accessed 11 Oct 2019
11. IOVisor: BPF Compiler Collection (BCC), github.com, para. 1, n.d. https://github.com/iovisor/bcc/blob/master/README.md
12. IBM: host byte order and host byte order. IBM, n.d. https://www.ibm.com/support/knowledgecenter/en/SSB27U_6.4.0/com.ibm.zvm.v640.kiml0/asonetw.htm. Accessed 11 Oct 2019
13. Plotly: Plotly Python Open Source Graphing Library, plotly.com, para. 1, n.d. https://plot.ly/python/. Accessed 11 Oct 2019
14. Lee, S., et al.: Relocatable service composition based on microservice architecture for cloud-native IoT-cloud services. In: Proceedings of the Asia-Pacific Advanced Network, vol. 48, pp. 23–27 (2019)

Recognition of Historical Characters by Combination of Method Detecting Character in Photo Image of Document and Method Separating Block to Characters

Liao Sichao and Hiroyoshi Miwa[✉]

Graduate School of Science and Technology, Kwansei Gakuin University,
2-1 Gakuen, Sanda-shi, Hyogo, Japan
{liaosichao,miwa}@kwansei.ac.jp

Abstract. There are vast amount of historical documents written in cursive writing style in Japan. However, characters written by the style cannot be read by modern people, because the style was taught no longer. Therefore, an efficient method to convert historical characters into modern characters automatically is required. Especially, since every page in a Japanese historical document is stored by a photo image, it is necessary to automatically recognize all characters in a photo image. However, it is difficult to recognize each historical characters separately, because they are written connected and because there are many types of shape of characters. In this paper, we propose a method combining a method using deep learning to detect characters in a photo image and a method separating a block into characters. The remained parts that cannot be recognized by the former method are separated into characters by the latter method. Thus, it is expected that the recognition ratio is improved. We evaluate the performance of the proposed algorithm by using photo images of actual documents.

1 Introduction

There are vast amount of historical documents written in cursive writing style in Japan. Especially, in the Edo period (1603–1868) that the level of education rose and the number of people who could read and write increased, a large number of books made by woodblock printing were on the market. General Catalog of National Books [1] estimates that 1.7 million books in total were published in Japan prior to 1867. Since there are many documents not registered in the catalog, the number of documents is much larger.

Since most of historical documents are written on papers, the documents face the risk of loss due to disasters and aging degradation. Therefore, the historical documents must be urgently recorded by media other than paper. Indeed, documents have been recorded by photo image page by page; however, a huge number of documents are not recorded yet.

L. Barolli et al. (Eds.): EIDWT 2020, LNDECT 47, pp. 469–477, 2020.
https://doi.org/10.1007/978-3-030-39746-3_48

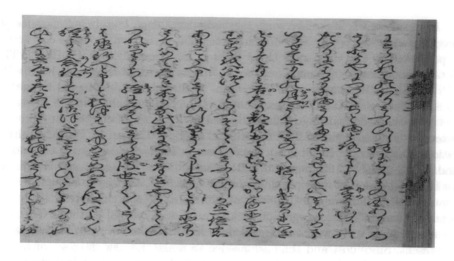

Fig. 1. Example of photo image

We combine the method of deep leaning based on the semantic segmentation and our previous proposed method. First, frequently appearing Kanji characters and Hiragana characters are recognized by the former method. Since Hiragana characters appear frequently in Japanese documents, a part of a document can be recognized by this method. The remained parts consist of blocks of characters. We apply our previous method [11] to these blocks. The block is divided to some pieces and these pieces are recognized by a single character recognizer by deep learning. This division is iterated by changing the separator lines, and the separator lines are determined so that the probability of recognition is large. Thus, it is expected that many characters are recognized in a photo image.

We describe the proposed method in detail as follows.

First, we describe the method using deep learning based on the semantic segmentation. We use the data set of kaggle [12]. The data set includes two types of data: one is the set of photo images (Fig. 1); the other is the set of characters and their positions.

We use the u-net model [10] of CNN (Convolutional Neural Network) for the semantic segmentation with Adam Optimizer [13] and the group normalization [14].

We make the training data for the semantic segmentation by assigning different colors to different characters (Fig. 2). We restrict the number of characters to be simultaneously recognized to six in order to reduce learning time to distinguish different characters. Although we tried more number of characters, the accuracy decreased rapidly. Applying the trained classifier to a photo image, the colors corresponding to the characters are painted in the photo image. Extracting the colors, we can identify the characters and their positions. We iterate this procedure of simultaneously six characters recognition, and eventually 72 Hiragana characters and frequently appearing 30 Kanji characters are recognized in total.

Fig. 2. Example of training data

Next, we describe the method to recognize the remained parts that cannot be recognized by the above method. We apply our previous method [11] to the remained parts. The block is divided to some pieces and these pieces are recognized by a single character recognizer based on deep learning. This division is iterated by changing the separator lines, and the separator lines are determined so that the probability of recognition is large. We describe the method in detail as follows. We made the single character recognizer by deep learning. We used the data set [5] including 684,165 images (581,540 images for learning data and 102,625 images for test data) of 4645 characters of Hiragana and Kanji. After an image is binarized and dirt and noise are removed, the image is converted to the image of 64 pixels × 64 pixels. We use VGG16 [15] as the model of CNN having 16 layers. The size of the input layer is 64 × 64, and the size of the middle layers is 4 × 4 × 512. The recognition ratio of this single character recognizer is 96% for 4645 characters of Hiragana and Kanji.

A block is divided to some pieces and these pieces are recognized by the single character recognizer. First, the number of characters in a block is estimated based on the ratio of the width and the height of a block. If the ratio is 1.8 or more, we assume that two characters are contained in the block; if the ratio is four or more, we assume that three characters are contained in the block. If a block contains two characters, we assume that the height of the first character is 35% to 65% of the height of the block; if a block contains three characters, we assume that the height of the first character is 20% to 40% of the height of the block and that the sum of the heights of the first and the second characters is 60% to 80% of the height of the block. This division is iterated by changing the separator

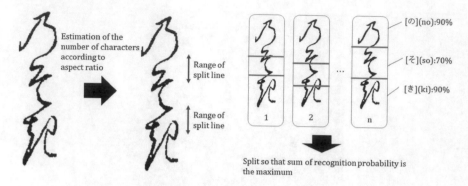

Fig. 3. Recognition of multiple characters in a block

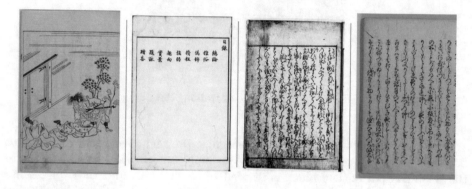

Fig. 4. Example of a page

lines, and the separator lines are determined so that the sum of the probability of recognition by the single character recognizer is the maximum (Fig. 3).

Thus, it is expected that the proposed method can recognize many characters and the positions in a photo image.

4   Performance Evaluation

In this section, first, we evaluate the performance of the semantic segmentation classifier, and then we evaluate the performance of the proposed method.

We use the training data 3,881 pages and the test data 4,150 pages in the data set of kaggle [12] for the semantic segmentation classifier. We show an example of a page in the data set (Fig. 4).

The semantic segmentation classifier that simultaneously recognizes six characters at once is trained in 15 epochs. It takes about 10 min per an epoch by using GPU (NVIDIA GTX1080Ti). We show an example of an image in which the characters and their positions are recognized (Fig. 5).

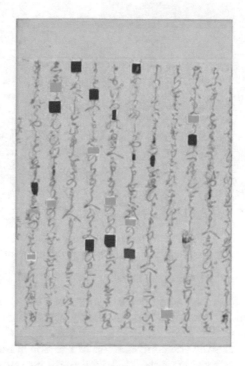

Fig. 5. Example of characters and positions recognized by semantic segmentation classifier

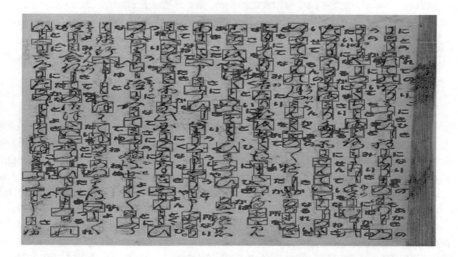

Fig. 6. Example of 102 characters and positions recognized by semantic segmentation classifiers

Fig. 7. Example of recognition by proposed method

We iterate this procedure of simultaneously six characters recognition, and eventually 72 Hiragana characters and frequently appearing 30 Kanji characters are recognized in total. We show an example of the result (Fig. 6).

The proposed method is the combination of the above method and our previous method that the division of the remained parts into one character and the recognition by a single character recognizer. We show an example of the result in Fig. 7.

The performance measure is the ratio of the number of correctly recognized characters to the number of all characters. We defined that a character is correctly recognized if and only if both the character and its position are recognized. As a result, the ratio is 61% for test data of 4,150 pages. This is better than our previous result of about 39% in ref. [11].

5   Conclusions

There are vast amount of historical documents written in cursive writing style in Japan. However, characters written by the style cannot be read by modern people, because the style was taught no longer in school. Therefore, an efficient method to convert historical characters into modern characters automatically is required. Especially, since every page in a Japanese historical document is stored by a photo image, it is necessary to recognize all characters in a photo image.

In this paper, we proposed a method to identify characters and positions in a photo image. Our proposed method detects characters as many as possible by the semantic segmentation method, a method of deep learning, and then detects characters by separating the blocks of the remained parts into characters by our previous proposed method.

Furthermore, we evaluated the performance of the proposed algorithm using photo images of actual documents. As a result, the proposed method achieved the recognition ratio of 61% for test data of 4,150 pages, which is better than our previous result of about 39%.

References

1. Shoten, I.: General Catalog of National Books. Iwanami Shoten (2002)
2. Shibayama, M., et. al.: Research on Higher Accuracy Document Character Recognition Systems (in Japanese), Grants-in-Aid for Scientific Research (B) (1) report on research results, pp. 33–49 (2005)
3. Yamamoto, S., Osawa, T.: Labor saving for reprinting Japanese rare classical books: the development of the new method for OCR technology including Kana and Kanji characters in cursive style. Inf. Manag. **58**(11), 819–826 (2016)
4. Hayasaka, T., Ohno, W., Kato, Y., Yamamoto, K.: Recognition of Hentaigana by deep learning and trial production of WWW application (in Japanese). In: Proceedings of IPSJ Symposium of Humanities and Computer Symposium, pp. 7–12 (2016)
5. The Dataset of Kuzushiji (Open Data Center for Humanities). http://codh.rois.ac.jp/pmjt
6. Tarin, C., Mikel, B., Asanobu, K., Alex, L., Kazuaki, Y., David, H.: Deep learning for classical Japanese literature. In: Proceedings of 2018 Workshop on Machine Learning for Creativity and Design (Thirty-second Conference on Neural Information Processing Systems), 3 December 2018
7. https://sites.google.com/view/alcon2017/prmu
8. Nguyen, H.T., Ly, N.T., Nguyen, K.C., Nguyen, C.T., Nakagawa, M.: Attempts to recognize anomalously deformed Kana in Japanese historical documents. In: Proceedings of 4th International Workshop on Historical Document Imaging and Processing, pp. 31–36, 10–11 November 2017
9. Clanuwat, T., Alex, L., Asanobu, K.: End-to-end pre-modern Japanese character (Kuzushiji) spotting with deep learning. In: Proceedings of IPSJ SIG Computers and the Humanities, pp. 15–20 (2018)
10. Ronneberger, O., Fischer, P., Brox, T.: U-Net: convolutional networks for biomedical image segmentation. In: Medical Image Computing and Computer-Assisted Intervention (MICCAI). LNCS, vol. 9351, pp. 234–241. Springer (2015)
11. Sichao, L., Miwa, H.: Algorithm using deep learning for recognition of Japanese historical characters in photo image of historical book. In: Proceedings of INCoS, Oita, Japan, pp. 5–7, September 2019
12. Kaggle. https://www.kaggle.com/c/kuzushiji-recognition
13. Kingama, D.P., Ba, J.L.: A method for stochastic optimization. In: Proceedings of ICLR, San Diego, 7–9 May 2015
14. Wu, Y., He, K.: Group normalization. In: Proceedings of ECCV, Munich, Germany, 8–14 September 2018
15. Simonyan, K., Zisserman, A.: Very deep convolutional networks for large-scale image recognition. In: Proceedings of International Conference on Learning Representations, San Diego, 7–9 May 2015

A Study for Semi-supervised Learning with Random Erasing

Yuuhi Okahana and Yusuke Gotoh[(✉)]

Graduate School of Natural Science and Technology, Okayama University,
Okayama, Japan
gotoh@cs.okayama-u.ac.jp

Abstract. Due to the recent popularization of various data classified by computer, machine learning is attracting great attention. A common method of machine learning is supervised learning, which classifies data using a large number of class labeled training data called labeled data. To improve the processing performance of supervised learning, it is effective to use Random Erasing in data augmentation. However, since supervised learning requires much labeled data, the cost of manually adding label information to an unclassified training case (unlabeled data) is very high. In this paper, we propose a method for achieving high classification accuracy using Random Erasing for semi-supervised learning using few labeled data and unlabeled data. In our evaluation, we confirm the availability of the proposed method compared with conventional methods.

1 Introduction

Due to the recent popularization of various data classified by computer, machine learning is attracting great attention. A common method of machine learning is supervised learning, which classifies data using a large number of class labeled training data called labeled data.

In supervised learning [3,4] using deep learning, we can obtain better accuracy than conventional methods using a network with a complex structure such as a convolutional neural network. In particular, Random Erasing [13], which efficiently performs supervised learning, can improve the classification accuracy by erasing a part of the image.

However, since supervised learning requires a large amount of labeled data, the cost of manually adding label information is very high. Therefore, many researchers have proposed semi-supervised learning methods [8,9] that combine labeled and training data without labeled data (unlabeled data). In semi-supervised learning using deep learning, several methods for classifying data achieve high classification accuracy [1,5,7,10].

Random Erasing [13] expands the data by erasing some images and improves the accuracy of image recognition. We can define a new semi-supervised learning by combining a semi-supervised learning method that uses the gradient for error as noise and Random Erasing that selects a region to be deleted at random.

© Springer Nature Switzerland AG 2020
L. Barolli et al. (Eds.): EIDWT 2020, LNDECT 47, pp. 478–490, 2020.
https://doi.org/10.1007/978-3-030-39746-3_49

In this paper, we propose a semi-supervised learning method based on Random Erasing used in supervised learning. In our proposed method, by combining the supervised and semi-supervised learning algorithm, we can use Random Erasing in semi-supervised learning. In addition, we evaluate the classification accuracy of the data in our proposed method.

Our contributions in this paper can be summarized as follows:

- We perform semi-supervised learning after erasing a part of the image based on Random Erasing.
- We classify image data as semi-supervised learning.
- Our proposed method has the same classification accuracy as such conventional methods as Virtual Adversarial Training (VAT) [8,9].

The remainder of the paper is organized as follows. Related works are explained in Sect. 2, and Random Erasing is introduced in Sect. 3. In Sect. 4, we explain the details of our proposed method in semi-supervised learning with Random Erasing. The performance of our proposed method is evaluated in Sect. 5. Finally, we conclude in Sect. 6.

2   Related Works

There are many semi-supervised learning methods. In this section, we describe Virtual Educational Training (VAT) [8,9] and Neighbor Embedding [5], both of which are semi-supervised learning methods using deep learning. In addition, we describe a graph structured method [11] for semi-supervised learning without deep learning.

2.1   Virtual Adversarial Training

We can derive the error of unsupervised learning by minimizing the amount of information, called the Kullback-Leibler (KL) divergence, between the correct input data and other input data with a small amount of added noise.

When the input data is x, the label for the input data is y, the parameter of the neural network is w, and the small vector is r. Then $\triangle KL$ is:

$$\triangle KL(r, x, w) = KL[p(y|x, w)||p(y|x + r, w)]. \tag{1}$$

In Eq. 1, r is not created randomly. In addition, r that maximizes $\triangle KL$ is defined:

$$r_{r-adv} = \arg\max_{r} \triangle KL(r, x, w). \tag{2}$$

Minimize $\triangle KL$, given by:

$$\triangle KL(r_{v-adv}, x, w) = KL[p(y|x, w)||p(y|x + r, w)]. \tag{3}$$

Randomly initialized unit vector d, parameters β and ε, and r_{v-asv} can be obtained by the following equations:

$$d = \nabla r \, \triangle \, KL(r, x, w)|_{r=\beta d} \ and \tag{4}$$

$$r_{v-adv} = \varepsilon d. \tag{5}$$

By minimizing the value of r_{v-adv} in the Eq. (5), we can learn the parameters suitable for classification. In addition, r_{v-adv} can be calculated using unlabeled data instead of labeled data.

Figure 1 shows the VAT configuration. First, we calculate $network(w)$ that is the output of the neural network for input x. After calculating r_{v-adv} using the calculation procedure described in Eqs. (1) to (5), we calculate the output $network(x)$ for input $x+r_{v-adv}$. Here supervised loss, which is error in supervised learning, is used to approximate the output $network(w)$ for input x to label y. On the other hand, unsupervised loss, which is error in unsupervised learning, is optimized so that the output $network(w)$ for $x + r_{v-adv}$ approaches x.

The supervised learning error can be calculated using only labeled data with known class labels. On the other hand, the unsupervised learning error can be calculated using labeled data whose class labels are unknown. Therefore, VAT can perform semi-supervised learning by combining both labeled and unlabeled data. VAT can be applied to supervised learning with high accuracy for an evaluation dataset.

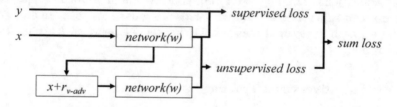

Fig. 1. Virtual Adversarial Training (VAT)

An image generated by VAT using MNIST is shown in Fig. 2. The top row is the original image, and the bottom row is the generated image. The clarity of the image generated by VAT is reduced due to the influence of noise.

2.2 Neighbor Embedding

Neighbor Embedding [5] is a semi-supervised learning method based on Euclidean distance. When x_l is labeled data, x_u is unlabeled data, and z_c is labeled data belonging to class c in x_l, we define an expression to minimize the error for the labeled data. Λ_L and λ_U are parameters that adjust the ratio of supervised error and unsupervised error:

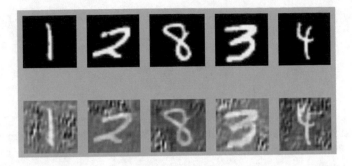

Fig. 2. Example of applying VAT to MNIST

$$L(x_l, x_u, z_1, \cdots, z_c) = \lambda_L L(x_l, z_1, \cdots, z_c)_L$$
$$+ \lambda_U L(x_u, z_1, \cdots, z_c)_U, \tag{6}$$

$$L(x_l, z_1, \cdots, z_c)_L = -log \frac{e^{||Net(x_l) - Net(z_k)||^2}}{\sum_{i=1}^{c} e^{||Net(x_l) - Net(z_i)||^2}}, \tag{7}$$

$$L(x, z_1, \cdots, z_c)_U = -\sum_{i=1}^{c} \frac{e^{||Net(x) - Net(z_i)||^2}}{\sum_{j=1}^{c} e^{||Net(x) - Net(z_j)||^2}}$$
$$*log \frac{e^{||Net(x) - Net(z_i)||^2}}{\sum_{j=1}^{c} e^{||Net(x) - Net(z_j)||^2}}. \tag{8}$$

$L(x_l, z_1, \cdots, z_c)_L$ in Neighbor Embedding indicates the error for the labeled data. In the Eq. (7), although the distance between the data with the same label becomes shorter, the distance between the data with different labels becomes longer. On the other hand, $L(x, z_1, \cdots, z_c)_U$ indicates the error for the unlabeled data. By minimizing this error, we can learn the mapping to a space suitable for identification. Neighbor Embedding achieves sufficiently high accuracy in MNIST using such a classifier as the k neighborhood method [5] after learning a mapping to a space suitable for classification.

2.3 Method Using Graph Structure

Semi-supervised learning with a graph structure can be classified into two stages: graph construction and label propagation. In graph construction, when all the data are set as vertices, k neighborhood graph is created based on the distance between each bit of data. Label propagation adds tags from the labeled data to the unlabeled data based on the graph created at the graph construction stage.

Next we explain Local and Global Consistency (LGC) [14], which adds labels to the unlabeled data using the adjacency matrix for the k neighborhood graph.

First, we calculate matrices D and S used for label propagation. D is Eq. (9) and S is the matrix indicated by Eq. (10):

$$D_{ii} = \sum_j A_{ij} \qquad (9)$$

$$S = D^{-\frac{1}{2}} A D^{-\frac{1}{2}}. \qquad (10)$$

The result of predicting to which label the data belong is shown in matrix F. Initial value $F(0)$ of matrix F is set to $F(0) = Y$. We show the formula for predicting labels:

$$F(t+1) = \alpha S F(t) + (1-\alpha)Y. \qquad (11)$$

Label propagation ends after iterating until Eq. (11) does not change. Conventional research [14] proved that $F(t)$ finally converges to Eq. (12). Label Propagation [15, 16] can be cited as a method other than LGC:

$$\lim_{t \to \infty} F(t) = (1-\alpha)(I - \alpha S)^{-1} Y. \qquad (12)$$

Unlike deep learning methods, semi-supervised learning that does not use deep learning, such as a scheme using graph structure [6, 11], can learn at high speed without such hardware as GPU. On the other hand, classification accuracy in semi-supervised learning without deep learning is lower than that with deep learning. Therefore, in recent years, many researchers have proposed methods using deep learning.

3   Random Erasing

Random Erasing [13] expands the data by erasing some images and improves the accuracy of image recognition. It also selects the size of the area to be randomly erased in the image and erases the selected area based on the algorithm. There

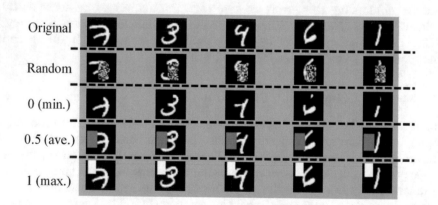

Fig. 3. Example of applying Random Erasing for MNIST

are four types of pixel-value setting methods in the erased area: random, zero, average pixel value, and maximum pixel value.

An example of applying Random Erasing to MNIST is shown in Fig. 3. The top row shows the original image. Next we set four types of pixel values in the area to be erased: Random, 0 (minimum), 0.5 (average), and 1 (max). In Fig. 3, when the value in the area to be erased is set at random, a certain area is erased by noise. Similarly, black is displayed when it is set to 0, white when it is set to 1, and gray when it is set as an average value, and certain areas are erased by noise.

Zhong et. al. [13] evaluated both data expansion and regularization. In terms of regularization, Random Erasing improves the classification accuracy more than Drop Out and Random Noise. In setting methods of pixels, the image classification accuracy is best when the pixel values in the area to be erased are set randomly.

4   Proposed Method

In this paper, we propose two types of methods for improving classification accuracy in semi-supervised learning. We define a new semi-supervised learning by combining a VAT that uses the gradient for error as noise and Random Erasing that selects a region to be deleted at random.

4.1   Virtual Adversarial Erasing

The first proposed method is Virtual Adversarial Erasing (VAE), which extends VAT in semi-supervised learning using deep learning. In VAE, the classification accuracy is improved by changing the process of adding noise to the image to the process of deleting a part of the image in the VAT processing procedure.

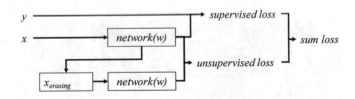

Fig. 4. Virtual Adversarial Erasing (VAE)

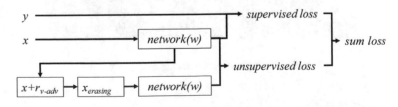

Fig. 5. VAT + VAE concept

Processing Procedure of VAE

Here is the procedure for erasing a region in VAE:

1. Based on the input, calculate the distribution that predicts the class label (predicted distribution).
2. Calculate the gradient of the input image for the predicted distribution.
3. Erase part of the image centering on the pixels with the largest absolute gradient value.

When the image after deleting the region in the above procedure is set to $x_{erasing}$, the unsupervised learning error is calculated:

$$\triangle KL(r,x,w) = KL[p(y|x,w)||p(y|x_{erasing},w)]. \qquad (13)$$

The VAE concept is shown in Fig. 4. In VAE, the processing of $x + r_{v-adv}$ shown in Fig. 1 is replaced with $x_{erasing}$, which is an image from which some areas were deleted. The other processing is the same as VAT.

Next we explain a method (VAT + VAE) that combines the conventional method VAT and the proposed method VAE. The concept of VAT + VAE is shown in Fig. 5. VAE + VAT improves the process of calculating $x_{erasing}$ after calculating $x_{v_a dv}$ compared to that in Fig. 4.

Generated Image by VAE

The image generated by VAE is shown in Fig. 6, where a region with a large gradient is erased, and a part of the number is erased. Next the image generated by VAT + VAE is shown in Fig. 7. By deleting the image area in addition to the noise from the gradient, a part of the image is erased and noise is generated in the entire image.

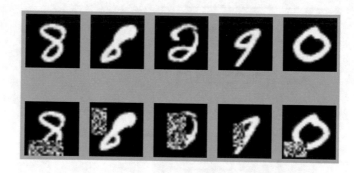

Fig. 6. Example of generating image by VAE

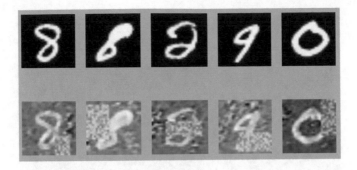

Fig. 7. Example of generating image by VAT + VAE

4.2 Random Erase Training

Our second proposed method is Random Erase Training (RET) in which semi-supervised learning using deep learning is combined with Random Erasing by extending VAE. In RET, the region to be erased using VAE is determined randomly instead of using the gradient. In this case, since it is not necessary to calculate the gradient for the input image, the calculation speed of RET is faster than VAE.

Figure 8 shows the RET concept. Unlike in Fig. 4, it is not necessary to calculate $network(w)$ when generating $x_{erasing}$. Therefore, RET can compute $x_{erasing}$ directly from input x.

5 Evaluation

5.1 Classification Accuracy in MNIST

MNIST [12] is a set of data with handwritten numbers from 0 to 9: 28×28 pixels and 70,000 images. An example of an image in the MNIST dataset is shown in Fig. 9. There are ten types of class labels from 0 to 9, and one of these numbers is stored in the dataset as an image. In our evaluation for 50,000 images, we used 100 labeled data and 49,900 unlabeled data. In addition, 10,000 images were used separately as test data for evaluating the classification accuracy.

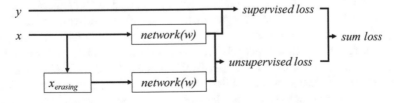

Fig. 8. Random Erase Training RET) concept

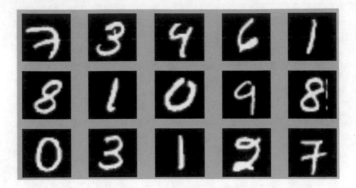

Fig. 9. Example of images in MNIST dataset

The proposed method uses VAE, RET, and VAT + VAE. In VAE, we evaluate the method of setting values in the erased area for the four types described in Sect. 3. The conventional method uses VAT and Convolutional Neural Network (CNN) [3]. In the evaluation, CNN is used as simple supervised learning using convolution. In CNN, since only labeled data are used for learning, classification accuracy is reduced compared to semi-supervised learning.

Table 1. Evaluation results of classification accuracy in MNIST

Method	Accuracy (%)	Semi-supervised/Supervised
VAE (Random)	98.80%	Semi-supervised
VAE 0	98.66%	Semi-supervised
VAE 1	96.83%	Semi-supervised
VAE Average	98.46%	Semi-supervised
RET Random	98.85%	Semi-supervised
VAT	98.99%	Semi-supervised
VAT+VAE	98.97%	Semi-supervised
CNN	83.70%	Supervised

The evaluation results for the classification accuracy using MNIST are shown in Table 1. The accuracy of the proposed method using semi-supervised learning was significantly improved compared to the conventional method using supervised learning. Also, in the pixel-filling method in the proposed method, the classification accuracy for random pixel filling with VAE and RET was higher than the others. When pixels are filled randomly in the proposed method, since there is no regularity in the values to be erased, the classification accuracy is higher than in other cases where the area to be erased is set at a constant value. Therefore, we confirmed that the classification accuracy is the highest when the pixels are filled randomly not only in supervised learning but also in semi-supervised learning.

When comparing our three proposed methods, VAE, RET, and VAT + VAE, with the conventional method VAT, the classification accuracy of the proposed method is slightly lower than VAT. In MNIST, the classification efficiency in VAT that adds noise to the whole image exceeds the other methods. However, since the difference between the proposed and the conventional methods is small, the classification accuracy of the former is almost the same as the latter.

In the proposed method, the classification accuracy of RET is about 0.05% higher than VAE (random). RET improves the classification accuracy by randomly setting the region to be erased instead of using the gradient for the input image. Therefore, we confirmed that the classification accuracy in semi-supervised learning by RET is the highest in MNIST.

5.2   Classification Accuracy in Fashion MNIST

FMNIST [2] is a dataset about such fashion items as clothing and shoes. The data size is 28 × 28 pixels, and the number of images is 70,000. An example of an image in the FMNIST dataset is shown in Fig. 10. Ten types of class labels are used in the calculation of the classification accuracy in FMNIST: t-shirts/tops, trousers, pullovers, dresses, coats, sandals, shirts, sneakers, bags, and ankle boots. In addition, the image classification by FMNIST is more difficult than with MNIST, and the classification accuracy is lower.

Case of Supervised Learning

By adding an semi-supervised learning error to supervised learning, the classification accuracy is improved compared to supervised learning. Table 2 shows

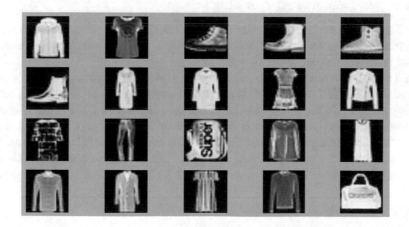

Fig. 10. Example of images in FMNIST dataset

Table 2. Classification accuracy using FMNIST (applying proposed method to supervised learning)

Method	Accuracy (%)	Semi-supervised/Supervised
VAE	93.12%	Semi-supervised
RET	93.24%	Semi-supervised
VAT	92.96%	Semi-supervised
CNN	92.64%	Supervised

Table 3. Classification accuracy using FMNIST (applying proposed method to semi-supervised learning)

Method	Accuracy (%)	Semi-supervised/Supervised
VAE	84.07%	Semi-supervised
RET	84.35%	Semi-supervised
VAT	83.96%	Semi-supervised
CNN	82.95%	Supervised

the evaluation results using FMNIST for the classification accuracy when the proposed method is applied to supervised learning.

In Table 2, the classification accuracy of VAE, RET, and VAT in semi-supervised learning is higher than CNN in supervised learning. We confirmed that it was improved using semi-supervised learning. In addition, the classification accuracy of RET is higher than that of VAE and VAT. With FMNIST, the method of erasing a part of the image is more efficient than the method of adding noise.

Case of Semi-supervised Learning

Table 3 shows the evaluation results using FMNIST for the classification accuracy when the proposed method is applied to semi-supervised learning. We used 100 labeled data and 49,900 unlabeled data.

In Table 3, the classification accuracy of the three semi-supervised learning methods, VAE, RET, and VAT, is improved compared to CNN, the supervised learning method. In addition, the classification accuracy in RET is the highest in the semi-supervised learning method. As in the evaluation in supervised learning, the method of erasing a part of the image is more efficient than the method of adding noise in semi-supervised learning. Therefore, we confirmed that the addition of noise complicates recognizing small changes in color and shape.

6   Conclusion

We proposed Virtual Adversarial Erasing (VAE), which determines the area of the image to be erased using gradients, and Random Erase Training (RET), which randomly erases the image. We extended Random Erasing, a data classification method for supervised learning, to semi-supervised learning. Our proposed method performs semi-supervised learning after erasing a part of the image. Our evaluation result confirmed that both our proposed methods classify image data as semi-supervised learning and have the same classification accuracy as the conventional method, Virtual Adversarial Training (VAT).

In the future, we will conduct an evaluation using more practical datasets.

Acknowledgement. This work was supported by JSPS KAKENHI Grant Number 18K11265. In addition, this paper is partially supported by Innovation Platform for Society 5.0 from Japan Ministry of Education, Culture, Sports, Science and Technology.

References

1. Rasmus, A., Valpola, H., Honkala, M., Berglund, M., Raiko, T.: Semi-supervised learning with ladder networks. In: Proceeding of the 35th Advances in Neural Information Processing Systems (NIPS 2015), vol. 2, pp. 3546–3554 (2015)
2. Han, X., Kashif, R., Roland, V.: Fashion-MNIST: a novel image dataset for benchmarking machine learning algorithms. arXiv preprint, arXiv: 1708.07747 (2017)
3. He, K., Zhang, X., Ren, S., Sun, J.: Deep Residual Learning for Image Recognition. arXiv preprint arXiv: 1512.03385 (2015)
4. Hoffer, E., Aillon, N.: Deep metric learning using triplet network. In: Proceeding of the 3rd International Conference on Learning Representations (ICLR 2015). arXiv preprint arXiv: 1412.6622 (2015)
5. Hoffer, E., Ailon, N.: Semi-supervised deep learning by metric embedding. In: Proceeding of the 5th International Conference on Learning Representations (ICLR 2017). arXiv preprint arXiv: 1611.01449 (2016)
6. Joachims, T.: Transductive inference for text classification using support vector machines. In: Proceedings of the 16th International Conference on Machine Learning, pp. 200–209 (1999)
7. Kingma, D., Mohamed, S., Rezende, D.J., Welling, M.: Semi-supervised learning with deep generative models. In: Proceeding of the 34th Advances in Neural Information Processing Systems (NIPS 2014), pp. 3581–3589 (2014)
8. Miyato, T., Maeda, S., Koyama, M., Nakae, K., Ishii, S.: Distributional Smoothing with Virtual Adversarial Training. arXiv preprint, arXiv: 1507.00677 (2016)
9. Miyato, T., Maeda, S., Koyama, M., Ishii, S.: Virtual adversarial training: a regularization method for supervised and semi-supervised learning. IEEE Trans. Pattern Anal. Mach. Intell. **41**(8), 1979–1993 (2019)
10. Park, S., Park, J.K., Shin, S.J., Moon, C.: Adversarial dropout for supervised and semi-supervised learning. In: Proceedings of the 32nd AAAI Conference on Artificial Intelligence, pp. 3917–3924 (2018)
11. Sousa, C., Rezende, S., Batista, G.: Influence of graph construction on semi-supervised learning. In: Proceeding of the European Conference on Machine Learning and Principles and Practice of Knowledge Discovery in Databases (ECML PKDD 2013), vol. 8190, pp. 160–175 (2013)
12. The MNIST handwritten digit database (2017). http://yann.lecun.com/exdb/mnist/
13. Zhong, Z., Zheng, L., Kang, G., Li, S., Yang, Y.: Random Erasing Data Augmentation. arXiv preprint, arXiv: 1708.04896 (2017)

14. Zhou, D., Bousquet, O., Navin, T., Weston, J., Scholkopf, B.: Learning with local and global consistency. In: Proceeding of the 24th Advances in Neural Information Processing Systems (NIPS 2004), pp. 321–328 (2004)
15. Zhu, X., Ghahramani, Z.: Learning from Labeled and Unlabeled Data with Label Propagation. Carnegie Mellon University CALD Technical Reports, CMU-CALD-02-107, pp. 1–17 (2002)
16. Zhu, X., Ghahramani, Z.: Semi-supervised learning using gaussian fields and harmonic functions. In: Proceedings of the 21th International Conference on Machine Learning (ICML 2003), pp. 912–919 (2003)

Hybrid Type DTN Routing Protocol Considering Storage Capacity

Kenta Henmi[(✉)] and Akio Koyama

Graduate School of Science and Engineering, Yamagata University,
4-3-16 Jonan, Yonezawa, Yamagata 992-8510, Japan
henmi@yamagata-cit.ac.jp, akoyama@yz.yamagata-u.ac.jp

Abstract. The conventional DTN routing protocol replicates messages ignoring storage state of replication node. However, it is necessary to take into account the number of replication in case of less space in storage. Therefore, in this paper, we propose a hybrid type DTN routing protocol. In this method, if there is enough free storage space, the proposed method selects the Epidemic base routing protocol which replicates a lot of messages. Meanwhile, if there is less free storage space, the proposed method selects the Spray and Wait base routing protocol which constrains the number of messages. In addition, regarding Spray and Wait base, we propose the method which dynamically set the number of replications, and the method which selects a replicated node by mobility. We verified by simulation that the proposed method has high packet delivery ratio and less overhead compared with conventional methods.

1 Introduction

Recently, network researchers are researching communication methods in poor communication environments. For example, the communication infrastructure fails due to a disaster, the network is congested, or the environment has a large delay. In such an environment, TCP/IP may not be able to communicate normally. Delay/Disruption Tolerant Network (DTN) [1, 2] is expected as a technology to solve this problem.

The origin of DTN began with the interplanetary Internet [3], and now it is not limited to the interplanetary Internet, but today it covers not only the interplanetary Internet but also poor communication environments. For example, there are various ecological surveys targeting zebras [4] and alternatives in places where it is difficult to establish infrastructure economically [5]. In DTN, the bundle layer [6] above the transport layer is responsible for processing to enable communication even in poor environments. In the bundle layer, data called a bundle is handled. If communication is impossible, the bundle is stored in the storage. When communication is possible, the bundle communication is performed.

In DTN routing, the node with data replicates the data to encountered nodes, and either node delivers it to the destination. By the routing, it is possible to communicate each other even a poor communication environment. Epidemic [7] which is typical routing protocol, replicates all messages that the encountered nodes do not have. In the Epidemic, the delivery ratio is good when there is enough space in the storage, but the disadvantage is that the number of replicates is larger than other routing protocols.

© Springer Nature Switzerland AG 2020
L. Barolli et al. (Eds.): EIDWT 2020, LNDECT 47, pp. 491–502, 2020.
https://doi.org/10.1007/978-3-030-39746-3_50

Spray and Wait statically sets the maximum number of replications into the sending message. Therefore, Spray and Wait can suppress excessive replication. However, it is difficult to set an appropriate maximum number of replications. In addition, since an appropriate replication node is not selected, replication to nodes that do not contribute to data arrival may occur. In DTN routing, if there are too many replications, storage overflow occur and messages must be deleted. As a result, the delivery ratio may decrease. Conversely, if there are too few replications, the probability of reaching the destination decreases and the delivery ratio may be decreased.

In this paper, we propose a hybrid DTN routing method which selects the Epidemic base method with many replications and the Spray and Wait base method with few replications according to the storage state of the nodes. Moreover, in Spray and Wait base method, we propose a method to set the number of replications by the network state and a method which select a node to replicate the messages by mobility. We evaluate the proposed method by simulation and confirm its effectiveness.

The structure of this paper is as follows. We introduce related works in Sect. 2. We explain the proposed method in Sect. 3. In Sect. 4, we describe the performance evaluation to verify the usefulness of the proposed method by simulation. Finally, conclusions of this paper is given in Sect. 5.

2 Related Works

DTN does not always have an established route, so a routing protocol for transmission of messages to the destination is important. Moreover, the performance of the DTN routing protocol depends greatly on the environment in which it is used. Therefore, it is most important to select a routing protocol suitable for the environment. In this section, we introduce five typical routing protocols, and a recovery method which improves storage utilization efficiency.

2.1 Epidemic

When a message occurs, Epidemic replicates the message to nodes within the communicable range regardless of the destination [7]. When nodes encountered each other, the node compares a Summary Vector which is a list of a bundle. Then, the node replicates unowned messages to the encountered node. The replicated node is repeated same things. So, it was named Epidemic because the messages spread in the network as if it were an infectious disease.

2.2 Spray and Wait

Unlike Epidemic, Spray and Wait has a limit on the maximum number of replications because the maximum number of replications N is defined for each message [8].

When a message is generated, the message is set to an integer N that indicates the maximum number of replications. The value of N determines whether to replicate. When a message is replicated, N value of the message is made to half. When N is 2 or more, replication is performed as usual. If N is 1, the message is sent only to the destination.

2.3 Spray and Focus

Spray and Focus [9] can replicate messages depending on the conditions even when N is 1. In the condition, the node checks whether the time when encountered to the destination node exceeds the threshold, and if the time is exceeded, the node replicates the messages.

2.4 PRoPHET

PRoPHET is a protocol that each node calculates the reaching probability to the destination and the node replicates the messages to encountered nodes with higher reaching probability [10]. The reaching probability is calculated when the node encounters to other nodes. The reaching probability becomes high, when the encountered time is close to the current time.

2.5 MaxProp

MaxProp is a protocol designed for vehicle to vehicle communication [11]. It defines the order in which messages are sent because the connection time for vehicle to vehicle communication is short. This order is also used when deleting messages. When there is no free space in the storage, messages are deleted in order of lower priority.

2.6 Recovery Method

A problem in DTN routing protocols is that replication may perform even after messages arrive.

The recovery method was devised as a method to solve this problem [12]. Because the messages that reaches the destination does not need to be replicated anymore, the destination node generates an anti-packet that requests the message to be deleted. Anti-packets suppress unnecessary replication.

3   Proposal Method

We propose a routing method which combines an Epidemic-based protocol and a Spray-and-Wait-based protocol according to the storage state. There are the following four functions. Functions (1) and (2) are common functions of Epidemic and Spray and Wait based protocols. However, the setting of the maximum number of replications in (3) and the selection of the replication destination in (4) are functions that are applied only to Spray and Wait based protocol.

(1) Judgment of storage state
    In order to select an appropriate routing method, each node judges its own storage status and sets an appropriate routing method for itself.

(2) Recovery method

The proposed method has the function of a recovery method. The recovery method is usually used to delete messages that have arrived destination, but it is also used to set a threshold for calculating the number of replications for each message.

(3) Setting of the number of replications

Spray and Wait has a fixed N for message replication times. However, the proposed method dynamically sets N according to the network conditions. Although there is a method to devise updating of the value of N after replicating a message [13], the proposed method conforms to the original Spray and Wait, and N after replicating is halved. In addition, we propose a method to replicate even if N is 1 only when a specific condition is satisfied.

(4) Selecting of the replication node

Since a combination method of Spray and Wait and PRoPHET has been proposed as a method for selecting the replication node [14], the proposed method replicates the messages to a node that has high mobility than self. This section describes details of these functions.

3.1   Judgment of Storage State

Each time, the node encounters to other nodes, it judges the storage state and sets a routing method suitable for itself. When the node encounters to other nodes, the node refer to the routing method of the encountered node and replicate the messages according to the routing method of the encountered node.

Table 1 shows parameters used for the judgment of storage state and Eq. (1) shows judgment condition.

Table 1. Parameter used for judgment of storage state.

F	Free storage space [byte]
R	Average received message size per node [byte]

$$\begin{cases} Epidemic\,Base \;\; \left(\frac{R}{F} \geq 3\right) \\ Spray\,and\,WaitBase \;\; (otherwise) \end{cases} \tag{1}$$

Here, we show calculation way of R. When the node encounters to other nodes, the node records the received total message size and the number of nodes. By dividing the received total message size by number of encountered nodes, average received message size for the node is calculated. Using the values of F and R, the protocol is switched by the Eq. (1). An example is shown in Fig. 1.

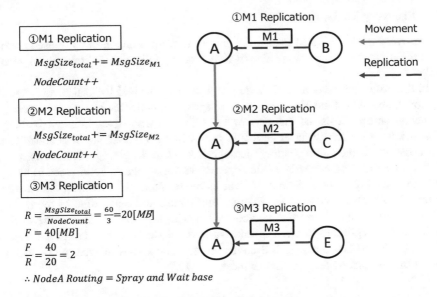

Fig. 1. Judgment of storage state.

(1) The node records the total received message size when the node encountered to other nodes. In Fig. 1, node A records the received message size at encountering to nodes B, C and E.

(2) The average received message size R is obtained by dividing the total received message size by the total number of encountered nodes.

(3) Using the parameters F and R, a node sets an appropriate protocol as follows. In Fig. 1, node A set a Spray and Wait base. Therefore, node E replicates to node A by a Spray and Wait base manner.

- Epidemic base

This is used when it is determined that there is enough space in the storage capacity at the replication node. In the proposed method, the number of replications N is set for all messages. However, the node replicates messages regardless the number N.

- Spray and Wait base

If Epidemic is used when free space in the storage is not much, messages may be deleted and the delivery ratio may decrease. Therefore, Spray and Wait base method replicates messages according to the number of replication. If the old message is deleted and the new message is replicated when free space is not much, the probability that the old message will reach the destination decreases. In this case, the proposed method does not replicate new messages. When a node replicates messages on a Spray and Wait base method, the node selects the replication node to avoid useless replication

3.2   Recovery Method

The recovery method is used to set the number of replications in the proposed method in addition to the suppression of arrival message replication introduced in related works.

In the proposed method, each node has a message arrival list that records the ID, hop count, and delay time of the arrived messages. The node that received the messages compares the arrival list of the other node with its own arrival list, and if there is information on the arrival message that is not in its own arrival list, it is added to the arrival list and if the corresponding message is exist in the storage, it is deleted.

Figure 2 shows an example. Node A generates a message M2 addressed to node C. The message M2 arrives at node C via node B. Node C receiving the message addressed to itself records ID, destination, hop count, and delay time of the message M2 in the arrival list. After that, an anti-packet with information on the arrival list is generated and replicated to Node B and Node A. Node B and node A receiving the anti-packet know that message M2 addressed to node C has arrived, and add information to their arrival list, and delete message M2.

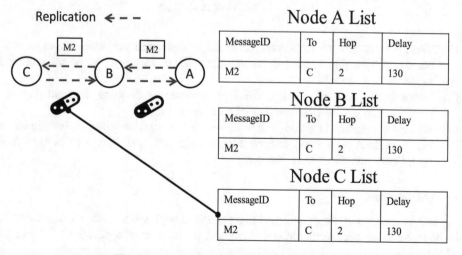

Node A List

MessageID	To	Hop	Delay
M2	C	2	130

Node B List

MessageID	To	Hop	Delay
M2	C	2	130

Node C List

MessageID	To	Hop	Delay
M2	C	2	130

Fig. 2.   Recovery method in proposed method.

3.3   Setting the Number of Replications

(a) Setting the number of replications

In Spray and Wait, the value of the number of replications N must be set in advance. The appropriate value of N varies depending on the network environment. If N is small, the number of message replications is small and the possibility of reaching the destination is low. If it is too large, useless replication occurs.

Therefore, we propose a method for dynamically changing N according to the network conditions.

In the proposed method, the average number of hops Hop_{avg}. of the 10 newest messages in this list is calculated, and Hop_{avg}. is applied to the following Eq. (2) to calculate a new N. The Eq. (2) is an equation for setting N so that the maximum number of replications per message is Hop_{avg}. Figure 3 shows an example. First, Hop_{avg} is calculated to set the number of replications N. Hop_{avg} is 3 from the average number of hops in the list. The value was applied to Eq. (2), and the number of replicas N was calculated to be 4. Then node A generates a message. The number of replications N is set to 4 for the message, and the destination is node D. Node A replicates the message to node F, N is 2, and the hop count is 1. Similarly, node F replicates the message to node H, and node H replicates the message to node D. Finally, when the message is reached to the destination node D, the number of hops became to 3, and the average number of hops calculated from the arrival list could be set as the maximum number of replications per message.

$$N = 2^{Hop_{avg}-1} \tag{2}$$

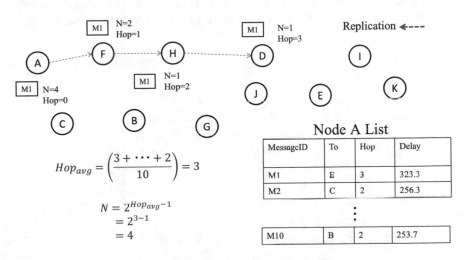

Fig. 3. Setting of the number of replications.

(b) Resetting of the number of replications
In setting of the number of replications, the arrival list is averaged, so the number of replications may be insufficient depending on the conditions such as the destination. Therefore, in the proposed method, it is allowed to reset the number of replicates if a specific condition is satisfied, as in Spray and Focus. However, in Spray and Focus, if the conditions are met, it is replicated unlimitedly. This may result in excessive replication. For this reason, the proposed method restricts the resetting of the number of replications. In order to reset the number of replications, it is necessary to satisfy Eqs. (3) and (4) at the same time.

$$T_{now} - T_{msgCreation} \geq Delay_{avg} \tag{3}$$

$$N_{init} \geq Hop_{msg} \tag{4}$$

Where, T_{now} is the current time, $T_{MsgCreation}$ is the message generation time, $Delay_{avg}$ is the average delay time of arrival list, N_{init} is the value of number of replications N when the message is generated, and Hop_{msg} is the current number of hops in the message.

Equation (3) represents the comparison between the elapsed time since the message was generated and the average delay time. Equation (4) is a conditional equation to prevent excessive replication.

Figure 4 shows an example of resetting the number of replications. Node A generates a message M1 addressed to node D and replicates it. When node H receives the message, N in the message is 1, so normally it will be replicated only to the destination. At this time, when the node H becomes communicable with the node E, it is confirmed whether M1 is applied to the Eqs. (3) and (4). In this case, it is assumed that both equations satisfy the conditions. Therefore, node H sets N of M1 as 2 again and allow replication to node E. Figure 4 shows that the message M1 reaches to the destination.

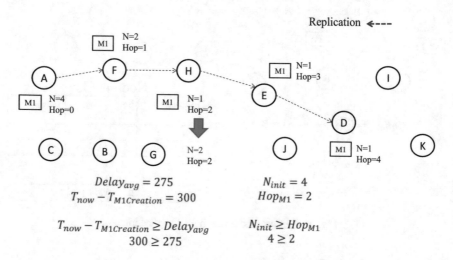

Fig. 4. Resetting of the number of replications.

3.4   Selection of Replication Nodes

Another disadvantage of Spray and Wait is that it doesn't select appropriate replication nodes.

Despite the protocol that limits the number of replications, useless replication may occur because appropriate replication nodes are not selected. For example, if the replication node remains stationary and there are no neighboring nodes, the possibility of reaching the destination is low, so the number of replications is wasted. In order to solve this problem, the proposed method defines the number of nodes encountered for 5 min as mobility as an indicator of whether it contributes to message arrival. When a node that owns a message encounters a certain node, if the other node's mobility is higher than own mobility, the message is replicated to the encountered node. On the other hand, if the encountered node's mobility is lower than own mobility, even if the node replicates, the probability of reaching the destination is low, so it will not be replicated.

4 Performance Evaluation

The comparison methods of evaluation are Spray and Wait, which is related to the proposed method, and MaxProp using the recovery method. Three protocols are evaluated by DTN simulator the ONE.

4.1 Evaluation Scenario

The evaluation scenario assumes an environment where infrastructure is destroyed by a disaster and safety information is exchanged between nodes by ad hoc communication of microcomputers with little storage capacity. In the simulation field, map information around the Faculty of Engineering, Yamagata University is used.

4.2 Evaluation Conditions

Three types of nodes with different mobility were prepared as simulation nodes. The outline is shown in Table 2. The simulation time is 1 h.

Table 2. Parameters for node mobility.

Node	Car	Bicycle	Walk
WaitTime[s]	0–120	0–120	120–240
Speed[m/s]	11.1–16.6	2.7–5.4	1.4–2.7

There are 40 nodes for each type of nodes with different mobility shown in Table 2, for a total of 120 nodes. The storage size of node is changed from 10 MB to 50 MB in increments of 10 MB. Also, one plot is the average value obtained by changing the seed 10 times. The movement model is the Shortest Path Map Based Movement model. The initial value of N for Spray and Wait and the proposed method is 10. In the scenario, all types of nodes shown in Table 2 are evaluated by changing the storage size from 10 MB to 50 MB in increments of 10 MB. Furthermore, the source and destination nodes are determined randomly, and the message size is also determined randomly between 500 kB and 5000 kB.

Under these conditions, the delivery rate, delay time, and overhead are evaluated.

4.3   Simulation Results

(a) Delivery ratio

Figure 5 shows the result of the delivery ratio.

The delivery ratio is good when Spray and Wait is low capacity and MaxProp is high capacity. The performance of conventional methods varies depending on the storage size. Compared with other methods, the proposed method has the same or better message delivery ratio for all storage sizes. The reason is that Spray and Wait is less prone to storage overflow when the storage capacity is small, and it has better performance than MaxProp. However, as the storage capacity increases, it cannot be replicated as much as necessary. Therefore, when the storage capacity is large, the performance of Spray and Wait decreases.

MaxProp uses a recovery method, but the operation is similar to Epidemic, because it replicates the messages to all encountered nodes. For this reason, when the storage capacity is small, MaxProp decreases the arrival ratio due to frequent overflow.

The proposed method works on a Spray and Wait basis that refrains from replication when storage capacity is low. In Spray and Wait base, the message arrival ratio is improved by selecting the replication nodes, setting the dynamic replication number N and using the recovery method. Futhermore, the proposed method improves the storage usage efficiency and resetting the replication number of messages that are difficult to reach the destination. Therefore, it is considered that the proposed method has good performance. Meanwhile, if there is enough storage capacity, all the messages are replicated to encountered nodes like MaxProp, so the result is almost the same value of MaxProp.

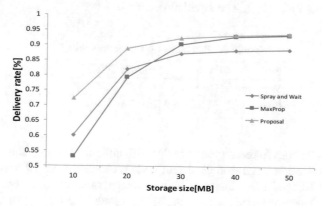

Fig. 5. Delivery ratio.

(b) Delay time

Figure 6 shows the results of delay time. When the storage size is small, messages are easily deleted. As the results, the time that messages remain in the storage is inevitably shortened. Therefore, MaxProp and Spray and Wait have shorter average delay time than the proposed method. Meanwhile, the proposed method is often replicated on a Spray and Wait basis when the storage is small, and since the recovery method is used,

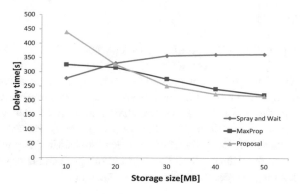

Fig. 6. Delay time.

the storage usage efficiency is good and the time that unreached messages can be stored in the storage is relatively long. Therefore, the delay time becomes longer, but the delivery rate is improved.

However, as the storage size increases, the number of replications increases in the proposed method, so the probability of reaching the destination quickly increases and the delay time tends to decrease.

(c) Overhead

Figure 7 shows the results of overhead. Since Spray and Wait has a fixed number of replications N as 10 in advance, there is no significant change in overhead. Since MaxProp is a protocol with many opportunities to replicate to encountered nodes, the overhead is large compared with other protocols. The overhead of the proposed method increases as the storage size increases. This is because the Epidemic base is often selected as an appropriate protocol as the storage size increases. As the results, the number of replications increases. The overhead of the proposed method is increased compared to Spray and Wait. However, the proposed method has the best delivery rate, and when the storage size is 40 MB and 50 MB, the delivery rate is the same as MaxProp, but the overhead can be greatly reduced.

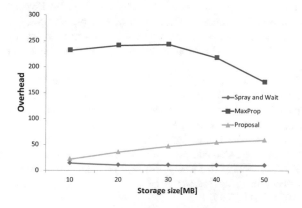

Fig. 7. Overhead.

5   Conclusion

In this paper, we proposed a DTN routing protocol that selects an appropriate protocol from the Epidemic based protocol with many replications and the Spray and Wait based protocol with few replications depending on the storage state. As a result, it was confirmed that the delivery ratio of the proposed method was improved compared with the conventional methods. Also, in a scenario that assumes a disaster, when the storage size is 30 MB or more, we confirmed that the proposed method reduces the overhead to less than half with the same delay time and delivery rate as MaxProp which can deliver the message fast. As the future work, To improve the performance, we would like to apply machine learning technique to the propose method.

References

1. Delay Tolerant Networking Research Group. http://www.dtnrg.org/
2. Cerf, V., et al.: Delay-Tolerant Network Architecture, IETF RFC 4838. http://www.ietf.org/rfc/rfc4838.txt
3. Cerf, V., Burleigh, S., Hooke, A., Torgerson, L., Durst, R., Scott, K., Travis, E., Weiss, H.: Interplanetary internet IPN: Architectural definition, IETF Internet Draft. http://tools.ietf.org/html/draft-irtf-ipnrg-arch-00
4. Juang, P., Oki, H., Wang, Y., Martonosi, M., Peh, L.-S., Rubenstein, D.: Energy-efficient computing for wildlife tracking: design tradeoffs and early experienceswith ZebraNet. ACM SIGOPS Operat. Syst. Rev. **36**, 96–107 (2002)
5. Pentland, A., Fletcher, R., Hasson, A.: DakNet: rethinking connectivity in developing nations. Computer **37**(1), 78–83 (2004)
6. Scott, K., Burleigh S.: Bundle Protocol Specification, IETF RFC 5050. http://www.ietf.org/rfc/rfc5050.txt
7. Vahdat, A., Becker, D.: Epidemic routing for partially connected ad hoc networks, Technical Report CS-200006, Duke University (2000)
8. Spyropoulos, T., Psounis, K., Raghavendram, C.S.: Spray and wait: an efficient routing scheme for intermittently connected mobile networks. In: Proceedings of ACM SIG-COMM Workshop on Delay-Tolerant Networking (WDTN), pp. 77–90 (2005)
9. Spyropoulos, T., Psounis, I.L., Raghavendra, C.S.: Spray and focus efficient mobility-assisted routing for heterogeneous and correlated mobility. In: Proceedings of the IEEE PerCom Workshop on Intermittently Connected Mobile Ad Hoc Networks (2007)
10. Lindgren, A., Doria, A., Schelen, O.: Probabilistic routing in intermittently connected networks. In: ACM SIGMOBILE Mobile Computing and Communications Renew, pp. 19–20 (2003)
11. Burgess, J., Gallagher, B., Jensen, D., Levine, B.N.: MaxProp: routing for vehicle-based disruption-tolerant networks. In: Proceedings of IEEE Infocom (2006)
12. Haas, Z.J., Small, T.: A new networking model for biological applications of ad hoc sensor networks. ACM/IEEE Trans. Netw. **14**(1), 27–40 (2006)
13. Maywad, D., Agrawal, C., Kumar Pal, S.: Improved spray and wait protocol for DTN networks. Int. J. Sci. Res. Comput. Sci. Eng. Inf. Technol. **2**, 441–444 (2017)
14. Yashaswini, K.N., Prabodh, C.P.: Spray and wait protocol based on prophet with dynamic buffer management in delay tolerant networks. Int. J. Recent Innov. Trends Comput. Commun. **5**, 1034–1037 (2017)

IntelligentBox for Web-Based VR Applications (*WebIBVR*) and Its Collaborative Virtual Environments

Yoshihiro Okada[1(✉)] and Taiki Ura[2]

[1] Innovation Center for Educational Resources (ICER),
Kyushu University Library, Kyushu University, Fukuoka, Japan
okada@inf.kyushu-u.ac.jp
[2] Graduate School of Information Science and Electrical Engineering,
Kyushu University, Fukuoka, Japan
ura.taiki.653@s.kyushu-u.ac.jp

Abstract. This paper treats an interactive 3D graphics software development system called *IntelligentBox* and its extended version for Web-based VR applications called *WebIBVR*. *WebIBVR* consists of two main components, a client side part and a server side part. The server side part is almost the same as the original *IntelligentBox*. Its difference from the original IntelligentBox is functionality of off-screen rendering of 3D scenes and communication with a client side part. The client side part is a JavaScript program runs on a Web-browser and displays the images of 3D scenes sent from the server side. The client side part provides several functionalities for VR applications, i.e., stereo view support and multi-angle view support for a Head Mounted Display (HMD) of a VR goggle with a smartphone. Multiple HMDs can share the same view image of a 3D scene generated by server side *IntelligentBox*. However, the server side *IntelligentBox* cannot generate multiple different view images of a 3D scene correspond to the orientation of each of multiple HMDs because it needs much time and the interactivity becomes worse. To overcome this, in this paper the authors proposes the use of multiple server side *IntelligentBoxes* to generate multiple different view images of a common 3D scene at once by communicating with each other using *RoomBox*, one of the special purpose components of original *IntelligentBox* used for building collaborative virtual environments. This paper also explains the performance improvement of image transmission from a server side to a client side.

Keywords: 3D graphics · Development systems · Component ware · Web contents · Virtual Reality

1 Introduction

The development of interactive 3D graphics software had made possible because of advances in 1990s' computer hardware technology, and then 3D graphics software had become in great demand. Then, we proposed an interactive 3D graphics software development system called *IntelligentBox* [1]. Using *IntelligentBox*, we have already

L. Barolli et al. (Eds.): EIDWT 2020, LNDECT 47, pp. 503–515, 2020.
https://doi.org/10.1007/978-3-030-39746-3_51

developed various applications those are 3D-CG animations [3, 4], Virtual Reality (VR) systems [5–9], interactive visualization tools [10–15], and so on. However, these applications could not be available on the web because originally *IntelligentBox* was realized as a development system for desktop 3D graphics applications. If these 3D graphics applications were available on the web, the usefulness of *IntelligentBox* would become higher than ever. Then, we extended *IntelligentBox* in order to make it possible to develop web-based 3D graphics applications [16–18]. This extended *IntelligentBox* is called *WebIB* as its web version. *WebIB* consists of two main components, a client side part and a server side part. The server side part is almost the same as the original *IntelligentBox*. Its difference from the original *IntelligentBox* is functionality of off-screen rendering of 3D scenes and communication with a client side part. The client side part is a JavaScript program runs on a Web-browser and displays the images of 3D scenes sent from the server side.

On the other hand, advances in recent computer hardware technologies of display and sensor devices have made possible the development of VR applications, and then their popularity has become higher than ever. Especially, VR applications available on the web have become more and more popular. If *WebIB* can be used for the development of web-based VR applications, its usefulness would become higher than now. Then, we also extended *WebIB* in order to make it possible to develop web-based VR applications [24]. This extended version of *WebIB* is called *WebIBVR*. The differences of *WebIBVR* from *WebIB* are several functionalities for VR applications, i.e., stereo view support and multi-angle view support for a Head Mounted Display (HMD) of a VR goggle with a smartphone. The use of a HMD can give higher immersion rather than a standard PC display. Multiple web-clients can share the same view image of a 3D scene generated by server side *IntelligentBox*. However, the server side *IntelligentBox* cannot generate multiple different view images of a common 3D scene correspond to the orientation of each of multiple HMDs because it needs much time and the interactivity becomes worse. To overcome this, this paper proposes the use of multiple server side *IntelligentBoxes* to generate multiple different view images of a common 3D scene at once by communicating with each other using *RoomBox* [2], one of the special purpose components of original *IntelligentBox* used for building collaborative virtual environments. By showing the several example use cases of *Room-Boxes*, we justify the use of *RoomBoxes* for building web-based VR collaborative applications. Furthermore, this paper also explains the performance improvement of image transmission from a server side to a client side.

The remainder of this paper is organized as follows: Sect. 2 describes related work about development tools and systems of 3D graphics applications including VR applications. After that, we explain essential mechanisms of *IntelligentBox* and its extended mechanisms of *WebIB* in Sect. 3. Section 4 explains extended functionalities of *WebIBVR* and introduce its web-based VR applications. In Sect. 5, we introduce *RoomBox* and its web-based collaborative VR applications. Finally, we conclude the paper in Sect. 6.

2 Related Work

Our research purpose on *IntelligentBox* is to propose a software architecture that makes it easier to develop 3D graphics applications including interactive web-based 3D contents and VR applications. Its related work includes researches on 3D graphics toolkit systems like Unreal Engine [19] and Unity 3D [20], and programming libraries like Open Inventor [21]. Unreal Engine and Unity 3D are very popular game engines. Open Inventor is an OpenGL (de facto standard 3D Graphics Library) based object oriented programming library. Some of them provide an authoring tool that allows developers to design interactive 3D graphics contents. Even if using such authoring tools for making 3D scenes, it is not easy to develop interactive 3D graphics applications because developers have to write any text-based programs for defining behaviors of primitive 3D objects. As development tools for interactive web 3D contents, there are library systems like WebGL [22] and Three.js [23]. WebGL (Web Graphics Library) is JavaScript API based on OpenGL for rendering interactive 3D graphics and 2D graphics within any compatible web browser without the use of any plug-ins. Three.js is also a library for the development of WebGL based 3D graphics contents. These are library systems so that developers have to write any text-based programs for making interactive web-based 3D graphics contents.

Our research system *IntelligentBox* and its web version *WebIB* provide various 3D software components called *boxes* represented as visible, manually operable, and reusable functional objects. They provide a dynamic data linkage mechanism called *slot connection*. These features make it easier for even end-users to develop 3D graphics applications including interactive web 3D applications. This is the main difference of our *IntelligentBox* and *WebIB* from the others. Moreover, we have already proposed *WebIBVR* as the extension of *WebIB* for the development of web-based VR applications. In this paper, we also propose the mechanism of *WebIBVR* with *RoomBox* for the development of web-based collaborative VR applications. By this mechanism, already existing 3D graphics applications actually developed using original *IntelligentBox* can be used as web-based collaborative VR applications without any modifications.

3 Original *IntelligentBox* and *WebIB*

This section explains essential mechanisms of original *IntelligentBox* and the extended mechanisms of *WebIB*.

3.1 Model-Display Object (MD) Structure

IntelligentBox is one of the component-ware whose components are called *boxes*. As shown in Fig. 1, each *box* consists of two objects, a model and a display object. This internal structure is called MD (Model-Display object) structure. MD structure allows the case that two different display objects share the same common model. This mechanism is called 'model sharing'.

A model holds state values of a *box* those are stored in variables called *slots*. A display object defines how the *box* appears on a computer screen and reacts to the user's manipulation. For instance, *RotationBox* of Fig. 1 has a *slot* named 'ratio' that holds a double precision number used as a rotation angle. Through the direct manipulation of a mouse device on the *box*, its *slot* value is changed. Furthermore, its visual image is also simultaneously changed, i.e. *RotaionBox* rotates. In this way, each *box* reacts to the user's manipulation according to its dedicated functionality.

3.2 Dynamic Data-Linkage Mechanism Called *Slot Connection*

IntelligentBox also provides a dynamic data-linkage mechanism called *slot connection*. Figure 2 shows the data linkage concept among three *boxes*. As explained in the previous subsection, each *box* has multiple *slots*. One of its *slots* can be connected to one of the *slots* of other *box*. This connection is called a *slot connection*. The *slot connection* is realized by three standard messages, i.e., a set message, a gimme (give me) message and an update message, when there is a parent-child relationship between any two *boxes*. The followings are their formats:

(1) Parent *box* set <slotname> <value>.
(2) Parent *box* gimme <slotname>.
(3) Child *box* update.

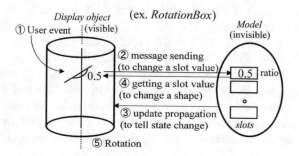

Fig. 1. *MD* structure of a *box* and its internal messages.

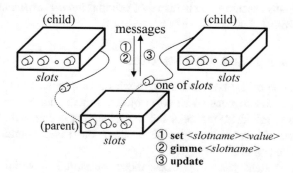

Fig. 2. Standard messages between a child and its parent *box*.

A <value> in a format (1) represents any value, and a <slotname> in formats (1) and (2) represents a user-selected slot of the parent *box* that receives these two messages. A set message writes a child *box* slot value into its parent *box* slot and a gimme message reads a parent *box* slot value to set it into its child *box* slot. Update messages are issued from a parent *box* to all of its child *boxes* to tell them that the parent *box* slot value has changed. By these three messages, the two slots of a child *box* and its parent *box* are connected and their two functionalities are combined.

3.3 Extended Mechanisms of *WebIB*

Figure 3 shows extended mechanisms of *WebIB*. Originally, *IntelligentBox* system uses OpenGL 3D graphics library which can generate off-screen rendered images of 3D scenes. As shown in the figure, *IntelligentBox* system runs on a web server, and its rendered image of a 3D scene is generated on the web server and transferred to a web browser through the Internet. On the web browser, besides a HTML program, a JavaScript program runs to manage user operation events, i.e., a mouse move, a mouse button click and so on. Such user operation events will be transferred to the web server using XMLHTTPrequest message through a CGI-program runs on the web server. The CGI-program is written in Perl.

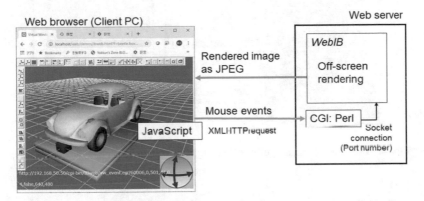

Fig. 3. Extended mechanisms of *WebIB*.

The CGI-program once receives the user operation events and applies them to *IntelligentBox* system. And then, *IntelligentBox* system will generate next off-screen rendered image of the 3D scene to be updated according to the received operation event. In this way, the user can interactively manipulate 3D contents of *IntelligentBox* that runs on the web server with looking at his/her web browser. *WebIB* supports most web browsers and then *WebIB* is available on any mobile device like iOS device and Android OS device besides a standard Desktop PC. However, when using iOS device or Android OS device, the touch interface is significant. Therefore, *WebIB* also supports the touch interface for both iOS and Android OS devices. As shown in Fig. 3, the touch interface image appears in the right lower part of the web browser. When the user do the tap on the image, the position/direction of the viewpoint will be changed according to the type of the image, i.e., there are three types for rotation, zoom and pan. Their details will be explain in the later section.

3.4   Mechanism of *WebIB* for Multiple Users

As shown in Fig. 4, *WebIB* can provide multiple users with a web-based collaborative environment based on the same mechanism of Fig. 3. *WebIB* has a System ID No. (SID No.), and by specifying it, each client user can access its corresponding *WebIB* that has the same SID No. through his/her web-browser. Using this data-linkage, communications between a teacher and a learner become possible in the cases of educational 3D contents as shown in Fig. 5. If the left user changes his/her viewpoint and direction, the

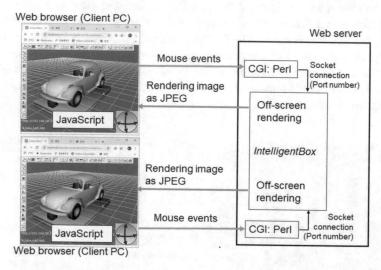

Fig. 4. Data-linkage among one *WebIB* (right part) and its two clients (left part).

Fig. 5. Collaboration case of a medical education content by sharing the same screen image.

viewpoint and direction of the right user are also changed similarly and simultaneously. Contrarily, when the right user changes his/her viewpoint and direction, the viewpoint and direction of the left user are also changed similarly and simultaneously. The collaboration like this becomes possible by sharing the same screen image.

4   *WebIBVR* for Web-Based VR

4.1   Extended Functionalities of *WebIBVR*

A head mounted display (HMD) is the most important feature for VR applications and the most portable/cheapest HMD is a set of a VR goggle and a smartphone as shown in Fig. 6. To use such a set as a HMD requires a stereo view support. Therefore, we extended *WebIB* to support the stereo view on a web browser. The left part of Fig. 6 is an image of a stereo view on a smartphone's web browser, the center is VOX+ 3DVR goggle and the right is Poskey blue-tooth gamepad. The blue-tooth gamepad is useful because the touch interface is not available when a smartphone put inside a google.

Fig. 6. The stereo view of *WebIBVR* on a smartphone's browser (left), VOX+ 3DVR goggle (center) and Poskey blue-tooth gamepad (right).

Besides stereo view support, we extended *WebIB* to support touch interfaces and device orientation/motion events as explained below.

(1) Stereo view support

Three.js [23], one of the most popular WebGL based 3D graphics library, supports a stereo graphics. However, JavaScript part of *WebIBVR* framework does not use any 3D graphics library. Therefore, a stereo view image of side-by-side should be generated by *IntelligentBox* system runs on a web server. Originally, *IntelligentBox* system supports a stereo view so that it generates a left eye image and a right eye image separately, and then merges them into one side-by-side stereo image. This stereo image will be transferred to the web browser on a smartphone. Then, the user can see the stereo image like the left image of Fig. 6.

(2) Touch interfaces

Originally, HTML5 supports JavaScript functions for touch interfaces like 'touchstart', 'touchend', 'touchmove' events and x and y positions of the user's touch fingers can be accessed by event.touches[*].pageX and event.touches[*].pageY, here, * means the indices of the user's fingers. As already explained in Subsect. 3.3, the touch interface part is included in the JavaScript program of *WebIB* framework and it provides the touch interface image that appears on the right lower part of a web browser. When the user do the tap on the image, the position/direction of a viewpoint will be changed according to the type of the image, i.e., there are three images for rotation, zoom and pan as shown in Fig. 7. X and Y values treated as mouse device moving values are calculated by the quadratic equation of the distance from the center position to the tapped position.

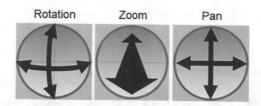

Fig. 7. Three images for the view controls, rotation, zoom and pan.

(3) Device orientation/motion interfaces

The device orientation/motion interfaces are simple but useful in a JavaScript program of HTML5 because there are 'deviceorientation' and 'devicemotion' events, and the device orientation/motion of the user's smartphone can be accessed by event. alpha, event.beta, event.gamma, event.acceleration.x, event.acceleration.y and event. acceleration.z, respectively. The JavaScript program of *WebIBVR* framework supports device orientation/motion events and their corresponding values will be sent to *IntelligentBox* system runs on a web server to change the position/direction of its viewpiont.

(4) Another extension

One of the problems of *WebIB* and *WebIBVR* is image transmission cost from a server side to a client side. To reduce this cost, we employed partial refreshing method that transfer only different parts of a present image with a previous image from a server side to a client side. Usually, if the view position/direction are not changed, the present view image and its previous view image are almost same and the different parts are very small. Therefore, image transmission cost can be drastically reduced.

4.2   VR Application of *WebIBVR*

There have been several VR applications actually developed using original *IntelligentBox* so far because it supports several VR devices by their dedicated software components, i.e., special purpose *boxes*. One of the VR applications is a dental training system [9] that supports a phantom device, one of the haptic devices, as a 3D input device with force-feedback. Using a phantom device, the user can feel force-feedback as operation sensation. Although the original dental training system of *IntelligentBox* had not supported a HMD, using *WebIBVR* framework, we can employ a set of a smartphone and a VR goggle as a HMD device of the dental training system as shown in Fig. 8. A side-by-side stereo graphics image of teeth and gum 3D models appears in the web browser on a smartphone. A trainee wears a VR goggle with a smartphone put inside, then he/she can see teeth and gum 3D models. Through IEEE 1394 interface, a phantom device is connected to the web server on which *IntelligentBox* system runs. The trainee can operate teeth 3D model using the phantom device by looking at the teeth and gum 3D models using a HMD with high immersion.

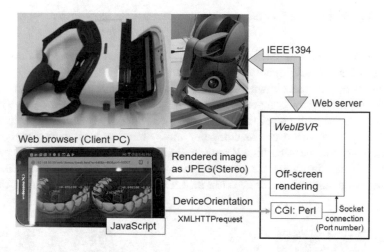

Fig. 8. Dental training VR application supporting a haptic device (phantom).

5 Collaborative Virtual Environments of *WebIBVR* with *RoomBox*

For building collaborative virtual environments, there is a special purpose component called *RoomBox* [2] of original *IntelligentBox*. This section explains the functionality of *RoomBox* and proposes the use of *RoomBox* for building web-based collaborative VR applications realized by *WebIBVR*.

As shown in Fig. 9, *RoomBox* has a slot named "event" which holds a current user-operation event operated on its descendant *boxes*. Some specific user-operation events generated in *RoomBox* are always stored in its slot until the next event is generated. Besides the model sharing mechanism explained in previous Sect. 3.1, original *Intel-ligentBox* provides a distributed model sharing mechanism concerning *RoomBox*. It enables multiple, distributed *RoomBoxes* to virtually share a common model with each other through messages passed via a network. When multiple *RoomBoxes* share a common model based on the distributed model sharing mechanism, they can share

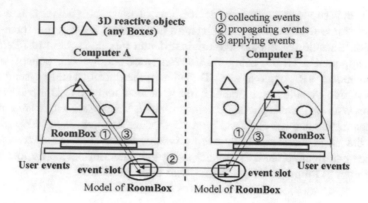

Fig. 9. Message flow between two *RoomBoxes* connected via a network.

user-operation events with each other. In this case, descendant *boxes* of *RoomBox* are treated as collaboratively operable 3D objects.

In the example case shown in Fig. 9, there are two *RoomBox* models existing separately on a different computer. These two models are kept in the same state by messages passed through a network. This linkage is built easily and rapidly by making a shared copy of *RoomBox* on one computer and by transferring it to another computer. The linkage for the message flow shown in Fig. 9 is automatically built. When a user operates one *box* in *RoomBox* on computer A, his/her operation event is sent to *RoomBox* model and subsequently set in its event slot. Further, this event is sent to the

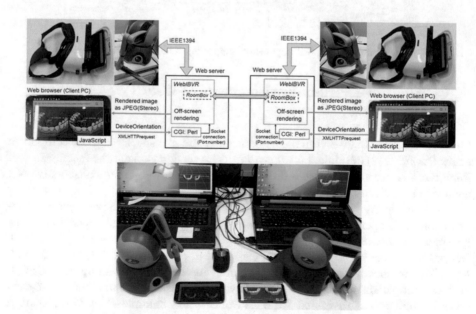

Fig. 10. Collaborative dental training system realized using *WebIBVR* with *RoomBox*.

model of the other *RoomBox* existing on computer B by a certain message via a network. After that, the operation event is applied to the corresponding *box* on computer B. In this way, by using distributed *RoomBoxes*, user-operation events are shared among multiple computers.

Since event data of a phantom device used in the dental training VR application shown in Fig. 8 can be also shared among several users using the event sharing mechanism of *RoomBox* [9], it is possible to provide collaborative environments of the dental training VR application as shown in Fig. 10. In this case, a teacher teaches how dental operations are performed to a learner even when the teacher and the learner are in a different place. Although the dental operations performed by each phantom device are shared between the teacher and the learner, the view direction of each of them is controlled by his/her own smartphone's device orientation individually because of the functionality of *WebIBVR*.

6 Conclusions

In this paper, we treated our *IntelligentBox*, an interactive 3D graphics software development system that provides various software components. Combining these software components by direct manipulations on a computer screen enables users to construct 3D graphics applications without writing any text-based program. This feature is significant for end-users who do not have any programming knowledge. Furthermore, we enhanced the usefulness of *IntelligentBox* by extending it to work as a development system for web-based 3D graphics applications. This is called *WebIB* as the web version of *IntelligentBox*. Recently, we also proposed *WebIBVR*, as the further extension of *WebIB* in order to make possible the development of web-based VR applications and introduced a dental training system as one of such VR applications.

Furthermore, in this paper, we introduced a special purpose component called *RoomBox* of original *IntelligentBox*. *RoomBox* enables standard 3D graphics applications developed using *IntelligentBox* to be collaborative 3D graphics applications. Then, this paper proposes the use of *RoomBox* for the development of web-based collaborative VR applications and showed a collaborative dental training system as one of such applications.

As future works, although we could reduce the cost of image transmission from a server side to a client by employing partial refreshing method, we will try to reduce the image transmission cost moreover to improve the interactive performance by employing sophisticated communication protocols such as WebRTC (Web Real-Time Communication). Also, we will develop more practical web-based single or collaborative VR applications to clarify the usefulness of *WebIBVR*.

Acknowledgments. This research was partially supported by JSPS KAKENHI Grant No. JP17H00773.

References

1. Okada, Y., Tanaka, Y.: *IntelligentBox*: a constructive visual software development system for interactive 3D graphic applications. In: Proceedings of Computer Animation 1995, pp. 114–125 (1995)
2. Okada, Y., Tanaka, Y.: Collaborative environments of *IntelligentBox* for distributed 3D graphics applications. Vis. Comput. **14**(4), 140–152 (1998)
3. Okada, Y.: Real-time character animation using puppet metaphor. In: Nakatsu, R., Hoshino, J. (eds.) IFIP First International Workshop on Entertainment Computing (IWEC 2002). Entertainment Computing Technologies and Applications, pp. 101–108. Kluwer Academic Publishers (2003)
4. Okada, Y.: Real-time motion generation of articulated figures using puppet/marionette metaphor for interactive animation systems. In: Proceedings of the 3rd IASTED International Conference on Visualization, Imaging, and Image Processing (VIIP 2003), pp. 13–18. ACTA Press (2003)
5. Okada, Y., Shinpo, K., Tanaka, Y., Thalmann, D.: Virtual input devices based on motion capture and collision detection. In: Proceedings of Computer Animation 1999, pp. 201–209. IEEE CS Press (1999)
6. Okada, Y.: 3D visual component based approach for immersive collaborative virtual environments. In: ACM SIGMM 2003 Workshop on Experiential Telepresence (ETP 2003), pp. 84–90 (2003)
7. Okada, Y.: 3D visual component based approach for effective telepresence systems. In: Proceedings of ACM SIGMM 2004 Workshop on Effective Telepresence: Toward Seamless Remote Interaction and Experience (ETP 2004), Demo Paper, pp. 46–47 (2004)
8. Okada, Y., Ogata, T., Matsuguma, H.: Component-based approach for prototyping of Tai Chi-based physical therapy game and its performance evaluations. ACM Comput. Entertain. **14**(1), 4:1–4:20 (2016)
9. Kosuki, Y., Okada, Y.: 3D visual component based development system for medical training systems supporting haptic devices and their collaborative environments. In: Proceedings of the 4th International Workshop on Virtual Environment and Network Oriented Applications (VENOA-2012) of CISIS-2012, pp. 687–692. IEEE CS Press (2012)
10. Akaishi, M., Okada, Y.: Time-tunnel: visual analysis tool for time-series numerical data and its aspects as multimedia presentation tool. In: Proceedings of 8th International Conference on Information Visualization (IV 2004), pp. 456–461. IEEE CS Press (2004)
11. Notsu, H., Okada, Y., Akaishi, M., Niijima, K.: Time-tunnel: visual analysis tool for time-series numerical data and its extension toward parallel coordinates. In: Computer Graphics, Imaging and Visualization as Proceedings of CGIV 2005, pp. 167–172. IEEE CS Press (2005)
12. Okada, Y.: Network data visualization using parallel coordinates version of time-tunnel with 2Dto2D visualization for intrusion detection. In: (WAINA 2013) IEEE 27th International Conference on Advanced Information Networking and Applications Workshops, pp. 1088–1093 (2013)
13. Okada, Y.: Parallel coordinates version of time-tunnel (PCTT) and its combinatorial use for macro to micro level visual analytics of multidimensional data. In: Modelling and Processing for Next Generation Big Data Technologies and Applications, 27 p. Springer (2014)
14. Tanaka, Y., Okada, Y., Niijima, K.: Treecube: visualization tool for browsing 3D multimedia data. In: Proceedings of 7th International Conference on Information Visualization (IV 2003), pp. 427–432. IEEE CS Press (2003)

15. Tanaka, Y., Okada, Y., Niijima, K.: Interactive interfaces of treecube for browsing 3D multimedia data. In: Proceedings of ACM The 7th International Working Conference on Advanced Visual Interfaces (AVI 2004), pp. 298–302 (2004)
16. Okada, Y.: *IntelligentBox* as development system for SaaS applications including web-based 3D games. In: Proceedings of the 9th Annual European GAMEON Conference, pp. 22–26 (2008)
17. Okada, Y.: Web version of *IntelligentBox* (*WebIB*) for development of web 3D educational contents. In: Proceedings of IADIS International Conference Mobile Learning 2011, pp. 251–255 (2011)
18. Okada, Y.: Web version of *IntelligentBox* (*WebIB*) and its integration with Webble World. In: Webble Technology as Proceedings of First Webble World Summit (WWS 2013). CCIS Series, vol. 372, pp. 11–20 (2013). ISSN 1865-0929, ISBN 978-3-642-38835-4
19. Unreal Engine, August 2019. https://www.unrealengine.com/
20. Unity 3D, August 2019. https://unity3d.com/
21. Open Inventor, August 2019. https://www.openinventor.com/
22. WebGL, August 2019. https://www.khronos.org/webgl/
23. Three.js, August 2019. https://threejs.org/
24. Okada, Y.: Web version of IntelligentBox (WebIB) and its extension for web-based VR applications – WebIBVR. In: Proceedings of the 14th International Conference on Broad-Band Wireless Computing, Communication and Applications (BWCCA-2019), pp. 303–314. Springer (2019)

Optimization Problem for Network Design by Link Protection and Link Augmentation

Hiroki Yano and Hiroyoshi Miwa[✉]

School of Science and Technology, Kwansei Gakuin University,
2-1 Gakuen, Sanda-shi, Hyogo 669-1337, Japan
{yano,miwa}@kwansei.ac.jp

Abstract. Information networks need high reliability. Although link failure is one of the causes of communication disconnection in networks, it is necessary to be able to continue communication even in such a failure. There are two approaches of link protection and link addition against link failure. Link protection is to make the failure probability of links sufficiently small by sufficient backup resource and rapid recovery system. Link addition is to increase network connectivity by adding new links to a network. Both link addition and link protection are methods to design a reliable network. Some network design methods either only by link addition or only by link protection have been investigated so far. However, it is possible to design a reliable network at lower cost by a combination of link addition and protection than a network designed either by only link addition or by only link protection. In this paper, we deal with a network design problem to realize a reliable network at lower cost by combining link addition and link protection. First, we formulate this network design problem as a decision problem and we prove the NP-hardness. Furthermore, we propose some polynomial-time algorithms for the problems of the optimization version under some restricted conditions.

1 Introduction

Information networks need high reliability. Although link failure is one of the causes of communication disconnection in networks, it is necessary to be able to continue communication even in such a failure. In other words, a network must be sufficiently connected, even if a link failure occur.

If we protect all links so that the failure probability of links is sufficiently small by sufficient backup resource and rapid recovery system, the resulting network is reliable. However, since link protection needs much cost, it is necessary to find the smallest number of links to be protected so that a network resulting from failures of any non-protected links can provide sufficient connectivity.

There is another method to improve the reliability of an information network. The edge-connectivity and the vertex-connectivity of the graph corresponding to the structure of an information network have been used as the measures for evaluation of the reliability of the network. As for a network design method to improve an information network that is lack of the connectivity, there is a method in which the connectivity is increased by adding links. The resulting network is reliable; however, since the addition

ⓒ Springer Nature Switzerland AG 2020
L. Barolli et al. (Eds.): EIDWT 2020, LNDECT 47, pp. 516–521, 2020.
https://doi.org/10.1007/978-3-030-39746-3_52

of links needs much cost, it is necessary to find the smallest number of links to be added so that a resulting network has a sufficient connectivity.

Both the addition and the protection of links are the methods to design a reliable network. Many network design methods either by link addition or by link protection have been extensively investigated so far. However, it is possible to design a reliable network with lower cost by a combination of link addition and link protection than a network designed either by only link addition or by only link protection.

We show an example. Figure 1 is a network that, if there is an edge in the network, a communication link corresponding to the edge can be later constructed. We cannot construct a link for an arbitrary pair of nodes in an actual information network because of geographical constraints and so on. We call this network an available network. An available network indicates the pairs for which the link can be constructed. An edge in an available network has the cost of addition and has the cost of protection. The cost of addition for edge e in an available network is the cost in order to construct actually a link on edge e; the cost of protection for edge e in an available network is the cost to protect the link already existing on e.

Figure 2 shows the initial information network. Note that the graph corresponding to the structure of the initial information network is the subgraph of the available network. Since the initial information network is a tree in this example, it is disconnected when only one link is broken. The purpose in this example is to make the information network robust so that it is always connected even if any one link is broken. If only link protection is used, the links of bold line are protected with the minimum cost of 19 so that node s and node t are connected even after any one link except the protected links is broken (See the left figure of Fig. 3). On the other hand, if only link addition is used, the links of dotted line are added with the minimum cost of 9 so that node s and node t are connected even after any one link except the protected link is broken (See the right figure of Fig. 3). However, if the combination of link protection and link addition is used, the information network can be constructed with the minimum cost of 8 so that node s and node t are connected even after any one link is broken (See Fig. 4).

In this paper, we address a network design problem to realize a reliable network with low cost by combining link addition and link protection. First, we formulate this network design problem as a decision problem and we prove the NP-hardness.

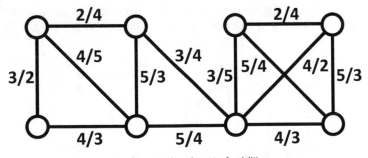

cost of protection / cost of addition

Fig. 1. Available network N

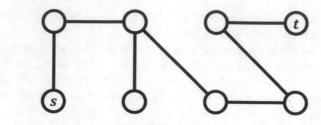

Fig. 2. Initial network G'

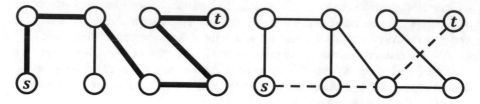

Fig. 3. Left: Only link protection, Right: Only link addition

Furthermore, we propose some polynomial-time algorithms for the problems of the optimization version under some restricted conditions.

2 Related Research

Many network design methods have been investigated, not only in the field of information networks, but also in the fields of graph and optimization theories. There is a problem for determining a set of edges to meet the required edge-connectivity or vertex-connectivity by adding edges to a graph. When the edge-connectivity (resp. vertex-connectivity) of a graph is k, the number of edge (resp. vertex) disjoint paths between two arbitrary vertices is at least k. Therefore, when a graph representing the structure of an information network has k-edge-connectivity (resp. vertex-connectivity), even if $k-1$ links (resp. nodes) are simultaneously broken, the information network is connected. The objective of this problem is to minimize the number of added edges. The problem of augmenting edge-connectivity can be solved in polynomial time [1,2]. On the other hand, the problem of augmenting vertex-connectivity is NP-hard [3].

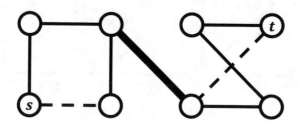

Fig. 4. Combination of link protection and link addition

Another measure of network reliability is the number of connected components when the information network is divided due to a failure [4]. Reference [4] addresses an optimization problem that determines a minimum broken node set maximizing the number of the connected components in the information network resulting from a failure. This problem is NP-hard, and a heuristic algorithm is proposed. However, from the viewpoint of link or node protection, it does not always give a set of protected nodes to keep small number of the connected components against any node failures, because this problem determines a node set whose failure maximizes the number of the connected components.

As for link protection, the group of the authors has investigated some problems so far. We addressed the network design problems that determine the protected links to keep the connectivity to a server in [5] and to avoid congestion in [6], and so on. The network design problem in [7,8] asks the smallest number of links to be protected so that the diameter of a network resulting from failures of non-protected links is less than or equal to a given integer. It was proved that this problem is NP-hard, and the approximation algorithms were presented to solve the problem in polynomial time when the number of broken links is restricted.

3   Problem of Augmenting Connectivity by Protecting and Adding Edges

Let $G = (V, E)$ where V is a set of vertices and E is a set of edges be a connected undirected graph representing the structure of an available network. An edge in an available network has the cost for edge addition and has the cost for edge protection in case that the edge already exists. We associate the costs with each edge by the cost functions $c_A : E \to \mathbb{R}_+$ and $c_P : E \to \mathbb{R}_+$. Let $c_A(F)$ be $\sum_{e \in F} c_A(e)$ and $c_P(F)$ be $\sum_{e \in F} c_P(e)$ where $F \subseteq E$. Let $Q = \{q_1 = \{s_1, t_1\}, q_2 = \{s_2, t_2\}, \ldots, q_r = \{s_r, t_r\}\}$ be a set of vertex pairs. When all vertex pairs are connected in a graph G, we say that Q is connected in G.

Problem for Network Design Improving Connectivity by Link Protection and Link Addition (LPLA)
INSTANCE: *Network* $N = (G = (V, E), c_A, c_B)$, *graph* $G' = (V, E')$ *where* $E' \subseteq E$, *set of vertex pairs* $Q = \{q_1, q_2, \ldots, q_r\}$, *a positive real number* B, *a positive integer* k.
QUESTION: *Is there edge sets* $E_P (\subseteq E')$ *and* $E_A (\subseteq E \setminus E')$ *that* Q *is connected in the resulting graph from the removal of any* k *edges except edges of* E_P *in the graph* $\tilde{G} = (V, E \cup E_A)$ *and that* $c_P(E_P) + c_A(E_A)$ *is less than or equal to* B? ☐

Network N corresponds to an available network, G' corresponds to an initial information network. The feasible solution of this problem, E_P and E_A, correspond to the set of protected links and the set of added links, respectively.

We show the computational complexity of LPLA.

Theorem 1. *LPLA is NP-hard.*

Proof: When an available network is the complete graph, the cost of addition is sufficiently large, and the cost of protection is one for all edges, LPLA is included in the problem in Ref. [9], because, in this case, only protection is used. Since the problem is NP-hard, LPLA is also NP-hard. ☐

The above definition is the formulation as the decision problem; however, we can formulate as the optimization problem whose objective function is to minimize the cost. In the rest of the paper, LPLA indicates the optimization version.

Since LPLA is NP-hard, we cannot expect a polynomial-time algorithm for LPLA; however, LPLA can be solved in polynomial time when the constraint is restricted. We prove the following theorems.

Theorem 2. *When an available network is the complete graph, the cost of protection is sufficiently large, and the cost of addition is one for all edges, LPLA can be solved in polynomial time.*

Proof: In this case, LPLA is the same as the problem augmenting edge-connectivity by edge addition [2], because, since the cost of addition is smaller than the cost of protection, the cost of the network made only by addition of edges is smaller. Since the problem augmenting edge-connectivity by edge addition can be solved in polynomial time and output the minimum number of added edges, LPLA can be also solved in polynomial time.   □

Theorem 3. *When an available network is the complete graph and the number of the protected edges is a constant integer, and the cost of addition is one for all edges, LPLA can be solved in polynomial time.*

Proof: For a set of protected edges, we apply the polynomial-time algorithm for the problem augmenting edge-connectivity by edge addition [2] to the graph whose protected edges are shrinked. We iterate the procedure for all sets of protected edges, and we choose the solution that the sum of the costs of protection and addition is the minimum. Since the number of the protected edges is a constant integer, the number of all the sets of protected edges is the polynomial order. Therefore, the total computational complexity is the polynomial order.   □

Theorem 4. *When the number of added edges is a constant integer, the cost of addition is one and the cost of protection is one for all edges, we can get the approximation solution with the approximation ratio of k in polynomial time.*

Proof: For a set of added edges, we make the network by adding these edges to the initial network G'. We can get a set of protected edges whose number is at most k times of the optimization solution by applying the algorithm in Ref. [9] to the network. We iterate the procedure for all the sets of added edges, and we choose the solution that the number of protected edges is the minimum. Thus, we can get the approximation solution with the approximation ratio of at most k. When the number of added edges is a constant integer, the number of all the sets of added edges is the polynomial order. Therefore, the total computational complexity is the polynomial order.   □

4   Conclusion

In this paper, we addressed a network design problem to realize a reliable network at low cost by combining link addition and link protection. Some network design methods either only by link addition or only by link protection have been investigated so far. However, it is possible to design a reliable network at lower cost by a combination of link addition and protection than a network designed either by only link addition or by only link protection.

First, we formulated this new network design problem LPLA, and we proved the NP-hardness of this problem. Furthermore, we proposed some algorithms for the problems under some restricted conditions. When an available network is the complete graph, the cost of protection is sufficiently large, and the cost of addition is one for all edges, LPLA can be solved in polynomial time. When an available network is the complete graph, the number of the protected edges is a constant integer, and the cost of addition is one for all edges, LPLA can be solved in polynomial time. On the other hand, when the number of added edges is a constant integer, the cost of addition is one and the cost of protection is one for all edges, we can get the approximation solution with the approximation ratio of k in polynomial time.

In the future work, it remains to design the algorithms for the problem when the ratio of the cost of link addition to the cost of link protection is small.

Acknowledgements. This work was partially supported by the Japan Society for the Promotion of Science through Grants-in-Aid for Scientific Research (B) (17H01742).

References

1. Frank, A.: Augmenting graphs to meet edge-connectivity requirements. In: Proceedings of the 31st Annual Symposium on Foundations of Computer Science, St. Louis, 22–24 October 1990
2. Ishii, T., Hagiwara, M.: Minimum augmentation of local edge-connectivity between vertices and vertex subsets in undirected graphs. Discrete Appl. Math. **154**(16), 2307–2329 (2006)
3. Kortsarz, G., Krauthgamer, R., Lee, J.R.: Hardness of approximation for vertex-connectivity network design problems. In: Jansen, K., Leonardi, S., Vazirani, V. (eds.) Approximation Algorithms for Combinatorial Optimization. Lecture Notes in Computer Science, vol. 2462, pp. 185–199. Springer, Heidelberg (2002)
4. Arulselvan, A., Commander, C.W., Elefteriadou, L., Pardalos, P.M.: Detecting critical nodes in sparse graph. Comput. Oper. Res. **36**(7), 2193–2200 (2009)
5. Imagawa, K., Miwa, H.: Detecting protected links to keep reachability to server against failures. In: Proceedings of ICOIN 2013, Bangkok, 28–30 January 2013
6. Noguchi, A., Fujimura, T., Miwa, H.: Network design method by link protection for network load alleviation against failures. In: Proceedings of International Conference on Intelligent Networking and Collaborative Systems (INCoS 2011), Fukuoka, Japan, 30 November–2 December 2011, pp. 581–586 (2011)
7. Fujimura, T., Miwa, H.: Critical links detection to maintain small diameter against link failure. In: Proceedings of INCoS 2010, Thessaloniki, Greece, 24–26 November 2010, pp. 339–343 (2010)
8. Imagawa, K., Miwa, H.: Approximation algorithms for finding protected links to keep small diameter during link failures. In: IEICE Technical Report, Miyazaki, Japan, 8–9 March 2012, vol. 111, no. 468, pp. 403–408 (2012)
9. Morino, Y., Miwa, H.: Network design method resistant to cascade failure considering betweenness centrality. In: Proceedings INCoS 2019, Oita, 5–7 September 2019, pp. 360–369 (2019)

Workflow Improvement for KubeFlow DL Performance over Cloud-Native SmartX AI Cluster

Yujin Hong and JongWon Kim[✉]

School of Electrical Engineering and Computer Science,
Gwangju Institute of Science and Technology (GIST), Gwangju, Republic of Korea
{hyj2508,jongwon}@nm.gist.ac.kr

Abstract. Cloud-native Kubernetes-based orchestration is widely adopted to take advantage of building large-scale resource pools by flexibly expanding the size of pools with the insertion of additional worker nodes. To meet the emerging demand for AI (Artificial Intelligence)-inspired HPC (High Performance Computing)/HPDA (High Performance Data Analytics) workloads, versatile AI clusters driven by open-source KubeFlow software have been rapidly developed by leveraging for various ML (Machine Learning)/DL (Deep Learning) tools and frameworks. However, since the current version of KubeFlow is not fully aware of underlying GPU (Graphics Processing Unit) resources, special attention should be made to smoothly execute the ML/DL workloads. Thus, in this paper, we explore tentative options to improve the ML/DL workflow under a KubeFlow-enabled AI cluster, which focus on GPU utilization efficiency with the assistance of Prometheus open-source monitoring.

1 Introduction

Recent advances in AI technology has increased the demand for ML/DL workloads for Machine Learning (ML) or Deep Learning (DL), HPDA (High Performance Data Analytics) workloads for big data processing, and HPC workloads for heavy computing jobs. In order to leverage the massive amount of resources necessary to operate these workloads, in several cases, multiple nodes are clustered together and utilized as a single large pool of resources. Thus, the adoption of dedicated aggregation of nodes into a cluster is rapidly rising to accommodate the increasing AI-inspired HPC/HPDA workloads [1].

Also, the cloud-native computing paradigm based on Kubernetes(K8S)-orchestrated containers, spearheaded by Cloud Native Computing Foundation (CNCF), is widely spreading. Motivated by this explosive expansion of cloud-native computing, it became quite interesting and future-proof to build AI-inspired HPC/HPDA clusters in a cloud-native way so that it could provide the flexibility to scale out by simply adding more nodes and building a single large-scale shared resource pool. Aligned with this cloud-native computing trend, open-source ML/DL software tools and frameworks are becoming de-facto in servicing the above AI-related workloads.

© Springer Nature Switzerland AG 2020
L. Barolli et al. (Eds.): EIDWT 2020, LNDECT 47, pp. 522–531, 2020.
https://doi.org/10.1007/978-3-030-39746-3_53

Among them, KubeFlow [2] is a leading open-source machine learning platform that supports cloud-native environment. This offers a variety of features that help you easily deploy and operate containerized ML/DL workloads on clusters built on Kubernetes [3]. KubeFlow includes several tool-style services to create and manage interactive Jupyter notebooks. With the web GUI (Graphical User Interface) of Jupyter notebooks, one can customize notebook deployments by preparing and utilizing necessary computing resources. Furthermore, KubeFlow provides effective ML/DL pipelines that comprehensively deploy and manage end-to-end ML/DL workflows. However, when executing ML/DL workloads in a KubeFlow cluster, resource efficiency is still limited. It is not straightforward to manage K8S-orchestrated multiple pods, which are dedicated to their workspaces. Also, with current KubeFlow usage via Jupyter notebook GUI, GPU is not automatically unmounted.

Thus, in this paper, in order to mitigate the above limitations, we first check a workaround solution without KubeFlow by directly utilizing the Linux shell script matched for SmartX AI Cluster [4]. We then suggest a resource-aware workflow for KubeFlow-based ML/DL workloads by leveraging Prometheus open-source monitoring [5], which is known for dimensional data model, flexible query language, efficient time series database, and modern alerting approach.

2 Cloud-Native SmartX AI Cluster

SmartX AI cluster targets cloud-native version of horizontal scalability by connecting several high-performance GPU nodes via high-speed inter-connecting switches. When executing containerized ML/DL workloads, we adopt cloud-native concept with open-source Kubernetes container orchestration and Docker container runtime. This enables the scalable and flexible execution of ML/DL workloads more controllable.

2.1 Hardware and Software of Cloud-Native SmartX AI Cluster

Figure 1 illustrates the current hardware configuration for the SmartX AI cluster, which consists of one intelligence center node for computing and storage masters, several types of intelligence worker nodes, and cloud-enabled storage nodes. In this paper, we are mainly utilizing one type of intelligence worker nodes, which respectively possess 2 Intel Xeon E5-2650v3 CPUs, 1 NVidia Titan V GPU, and 128 GB RAM. Also, for storage support, we leverage both internal storage of individual worker node and Ceph-driven cloud-enabled storage nodes. Note that there are four storage nodes and one of them is an all-flash storage node with large-capacity next-generation NF1 all flash drives. Finally, the data plane of the SmartX AI cluster is constructed through Mellanox SN2100 100 Gbps converged Ethernet switch, where either dual 100G-ports or dual 25G-ports of Mellanox smart NICs is used to connect nodes to the switch fabric.

Figure 2 illustrates the software specification for cloud-native SmartX AI cluster. Kubernetes 1.15.3 is used for container orchestration. In each worker node, Docker-CE 18.03 container runtime is running. Also, Project Calico 3.8 is used for container networking. Finally, KubeFlow 0.6.2 is used for ML/DL pipeline support. Besides, various types of shared storage modes, based on BeeGFS [6] and CephFS [7], are configured.

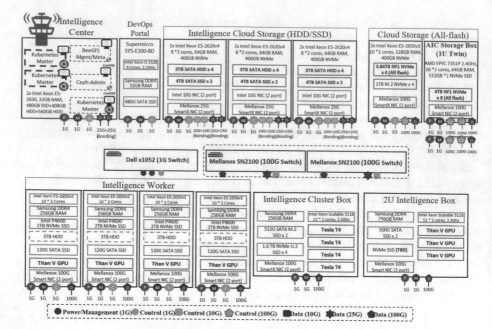

Fig. 1. Hardware specification for SmartX AI cluster.

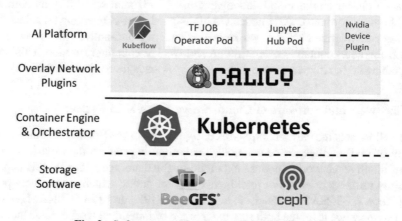

Fig. 2. Software specification for SmartX AI cluster.

2.2 Typical Usage Pattern for the SmartX AI Cluster

- **KubeFlow GUI for Jupyter Notebooks:** With KubeFlow GUI, multiple users can make various Jupyter notebooks by selecting memory sizes, the number of CPU cores, the number of GPUs, and the container image supporting ML/DL framework. They can simply write scripts and execute ML/DL workloads. Figure 3 shows the screen of KubeFlow GUI for Jupyter Notebooks.

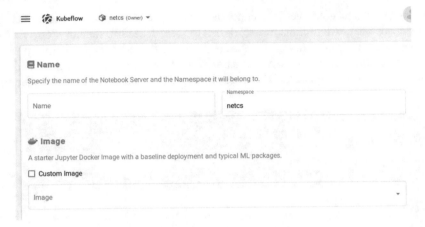

Fig. 3. KubeFlow GUI to create customized workspace.

- **Using CLI Based on Kubernetes:** Users create a new Docker image by making a dockerfile based on the ML/DL workload. Based on this Docker image and the size of resources to be allocated, write the Kubernetes manifest in YAML format. By applying this manifest to the K8S master node, they can deploy containerized ML/DL workloads in the targeted cluster environment.

3   Efficiency Problems with KubeFlow-Based DL Workloads

3.1   Resource Efficiency Limitation of KubeFlow-Based AI Cluster

KubeFlow based AI Cluster provides each user with an account and central dashboard, supporting multiple users to use AI Cluster. Furthermore, intuitive KubeFlow GUI allows users to easily select the desired resource and container images for their own environment. This makes a user have a bunch of environments suitable for various ML/DL workloads. Though, it shows some limitations in terms of sophisticated resource management.

- **Inefficient pod management:** In KubeFlow, an ML/DL workload is considered as only a part of a workspace. When a user makes workspace for a specific docker image, a pod for the workspace is created. Although there is no file, nor no work is done in the workspace, the pod is still managed by Kubernetes. Suppose multiple users make their own workspaces in KubeFlow and only the users whose pods are allocated in *worker-node1* try to run ML/DL workloads. The users with *worker-node1* will struggle with broken workspaces due to GPU memory shortage. GPU memory except *worker-node1* is idle but cannot be allocated to the ML/DL workloads since pods are already allocated to *worker-node1*.
- **GPU Unmount Problem:** In KubeFlow, GPU unmount is not performed automatically. Although the workload is stacked and the first workload is already completed,

the first workload continues to occupy the GPU of the worker node, making the other workloads fail or wait for the first workload to release its GPU memory. This problem is depicted in Fig. 4. This means that if there are 4 worker nodes, only up to four ML/DL workloads can be run at the same time. In order to run more workloads than the number of worker nodes, it is necessary to delete finished workloads manually.

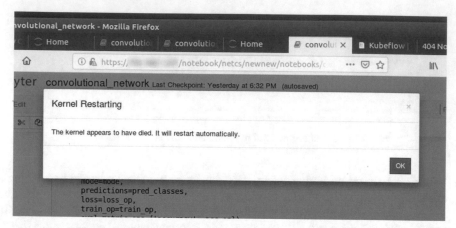

Fig. 4. Failed ML/DL workload due to GPU unmount problem

3.2 A Workaround Workflow Solution Without KubeFlow

KubeFlow makes users to conveniently set up the environment for ML/DL workloads, but resource efficiency limit occurs. We present a workflow that makes it easier for users to create an environment and to perform ML/DL workloads while using resources cleverly. This workflow does not use KubeFlow, and it is implemented through the basic functions of Kubernetes and the Linux shell. This workflow receives ML/DL workloads, the docker image used to run ML/DL workloads and identification value for the work as input. As a result, it automatically generates, deploys, and even logs ML/DL workloads using Kubernetes *job*.

- **Deployment of ML/DL workload:** KubeFlow creates a pod for each workspace. However, Kubernetes does not have to manage empty workspaces. Kubernetes provides a Job, which exterminates its pod when a workload finished. If we use a *Job* for workloads, resources leveraged are unmounted automatically and make Kubernetes manage multiple workloads in a workspace flexibly. In other words, *Job* resolves not only the GPU unmount problem but also Kubernetes' inefficient pod management.
- **Make local workload operate in the container:** To containerize a local ML/DL workload into the container, we use *kubectl cp* commands. After deploying a Kubernetes manifest, the Linux shell file waits for a pod to be created. As soon as a pod is created, the Linux shell file uses *kubectl cp* commands to copy the local ML/DL workloads into a container.
- **Logging for ML/DL workload:** While the ML/DL workloads execution, a user can check the workloads' output using *kubectl logs* commands in real-time. For the user's

convenience, the Linux shell file automatically saves workloads' log as a file so that the user does not have to type any commands.

4 Proposed Workflow for KubeFlow-Based DL Performance

Although KubeFlow has resource efficiency limitations, there are many advantages that cannot be covered by a simple Linux shell file. Furthermore, because KubeFlow, the open-source project, would provide more and more useful functions, the workflow solving resource efficiency limitation with KubeFlow-based DL workload is more stable and future-oriented methodology. In this workflow, the GPU status of the worker nodes is continuously monitored by Prometheus, the open-source monitoring tool. It also checks for processes that occupying GPU memory abnormally despite the completion of the workload. If such processes are detected, kill them immediately. This workflow mitigates resource efficiency limits that require other processes to wait for GPU memory until users arbitrarily delete already completed workloads.

4.1 Worker Node's GPU Status Retrieval in Real-Time

To monitor each worker nodes' GPU status, *nvidia-smi* command can be used. In order to make the master node monitor worker nodes' GPU status in real-time, we use Prometheus. We use GO language [8] to monitor each nodes' status and display a Prometheus exporter output at [Each node's IP Address: 9101] in real-time as shown in Fig. 5. Metrics collected by the Prometheus exporter is shown in Table 1.

Table 1. Metrics collected by the Prometheus exporter

Label	Description
temperature_gpu	Temperature of GPU (unit: °C)
utilization_gpu	Percent of time over the past sample period during which one or more kernels was executing on the GPU
utilization_memory	Percent of time over the past sample period during which global (device) memory was being read or written
memory_total	Total installed GPU memory
memory_free	Free GPU memory capacity
memory_used	Total memory allocated by active contexts
gpu_using_pid	Process id which mounts in GPU

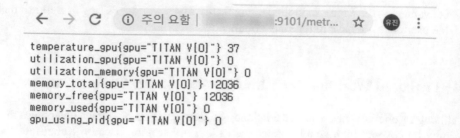

```
temperature_gpu{gpu="TITAN V[0]"} 37
utilization_gpu{gpu="TITAN V[0]"} 0
utilization_memory{gpu="TITAN V[0]"} 0
memory_total{gpu="TITAN V[0]"} 12036
memory_free{gpu="TITAN V[0]"} 12036
memory_used{gpu="TITAN V[0]"} 0
gpu_using_pid{gpu="TITAN V[0]"} 0
```

Fig. 5. Monitoring GPU status of worker-node1 in real-time

4.2 Monitoring the Node's GPU Status from Master

After confirming that the GPU usage of all nodes is monitored normally, we add their IP addresses and port numbers to the Prometheus configuration file on the master node.

After restarting the Prometheus service, the status of GPUs can be found in the Prometheus dashboard as shown in Fig. 6. Since Prometheus supports PromQL (Prometheus Query Language), a flexible query language to leverage this dimensionality, we can check GPU monitoring results from the command line interface.

Fig. 6. Data from GPU exporters in nodes displayed in the Prometheus Panel

4.3 Abnormally Mounted GPU Processes

memory_used having a non-zero value while *utilization_gpu*'s value having zero, as shown in Fig. 7, is abnormal. In general, processes in GPU memory are utilized with a GPU. However, the value of *utilization_gpu* can be zero instantly while a process is running. Therefore, instead of searching for a brief moment when the utilization_gpu value goes to zero, when the value of utilization_gpu stays zero for ten seconds, we determine that some processes are abnormally occupying the GPU.

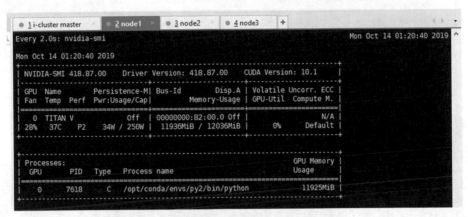

Fig. 7. Process that abnormally mounting GPU

4.4 Abnormal Process Termination

We can get the IP addresses and the *gpu_using_pid* value of nodes whose processes are abnormally mounted to a GPU using Python with PromQL. Then we exterminate the abnormal processes in particular nodes. Since Kubernetes does not support ssh commands to remotely access nodes, we use Paramiko Python open source library [9] which provides SSHv2 protocol for both client and server functionality.

4.5 Monitoring Without Intermission

It is necessary for our workflow to monitor GPU status and kill abnormal processes continuously. Otherwise, the remaining jobs could be stalled until the GPU is unmounted by the first job. To ensure continuous workflow execution, we deploy our workflow as Kubernetes deployment and service.

4.6 GPU Memory Comparison

In Fig. 8, a graph with an X-axis of 40 or higher is what happens when a service consisting of the workflow above is stopped. When the service is stopped, GPU memory is still occupied even though the workload is finished, let other workloads cannot be run. However, the output when we start a service is depicted in Fig. 9. The graph shows a short and repetitive appearance because the workflow is configured to occupy the memory only when the workload is actually running. In conclusion, it has been verified that multiple workloads can be efficiently managed for GPU memory with the KubeFlow when we use the proposed workflow.

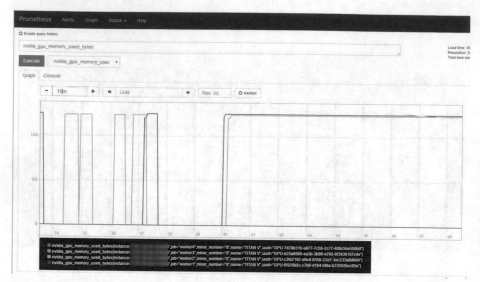

Fig. 8. GPU memory bytes in use on each node (Y-axis) in accordance with time elapsed (X-axis) collected by Prometheus NVidia GPU exporter when a service consisting of the workflow above is stopped

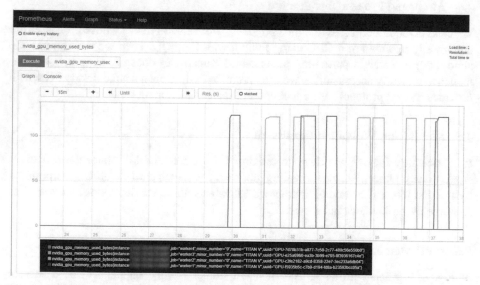

Fig. 9. GPU memory bytes in use on each node (Y-axis) in accordance with time elapsed (X-axis) collected by Prometheus NVidia GPU exporter when a service with the proposed workflow executed

5 Conclusion

This paper proposes methods to resolve the limitations in the current version of KubeFlow that arises from the lack of the underlying GPU resources awareness. One of them is inefficient pod management, which is KubeFlow making their pod for each workspace. Another one is the GPU unmount issue with the Jupyter Notebook GUI included in KubeFlow. As a resolution, we propose a user-friendly workaround workflow solution without KubeFlow using a Linux shell file. Furthermore, we also propose the resource-aware workflow for KubeFlow-based DL Performance, which is more stable and future-oriented methodology, using GPU status monitoring service based on Prometheus.

Acknowledgments. This work was supported by 2019 GIST Research Institute (GRI) grant funded by GIST.

References

1. Gannon, D., Barga, R., Sundaresan, N.: Cloud-native applications. IEEE Cloud Comput. **4**(5), 16–21 (2017)
2. Documentation. https://www.KubeFlow.org/docs/. Accessed 13 Oct 2019
3. Brewer, E.A.: Kubernetes and the path to cloud native. In: Proceedings of the Sixth ACM Symposium on Cloud Computing (ACM), p. 167 (2015)
4. Kwon, J., Kim, N.L., Kang, M., Kim, J.: Design and prototyping of container-enabled cluster for high performance data analytics. In: International Conference on Information Networking (ICOIN), pp. 436–438 (2019)
5. Getting started | Prometheus. https://prometheus.io/docs/prometheus/latest/getting_started/. Accessed 13 Oct 2019
6. BeeGFS - The Leading Parallel Cluster File System. https://www.beegfs.io. Accessed 13 Oct 2019
7. Ceph. https://ceph.com/. Accessed 13 Oct 2019
8. How to Write Go Code - The Go Programming Language. https://golang.org/doc/code.html. Accessed 13 Oct 2019
9. Welcome to Paramiko! — Paramiko documentation. http://www.paramiko.org/. Accessed 13 Oct 2019

DevOps Portal Design for SmartX AI Cluster Employing Cloud-Native Machine Learning Workflows

GeumSeong Yoon, Jungsu Han, Seunghyung Lee,
and JongWon Kim[✉]

School of Electrical Engineering and Computer Science,
Gwangju Institute of Science and Technology (GIST),
Gwangju, Republic of Korea
{gsyoon, jshan, shlee, jongwon}@nm.gist.ac.kr

Abstract. This paper introduces DevOps Portal for AI multi-cluster environment and management using Kubernetes (K8S), a representative container orchestration technology based on cloud-native. Specifically, we propose and verify the concept of DevOps Portal that provides the management function for multi-cluster operators in DevOps aspect and enables the operation of multiple clusters through the support of cluster selection from the developer's point of view. In addition, after using DevOps Portal, developers can create an environment in which Machine Learning (ML) workflows can be performed according to the processing of data through a web dashboard. This allows partial validation of cloud-native based HPC/HPDA/AI workloads.

1 Introduction

As the demand for cloud-native computing increases with the micro-service architecture, container virtualization, one of the core technologies of cloud-native, is drawing attention. Unlike traditional virtual machine (VM)-based virtualization, containers can be deployed quickly on a per-application basis, and because they are independent structures, they do not affect other containers. Orchestration technologies are being applied to various ICT applications, centering on Kubernetes (K8S), a container orchestration technology that enables the efficient management of multiple containers. Among them, the number of cases applied to workloads of AI technologies such as machine learning continues to increase.

Furthermore, open-source software that supports the machine learning tools and frameworks used to run these AI workloads and provides the ability to run in container environments is rapidly being developed and utilized. In addition, advances in AI technology have increased the demand for a variety of cost-effective AI-inspired HPC/HPDA workloads.

However, the above workload requires a large resource configuration. To solve this problem, K8S, which uses a cluster structure by tying up multiple nodes, is getting more attention. K8S-based clusters make a large pool of resources into one pool and have the flexibility to expand the pool size by adding nodes. The above structure can

© Springer Nature Switzerland AG 2020
L. Barolli et al. (Eds.): EIDWT 2020, LNDECT 47, pp. 532–539, 2020.
https://doi.org/10.1007/978-3-030-39746-3_54

satisfy the 3Vs (Velocity, Volume, and Variety) due to the attribute of rapidly changing data in ICT application service. Furthermore, building a K8S cluster allows you to do whatever you need for HPDA workloads that handle big data, HPC workloads for high-performance computing, and AI workloads for ML or Deep Learning (DL). In some cases, roles are categorized for each K8S cluster to improve performance by focusing on specific workloads. Running multiple workloads requires building multiple clusters. This has given rise to the concept of multi-clusters that combine and operate a single K8S cluster deployed in multiple locations. However, in order to operate K8S-based multiple clusters, unlike a single cluster, a method of managing multiple clusters in a location capable of managing multiple clusters is required. In addition, users should be able to freely access the environment where multiple clusters capable of running HPC/HPDA/AI workloads are constructed.

Therefore, this paper introduces SmartX AI cluster operated by our research team. This approach, unlike the traditional way of operating a single cluster, suggests the concept of managing multiple clusters and allowing multiple users to be used simultaneously in an entity called DevOps Portal [1]. It also performs partial validation of AI workloads among cloud-native computing workloads (HPC, HPDA, AI) by demonstrating several machine learning workflows through the DevOps Portal of the SmartX AI cluster. Furthermore, Users can freely access SmartX AI clusters through a web dashboard and use a common UI. It shows that DevOps Portal manages AI SmartX cluster to be used only by authenticated users, rather than indiscriminately while ensuring an independent environment.

2   SmartX AI Cluster and DevOps Portal

2.1   Introduction of SmartX AI Cluster and DevOps Portal

Cloud-native based SmartX AI cluster is a collection of general single clusters. In this paper, we introduce the hardware configuration of SmartX AI cluster with high-end devices that support AI computing, as shown in Fig. 1.

It is noteworthy in the hardware configuration that the network fabric consists of 100Gbps high-performance switches to cope with sudden data fluctuations. Moreover, there are three clusters, each with a different type of high-performance GPU that can perform high-efficiency tasks depending on the AI workloads to be performed. And as secondary storage, cloud storage can be deployed separately from K8S clusters, allowing SATA HDD and SSD (SATA, NVMe) resources to be combined into a single resource pool. It can connect with clusters through the network fabric to digest the massive data used for AI workloads.

Intelligence Center (I-Center) manages K8S and storage functions as hypervisor-based virtual machines in VMware ESXi. Specifically, there are BeeGFS parallel file systems that tie internal NVMe SSDs and Ceph to use cloud storage [2]. Also, there are K8S to run a workload in container form, Kubeflow supporting various machine learning tools and frameworks in a clustered environment, and Prometheus to provide visibility through the collection of cluster resources. Through Kubeflow's dashboard, users can access the Jupyter development environment and do what they want.

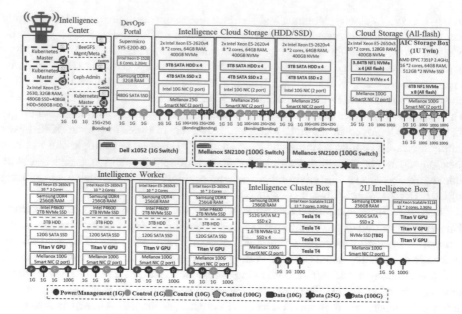

Fig. 1. SmartX AI cluster hardware configuration

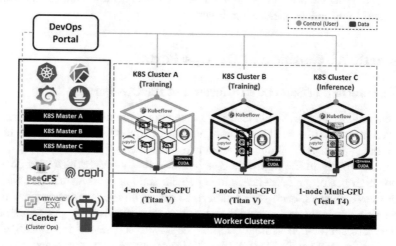

Fig. 2. SmartX AI cluster and DevOps Portal software configuration

The software configuration representing the cloud-native based software tools is shown in Fig. 2.

The DevOps Portal is an entity that enables the simultaneous use of SmartX AI clusters with the above hardware and software configurations [3]. From a DevOps perspective, developers use the DevOps Portal to make sure they have the resources available to perform AI workloads. Developers can use the visibility tools built into the DevOps Portal to check the status of multiple clusters to help them choose a development environment. Developers must also create an account in the DevOps Portal to access the Kubeflow dashboard to ensure an independent development environment. On the other hand, for the operator, the software is managed so that the developer can receive only the features that are needed only in the DevOps Portal. This ensures a consistent operating environment that only requires operating from the DevOps Portal in terms of multi-cluster management. This configuration has the advantage of separating developers and operators who use multiple clusters by restricting access to DevOps Portal.

2.2   Partial Validation of SmartX AI Cluster DevOps Portal

In this paper, we partially verified the functions provided by DevOps Portal in SmartX AI cluster. The conceptual diagram of DevOps Portal including the DevOps Portal's functions is shown in Fig. 3.

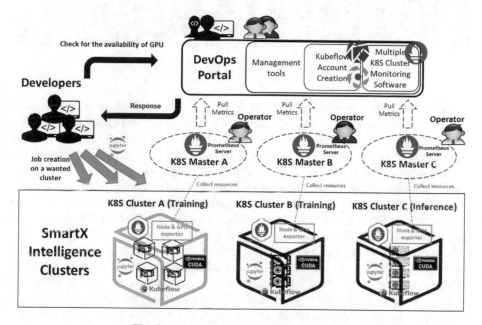

Fig. 3. Proposed concept of DevOps Portal

Prometheus collects resource usage information such as CPU, RAM, and GPU by distributing Prometheus exporter in pod form to each cluster node [4]. Resources of the entire node are collected from the exporter, and the status of resources used as pod units of K8S can be checked by requesting Kube-api server provided by K8S itself. There is only one Prometheus server per K8S master that collects information from exporters deployed on K8S nodes. Therefore, to check the resource status of multiple clusters, it is necessary to get information from the Prometheus server at the location where single K8S masters are managed. According to the above demand, cluster resources are collected and devised for resource monitoring in DevOps Portal, a multi-cluster management entity. Developers can access the Portal to understand the resource availability of multiple clusters running, and access Kubeflow dashboards in their environment to perform AI workloads.

Figure 4 shows the data collected from the Prometheus exporter deployed on the node. Along with the screen displaying the collected metrics, the data can be used by users and displayed on the screen to confirm that the monitoring function is provided.

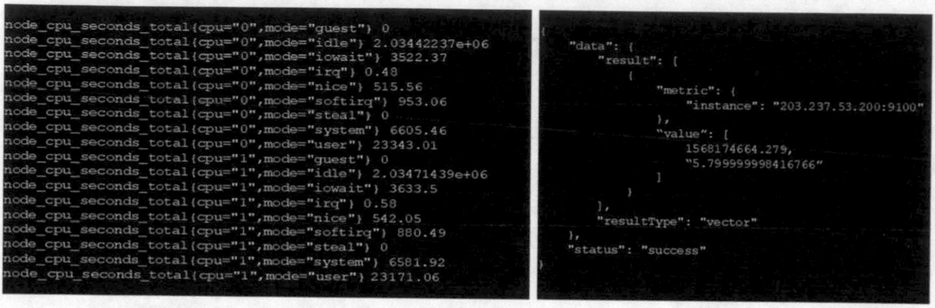

Fig. 4. Data collected from Prometheus exporters

In addition, user authentication is required to access the Kubeflow dashboard to ensure user multi-tenancy. There are various methods of authentication used in K8S. Among them, DevOps Portal uses K8S user authentication through OpenID Connect (OIDC). OIDC is a feature of OAuth2 that is supported by some providers. OAuth2 is a general-purpose framework for managing authentication and authorization of third-party applications. In addition, OAuth-based services use an access token called an ID token. And to manage the OAuth token, an intermediate server is needed. We used Dex, an authentication system that uses OIDC. In other words, Dex, an intermediary with a third-party, performs authentication using ID token in K8S, and makes it easier to issue, store, and manage tokens. Dex, which acts as a gateway, supports several authentication methods in K8S. First, the method for static users. This is a method of directly adding authentication information through the configuration of a file in the back-end such as K8S. Second, Dex also supports methods for LDAP (Lightweight Directory Access Protocol)/Active Directory. To add an existing LDAP/Active Directory, the user adds the connectors option used as an authentication tool of another identity provider in the ConfigMap section where they can configure Dex, and modify the necessary settings. It also includes support for External Identity Providers (IdPs) such as Google, LinkedIn, and GitHub [5].

In this paper, partial verification of user authentication using Dex is performed in K8S. Dex performs vendor-neutral authentication and authenticates static users. The above authentication is performed by adding a user account through Dex's Configmap YAML file setting. If you enter an account created using the above method into Kubeflow's dashboard, it will have its namespace. Each user has their own Jupyter Notebook development environment for each namespace, and the work environment is separated from other users [6].

The left side of Fig. 5 shows the screen that requires user authentication when accessing the Kubeflow dashboard. On the right, you can see that multi-tenancy is supported by separating the development environment by namespace according to the account.

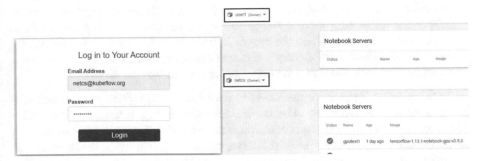

Fig. 5. Accounts and namespaces for using Kubeflow

3 Conducting ML Workflows Over DevOps Portal

In this paper, we show the ML workflows performed in the DevOps Portal to prove that the AI workloads of the HPC/HPDA/AI workloads are executed in the multi-cluster. ML workflows are divided into data preprocessing, data training, and data inferencing.

- **Data preprocessing:** The data preprocessing part of preparing data for data training is the associated concept for HPDA workloads. The main tool used for HPDA workloads is Apache Spark, which is an analysis tool for processing large amounts of data. Extract, Transform, Load (ETL) and analyze large volumes of data in real-time or in batches. Spark has various execution environments. Among them, K8S uses Spark Operator, which is a K8S operator dedicated to Spark. With Spark Operator, Spark applications can run inside a K8S cluster by default, so they run just as easily as any other common application. Custom resources also enable declarative application specification and management of applications. In addition, applications can be managed using kubectl, a basic command configuration tool of K8S, not spark-submit. Spark's job can be executed by creating a YAML file, a declaration script that describes the application, and by using the kubectl command [7].
- **Data training:** Data training is the process of training a specific model through data and datasets processed in ML workflows. The data training part is a concept associated with AI workload among the above workloads. This paper shows that users authorized through the DevOps Portal can perform training jobs through Kubeflow, which supports Tensorflow, an open-source library used for ML. Users connected through the Kubeflow UI can create a working environment by creating a Jupyter notebook server. Run through a web browser, the Jupyter notebook is easy for data visualization, code sharing, and real-time conversation, and can be created with simple manipulations in the Kubeflow UI. In the creation process, you can

select a Tensorflow image of a container to create a pod for a new server, and set up data and workspace volumes, memory size, and the number of CPUs and GPUs. The Jupyter notebook server, created as a pod of K8S, can be easily removed using Kubeflow UI.

- **Data inferencing:** As with data training, data inferencing workflows belong to the category of AI workload. Inferencing servers that generate results based on trained models need to be certified and run on the Jupyter notebook through the DevOps portal. The inferencing server running on the pod of each user's Jupyter notebook server waits for input. When the proper input data is delivered to the server, the trained model determines the input data and derives the result value. The resulting value derived from the above process can be used in various services.

In this paper, we introduce the Smart Energy IoT-Cloud service as an example of the data inferencing workflow. The service functions in Fig. 6 are programs with a microservices structure. The data collection function uses IoT pattern logic to measure the temperature and humidity inside the server room using sensors on the Raspberry Pi and collect information about the weather outside. DataPond leverages MongoDB to store collected data from the IoT pattern. To interconnect each feature of the Smart Energy IoT-Cloud service using EdgeX, an open-source edge computing framework, each feature can provide connectivity to each application feature via port forwarding. Kafka is used to reliably deliver large amounts of sensing data collected through Raspberry Pi to cloud storage. Through this Kafka relaying, the data is finally stored in a DataLake using InfluxDB, a time series database. Training is performed using the stored data to provide the recommended air conditioner temperature function considering energy efficiency. The inferencing function is applied to the ML inference server and the inference server returns the recommended data when it received the request. Inference calculates and returns the appropriate value considering the temperature of the outdoor and server room. Through the above service, it is easy to manage the equipment of the server room which needs to maintain a constant temperature [8].

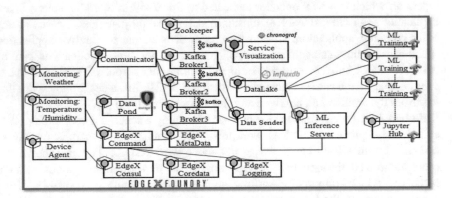

Fig. 6. Implemented Smart Energy IoT-Cloud service function diagram

4   Conclusion

In this paper, we talked about the Cloud-Native based SmartX AI cluster environment and DevOps Portal with the networking/computing/storage resources to perform AI workloads. In detail, DevOps Portal manages multiple clusters by providing a monitoring function to identify the resource status of clusters and a multi-tenancy function through an authenticated user account. In addition, we confirmed that developers and operators can play the role of working DevOps. And by performing partial ML workflows in the DevOps Portal, we showed that the clusters built internally support AI workloads.

Acknowledgments. This work was supported by GIST Research Institute (GRI) grant funded by the GIST in 2019 and Institute of Information & communications Technology Planning & Evaluation (IITP) grant funded by the Korea government (MSIT) (No. 2015-0-00575, Global SDN/NFV Open-Source Software Core Module/Function Development).

References

1. Kwon, J., Kim, J.: Supporting machine learning functionality over SmartX AI cluster for smart IoT-cloud services. In: Proceedings of Symposium of the Korean Institute of communications and Information Sciences, pp. 540–541 (2018)
2. Kwon, J., Kim, N.L., Kang, M., Kim, J.: Design and prototyping of container-enabled cluster for high performance data analytics. In: 2019 International Conference on Information Networking (ICOIN), pp. 436–438 (2019)
3. Jeon, I.: Integrated management of development operation organization in the non-stop environment considering security. In: Review of Korea Institute of Information Security and Cryptology (KIISC), pp. 47–52 (2015)
4. Kim, K., et al.: Kubernetes architecture for cloud services. J. Korean Inst. Commun. Sci. **35** (11), 11–19 (2018)
5. Dex. https://github.com/dexidp/dex/blob/master/Documentation/kubernetes.md. Accessed 13 Oct 2019
6. Multi-user isolation in Kubeflow. https://www.kubeflow.org/docs/other-guides/multi-user-overview/. Accessed 14 Oct 2019
7. Piotr, M.: Scaling cloud-native Apache Spark on Kubernetes for workloads in external storages. EECS, KTH, Stockholm (2018)
8. Lee, S., Han, J., Kwon, J., Kim, J.: Relocatable service composition based on microservice architecture for cloud-native IoT-cloud services. Proc. Asia-Pac. Adv. Netw. (APAN) **48**, 23–27 (2019)

A Management System for Electric Wheelchair Considering Agile-Kanban Using IoT Sensors and Scikit-Learn

Takeru Kurita[1], Keita Matsuo[2(✉)], and Leonard Barolli[2]

[1] Graduate School of Engineering, Fukuoka Institute of Technology (FIT),
3-30-1 Wajiro-Higashi, Higashi-Ku, Fukuoka 811-0295, Japan
mgm19103@bene.fit.ac.jp
[2] Department of Information and Communication Engineering,
Fukuoka Institute of Technology (FIT),
3-30-1 Wajiro-Higashi, Higashi-Ku, Fukuoka 811-0295, Japan
{kt-matsuo,barolli}@fit.ac.jp

Abstract. High quality electric wheelchairs can support handicapped persons to move freely in home, work, or hospital settings. However, the electric wheelchairs require a high standard of maintenance for safety. For this reason, we propose a wheelchair management system that can manage electric wheelchairs using Agile-Kanban. Agile is a technique to develop software and manage work efficiency. Kanban is a method to support Agile development. In this work, we design and implement a Wheelchair management system considering Agile-Kanban using IoT sensors and Scikit-learn. Our proposed system can manage the electric wheelchairs efficiently. We present the design and implementation of the proposed system and show that it can measure a wheelchair's states and activities. In addition, we investigate the timing of changing batteries for the electric wheelchair with Scikit-learn.

1 Introduction

The communication systems have been developed in order to better support health, wellness, and convenience. IoT technologies, IoT devices and sensors, in particular, can provide many assistive services. IoT sensors can be embedded in electronic devices. For instance, a refrigerator with IoT sensors can access the Internet to inform the user of what ingredients are needed for a recipe, any time and from anywhere. These systems are also used to control air conditioner, and other such appliances as televisions, cooking appliances, microwaves, bath-units, toilet-units, lights and so on can provide new services through IoT technologies.

Vehicles as well can be connected to the Internet. Toyota uses the phrase "Connected Car System". This system can collect the data using mobile network and sensors in the car. These include different kinds of data such as the data taken by the car when it is moving, car speed, braking or acceleration, weather conditions and traffic jam information. These data are gathered to the datacenter

© Springer Nature Switzerland AG 2020
L. Barolli et al. (Eds.): EIDWT 2020, LNDECT 47, pp. 540–551, 2020.
https://doi.org/10.1007/978-3-030-39746-3_55

through the network by using Internet technologies. The data can be analyzed and used for solving some traffic problems. There are also other applications using IoT such as the use of IoT technologies for agriculture, factories, offices, schools, hospitals and so on.

We consider the use of Agile and Kanban (from the Japanese word 'Sign Board') for electric wheelchair management system. Agile is a software developing method that implies "the quality of being agile, readiness for motion, nimbleness, activity, dexterity in motion". The software development methods can offer the answer to the business community asking for lighter weight paired with faster and nimbler software development processes [1,5]. Kanban is one of methods that can support the Agile process [11].

The structure of this paper is as follows. In Sect. 2, we introduce the IoT and Kanban. In Sect. 3, we describe the Agile-Kanban, In Sect. 4, we present our proposed wheelchair management system and in Sect. 5, we show experimental results. In Sect. 6, we show the conclusions and future work.

2   IoT and Kanban

2.1   IoT

Our life styles are drastically changing with IoT technologies. IoT can connect various things to the Internet, such as household electronics appliances, vehicles, robots, communication devices, and some application softwares. These advancements can improve our quality of life. In Fig. 1, we show the image of an IoT environment. An IoT system has many sensors, which can be used for operating factories, farms, offices, houses, and so on. However, given the fact that most IoT sensors are resource limited and operate on batteries, the power consumption and life time of sensors are important issues for the design of wireless IoT sensor networks [9,12].

There are some approaches for decreasing the number of packets in the networks. The Opportunistic Networks (OppNets) is one such approach. It can provide an alternative way to support the diffusion of information in special locations in a city, particularly in crowded spaces where current wireless technologies can exhibit congestion issues [8].

2.2   Kanban

Kanban is a system produced by Toyota. The system can efficiently manufactures vehicles. The main goal of Kanban was to manufacture cars at a similar or lower price than competitors. This Kanban system is called Toyota Production System (TPS). This system is also known as Just in Time (JIT) manufacturing and the basic principle is to produce "only what is needed, when it is needed and in the amount needed". Currently, there are many research projects using Kanban for developing software [6,7,14]. Initially, Kanban was used in manufacturing processes; however, its applications in other areas is continuously growing due to its proven success.

We have shown a schematic illustration of the Kanban system in Fig. 2. The system has two Kanbans. The Kanban system has 3 sections: Upstream, Downstream and Store. The Upstream section role is manufacturing parts for making some blocks of products. The Downstream section is used for assembling parts and completing the products. The Store section is used for keeping some parts needed for working operation. If the amount of stocking parts increases, the manufacturing efficiency decreases. In order to solve this problem, Kanban system uses Production-Kanban and Withdraw-Kanban.

The upstream section produces parts and stocks them in the store section. While, the downstream section uses the parts in the store to assemble the products. The Production-Kanban can only move between upstream and store, while the Withdraw-Kanban moves between store and downstream, as shown in Fig. 2.

We describe the movements of the Kanbans in Fig. 3. If there is a shortage of parts in the downstream section, the Withdraw-Kanban moves from downstream section to store section. Then, the Withdraw-Kanban informs the number of the shortage to the Production-Kanban. After that, the Production-Kanban instructs the upstream section to manufacture the number of parts requested by the Withdraw-Kanban. The Production-Kanban will put that number in the production queue. Then, the production of parts will begin. Thus, the Kanban system can control the supply and demand of parts, which leads to an efficient manufacturing.

Recently, the Kanban system combined with Agile has been used for software development. Some papers use Kanban and Agile for collaborative work [3,10]. In another paper, the authors used Agile approach with Kanban for managing the security risk on e-commerce [2].

3   Agile-Kanban System

In this section, we present the IoT sensor management system considering Agile-Kanban (see Fig. 4). This system uses Kanboard. Kanboard requires a web server, database and PHP. Kanboard offers 4 kinds of Kanbans; Backlog, Ready, Work in progress and Done. Each Kanban description is shown in Table 1.

Kanboard is free and open source software [4]. The Kanboard user interface is shown in Fig. 5. Kanboard focuses on simplicity and minimalism. The number of features are limited. Kanboard makes it easy to know the current status of a project because it is visual. It is very easy to understand and there is no need of special training.

Kanboard has a number of features as follows.

- Action Visualization,
- Limit work in progress to focus on the goal,
- Drag and drop tasks to manage the project,
- Self-hosted,
- Simple installation.

In Fig. 5, we present a user interface of Kanboard, which shows 4 Kanbans: Backlog, Ready, Work in progress and Done.

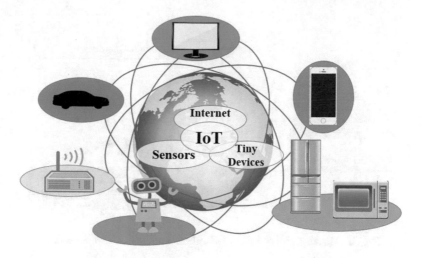

Fig. 1. Image of IoT environment.

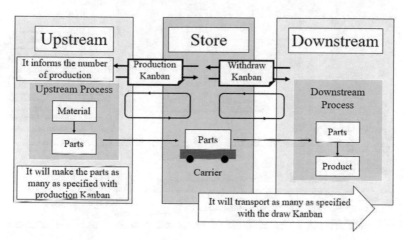

Fig. 2. Structure of Kanban system.

Table 1. Description of Kanban names.

Name of Kanban	Description
Backlog	The sensor needs maintenance
Ready	The sensor is ready for using
Work in progress	The sensor is working
Done	The sensor finish the work

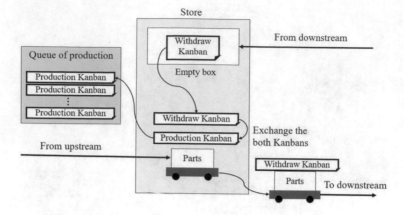

Fig. 3. Moving of Production-Kanban and Withdraw-Kanban.

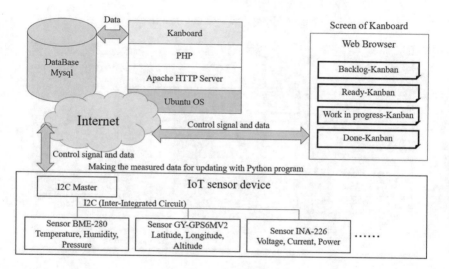

Fig. 4. Agile-Kanban system.

Fig. 5. User interface of Kanboard.

4   Proposed Electric Wheelchair Management System

With the rapid development of science and technology, traditional hand-propelled wheelchairs have gradually evolved into electric wheelchairs that can be operated by individuals with mobility impairments [13]. Handicapped persons require higher quality electric wheelchairs to move freely in the home, at work, or in hospitals. These electric wheelchairs will require through maintenance checks for its overall condition as well as more focused checks on the batteries and current, voltage and temperature of the motor.

Hospitals, nursing, or elderly care facilities need many electric wheelchairs and consequently will require a great deal of time and manpower to manage and maintain them. Therefore, we propose an electric wheelchair management system that can manage many electric wheelchairs using Agile-Kanban.

Figure 6 shows the user interface of the proposed management system and Fig. 7 shows its schematic illustration. The Kanbans can move in different states, such as Backlog, Ready, Work in progress and Done. One Kanban corresponds to one wheelchair.

The Kanban can change some states with drag and drop. For example, when there is a Kanban in the state of Backlog, it means that the wheelchair requires maintenance. When Kanban is in the state of Ready, it means that the wheelchair is ready to be used. Work in progress means the wheelchair is working. In this case, the measurement data of wheelchair by sensors will be uploaded to the database. When Kanban is in the state of Done, it means the end of work.

The proposed system has used Scikit-learn to detect some states of wheelchair. Scikit-learn is an open source software for machine learning. Scikit-learn has a number of algorithms that support vector machine, random forest, k-means clustering and neural networks. We use neural networks for predicting the states of wheelchair.

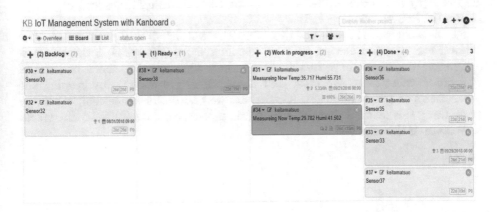

Fig. 6. User interface of wheelchair management system.

Figure 8 shows a diagram of machine learning and Fig. 9 shows a diagram of using neuron that has two medium layers. We can predict a battery's states using this model. When if the battery is in bad condition the corresponding wheelchair's Kanban moves to the Backlog section automatically. After that the wheelchair's battery will get recharged or replaced.

We implemented the IoT device in order to measure the states of the wheelchair as shown in Fig. 10. The device can track the wheelchair's states, including the battery's voltage, the motor's current and so on.

When the Kanban is in the "Work in progress" state, the measured data of the voltage, current and temperature of the battery are shown in Fig. 11. Figure 12 shows the image of a moving Kanban on the wheelchair management system using Kanboard.

The benefits of the wheelchair management system are as follows:

- The system can manage many wheelchairs.
- When the wheelchairs are not used, the management system can cut down the power consumption to save the wheelchair's battery.
- The system is able to analyze the wheelchair's status such as wheelchair's working time, frequency of use and activity.
- The system can be installed in other factory machines, vehicles, robots, offices or school facilities, home electrical appliances and so on.

5   Experimental Results

In order to check the states of wheelchair's battery, we used Scikit-learn (see Figs. 13, 14 and 15). Figure 13(a) shows the validation result for 1000 learning times and Fig. 13(b) shows the loss function. In Fig. 14(a) are shown the validation result for 3000 learning times and in Fig. 14(b) is shown the loss function. Figure 15(a) shows the validation result for 10000 learning times and Fig. 15(b) shows the loss function.

Table 2 shows the results of detecting the wheelchair's battery states. The highest accuracy is achieved for 3000 learning times. We see that the result of 10000 learning times have less accuracy than 3000 learning times because of a short learning time. We must find the optimal parameters for keeping the wheelchairs in good condition. After that, we can decide wheelchair's battery state using this method. We think that this process can be applied to any trackable state of the wheelchair.

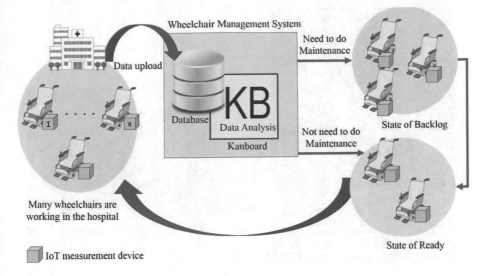

Fig. 7. Schematic illustration of proposed wheelchair management system.

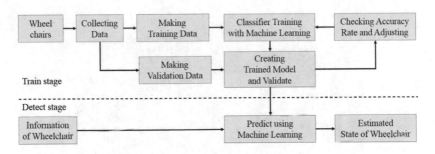

Fig. 8. Diagram of machine learning.

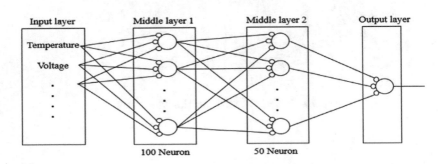

Fig. 9. Diagram of using neuron.

Fig. 10. IoT device for wheelchair management system.

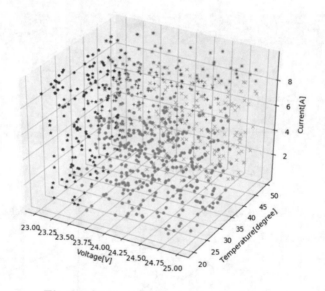

Fig. 11. Measured data on the wheelchair.

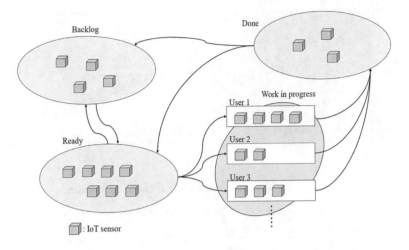

Fig. 12. Different states of Kanban moving.

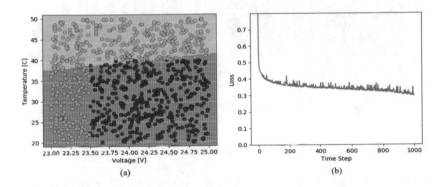

Fig. 13. Validation result with 1,000 times of learning.

Table 2. Results of detecting wheelchair's battery states.

Number of learning times	Accuracy rate
1000	84.05%
3000	96.52%
10000	93.17%

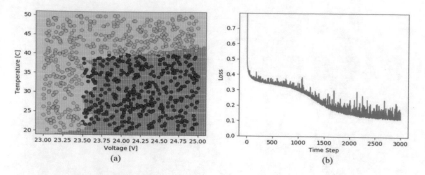

Fig. 14. Validation result with 3,000 times of learning.

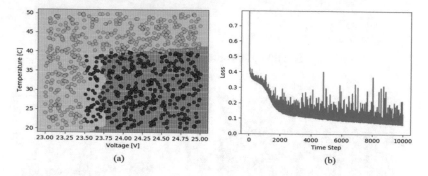

Fig. 15. Validation result with 10,000 times of learning.

6   Conclusions and Future Work

In this paper, we introduced Agile-Kanban. We presented in details the Kanban system and Kanboard. The proposed system can manage many wheelchairs. Furthermore, the proposed system can measure the electric wheelchair's states and activities using Scikit-learn. Through efficient management, the wheelchair's battery life time can be increased. In addition, we implemented a IoT measurement device using Agil-Kanban and Scikit-learn.

In the future work, we would like to improve the proposed system and carry out additional experiments.

References

1. Abrahamsson, P., Salo, O., Ronkainen, J., Warsta, J.: Agile software development methods: review and analysis. arXiv preprint arXiv:1709.08439 (2017)
2. Dorca, V., Munteanu, R., Popescu, S., Chioreanu, A., Peleskei, C.: Agile approach with Kanban in information security risk management. In: 2016 IEEE International Conference on Automation, Quality and Testing, Robotics (AQTR), pp. 1–6. IEEE (2016)

3. Hofmann, C., Lauber, S., Haefner, B., Lanza, G.: Development of an agile development method based on Kanban for distributed part-time teams and an introduction framework. Procedia Manuf. **23**, 45–50 (2018)
4. Kanboard: Kanboard web page. https://kanboard.org/
5. Kiely, G., Kiely, J., Nolan, C.: Scaling agile methods to process improvement projects: a global virtual team case study (2017)
6. Kirovska, N., Koceski, S.: Usage of Kanban methodology at software development teams. J. Appl. Econ. Bus. **3**(3), 25–34 (2015)
7. Maneva, M., Koceska, N., Koceski, S.: Introduction of Kanban methodology and its usage in software development (2016)
8. Miralda, C., Donald, E., Kevin, B., Keita, M., Leonard, B.: A delay-aware fuzzy-based system for selection of IoT devices in opportunistic networks considering number of past encounters. In: Proceedings of the 21th International Conference on Network-Based Information Systems (NBiS 2018), pp. 16–29 (2018)
9. Nair, K., Kulkarni, J., Warde, M., Dave, Z., Rawalgaonkar, V., Gore, G., Joshi, J.: Optimizing power consumption in IoT based wireless sensor networks using Bluetooth low energy. In: 2015 International Conference on Green Computing and Internet of Things (ICGCIoT), pp. 589–593. IEEE (2015)
10. Padmanabhan, V.: Functional strategy implementation-experimental study on agile Kanban. Sumedha J. Manag. **7**(2), 6–17 (2018)
11. Petersen, J.: Mean web application development with agile Kanban (2016)
12. Sheng, Z., Mahapatra, C., Zhu, C., Leung, V.: Recent advances in industrial wireless sensor networks towards efficient management in IoT. IEEE Access **3**, 622–637 (2015)
13. Tseng, T.H., Liang-Rui, C., Yu-Jia, Z., Bo-Rui, X., Jin-An, L.: Battery management system for 24-V battery-powered electric wheelchair. Proc. Eng. Technol. Innov. **10**, 29 (2018)
14. Yacoub, M.K., Mostafa, M.A.A., Farid, A.B.: A new approach for distributed software engineering teams based on Kanban method for reducing dependency. JSW **11**(12), 1231–1241 (2016)

An Approach of Usability Testing for Web User Interface Through Interaction Flow Modeling Language (IFML) Models

Muhammad Talha Riaz[1], Farooque Azam[1], Nazish Yousaf[1,2(✉)],
Muhammad Waseem Anwar[1], and Adil Aziz[1]

[1] Department of Computer and Software Engineering, College of Electrical and Mechanical Engineering, National University of Sciences and Technology (NUST), Islamabad, Pakistan
{talha.riaz18,nazish.yousaf15,adil.aziz18}@ce.ceme.edu.pk,
{farooq,waseemanwar}@ceme.nust.edu.pk
[2] Department of Computer Science, University of Wah, Wah Cantt, Pakistan

Abstract. Usability of the web User Interface (UI) is totally dependent on the user's interaction with system. Usability is to check how efficiently and easily user utilizes the system. To design a clear, well-managed UI that depicts all the elements of web to end user is a very complex task. As the creation of the UI being a complex and cost-effective task, most of the organizations squander most of their budget on it. In this research, we have proposed a usability testing model for the web UI. This model creates a checklist for the usability testing of web UI, then transforms this checklist into IFML constructs. Using this checklist, we have generated test cases to validate the web UI. This checklist can be used to design a new front-end or to test an existing UI. We have applied this checklist on MobileWeb case study. The proposed approach is validated using a UML Domain Model Diagram and IFML models. Finally, the comparison of general checklist items of MobileWeb is performed with IFML constructs.

Keywords: Usability testing of web user interface · UI testing · IFML · Interaction flow modeling language · IFML modeling constructs · Usability checklists

1 Introduction

Web applications are ubiquitously increasing with their cumulative adoption in market. The demand of web sites is increasing day by day, but there exist quality, performance and security problems. Web sites are widely used in all aspects of life and users demand proper well-managed, well-behaved and better performance of web sites. User interface (UI) is an important aspect of acceptance of the system as well as satisfaction of the systems' behavior. The usability of web UI is attention focus and most important factor of evaluating the system behavior. Usability testing is analysis to check users' needs and preferences by observing their reactions as they use the product. Usability testing helps to evaluate the efficiency of the web UI that how easily web interface fulfills the user needs.

© Springer Nature Switzerland AG 2020
L. Barolli et al. (Eds.): EIDWT 2020, LNDECT 47, pp. 552–563, 2020.
https://doi.org/10.1007/978-3-030-39746-3_56

End user satisfaction is very important to ensure a quality product. Usability testing of websites is very important in order to maintain QoS (Quality of service). International Organization for Standardization (ISO) has proposed different models to describe and to measure the usability [1]. Usability of web UI depends on how the user interact with the site at peak time under different fluctuating situations. Different tools and techniques are available to check usability of web UI. Some specific tools are defined for UI testing according to Web Services Description Language (WSDL) standard.

In Model Driven approach, UML is widely used for modeling and design purpose. However, UML cannot capture various specifications related to UI and user interaction. Web Modeling Language (WebML) was introduced to fill this gap using visual notations and principles. WebML was later evolved into Interaction Flow Modeling Language (IFML) to cover a broad range of front-end interfaces. IFML is a visual modeling language defined by Object Management Group (OMG) to design user interaction visually and to control behavior when user interact with the system. IFML can be integrated in mobile and web applications. IFML supports platform independent description of GUI for application. A typical IFML model consists of container, component, events, actions, navigation flow and parameters. IFML model has two main packages. Core package include IFML concepts and extension package contains some enhanced characteristics to make the interface more interactive. IFML model can be used to design the front-end UI and different template test cases are generated to test this UI.

There is a need to validate the UI requirements. Therefore, we propose an approach for validation of UI requirement. In this approach, a checklist is defined which is used to validate the web UI. This general checklist is transformed into IFML constructs. Using IFML constructs, validation of UI requirements is carried out (Sect. 3). Overview of

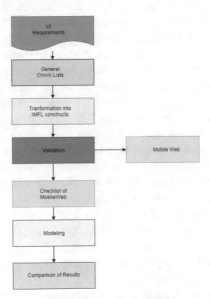

Fig. 1. Overview of research

research is shown in Fig. 1. The proposed approach has been validated on a case study of MobileWeb (Sect. 4).

2 Literature Review

Yousaf et al. in their study about model-based web UI test case generation proposed an approach for automated model-based testing of UI from IFML models at platform independent level. Authors also presented a fully automated MBUITC generator tool which takes domain and IFML model of web application as input and generates all main testing artifacts covering navigation and functional testing [2].

Conte et al. in their study provided usability evaluation method segmented into two divisions. First is usability inspection – based on expert analysis and second one is evaluating through user participation such as laboratory studies and cooperative evaluation. When utilizing evaluation method which is based on user's involvement. Usability problems are found out by observation and interaction with users. Users perform tasks and provide suggestion about design and ease of use. In usability inspection, problems are extracted by expert by using inspection technique [3]. Authors in [4] proposed different inspection techniques like, Guidelines and Checklists, Cognitive Walkthrough, and Heuristic Evaluation.

Some principles and tools have suggested to enhance the usability of UI. Leiva et al. [5] describe in their research that other commercial web analytics services allow website to track the user movement. ClickTale, MouseFlow, Mpathy, LuckyOrange and Clixpy allow web developers to track users' mouse cursor movement as user is performing scrolling, pointing and clicking on the website and developers have ability to reply cursors movement in a given session [6]. A CLICKHEAT tool is available to point out the difficult usability situation in websites through Heatmap [7].

Li et al. [8] presented MetroWeb tool which helps the designers to gather guidelines coming from different sources and organizes them in a structured form. This helps the designer to minimize the usability errors of websites. A paper prototype is discussed for the usability testing of web site design. Paper prototype is low cost, low tech but highly effective usability testing technique having the numerous benefits. Paper prototypes have three benefits. Paper prototypes are truly "hands on", portable and paper prototype seems less finished or real than electronic prototype [9].

Dias et al. [10] performed a pattern-based usability testing which is able to create test models. Usability tests patterns can be produced from these models and automatically executed over the websites without user interventions. Another usability test pattern is specified as "Reachability Test Pattern". This test pattern is executed after the analysis of the websites and by creating a tree whose vertex correspond to links and edges correspond to possible transition between link. This tree is created using Breath-First search algorithm [10]. Grigera et al. [11] presented a tool Kobold that discerns the usability problems of the working UI events and fixes them automatically or suggests some solution by using refactoring technique.

Testing the user interface is making the whole system safer and more robust. More user-friendly interface is difficult to test as it hides some of the complexities. Any complexity in the user interface is because the source code of the website contains bugs and is not properly tested. TestNG- Abbot is a developer friendly open source tool which provides a simple API that facilitates the Test-Driven Development technique to Java GUI. It provides fixture to most basic Java Swing component [12].

Pinho et al. [13] designed a UI of an existing case study on the bases of some usability principles, in order to make sure that all the end users take advantage of the application in an effective and efficient manner. As a substitute approach to conversion-based Split Testing another approach Usability-based Split Testing is introduced for the assurance of the web interface quality along with a WaPPU tool. Based on the code of conduct of Usability-based Split Testing, WaPPU tool derived an interaction-based heuristic and applied it to search engine result pages [14].

Usability of user interface is critical condition of e-commerce websites to attract the customers and provide good user experience throughout the complete process. Yang et al. [15] proposed a few guidelines to design user interface of e-commerce websites which included all the major sections of design format like overall page, clear and well aligned navigations, catalogs, forms input output process, personalization, date and numeric fields, abbreviation inconsistencies, images alignment, shopping cart, search function and customer services and complaints.

IFML can be used for the UI modeling of web mobile application (E-commerce is one of the popular domains) [16]. IFML model is widely used in UI modeling of web application. It elaborates the description of content and behavior of user interface. IFML can be integrated with other modeling languages like Unified Modeling Language (UML), Web Modeling Language (WebML) and Ontology Definition Meta-Model (ODM) [17].

Kaur et al. performed a comparative study to assess the parameters of web user interface. The study involved the evaluation of compatibility with mobile phones, check the complaint system, feedback forms, testing techniques during development phases, network performance, security algorithms, programing language used, scalability and availability of web sites. One more method of usability testing of websites is introduced by the combination of some existing methods named CARE (Cheap, Accurate, Reliable, and Efficient Testing) to control the limitation of cost overrun, resources, imprecise results, uncertainty and inefficient testing [23].

3   Methodology

This section explains the proposed standardized model for usability testing of web UI using IFML (Interaction Flow Modeling Language). The proposed approach is based on two key steps. First step is to design a checklist keeping in mind user's psychology, to achieve maximum usability of UI. A checklist is an organized method of applying usability testing to improve a website. Second part of the proposed approach is transformation of the general checklist into IFML constructs. This checklist can be used to design a new website or testing of an existing site can be performed using recommendation of specific intensification [18].

3.1 Usability Checklist Items

This checklist is designed to assess quantifiable features of the website [18, 19]. The usability index of a website is measured by using checklist which contains the following items.

- One Huge Descriptive Title [18, 20]
- Short and descriptive sentences and paragraphs
- One list of Object
- One CTA per screen clearly visible [19]
- One primary task
- Form, list, Details
- Content [20, 21]
- Buttons, Event [20]
- Graphics and Images [20]
- Website render correctly in different screen resolutions [20]
- Check page loading time (Quick loading).
- Response time to query [19, 20]
- Graphics and Images [20]
- Menu [18]
- Feedback form [18]
- Logo links home [20]
- Progress Bar and error messages [18]
- Provide error message [20]
- Compatible with cross platform [20]
- Format list for ease. Place important items at top of list [20]
- Site Map [20]
- Date and Numeric Field [20]
- Visual Clues for links
- Error messages [12]
- A glossary and linked is given if needed [10]

3.2 Transformation of General Checklist to IFML Constructs

Comparison of general usability checklists is performed with IFML core and extended constructs in Table 1.

Table 1. Comparison of general usability checklists with IFML metaclass.

Checklist in natural language	IFML constructs [22]
One Huge Descriptive Title [18, 20]	IFML::Core:: HEADElement
Navigation [20]	IFML::Core::NavigationFlow
Container [20]	IFML::Core::ViewContainer
Short and descriptive sentences and paragraphs	IFML::Core::Context

(continued)

Table 1. (*continued*)

Checklist in natural language	IFML constructs [22]
Form, list, Details	IFML::Core::ViewComponent
Content [20, 21]	IFML::Core::ContentBinding
Buttons, Event [20]	IFML::Core::Event
Website render correctly in different screen resolutions [20]	IFML::Core::ViewContainer
Graphics and Images [20]	IFML::Core::ViewContainer
Menu [18]	IFML::Extensions::Menu
Check page loading time (Quickly loading) Response time to query [19, 20]	IFML::Extensions::OnLoadEvent
Feedback form [18]	IFML::Extensions::Form
Logo links home [20]	IFML::Extensions::OnSelectEvent
Progress Bar and error messages [18]	IFML::Extensions::ValidationRule
Provide error message [20]	IFML::Extensions::Details
Compatible with cross platform [20]	IFML::Extensions::Device
Format list for ease. Place important items at top of list [20]	IFML::Extensions::List
Site Map [20]	IFML::Extensions::Position
Date and Numeric Field [20]	IFML::Core::BooleanExpression

4 Validation

4.1 MobileWeb Case Study

MobileWeb is an informational website which gives information related to new mobiles. Mobile prices, coming dates, specification is listed. Users can search the latest or upcoming mobiles. This site includes the latest reviews and blogs related to mobiles given by different bloggers or by staff. Users can also search the outlets of mobile brands. Users can give feedback related to complaints, improvements or for suggestions.

Home Page. MobileWeb should have a main master page which should be accessible through all other pages of the web. Logo of the web page redirects to its main page. Main page contains a one descriptive title. Each content on this page have specific heading. The main page contains a section of new arrival mobiles. In this section mobile dummy picture, price and available date of mobile is mentioned. This section contain mouse over button navigate to mobile details page. This page contains details of the mobile. Main page also contains another section of list of mobile brands. In this list all the brands names are mentioned. On selecting the brand name, the page is navigated to specific brand page. This page contains a section having list of all the available mobile phones of a specific brand.

The main page contains the navigation bar. This navigation bar contains the Home page, reviews, news, outlets and contact us button.

Reviews. This page contains a section having videos, links and blogs of reviews of different mobiles. Each content on this page have specific heading. These reviews are provided by different third-party users. This page also contains another section of list of mobile brands. Click on the brand name, the page is navigated to specific brand page. This page contains a section having list of all the available mobile phones of a specific brand.

News. This page contains a section having latest news about the mobiles. Each content on this page have specific heading. Coming dates, official announcements, official auctioning and pros and cons of mobiles.

Outlets. Outlet page contains the section having two drops down menu to search the official outlets of brands. 1st drop down list shows all the available name of the brands and 2nd drop down shows the city name in which the outlet is located. Search button is attached in this section. Search navigates to outlet details page containing the details of the outlets related to your input. This page contains a section in which outlet details are showing.

Contact Us. Contact us page contains a section with feedback form containing credentials of contact us. No registration or sign in is required for this form. Just a valid email is required to send the feedback form. Send button is attached with this form.

4.2   Checklist of MobileWeb Case Study

- One Huge title
- Heading
- Logo navigates to Home page
- Home Page contains container
- Search bar
- Navigation Bar
- Navigation bar contains the Home page, reviews, news, outlets and contact us menu button.
- List of mobiles.
- Identifying links.
- Name, price, date and Image of mobile.
- Mouse over button navigates to another page.
- Error Messages
- Reviews page have video
- Buttons and Event
- Feedback Form

4.3   Modeling

Domain Model. Domain model of the MobileWeb case study is designed as UML class diagram in papyrus editor using Eclipse. Eclipse provides public integrated development environment and IFML.editor is freely available plugin in eclipse. Domain model illustrating the behavior of MobileWeb is shown in Fig. 2.

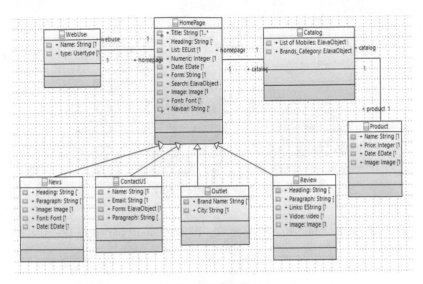

Fig. 2. Domain model

IFML Model. IFML model depicts the MobileWeb behavior and flow. IFML model requires UML model as pre-requisite to draw IFML model diagram. Below are the model diagrams of case study.

HomePage. IFML model depicts the navigation of homepage. Homepage navigates to Brand Category and list of Mobile. And logo links to Homepage. IFML model diagram of HomePage depicts the behavior and content of the Homepage UI as shown in Fig. 3.

Review. Homepage navigates to reviews page by clicking reviews button in navigation bar as shown in Fig. 4.

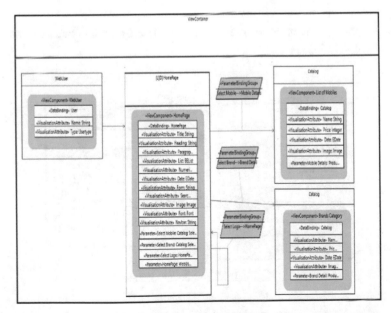

Fig. 3. IFML model of HomePage

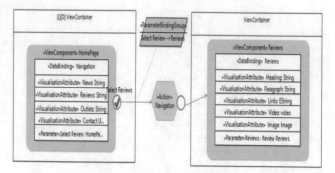

Fig. 4. IFML model of Reviews

News. Homepage navigates to news page by clicking news button in navigation bar as shown in Fig. 5.

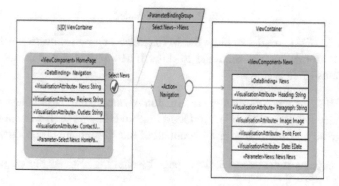

Fig. 5. IFML model of News

Outlets. Homepage navigates to outlets page by clicking outlets button in navigation bar as illustrated in Fig. 6.

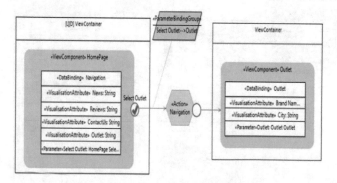

Fig. 6. IFML model of Outlets

Contact Us. Homepage navigates to contact us page by clicking contact us button in navigation bar as shown in Fig. 7.

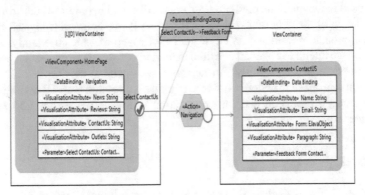

Fig. 7. IFML model of Contact Us

4.4 MobileWeb Checklist Comparison with IFML Constructs

General checklist items are validated using the IFML constructs given in Table 2. We validate that these items are included in the web site using IFML model. It is validated that all the pre-required items are present in the web site ensuring that usability testing is also checked.

Table 2. MobileWeb checklist comparison with IFML constructs

Checklist in natural language	IFML constructs
One Huge title	IFML::Core:: Element
Heading	IFML::Core::NamedElement
Name, price, date and Image of mobile	IFML::Core::ViewComponent IFML::Core::NamedElement
Logo navigates to Home page	IFML::Extensions::OnSelectEvent
Home Page contains section	IFML::Core::ViewComponent
Search bar	IFML::Core::ViewComponent
Navigation Bar	IFML::Core::NavigationFlow
Navigation bar contains the Home page, reviews, news, outlets and contact us menu button	IFML::Extensions::Menu IFML::Extensions::OnSelectEvent
List of mobiles	IFML::Extensions::List « List»
Identifying links	IFML::Core::Element
Mouse over button navigates to another page	IFML::Extensions::OnSubmitEvent
Error Messages	IFML::Core::CatchingEvent
Reviews page have video	IFML::Core::ViewComponent IFML::Core::NamedElement
Buttons and Event	IFML::Extensions::OnSelectEvent
Feedback Form	IFML::Extensions::Form
Response time to query	IFML::Extensions::OnLoadEvent

5 Conclusion

In this paper, the usability testing of UI is carried out using IFML constructs. IFML model is used to evaluate the content, user interaction and behavior of the front-end UI. Transformation of the general usability checklist to IFML constructs is performed. The proposed approach is validated on UI of the MobileWeb website case study.

In future we intend to cover all core and extended constructs from IFML specification using a more extended case study. Furthermore, we intend to develop a tool which will implement the proposed model in order to provide automated web UI testing to ensure the consistency. Automated test case generation will save more time and reduce the development cost.

References

1. Atzeni, A., Faily, S., Galloni, R.: Usable security. In: Advanced Methodologies and Technologie. System Security, Information Privacy, and Forensics, pp. 348–359. IGI Global (2019)
2. Yousaf, N., Azam, F., Butt, W.H., Anwar, M.W., Rashid, M.: Automated model-based test case generation for web user interfaces (WUI) from interaction flow modeling language (IFML) models. IEEE Access 7, 67331–67354 (2019)
3. Conte, T., Massolar, J., Mendes, E., Travassos, G.H.: Web usability inspection technique based on design perspectives. IET Softw. 3, 106–123 (2009)
4. Omar, K., Rapp, B., Gómez, J.M.: Heuristic evaluation checklist for mobile ERP user interfaces. In: Proceedings of 7th International Conference on Information and Communication Systems (ICICS), Irbid, Jordon, pp. 180–185 (2016)
5. Leiva, L.A., Huang, J.: Building a better mousetrap: compressing mouse cursor activity for web analytics. Inf. Process. Manag. 51, 114–129 (2015)
6. Martín-Albo, D., Leiva, L.A., Huang, J., Plamondon, R.: Strokes of insight: user intent detection and kinematic compression of mouse cursor trails. Inf. Process. Manag. 52, 989–1003 (2016)
7. Danilov, N., Shulga, T., Frolova, N., Melnikova, N., Vagarina, N., Pchelintseva, E.: Software usability evaluation based on the user pinpoint activity heat map, Cham, pp. 217–225 (2016)
8. Li, P.: Modular and flexible coordination for web-based applications. In: Proceedings of 2nd International Conference on Computer and Communication Systems (ICCCS), Krakow, Poland, pp. 6–10 (2017)
9. Van Eck, M.L., Markslag, E., Sidorova, N., Brosens-Kessels, A., Van der Aalst, W.M.: Data-driven usability test scenario creation. In: Proceedings of International Conference on Human-Centred Software Engineering, Sophia Antipolis, France, pp. 88–108 (2018)
10. Dias, F., Paiva, A.C.: Pattern-based usability testing. In: Proceedings of IEEE International Conference on Software Testing, Verification and Validation Workshops (ICSTW), Tokyo, Japan, pp. 366–371 (2017)
11. Grigera, J., Garrido, A., Rossi, G.: Kobold: web usability as a service. Presented at the Proceedings of the 32nd IEEE/ACM International Conference on Automated Software Engineering. Urbana-Champaign (2017)
12. Fucci, D., Erdogmus, H., Turhan, B., Oivo, M., Juristo, N.: A dissection of the test-driven development process: does it really matter to test-first or to test-last? IEEE Trans. Softw. Eng. 43, 597–614 (2017)

13. Pinho, R., Sousa, A., Restivo, A.: Applying usability principles to the design of a web interface for the iLab - inventory manager for electronics laboratory. In: Proceedings of 5th Iberian Conference on Information Systems and Technologies, Santiago de Compostela, Spain, pp. 1–6 (2010)
14. Bakaev, M., Mamysheva, T., Gaedke, M.: Current trends in automating usability evaluation of websites: can you manage what you can't measure? In: Proceedings of 11th International Forum on Strategic Technology (IFOST), Novosibirsk, Russia, pp. 510–514 (2016)
15. Yang, Z., Shi, Y., Yan, H.: Scale, congestion, efficiency and effectiveness in e-commerce firms. Electron. Commer. Res. Appl. **20**, 171–182 (2016)
16. Laaz, N., Mbarki, S.: A model-driven approach for generating RIA interfaces using IFML and ontologies. In: Proceedings of 4th IEEE International Colloquium on Information Science and Technology (CiSt), Tangier, Morocco, pp. 83–88 (2016)
17. Hamdani, M., Butt, W.H., Anwar, M.W., Azam, F.: A systematic literature review on interaction flow modeling language (IFML). In: Proceedings of 2nd International Conference on Management Engineering, Software Engineering and Service Sciences, Wuhan, China, pp. 134–138 (2018)
18. Keevil, B.: Measuring the usability index of your web site. In: Proceedings of the 16th Annual International Conference on Computer Documentation, Quebec, Canada, pp. 271–277 (1998)
19. Kumar, M., Emory, J., Choppella, V.: Usability analysis of virtual labs. In: Proceedings of IEEE 18th International Conference on Advanced Learning Technologies (ICALT), Bombay, India, pp. 238–240 (2018)
20. Bevan, N., Spinhof, L.: Are guidelines and standards for web usability comprehensive? In: Proceedings of International Conference on Human-Computer Interaction, Beijing, China, pp. 407–419 (2007)
21. Da Costa, S.L., Neto, V.V.G., De Oliveira, J.L.: A user interface stereotype to build web portals. In: Proceedings of 9th Latin American Web Congress, Ouro Preto, MG, Brazil, pp. 10–18 (2014)
22. Brambilla, M., Fraternali, P.: Interaction Flow Modeling Language: Model-Driven UI Engineering of Web and Mobile Apps with IFML. Morgan Kaufmann, Boston (2014)
23. Kaur, R., Sharma, B.: Comparative study for evaluating the usability of web based applications. In: Proceedings of 4th International Conference on Computing Sciences (ICCS), Phagwara, India, pp. 94–97 (2018)

Blockchain for IoT-Based Digital Supply Chain: A Survey

Haibo Zhang$^{(\boxtimes)}$ and Kouichi Sakurai

Kyushu University, Fukuoka, Japan
{zhang.haibo,sakurai}@inf.kyushu-u.ac.jp

Abstract. This exploratory investigation aims to discuss current network environment of digital supply chain system and security issues, especially from the Internet world, of digital supply chain management system with applying some advanced information technologies, such as Internet of Things and blockchain, for improving various system performance and properties. This paper introduces the general histories and backgrounds, in terms of information science, of the supply chain and relevant technologies which have been applied or are potential to be applied on supply chain with purpose of lowering cost, facilitating its security and convenience. It provides a comprehensive review of current relative research work and industrial cases from several famous companies. It also illustrates the IoT enablement and security issues of current digital supply chain system, and existing blockchain's role in this kind of digital system. Finally, this paper concludes several potential or existing security issues and challenges which supply chain management is facing.

1 Introduction

A traditional supply chain system can be regarded as a network of all individuals, organizations, resources, activities and technologies involved in the creation, delivery and sale of a product [1]. A supply chain is linked together through physical flows, which involves the production, transportation, movement, and storage of goods and materials, as well as information flows, which allows the various supply chain members to coordinate their long-term plans and control the daily flow of goods and materials up and down the supply chain [2].

While the traditional supply chain system can provide services to human's life within a relatively safe environment, it's difficult to satisfy ever-increasing diversified goods and complicated customers' demands which require supply chain to high-efficiently and less-costly work within a more complex consuming information network. Moreover, all supply chain participants should have the ability to share and gather product's data efficiently cross the whole supply chain network for making decisions coordinately in a timely way. By enabling technical methods to traditional supply chain's daily management work, for gathering, processing, analyzing, storing and sharing large amounts of information in a real-time, information technologies have become necessary components for information collaborating and performance improving [3].

© Springer Nature Switzerland AG 2020
L. Barolli et al. (Eds.): EIDWT 2020, LNDECT 47, pp. 564–573, 2020.
https://doi.org/10.1007/978-3-030-39746-3_57

Digital supply chain means applying advanced information technologies to traditional supply chain, i.e. internet of thing, cloud computing and blockchain technology. Digital supply chain is designed to prove the authenticity of goods received by customers, and efficiently track transportation status or items' condition (especially factors like temperature, power and light for cold or food chain) in real-time. Digital supply chain does not have an explicit starting time in its development history, that could be regarded as starting from the first application of information technology on supply chain system which might be the first network of smart devices, CMU's 'Only' coke machine in 1982 [4].

In this paper, we discuss how internet of things technology can work with supply chain system to constitute an advanced information system, what kinds of issues or challenges would appear, and how blockchain technology can address them.

2 Supply Chain with the Internet of Things

2.1 Internet of Things

The first idea of 'network of smart devices' was applied on Carnegie Mellon University's modified Coke machine which was able to report its inventory and whether the loaded drinks were cold or not in 1982. However, the first proposal of 'Internet of Things (IoT)' was formally proposed by MIT's Auto-ID Center in 1999 [4].

IoT is defined as a network of objects within which all objects are able to be identified by certain trustful mechanisms and connected, either with each other internally or to the internet externally through combining with IoT's necessary technologies like Radio Frequency Identification Devices (RFID), sensors, GPS chips and mobile phone to provide integrated services [5]. IoT system generally consists of three layers, including perception layer, network layer, and application layer. Figure 1 shows the general architecture of IoT system and its data transmission flows.

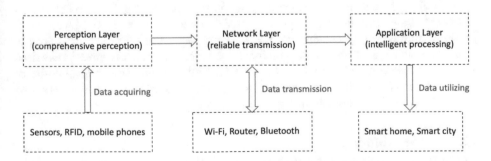

Fig. 1. IoT architecture.

2.1.1 Perception Layer

The task of perception layer is to acquire useful data, including light, sound, heat, biology, location, etc, from relative environment by using sensing technologies, i.e. RFID. This layer is for detecting, collecting and pre-processing the data, and then transmitting to the network layer for follow-up processing [6].

2.2 Network Layer

The network layer of IoT system could be regarded as a middle-ware layer which is for data aggregating, filtering, routing and transmission, through the network with some platforms, such as Internet gateways, switching and routing devices, to various IoT devices in application layer [6].

2.2.1 Application Layer

The application layer could achieve more intelligent data processing to guarantee IoT system's purpose and constitute a smart environment.

2.3 Working with Supply Chain (Benefits)

2.3.1 Trust Management

Traditional tracking methods, such as periodic bar code scanning and check points, providing segment information is incomplete. IoT technology is a kind of new technology which can improve the performance of supply chain's traceability with complete information that traditional information technology cannot achieve.

In recent years, IoT technologies, such as RFID, enables an automatic supply chain tracking capability with the lowest operational cost. For example, many companies have started using RFID technology to track real-time inventory information and to monitor human resource activities [7]. Moreover, RFID could be used by retailers to facilitate the speed of returns, to manage warranty claims by manufacturers and improve the performance of post-sales support.

Especially in pharmaceutical supply chain management system, RFID could cut down the counterfeiting of pharmaceutical drugs and insure the integrity of products purchased by consumers. RFID could also be used in the food supply chain to ensure that the foods are fresh by tracking food products' real-time status and condition. Consumers can use RFID information to check all nodes of supply chains, especially in the cool supply chain. That is to say, goods attached by RFID would be traceable in the supply chain [8].

2.3.2 Traceability

One of the most necessary and important properties for digital supply chain is to achieve its high-efficiency traceability in some fields, i.e. cold chain, food supply chain, health care, and pharmaceutical industries, for preventing against some transportation or production's condition accidents like food poisoning which could cause some serious health effects on consumers [9]. High-efficient traceability was able to allow supply chain members to make decisions and coordinate plans flexible and fast. A high-efficient traceability can also help organizations to manage their whole transportation network with lower cost.

IoT plays an important role on combining supply chain management system with information technologies regarded as a basic platform providing tracking and internet-connecting services. Thus, early research works should focus on facilitating the convenience and traceability by utilizing IoT technologies. The combination between IoT and supply chain would enhance the traceability especially in cold chain and food manufacturing.

2.3.3 Data-Provenance

Data-provenance have been worked in many research domains such as healthcare management for information navigating, tracing, monitoring and management. Data-provenance issues can also be researched with other related technologies such as IoT, blockchain and working in cloud environment, in various research or industrial domains as tracing evidences of objects or processes, especially working on food supply chains.

Some researchers have researched on data-provenance with related technologies but not focus on supply chain system, which would have research space for further research work by extending them to supply chain systems. Ramachandran and Kantarcioglu researched on developing a system which was for recording immutable data trails according to smart contracts and open provenance model, to facilitate the collection, verification and management of trustworthy data provenance based on using blockchain as a platform [10]. Javaid et al. provided a system architecture which could enforce data provenance and data integrity to work in IoT environment based on using Physical Unclonable Functions and Ethereum, as well as a blockchain variant with smart contracts [11].

Data-provenance can work with IoT technology as a kind of history and evidence record for tracking products and processes' activities. Unlike using RFID's traceability in terms of environment or conditions, the data-provenance could be produced by processing information, i.e. all participants' activities, physical and information flow, captured by IoT sensors.

The application of using IoT to trace data-provenance in supply chain system indicates the ability of products' information safe protection, especially in food supply chain for food's safety protection. IoT can also work with some existing mature data-provenance security methods such as applying its hash/signature chain architecture for integrity protection and access control policies for confidentiality protection, to solve those security issues come with IoT-enabled supply chain for achieving other efficient management features.

2.4   Related Work

2.4.1   Academic Study

Yan et al. defined objectives of trust management on IoT systems, which could be applicable on IoT-enabled supply chain system, with regard to trust relationship and decision, data perception trust, privacy preservation, data fusion and mining trust, data transmission and communication trust, quality of IoT services, system security and robustness, human-computer trust interaction and identity trust [12].

To provide a trust environment of a supply chain management system, IoT technologies and services enabled on supply chain should be enforced as many as better to achieve above objectives as standard measurement of IoT trust management systems.

Zhang et al. proposed a smart sensor data collection strategy for IoT, which aims to improve the efficiency and accuracy of provenance with the minimized size of data set at the same time by modeling IoT system structure for food supply chains, as well as algorithms from big data and self-correction strategies [13].

Mousavi et al. proposed a practical system framework, which could trace the whole process of meat producing from the individual animal to individual prime cuts in the boning hall, with technologies such as bar code scanning and RFID [9].

Regattieri et al. provided a practical system framework for food manufacturing supply chain traceability with tracing functions, which could identify products' self-information, track products' status and data by utilizing traceability tools and route related information [14].

Abad et al. demonstrated an example of using RFID tags for tracking products' status and conditions in the fresh fish cold chain in which multiple sensors were enabled to capture the real-time information in terms of temperature, humidity, power and light. The information thus collected are stored and can be further analyzed [15].

2.4.2   Industrial Applications

Cisco began to process its supply chain digitization initiative several years ago. They focused on their systems and processes foundation, upgraded their enterprise resource planning and product data management systems. They used technology such as internet of things, mobility, big data and cloud services, to provide real-time visibility.

Walmart has become the leading retailers because of having a powerful decision-making system that relied on data analysis through a bar-code scanning system, a point-of-sale system, and real time data collection through current RFID technology. They enabled some advanced inventory technology like automated recording system for real-time data recording to the database. When combining with the use of bar-codes, employees were able to gather and analyze real-time inventory information.

IBMs Watson Supply Chain applied artificial intelligence on their supply chain system, and trained in supply chain through machine learning, to provide comprehensive, end-to-end visibility and insights. Users can use Watson to rapidly assemble the right team to collaborate and manage incidents and resolve disruptions quickly. Watson provides an open integration platform for connecting and harmonizing disparate data, silos and systems to provide deeper visibility and insights.

3   Security Issues

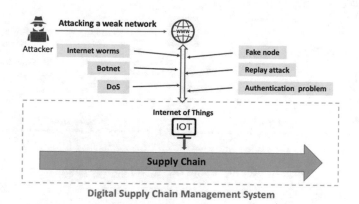

Fig. 2. Attacking methods on digital supply chain.

While the enablement of IoT can provide an efficient information collaboration for supply chain networks, it also reduced or removed the protection barrier of traditional supply chain system. Traditional supply chain system was separated from external unstable cyber attacks environment due to its disconnecting with the internet. Connecting with the internet would allow various external or internal risks to enter into supply chain system resulting in unexpected problems.

Figure 2 shows that attackers could infiltrate a weak network by some malicious methods, such as injecting Internet worms and counterfeiting nodes, to cause cyber attacks like botnets, denial of service attack, replay attack and authentication problems.

One of the most important part of preventing against security issues is how to preserve necessary security properties such as confidentiality, integrity, availability (CIA), authenticity and authority which could be vulnerable to diversified cyber attacks, like virus, malicious code injection, distributed denial of service, covert channel, malicious intrusion and misuse.

Although enabling IoT technologies could improve the performance of information sharing and reduce supply chain risks by facilitating its information collaboration, risk sources could be exposed at the same time and increase supply chain risks. In this way, system flaws and vulnerabilities, especially from devices provided by third party providers like IoT devices or cloud computing server, would be exploited by cyber attackers to also increase supply chain risks. Those attacks would be dangerous for system assets, including both system data and processes which could be regarded as supporters for information sharing.

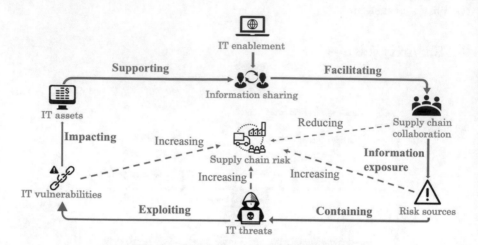

Fig. 3. Security issues on digital supply chain.

Smith et al. defined a supply chain information security risks flow model, shows as Fig. 3, to illustrate relationships, in terms of increasing, facilitating, supporting, reducing and impacting, between information technology enablement and security risks [3]. Information technology enablement could be regarded as the source of a series of security impacts and actions.

4   Blockchain's Role

4.1   Blockchain

Haber and Stornetta firstly proposed the idea of cryptographically secured chain architecture for their tamper-proof timestamp mechanism in 1990 [16]. Blockchain was formally proposed by a person or a group named Nakamoto in 2008 as a core component of his cryptocurrency system "bitcoin" [17].

Blockchain refers to a decentralized architecture consists of increasing numbers of cryptographically linked blocks, each of which stores the hash value of previous block [18]. Blockchain was able to protect the information among the network against any breached or vulnerable device through verifying identities and rejecting malicious parties by other members. The data stored in a block is immutable which is extremely hard to be tampered (tamper-proof), that means hackers have to manipulate all blocks' data until the head block to achieve their malicious attacks, which is impossible in a real blockchain world.

4.2 Blockchain's Role

Blockchain as one of the hottest research topic, its trustworthy architecture with distributed and decentralized ledger, as well as cryptographically linked blocks, can also provide an accurate way for measuring products quality during whole transportation on supply chain. For example, stake-holders in a supply chain can gather the location information about whether the product was in a wrong place or the whole journey from source to destination by analyzing collected data on the travel path and duration. Other utilizing cases of this kind of capability are applying blockchain technology on cold supply chain for food products environment monitoring, especially for temperature, and on food supply chain for food healthcare which could lead to serious health risks without enough attention.

Kshetri [19] demonstrate blockchain's roles in achieving the various strategic supply chain objectives. For general performance dimensions of supply chain, for instance, blockchain can reduce or zero costs of enterprises if technologies such as IoT have already been applied to detect, measure, and track key supply chain management processes; the response speed can be increased by digitizing physical process and reducing interactions as well as communications according to blockchain's digital signature storage and transmission which can validate the identities of individual and assets to minimize the needs of physical interactions and communications.

Blockchain can also play an important role on improving the performance of trust management by reducing the risk possibility since blockchain need to validate the identities of individual participating in transactions, which means only members who are mutually accepted in the network can engage in transactions.

Blockchain's architecture of distributed ledger can ensure a decentralized and transparent transaction mechanism in supply chain management system in industrial and business. Blockchain can help supply chain organizations and consumers to track products origins and whole processes during the whole transportation. Abeyratne and Monfared [20] proposed an architecture about how blockchain can manufacture supply chain system with factors of registrars, standards, certifiers, producers and consumers.

Applications of blockchain in supply chain has been employed in industrial area widely by business companies. Alibaba worked with AusPost, Blackmores and PwC to explore the ability of combining blockchain with food supply chain for food fraud fighting such as selling low-quality foods. The purpose of their team is to develop a 'Food Trust Framework' to improve the integrity and traceability on the global supply chains [19]. Walmart built a system for providing a service to monitor the pork production in the U.S. and China with blockchain enabling the digital tracking on individual pork products in a few minutes compared to many days taken in the past.

5   Challenges

For current digital supply chain management systems, an important security challenge is how to face those security risks which are also rigorous for enabled IoT technologies. While enabling above technologies on supply chain systems, some inherent security problems within themselves or between collaborations with each other would be long/short-term challenges for digital supply chain system in terms of improving efficient trust management.

IoT devices still face the unstable or untruthful internet environment with unexpected cyber attacks, so the information collaboration between IoT and data-provenance would be vulnerable to such risks. In addition, how to preserve the integrity and confidentiality of data-provenance in terms of producing and storing to ensure IoT's traceability needs more deeply research work.

The limitations of blockchain would limit the development on supply chain as well, i.e. the global supply chain operates in a complicated environment which requires various parties to comply with diverse laws, regulations and institutions. Increasing storage space for blockchain technology is also an emergent problem. Many researchers have worked on combining blockchain with cloud storage service to achieve storage space reducing and information accessing timely. However, this kind of application reduced blockchain's security due to cloud service's centralized architecture.

6   Conclusion

This paper discusses current situation and environment of supply chain management systems with advanced IoT enablement for improving the performance corresponding to diversified demands. While working advantages of those information technologies, supply chain has to face various security risks and challenges which are needed to be solved in the future research work. That would not be a short-term work to enhance the trust management of supply chain management system.

Acknowledgements. This research was partially supported by Collaboration Hubs for International Program (CHIRP) of SICORP, Japan Science and Technology Agency (JST).

References

1. Janvier-James, A.M.: A new introduction to supply chains and supply chain management: definitions and theories perspective. Int. Bus. Res. **5**(1), 194 (2012)
2. Zhou, W., Piramuthu, S.: IoT and supply chain traceability. In: International Conference on Future Network Systems and Security, pp. 156–165. Springer, Cham, June 2015
3. Smith, G.E., Watson, K.J., Baker, W.H., Pokorski Ii, J.A.: A critical balance: collaboration and security in the IT-enabled supply chain. Int. J. Prod. Res. **45**(11), 2595–2613 (2007)

4. Madakam, S., Ramaswamy, R., Tripathi, S.: Internet of Things (IoT): a literature review. J. Comput. Commun. **3**(05), 164 (2015)
5. Benabdessalem, R., Hamdi, M., Kim, T.H.: A survey on security models, techniques, and tools for the Internet of Things. In: 2014 7th International Conference on Advanced Software Engineering and Its Applications (ASEA), pp. 44–48. IEEE, December 2014
6. Mahmoud, R., Yousuf, T., Aloul, F., Zualkernan, I.: Internet of Things (IoT) security: current status, challenges and prospective measures. In: 2015 10th International Conference for Internet Technology and Secured Transactions (ICITST), pp. 336–341. IEEE, December 2015
7. Zhou, W., Piramuthu, S.: IoT and supply chain traceability. In: International Conference on Future Network Systems and Security, pp. 156–165. Springer, Cham, June 2015
8. Shen, G., Liu, B.: The visions, technologies, applications and security issues of Internet of Things. In: 2011 International Conference on E-Business and E-Government (ICEE), pp. 1–4. IEEE, May 2011
9. Mousavi, A., Sarhadi, M., Lenk, A., Fawcett, S.: Tracking and traceability in the meat processing industry: a solution. Br. Food J. **104**(1), 7–19 (2002)
10. Ramachandran, A., Kantarcioglu, D.: Using blockchain and smart contracts for secure data provenance management. arXiv preprint arXiv:1709.10000 (2017)
11. Javaid, U., Aman, M.N., Sikdar, B.: BlockPro: blockchain based data provenance and integrity for secure IoT environments. In: Proceedings of the 1st Workshop on Blockchain-Enabled Networked Sensor Systems, pp. 13–18. ACM, November 2018
12. Yan, Z., Zhang, P., Vasilakos, A.V.: A survey on trust management for Internet of Things. J. Netw. Comput. Appl. **42**, 120–134 (2014)
13. Zhang, Q., Huang, T., Zhu, Y., Qiu, M.: A case study of sensor data collection and analysis in smart city: provenance in smart food supply chain. Int. J. Distrib. Sens. Netw. **9**(11), 382132 (2013)
14. Regattieri, A., Gamberi, M., Manzini, R.: Traceability of food products: general framework and experimental evidence. J. Food Eng. **81**(2), 347–356 (2007)
15. Abad, E., Palacio, F., Nuin, M., De Zarate, A.G., Juarros, A., Gómez, J.M., Marco, S.: RFID smart tag for traceability and cold chain monitoring of foods: demonstration in an intercontinental fresh fish logistic chain. J. Food Eng. **93**(4), 394–399 (2009)
16. Haber, S., Stornetta, W.S.: How to time-stamp a digital document. In: Conference on the Theory and Application of Cryptography, pp. 437–455. Springer, Heidelberg, August 1990
17. Nakamoto, S.: Bitcoin: a peer-to-peer electronic cash system (2008)
18. Bocek, T., Rodrigues, B.B., Strasser, T., Stiller, B.: Blockchains everywhere-a use case of blockchains in the pharma supply chain. In: 2017 IFIP/IEEE Symposium on Integrated Network and Service Management (IM), pp. 772–777. IEEE, May 2017
19. Kshetri, N.: 1 blockchaina's roles in meeting key supply chain management objectives. Int. J. Inf. Manag. **39**, 80–89 (2018)
20. Abeyratne, S.A., Monfared, R.P.: Blockchain ready manufacturing supply chain using distributed ledger. Int. J. Res. Eng. Technol. **05**(09), 1–10 (2016)

Author Index

L. Barolli et al. (Eds.): EIDWT 2020, LNDECT 47, pp. 575–576, 2020.
https://doi.org/10.1007/978-3-030-39746-3

Printed in the United States
By Bookmasters